Basic Structure and Evolution of Vertebrates

Basic Structure and Evolution of Vertebrates

Volume 2

ERIK JARVIK

Naturhistoriska Riksmuseet,
Sektionen för Paleozoologi,
Stockholm, Sweden

1980

ACADEMIC PRESS

A Subsidiary of Harcourt Brace Jovanovich, Publishers

London · New York · Toronto · Sydney · San Francisco

ACADEMIC PRESS INC. (LONDON) LTD.
24/28 Oval Road, London NW1 7DX

United States Edition published by
ACADEMIC PRESS INC.
111 Fifth Avenue, New York, New York 10003

British Library Cataloguing in Publication Data

Jarvik, Erik
Basic structure and evolution of vertebrates.
Vol. 2
1. Vertebrates, Fossil 2. Vertebrates—
Anatomy 3. Vertebrates—Evolution
I. Title
566 QE841 80–40244

ISBN 0–12–380802–2

Text set in 11/12 pt Bembo
Printed in Great Britain by W & J Mackay Limited, Chatham

Preface

A goal common to students in all the various fields of biology is that of unravelling the evolution of life from its beginnings to the diversified fauna and flora of today. The time perspective furnished by the geological record gives palaeontology a central place in these efforts and the intricate skeletal structures of vertebrates are especially well suited for the pursuit of long phylogenetic histories.

Vertebrates appear in the fossil record nearly 500 million years ago at the end of the Cambrian but it was not until the late Silurian and the Devonian that they become common. This early Palaeozoic vertebrate fauna is of special interest because it included the ancestors of all vertebrates appearing later, including those living today. Except for the oldest known tetrapods, the ichthyostegalians, all these creatures are either cyclostomes or fishes. A profound knowledge of these most ancient vertebrates is of the greatest importance for students of vertebrate evolution and may also be of interest to other evolutionary biologists. One of the main aims of this book is to inform researchers in the life sciences of the results of work done during the last few decades on the earliest vertebrates; this work has included detailed anatomical studies and comparisons with living forms, both larval and adult.

The fossil record is incomplete and most of the early vertebrates are still imperfectly known. However, several well-preserved specimens have been discovered, and this institute is in the fortunate position of having at hand a large (and in many respects quite unique) collection of early cyclostomes and fishes, as well as fossil material of the oldest known tetrapods. Using advanced methods of investigation including Sollas's grinding method for the construction of wax models of anatomical structures on this fossil material, it has been possible to make out essential features of the internal anatomy of members of the principal groups of early vertebrates.

Interpretation of fossil structures must rest on comparisons with recent forms, and to provide a solid basis for such comparisons a review of the relevant anatomical features in the extant ganoid fish *Amia calva* is given. Thanks to studies of the excellent embryological material stored in the Zoological Institute of the University of Stockholm, many new and important ontogenetic data have been incorporated in this review. The review also aims to give an idea of the complexity of the anatomical details with which a palaeozoologist has to deal. While preparing the review, I soon realized that many of the figures of *Amia* and other extant vertebrates in the literature were either incorrect or insufficient for my purposes. Thus, I have prepared new figures based on photographs and dissections not only of *Amia*, but also of *Neoceratodus* and other living vertebrates.

Among early fossil vertebrates the Devonian fish *Eusthenopteron foordi* is outstanding in several respects. This fish, dealt with in a special part, is the only Palaeozoic vertebrate in which the skeleton of the head is completely known and its postcranial skeleton is also well preserved. Since this is the first time that it has been possible to give a full description of the structure of a Palaeozoic vertebrate, this account and the elaborate figures which accompany it should be of use and interest to all students of palaeozoology and comparative anatomy.

Eusthenopteron belongs to the group of fishes, the Osteolepiformes, from which the majority of the tetrapods (including man) has evolved. I have also studied the material of *Glyptolepis groenlandica* and other excellent material of the Porolepiformes, the group of fishes from which urodeles originate, as well as all the known material of the ichthyostegalians. Thus it has been possible to discuss the origins and evolution of the tetrapods from the widest possible palaeontological base while also using a very substantial body of data from embryological studies.

The problems of the origins of paired fins, the tetrapod limbs, and of the tetrapod tongue are also thoroughly discussed here. Moreover, *Eusthenopteron* is one of our closest relatives among fishes, and direct comparisons with man have been used to elucidate how some structures in the human body came about. The similarities between the pectoral and pelvic fins in *Eusthenopteron* and the embryonic human hand and foot are discussed, and a new theory of the origin of mammalian ear ossicles based on a comparison of *Eusthenopteron* and man is presented. Both these discussions and the detailed description of one of our Devonian ancestors should be of interest to anthropologists and other students of human biology who are interested in man's earliest origins.

When the grinding series of *Eusthenopteron* was completed in the early 1950s and the elaborate wax models of the cranial structures had been made, it became evident that the vertebrate cranium is composed of a modified portion of the vertebral column to which parts of the prootic visceral arches have been added. I have long wished to pursue the implications of these ideas. Now, thanks to increased knowledge of *Latimeria* and to many new and important embryonic data, it has been possible to present a coherent theory of the origin and composition of the vertebrate head. Metamerism (segmentation) is shown to play a much more dominant role in the morphogenesis of the head than was previously supposed, and I suggest that not only the somitic derivatives and the cranial nerves, but also the visceral arches, the sensory line system, and the dermal bones pertain to the metameric system.

Studies of the earliest fossil vertebrates and of extant vertebrates have led to important new results concerning their phylogenetic relationships and have shed new light on many anatomical and embryological problems. However, when tackling such problems in palaeozoology, we must be sure that we have a detailed and accurate knowledge of both the fossil material and the anatomy and embryology of the extant forms we use for comparison. It is only through painstaking study and comparison of the fossil material that it is possible to achieve lasting results; theoretical models and philosophical speculations may have their place in the study of vertebrate evolution, but the morphological evidence is primary. It is never enough to trust in authority, however eminent, or to rely on current opinions in textbooks; we must always check the evidence at first hand. In doing so we sometimes find opinions which have been long accepted without question resting on very scanty factual bases. I have tried to maintain such a critical attitude towards preconceived opinions and traditional views and to couple that attitude with a broad anatomical and embryological approach throughout this book.

This volume covers a rather wide subject, and many of the results and conclusions reached here rest on studies of unique material of early fossil vertebrates and of embryological materials curated in this institute. However, I have also undertaken a comprehensive review of the literature, and most of that literature is cited in the bibliography. There will inevitably prove to be omissions in a bibliographical list of this size, and I apologize in advance for such ommissions. In particular, I regret that I have been unable to read the imposing papers published in the Russian language which have recently appeared.

Many intricate problems are discussed here, but I have tried throughout to use a simple and easily readable style; I hope that the book will be useful even to scientists whose first language is not English. In order to facilitate the book's use and to more clearly demonstrate the points being discussed, I have included a large number of figures; my aim has been to illustrate all the topics discussed. Nearly all the figures used here have been redrawn for this volume, and with few exceptions they show the animal (or part of it) either from the left side or with the rostral part(s) directed upward.

It has been a great privilege for me to work at the Paleozoological Section of the Swedish Museum of Natural History and to be able to use its outstanding collections of early fossil vertebrates as well as its grinding series and wax models. For this I would first of all thank my teacher and old friend Professor Erik Stensiö who already, when I started my palaeozoological studies in the mid 1930s, generously placed excellent fossil material at my disposal and then through the years has supported me in multifarious ways. I also want to express my sincere gratitude, for splendid working facilities, continuous help and encouragement, to my friend and colleague Professor Tor Ørvig who since my retirement in November 1973 is the director of this institute.

A constant friend and critic in all aspects of the work that culminated in this volume has been my colleague Dr Hans Christian Bjerring, Stockholm. His wealth of ideas and his many valued suggestions have been an incessant source of information. Also I have derived much benefit from his embryological and anatomical knowledge as well as from the elaborate wax models of various embryos that he has made. I am further indebted to Bjerring, to Professor Tor Ørvig, Stockholm, to Professor Orvar Nybelin, Gothenburg and Dr Philippe Janvier, Paris, for the permission to use figures and data not yet released.

I am under great obligations to the late Dr Lauge Koch, Copenhagen, for giving me the opportunity to join eight summer expeditions to East Greenland in order to collect fossils in the Devonian, and I am much obliged to the authorities of Geologisk Museum, Copenhagen, for permitting me to study those specimens of this material which have been of special interest for my work. Thanks are also due to Professor Lars

Silén and Professor Ragnar Olsson at the Zoological Institute of the University of Stockholm for the allowance to use the embryological material under their care as well as to the many persons (Professor G. Haas, Jerusalem, Dr P. H. Greenwood, London, Dr M. Jollie, De Kalb, and others) who have provided me with material from recent animals.

The illustrations are an important part of this book. Many of the drawings were done earlier at this institute by Mr S. Ekblom, Mrs S. Samson, Mrs M. Liepina, Mr C. Salgueiro, Mr B. Bergman, and other artists. Most of these figures plus those obtained from other sources have been redrawn and rearranged for this book and many new ones have been added. This artwork has been carried out mainly by Mr Bertil Blücher and Mr Lennart Andersson. To these skilled artists grateful thanks are due for their never-failing patience in the execution of their arduous task.

Most of the photographs in this book together with the many photographs which have served as a basis for drawings have been taken by Mr Uno Samuelson. The mechanical preparation of fossil specimens has been performed by Miss Agda Brasch and Miss Eva Norrman. Mrs Sif Samuelson has typed earlier versions of the manuscript and she and Mrs Kamlesh Khullar have typed the final draft. Stylistic corrections were made by Dr John Reed, Berkeley and Dr Kubet Luchterhand, Chicago. To all these collaborators I extend my unstinting thanks and appreciation.

The work has been financially supported by the Swedish Natural Science Research Council.

June 1980
Stockholm

Erik Jarvik

Contents

III THE ORIGIN OF THE TETRAPODS

IV RECAPITULATION AND COMMENT

Contents of Volume 1

I Basic Structure and Composition of Vertebrate Head

1 Notes on the Development of the Vertebrate Embryo

In this and the following sections we will deal firstly with three principal complexes of problems: the composition of the vertebrate head, the origin of the paired fins and the origin and evolution of the tetrapods. Students of evolution often use an intricate terminology. For our purposes it may be sufficient to distinguish between primitive (or basic) and advanced (or specialized) characters. It is to be emphasized, however, that these terms are relative and have a different meaning dependent on which level in the vertebrate pedigree is concerned. Thus, with basic vertebrate characters we mean not only the primitive characters which are shared with the Acrania (Cephalochordata) and are common to all the Euchordata*. We also include all the advanced characters which distinguish the Vertebrata (Craniota) from the Acrania and are common to all vertebrates. Such basic common characters can be established by comparisons of adult animals; and in particular comparisons of extant forms with well known representatives of the early Palaeozoic vertebrates may provide important information. However, the early fossil vertebrates were already highly specialized (Jarvik, 1959; 1964) and to a considerable extent we have to rely on studies of the ontogenetic development. In fact embryology is also an important tool for the palaeozoologist and omitting the earliest stages (fertilization, egg cleavage, blastula, gastrulation) we may start with a brief review of the early ontogenetic development.

The process of gastrulation in the vertebrate embryo as well as in the acraniate *Branchiostoma* (*Amphioxus*) results in the formation of three germ-layers (Fig. 1): an outer, the ectoderm, and two inner, the mesoderm and the endoderm (Brachet, 1935; Dalcq and Pasteels, 1954; Balinsky, 1970; Romer, 1970; Torrey, 1971).

The ectoderm forms important parts of the nervous system and the skeleton. In the early embryo, the neurula (Fig. 1A, C), the part of the ectoderm (the neurectoderm) destined to form the brain and the spinal cord appear as a depressed area, the medullary or neural plate. On each side this plate is bounded off from the lateral or skin ectoderm by a thickening, the neural fold (ridge). Very soon, by processes known as neurulation, the neural folds rise, approach each other and fuse in the median line, leaving an opening, the neuropore, most anteriorly (Fig. 1B, D). Then the outer layer of the fold formed by skin ectoderm separates from the neurectoderm which features the neural

* p. 238.

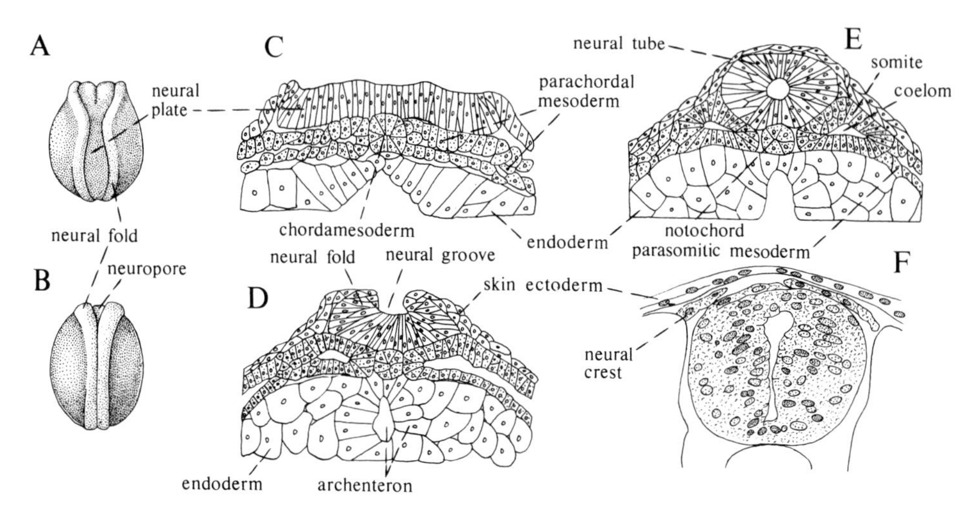

Fig. 1. Embryos of urodele (*Triturus*). A–D, neurulae in dorsal aspects and transverse sections. E, transverse section of embryo with closed neural tube and well developed somites. F, transverse section of neural tube and neural crest. A, B, from Glaesner (1925); C–E, from Hertwig (1903); F, from Neumayer (1906).

tube (Fig. 1E), inside the skin ectoderm. Between that tube and the overlying skin ectoderm the early embryo shows free cells often arranged in a paired longitudinal ridge along the dorsal side of the neural tube (Fig. 1F). This ridge is termed the ganglionic or neural crest (Hörstadius, 1950; Weston, 1970; Chibon, 1974), but if the material of this structure originates from neurectoderm, from skin ectoderm, from both or from a separate primordium (Raven, 1931) still seems to be an open question (Fig. 2). The neural crest cells have the capacity to migrate. The migrating cells are initially arranged in continuous columns or streams following rather precise paths, but are soon dispersed through the vertebrate embryo, forming important structures such as the visceral endoskeleton, ganglia and pigment cells.

For these migrating neural crest cells Hörstadius (1950) introduced the term ectomesenchyme replacing the terms mesectoderm or ectomesoderm used by previous writers. It should be observed, however, that the thickenings of the skin ectoderm known as placodes also produce migrating mesenchyme cells. We have therefore to distinguish between neural crest and placodal ectomesenchyme. The migratory mesenchyme cells derived from the mesoderm may be referred to as mesomesenchyme (endomesoderm, Hörstadius, 1950) and those from the endoderm as endomesenchyme.

In *Amphioxus* (Fig. 3A) the neural tube arises in much the same way as in vertebrates, but there is no neural crest. Moreover, the neural tube is straight, whereas a characteristic feature of the vertebrate embryo is that the neural tube is bent strongly downwards in its anterior part (Fig. 3B).

This pronounced cephalic flexure in the vertebrate embryo together with the development of the otic vesicle strongly modifies the underlying anterior part of the mesoderm, a condition which causes considerable disagreement as to the development events and has created confusion in the terminology. However, judging from

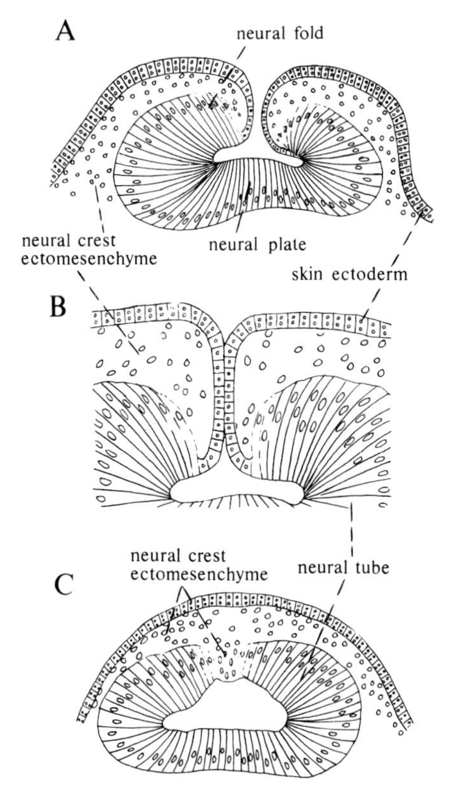

Fig. 2. Three stages in closure of neural tube and formation of neural crest in embryos of *Ambystoma* (axolotl). Drawings after photographs in Raven (1932).

the comprehensive literature (*Amphioxus*, Hatschek, 1881; von Kupffer, 1894; Conklin, 1932; petromyzontids, Sewertzoff, 1916–17, Damas, 1944; 1951; sharks, Balfour, 1878; 1880–1881; Dohrn, 1884; 1885; 1886; 1904; van Wihje, 1882, 1882a; 1905; 1922; Platt, 1891; Hoffmann, 1896–1897; Neal, 1898; Scammon, 1911; Goodrich, 1918; 1930; de Beer, 1924; 1947; Holmgren, 1940; Bjerring, 1967; 1968; 1971; 1977; rays, Sewertzoff, 1899; urodeles, Vogt, 1929; Adelmann, 1932) it seems likely that the early differentiation of the mesoderm takes place somewhat as follows.

In early stages, when the neural tube has not yet arisen (Fig. 1C, D) the mesoderm forms a thin plate between the presumptive neurectoderm above and the endodermal primitive gut, the archenteron, below. In this plate a median band of somewhat modified cells is discernible. This band, the chordamesoderm, is the primordium of the notochord, but is of interest also because it acts as an inductor which determines the development of the neuroectoderm (Fig. 11A; Saxén and Toivonen, 1962; Balinsky, 1970). The chordamesoderm extends to the anterior end of the embryo and divides the mesodermal plate into two equal lateral parts. This paired part of the mesoderm in the now bilateral symmetric embryo will be referred to as the parachordal mesoderm.

In vertebrates the notochord ends close behind the hypophysis and because no notochord develops in the most anterior part of the mesodermal plate, this part has generally been distinguished as some kind of transition area between mesoderm and endoderm, called the prechordal plate (Figs 3B, C, E, 6D, 11A). This distinction, which is clearly artificial, has for a long time hampered our understanding of the morphogenesis of the vertebrate embryo. In fact it was not until quite recently that it

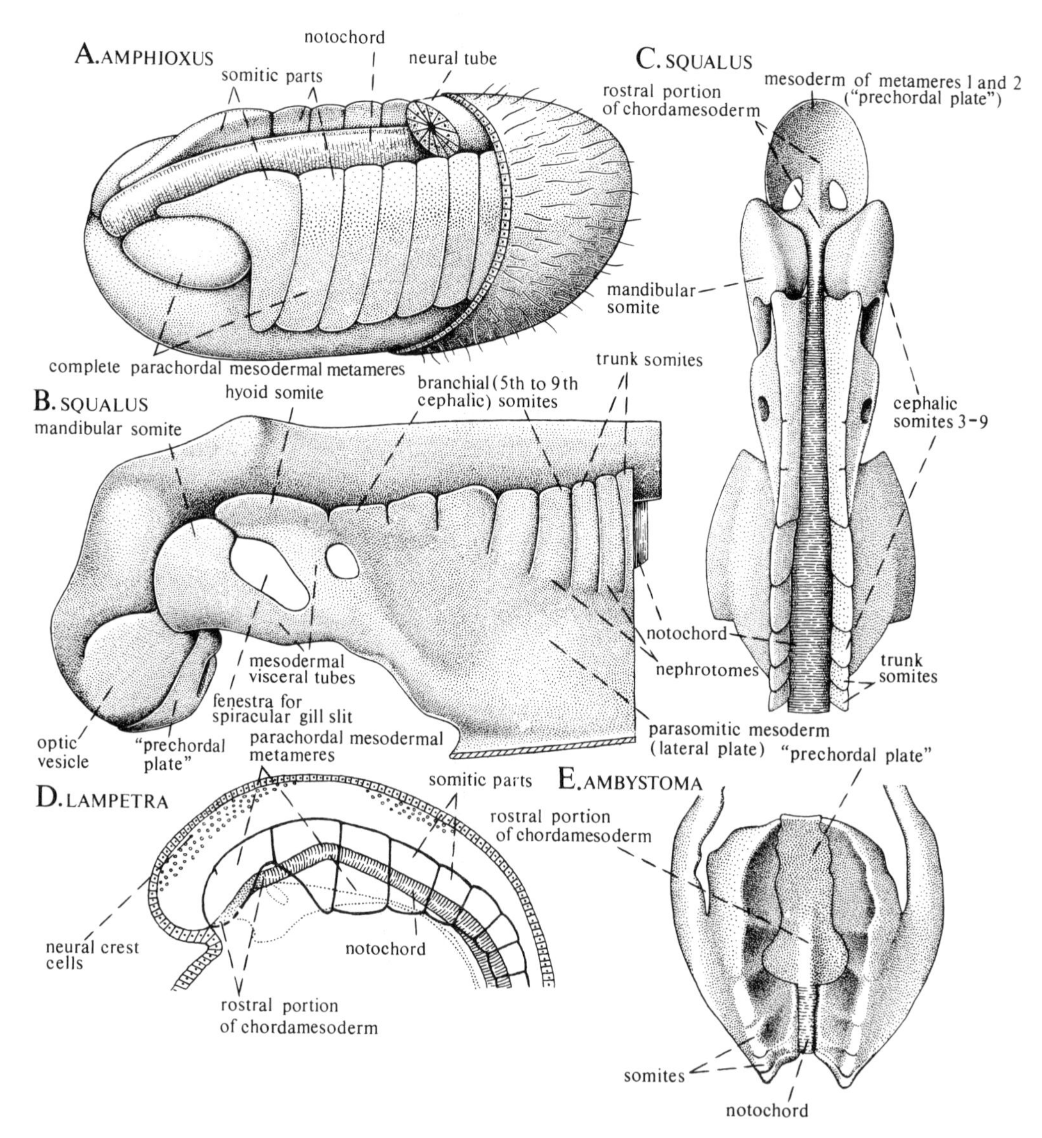

Fig. 3. A, young embryo of *Branchiostoma lanceolatum* (*Amphioxus*) in oblique anterodorsal aspect. From Bjerring (1971). B–C, *Squalus acanthias*. Drawings of model made by H. C. Bjerring of cephalic mesoderm (in B with brain) of embryo 4·5 mm in lateral and dorsal aspects. D, *Lampetra fluviatilis*. Part of embryo to show extension forward of chordamesoderm. From Damas (1944). E, anterior mesodermal part of embryo of urodele (*Ambystoma*) to show "prechordal plate". From Adelmann (1932).

was demonstrated (Bjerring, 1967; 1971) that the notochord in vertebrates originally continued in front of the hypophysis and extended to the anterior end of the embryo, as do the neural tube and the archenteron. This means that the vertebrates in this respect agree with *Amphioxus* (Fig. 3A). However, in vertebrates, probably as a consequence of the strong cephalic flexure of the brain, the notochord formation has been suppressed in the anterior part of the chordamesoderm situated in the "prechordal plate". This anterior part of the chordamesoderm where no notochord is formed has been distinguished as its rostral portion (Fig. 4).

That the notochord in vertebrates originally reached the anterior part of the embryo as it does in *Amphioxus*, is, apart from the relations to the somites,* suggested by the following facts (cf. Bjerring, 1971).

(1) As shown by Dohrn (1904, p. 117) in sharks the median band of specialized cells, the "Chordaanlage", continues—before the hypophysis has arisen—forwards through the "prechordal plate" almost to the anterior end of the embryo. Similar conditions have been described by Vogt (1929) and Adelmann, (1932) in urodeles (Figs 3E, 4B–D); and in the larval lamprey (*Ammocoetes*) the median band of columnar cells which represents the chordamesoderm continuous far in front of that part which forms the notochord (Fig. 4A; Damas 1944).

(2) By cultivating pieces of the material destined to form the "prechordal plate" it had been shown (Holfreter, 1938; Takaya, 1953) that this material, before its invagination at the gastrulation process, possesses a strong notochordal-forming potency. These experiments show that also the rostral portion of the chordamesoderm is capable of forming true notochordal tissue.

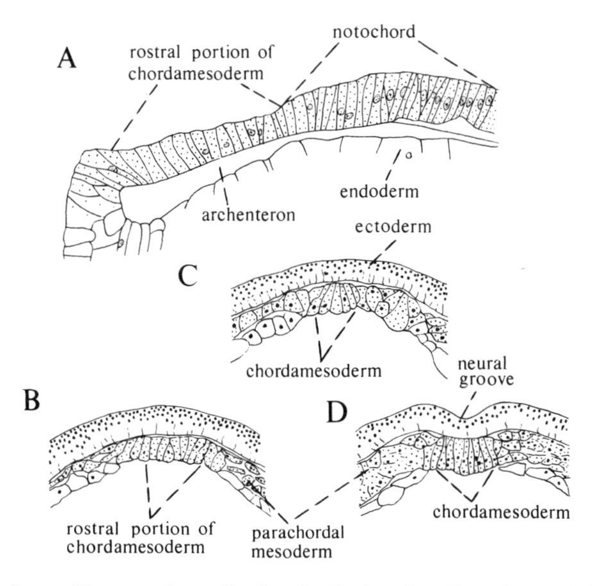

Fig. 4. A, part of early embryo of *Lampetra* in median longitudinal section. From Damas (1944). B–D, parts of early embryos of urodele (*Pleurodeles*) in transverse sections. B, through "prechordal plate"; C at transition between notochord and rostral portion of chordamesoderm; D, through trunk. Drawings after photographs in Vogt (1929).

* p. 11.

(3) As also shown experimentally (Saxén and Toivonen, 1962; Balinsky, 1970) the inducing capacity of the chordamesoderm is not restricted to that part which later forms the notochord. The rostral portion in the "prechordal plate" also exerts a strong inductive acitivity (Fig. 11A) and initiates the development of the anterior division (archencephalon) of the embryonic brain (archencephalic inductor).

(4) Remarkable is also that the "prechordal plate" in the human embryo (George, 1942; Gilbert, 1952; Bergquist, 1953) is pierced by an irregular space which appears to be an anterior continuation of the central canal of the notochord.

The notochord soon assumes a cylindrical shape and separates from the parachordal mesoderm which forms a thin descending sheet on each side of the archenteron (Fig. 1D, E). Simultaneously with this separation transverse crevices appear in the parachordal mesoderm dividing it into a series of portions. These portions are generally referred to as segments but will—in order to avoid confusion (e.g. with the vertebral segments)—be called metameres. This is the first indication of the metamerism ("segmentation") of the vertebrate body. In larval lampreys (*Ammocoetes*) the subdivision of the parachordal mesoderm into metameres is complete (Fig. 3D) and since this is true also of *Amphioxus* (Fig. 3A) this condition may be regarded as primitive. In gnathostomes, by contrast, we have to distinguish between a dorsal somitic portion divided by crevices into metameric somites and a ventral incompletely divided portion, the parasomitic mesoderm (Fig. 3B, C). The parasomitic mesoderm becomes split into two layers, an outer which altogether with the epidermis forms the somatopleure, and an inner which together with the wall of the gut constitutes the splachnopleure (Fig. 5). The space between the two mesodermal layers is the coelom. The somites develop into thick-walled vesicles situated along the notochord and neural tube. Characteristic of somites is the fact that they become subdivided into three

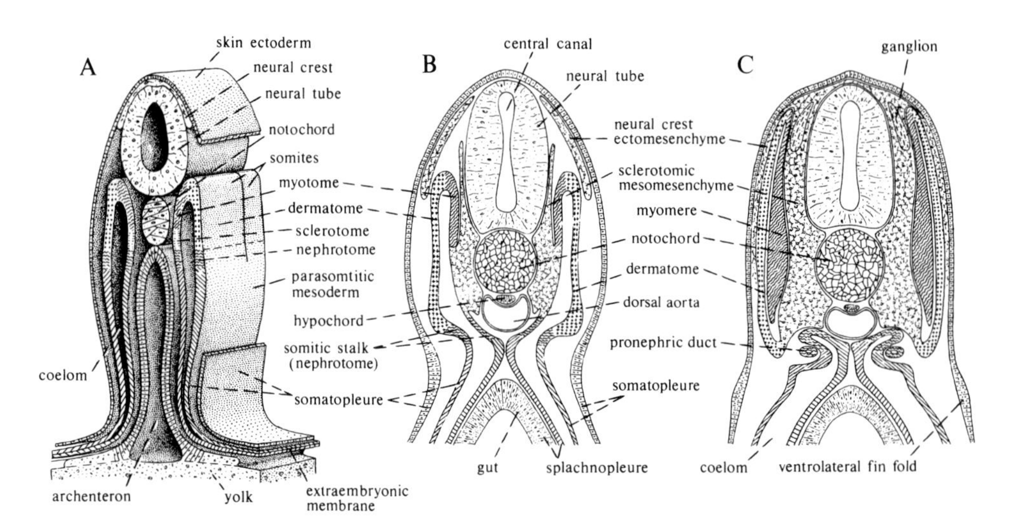

Fig. 5. Squalus acanthias, early embryos (3–4, 4·5 and 9 mm). A, diagrammatic thick transverse section through trunk shortly after closure of neural tube. Posterolateral view. B, C, transverse sections through trunk of later stages. From Bjerring (1977, and unpublished).

organ-forming parts: one lateral, the dermatome, formed by the lateral wall of the vesicle; one ventromedial, the sclerotome; and one dorsomedial, the myotome. The latter forms the somatic musculature but also several skeletal elements (myotomic skeleton*). The sclerotome disintegrates into migratory mesomesenchyme cells which are dispersed around the notochord and the neural tube and form the axial skeleton and certain vessels. The dermatome also proliferates migrating mesomesenchyme cells but because these cells are mixed with neural crest ectomesenchyme cells their fate is not known, and to which extent the dermatomes contribute to the formation of the inner mesenchymatic layer of the skin (dermis) is difficult to say.

In his outstanding pioneer work on the development of the elasmobranch fishes Balfour (1878) demonstrated that the head mesoderm is subdivided into somites (called sections of the head cavity) similar to and continuous with those of the trunk. Balfour also made out the relations to the cranial nerves and showed that the cephalic somites ("sections") are continued by "muscle plates" forming the visceral musculature of the visceral arch belonging to the same metamere as the somite (which means that branchiomerism corresponds with the metamerism†). In view of these facts Balfour (1881) suggested the term mandibular for the somite associated with the "muscle plate" in the mandibular arch. The somite next in front, in his opinion the foremost of the cephalic series, was termed the premandibular and that following behind the hyoid (Fig. 6A). The remaining cephalic somites, one to each branchial arch, were termed branchial. Since the shark (*Scyllium*) studied by Balfour, like most other fishes, has five branchial arches there are most often five branchial somites which together with the three anterior ("prootic") somites admitted by Balfour make eight cephalic somites in all.

Balfour's papers stimulated further researches and in the years around the turn of the century a great number of papers on the metamerism ("segmentation") of the vertebrate head appeared (for bibliography see Goodrich, 1930; de Beer, 1937). Disagreement soon arose as to the number and fate of the cephalic somites, the position of the posterior limit of the head, the relations of the gill arches and gill slits to the metameric somites, and so forth. However, after much unnecessary confusion and many heated debates we can now say that Balfour's main conclusions have been fully confirmed (Goodrich, 1911; 1918; 1930; de Beer, 1924; 1937; Holmgren, 1940; 1943; Damas, 1944; Jarvik, 1954; 1960; 1972; Bertmar, 1959; Bjerring, 1967; 1968; 1971; 1977). The head and trunk somites arise in a similar way; they form a continuous series; no cephalic somites have disappeared, and in the cephalic somites myotome, sclerotome and dermatome may be distinguished. However in spite of the agreement in these regards that long has been prevalent it now seems called for to add a new somite (the terminal) in front of the foremost one (the premandibular) distinguished by Balfour (Bjerring, 1973; 1977). This increases the total number of cephalic somites in most fishes to nine.

From the mandibular somite backwards the somites in each pair are separated by an intervening portion of the notochord. The two premandibular somites, in contrast, are products of the parachordal mesoderm of the "prechordal plate" and because no

* Volume 1, p. 43; † p. 12.

notochord is formed in the intermediary rostral portion of the chordamesoderm the premandibular somites are connected by a transverse commissure (Figs 6B, 16D). This premandibular commissure situated between the anterior tip of the notochord and the infundibular evagination of the brain and probably formed by chordamesoderm, later disintegrates into mesomesenchyme cells which may form skeletal elements (exchordals, Bjerring*). In front of the premandibular somite in *Squalus* Platt (1891) and Zimmermann (1891) independently of each other discovered a similar vesicle, the anterior head cavity (Platt), which they regarded as a true somite. However, since this vesicle, also known as Platt's vesicle (Figs 6A, 8A, 9), according to de Beer (1937) and others, besides in *Squalus*, has been recorded only in *Galeus* (van Wijhe, 1882) its somitic nature has been doubted. Thus Holmgren (1940) regards it as a part of the premandibular somite, whereas de Beer states that it is a structure of no

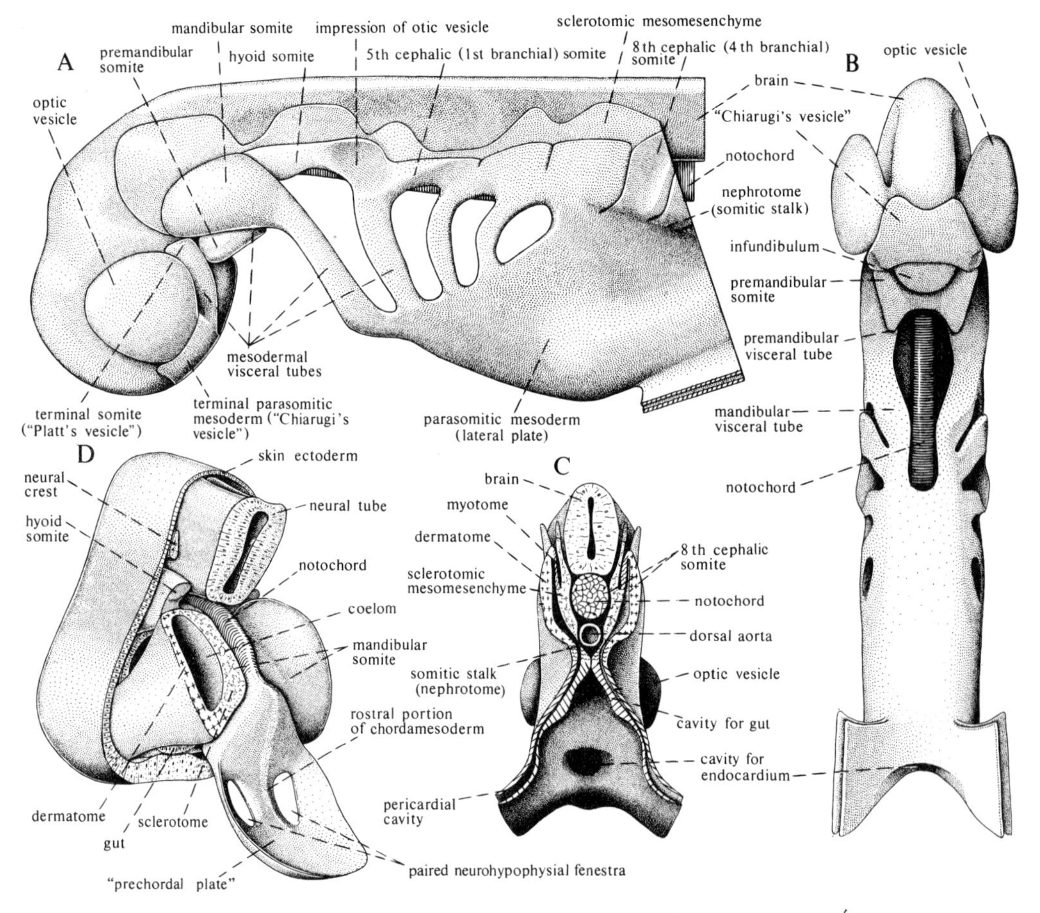

Fig. 6. Squalus acanthias. A–C, drawings of model made by H. C. Bjerring of cephalic mesoderm and brain of embryo 5·5 mm in lateral, ventral and posterior aspects. D, drawing by H. C. Bjerring (unpublished) of same model (embryo 4·5 mm) as in Fig. 3 *B*, *C* in oblique anterodorsal aspect. Anterior part of neural tube and dorsal part of right mandibular somite removed.

* p. 71.

morphological significance which soon disappears. This last statement has been proved to be incorrect, and moreover, the fact that Platt's vesicle has also been described in a number of other sharks (*Heptanchus*, *Pristiurus*, *Scyllium*, *Scymnus*, Dohrn, 1904) has been overlooked (cf. also *Ambystoma* Fig. 8D).

As shown by Holmgren (1940) Platt's vesicle disintegrates into mesenchyme which forms both a somatic muscle (m. obliquus inferior) and skeletal elements (parts of the scleral capsule and the "trabecle"). Accordingly Platt's vesicle, like accepted somites, includes myotome, sclerotome and probably dermatome as well. That it is a true somite, maintained not only by Platt and Zimmermann, but also by Hoffmann (1896–1897), Neal (1898), Lamb (1902), Scammon (1911) and Bjerring (1973; 1977), is further evidenced by the following facts: as indicated by the term Platt's vesicle, the foremost or terminal somite (Bjerring) contains a coelomic lumen. Moreover it arises in front of the premandibular somite, along the most anterior part of the chordamesoderm in the "prechordal plate". Judging from the description given by Hoffmann (1896–1897; see also Holmgren, 1940) the foremost part of the "prechordal plate" in early embryonic stages of *Squalus*, together with the underlying part of the archenteron, forms a broad cellular band extending forwards to the neuropore. Very soon this band divides into a paired lateral part which is the terminal somite (Platt's vesicle) and a median portion which includes the anterior parts of the chordamesoderm and the archenteron. Moreover, the parasomitic mesoderm which forms a continuation of the terminal somite is developed much as the parasomitic mesoderm in other metameres.*

Since the m. obliquus inferior in other vertebrates than *Squalus* and some other sharks arises from what is generally called the premandibular somite it is evident that this somite in most vertebrates is complex including also the terminal somite. The fact that the separation of the terminal somite most often fails to take place is not surprising with regard to its position in the foremost low part of the embryo and the modifications (cephalic flexure, development of mouth) which have occurred in this part and which have contributed to the suppression of the anterior part of the notochord.

That the terminal and the so called prootic somites (premandibular, mandibular and hyoid) are true somites which has been doubted (Starck, 1955; 1963) is evidenced also by their relations to the parasomitic mesoderm.

As pointed out above the mesoderm both in *Ammocoetes* and in *Amphioxus* is completely divided into metameres. In gnathostomes, in contrast, the subdivision of the parasomitic mesoderm is incomplete. This is true in particular of the trunk where the metamerism in the parasomitic mesoderm is confined to its dorsal part. This part is divided into metameric hollow tubes, the somitic stalks (mesomeres), which connect each of the trunk somites with the undivided main part (the lateral plate) of the parasomitic mesoderm (Figs 5, 6, 9). The somitic stalks soon separate from the somites and develop into embryonic excretory organs, nephric units or nephrons. The somitic stalks in the trunk are therefore often called nephrotomes. The most anterior nephrons which open into the pronephric duct constitute the pronephros. The extension forwards of the pronephric series is subject to variation but in *Squalus* (Fig. 6A), at

* p. 14.

any rate, it reaches into the head, the somitic stalks (nephrotomes) of the two hindmost cephalic somites (the eighth and ninth) forming pronephric nephrons (Bjerring, 1967).

In the head the subdivision of the parasomitic mesoderm is more complete than in the trunk. Due to the formation of the gill slits (see below) the parasomitic mesoderm early becomes subdivided into hollow tubes enclosing a detached part of the coelom (Figs 6, 7). These tubes, the mesodermal visceral tubes, give rise to the visceral musculature and have therefore generally been called muscle plates (Edgeworth, 1926; 1928; 1935) or muscular processes (Holmgren, 1940; Bertmar, 1959). Ventrally these tubes are, from the mandibular metamere backwards, connected with the undivided part of the parasomitic mesoderm. Dorsally each of the tubes is continuous with a somite (van Wijhe, 1882; Ziegler, 1908; Goodrich, 1918; Holmgren, 1940; Bertmar, 1959) and there is one tube to each cephalic somite. Accordingly the mesodermal visceral tubes (mandibular, hyoid and most often, five branchial) are metameric structures (as regards the terminal and premandibular somites see below). However, already early in ontogeny the dorsal parts of the visceral tubes become constricted and separate from the adjacent parts of the somites. These constricted parts are suggestive of, and most likely serially homologous with the somitic stalks in the trunk. Because the visceral tubes in the branchial metameres separate earlier than in the mandibular and hyoid metameres these tube for some time appear as a series of ascending rods without connection with the somites. Figures of such stages are often reproduced (Edgeworth, 1935, fig. 86; Starck, 1963, fig. 12; Romer, 1970, fig. 185) and have contributed to the widespread but erronous view (Edgeworth, 1935; Romer, 1970; 1972; Torrey, 1971) that there is no correspondance between metamerism and "branchiomerism". The reason for this misconception is no doubt—besides the early disintegration of the cephalic somitic stalks—that these stalks, the two most posterior in *Squalus* excepted, do not form nephrons. A remarkable fact to be mentioned in this connection is that the series of nephrons in continued forwards in the head by the series of thymus glands which are also metameric organs. The thymus glands (Dohrn, 1884; Hoffmann, 1896–1897; Hill, 1935; Rauther, 1937; Romer, 1970) arise as epithelial thickenings in the endodermal anterodorsal part of the walls of the gill slits, that is not far from the cephalic somitic stalks. The thymus glands also include mesenchymatic material but if this is derived from the somitic stalks as might be a possibility or has another origin is unknown.

The gill slits are formed by pouches of the endodermal gut which meet similar ectodermal pouches (Fig. 7). The intervening branchial membrane soon disappears and a slit is formed bounded by epithelial walls, endodermal in their medial and ectodermal in their lateral parts. A slit together with its epithelial walls constitutes a branchiothyrium (Bjerring, 1977). As a consequence of this mode of formation of the gill slits each mesodermal (mesomesenchymatic) visceral tube becomes surrounded by an outer tube, the medial wall of which is formed by endoderm and the lateral by ectoderm. In gnathostomes the ectodermal part of the wall of the outer tube forms the gills (Dohrn, 1884; Goette, 1901; Moroff, 1904; Sewertzoff, 1923; Goodrich, 1930), whereas the inwardly directed gills of the cyclostomes are derived from the endodermal part of the outer tube. The space between the outer and inner tubes is filled out by

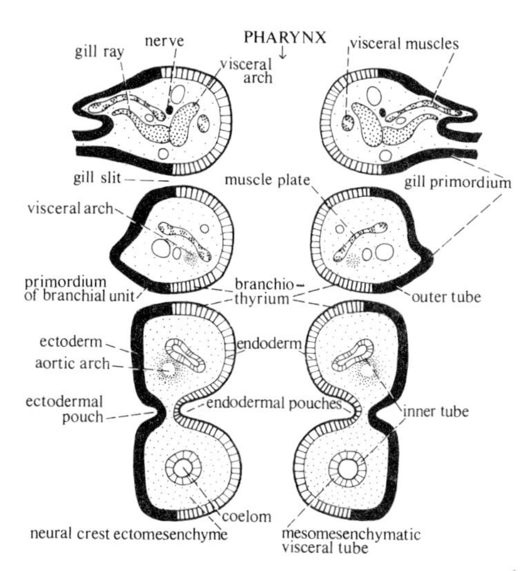

Fig. 7. Diagram illustrating formation and early differentiation of branchial units in shark embryo. Branchial units shown in horizontal sections.

migratory neural crest ectomesenchyme. The complex structures thus formed are the primordia of the branchial units. Since each of the mesodermal visceral tubes originally is continuous with a somite (Fig. 9) the primordia of the branchial units and their derivatives are truly metameric structures. The gill slits are intermetameric in position.

The coelomic lumen of the mesodermal visceral tube which is an isolated part of the coelom of the parasomitic mesoderm and originally continuous with the coelom in the somite, soon disappears. Simultaneously with this process the walls of the tube are converted into muscular tissue which eventually differentiates into the visceral musculature of the unit. The visceral arches are formed by the neural crest ectomesenchyme and judging from figures by Dohrn and Edgeworth (1935) this seems to be true also of the aortic arches (most other vessels are probably mesomesenchymatic in origin, Balinsky, 1970).

In the premandibular and terminal metameres the parasomitic mesoderm, at least partly due to the development of the stomodeal invagination, has been modified and reduced. However, in *Squalus* the premandibular somite (Fig. 6A, B) is continued by a process running along the mandibular visceral tube, and as suggested by Holmgren (1940) this process most likely represents the premandibular visceral tube. That this is so is strongly supported by the conditions in *Pristiurus* (Fig. 8B; Dohrn, 1885; Zimmermann, 1891). In this form the process is longer and moreover it joins the mandibular tube in the same way as the latter joins the hyoid tube. As suggested by Dohrn this process forms the visceral musculature in the premandibular branchial unit, a suggestion which is supported by the conditions in *Acipenser* and urodeles (Fig. 51). An important fact is also that the presumed premandibular tube is accompanied by a distinct strand of neural crest ectomesenchyme (Fig. 8A), and similar strands have been observed also in other sharks and in rays (Hoffmann, 1896–97; Neal, 1898;

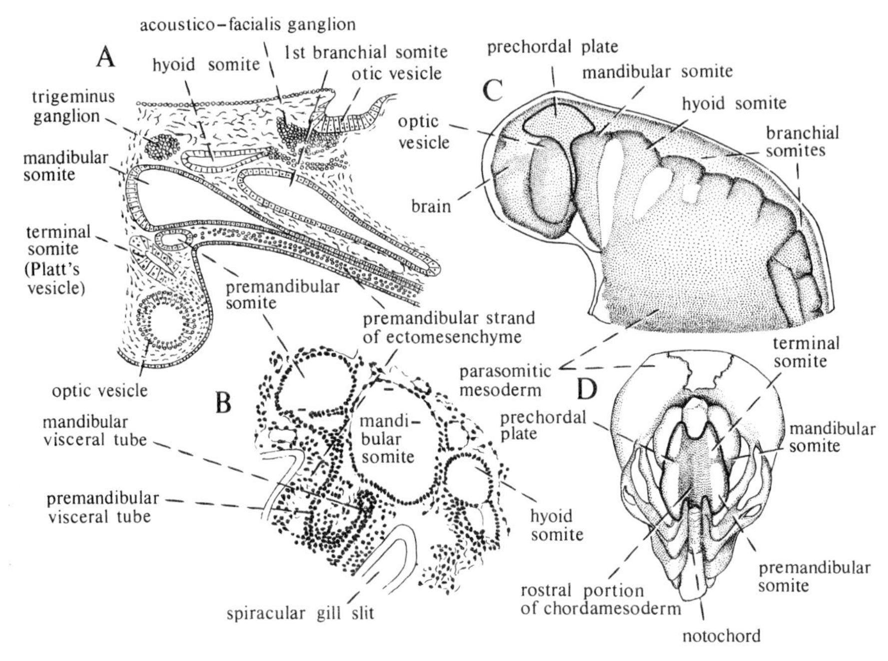

Fig. 8. A, *Squalus acanthias*, embryo 8 mm. Longitudinal section through part of head. From Hoffmann (1896). B, similar section of embryo of *Pristiurus* (9 mm) with well developed premandibular visceral tube joining mandibular tube. From Dohrn (1904). C, D, *Ambystoma punctatum*, embryo. Model of mesoderm (in C with brain) in lateral and anterodorsal aspects, to show anterior cephalic somites and "prechordal plate". From Adelmann (1932).

Dohrn, 1904). Because this strand is very suggestive of the ectomesenchymatic strand in the mandibular branchial unit which forms the mandibular visceral arch it is evident that the strand associated with the premandibular visceral tube must be the primordium of the premandibular visceral arch as indicated already by Zimmermann (1891).

The parasomitic mesoderm in the terminal metamere is much modified being compressed between the developing forebrain with the optic vesicle above and the stomodeal invagination below. However judging from the models of early embryos of *Squalus* (Figs 3, 6, 9; Bjerring, 1967; 1977) and the descriptions given by Dohrn (1904) and Holmgren (1940) the terminal somite ("Platt's vesicle") is continued by a rodlike descending structure with a radiating arrangement of the cells much as in the mesodermal visceral tubes in other cranial metameres. These latter tubes ventrally merge into the ventral undivided part of the parasomitic (lateral plate) mesoderm which grows downwards and fuses below the gut with the corresponding mesoderm of the other side. The presumed terminal mesodermal visceral tube in the same way merges into a thin plate of mesomesenchyme which is continuous with that of the other side. The fate later in ontogeny of the continuous plate thus formed (Chiarugi's vesicle; Chiarugi, 1898; Dohrn, 1904; Holmgren, 1940) and of the presumed terminal visceral tube is unknown, but the fact that the latter, like other visceral tubes is surrounded by neural crest ectomesenchyme indicates that this mesenchyme may

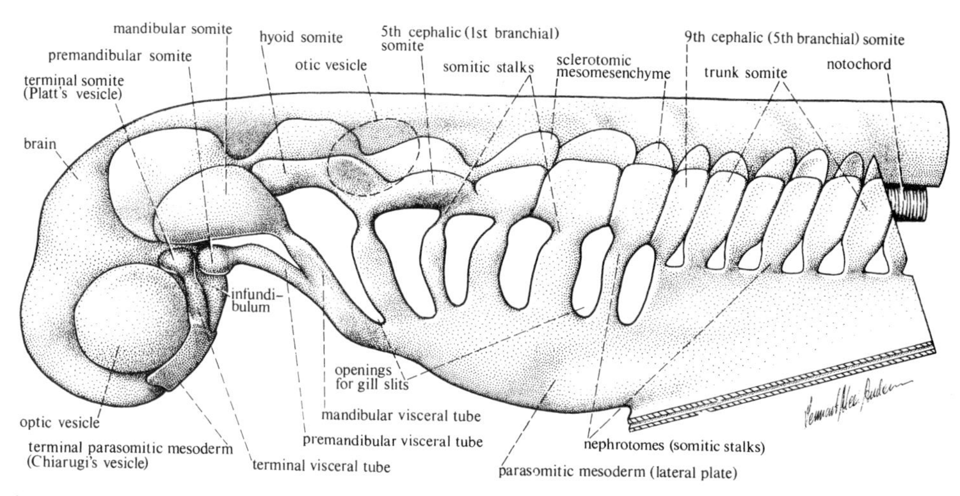

Fig. 9. Diagrammatic representation of mesoderm and neural tube of shark embryo a little more advanced than that shown in Fig. 6A.

develop into visceral skeletal elements belonging to the terminal visceral arch.*

Summary. The embryologic facts presented above justify the following conclusions.

(1) The notochord originally extended forwards to the anterior end of the embryo as do the archenteron and the neural tube. In these regards the vertebrates agree with *Amphioxus*.

(2) In vertebrates the most anterior part of the notochord has become suppressed but also in the "prechordal plate" the chordamesoderm induces the formation of the overlying part of the neural tube (archencephalic inductor) and may, if cultivated *in vitro*, form true notochordal tissue.

(3) Moreover true somites (the terminal and the premandibular) are formed also along the foremost rostral ("prechordal") portion of the chordamesoderm. Like other somites these two foremost somites have a coelom, they differentiate into sclerotome, myotome and (probably) dermatome, and they have similar relations to the parasomitic mesoderm as have other somites.

(4) The metamerism of the parasomitic mesoderm is, in the trunk, restricted to its most dorsal part adjacent to the somites, where a series of metameric somitic stalks or nephrotomes are formed. In the head the subdivision of the parasomitic mesoderm is more complete. Due to the formation of intermetameric gill slits it is divided into metameric mesodermal visceral tubes, each connected with a somite by a somitic stalk. Those somites which are connected with visceral tubes are regarded as cephalic or head somites. In addition to the four "prootic" somites (terminal, premandibular, mandibular and hyoid) there are in most fishes five branchial somites, thus nine cephalic somites in all. The somites of the trunk and head form a continuous series. Some of the cephalic somites have been modified by the development of the otic

* p. 86.

vesicle, the flexure of the brain, and the formation of the mouth, but none has disappeared and none has been added.

(5) Ventrally the mesodermal visceral tubes merge into the ventral undivided part of the parasomitic mesoderm, which below the gut fuses with that of the other side. A corresponding fusion has occurred also in the terminal metamere where the fused parts form a thin preoral cell lamina (Chiarugi's vesicle).

(6) Mesodermal visceral tubes situated between two gill slits are surrounded by an outer tube, partly ectodermal and partly endodermal. These two tubes, the outer and the inner, together with the strand of neural crest ectomesenchyme which is found in the space between them constitute the primordium of a branchial unit. Also most anteriorly and most posteriorly were no distinct gill slits and no complete outer tubes are formed the visceral tube is accompanied by a strand of ectomesenchyme. Since each of the visceral tubes, generally nine in number as the cephalic somites, in early ontogenetic stages is connected with a somite it is evident that the branchial units and their derivatives are truly metameric structures. It is therefore incorrect to speak of "branchiomerism" as opposed to metamerism. The subdivision into visceral tubes and branchial units is a part of the metamerism of the head.

(7) The early ontogenetic development of the vertebrates indicates that their common ancestor, the protovertebrate, was a free-swimming bilateral symmetric animal with only short or no trunk, in several regards suggestive of the larva of *Amphioxus* (Fig. 3A; Sewertzoff, 1899; Jarvik, 1972).

(8) The pronounced metamerism displayed by the early vertebrate embryo to a considerable extent determines the further development, a condition which is of fundamental importance for the understanding of the composition of the vertebrate head.

2 Nervous System and Sense Organs

Neuromery. In the early vertebrate embryo of lampreys as well as man, the neural folds and the neural tube show a series of bulges separated by constrictions (Locy, 1895; Neal, 1898; 1918; von Kupffer, 1905; Bergquist, 1952; Bergquist and Källén, 1954; Vaage, 1969; Kuhlenbeck 1973). The importance of these bulges or neuromeres (Fig. 15C, D) has been much discussed, but as far as may be judged at present there is no clear correspondance between neuromery and metamerism. According to Bergquist the bulges mark the position of budding centres of brain nuclei (migration areas) and are therefore of greater importance for the understanding of the further development of the brain tube than the two-, three- and five-vesicle stages of von Kupffer (1905).

SUBDIVISION OF BRAIN AND ROSTRAL PROLONGATION

The five vesicles of von Kupffer (Fig. 10) form, in the adult, the five well known divisions of the brain: telencephalon, diencephalon, mesencephalon, metencephalon and myelencephalon or medulla oblongata. This terminology has become generally accepted despite the fact that in certain respects it is unsatisfactory. This is true especially of the terms mesencephalon, metencephalon and medulla oblongata. The ventral part of the mesencephalon (tegmentum) may most properly be regarded as a part of the medulla,* whereas its dorsal part, the tectum opticum or optic lobe, belongs with the visual sense organs.† The ventral part of the metencephalon, too, clearly belongs with the medulla. Its dorsal part is the cerebellum. The names applied by von Kupffer to the divisions of the brain in the two- and three-vesicle stages are also frequently used. The two divisions in the early embryo (Fig. 10A), the archencephalon and the deuteroencephalon, are separated by a fold, the plica encephali ventralis, which is the first indication of the cephalic flexure. As emphasized by von Kupffer the deuteroencephalon and its posterior continuation, the medulla spinalis (spinal cord in the adult), arise dorsal to the notochord, whereas the archencephalon is "prechordal" in position which means that it lies dorsal to the "prechordal plate".

* p. 30; † p. 20.

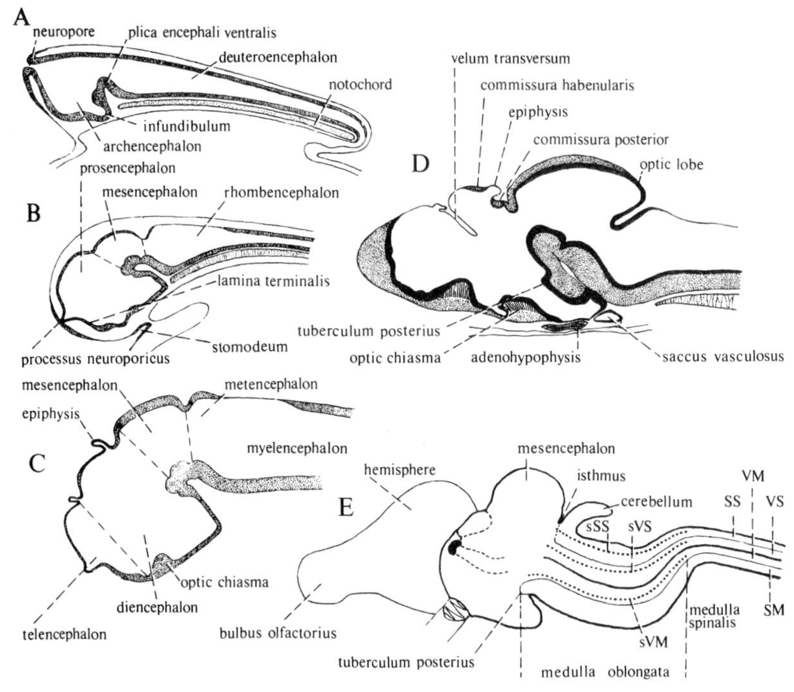

Fig. 10. A–C, three stages in development of vertebrate brain. A, two vesicle stage; B, three vesicle stage; C, five vesicle stage. After von Kupffer 1905. D, *Amia calva*, larva four weeks old. Brain in median section. From Dean (1896). E, diagram of brain and anterior part of spinal cord (medulla spinalis) to show the four columns of His-Herrick and extent of medulla oblongata. Position of special sensory and motor columns in medulla indicated by dotted lines. Mainly after Senn (1970.

sSS, special somatic sensory (acoustico-lateralis area); SS, somatic sensory; sVS, special visceral sensory; VS, visceral sensory; sVM, special visceral motor; VM, visceral motor; SM, somatic motor.

This subdivision has found some support in the experiments made by F. E. Lehmann, S. Toivonen and others (see Bergquist and Källén, 1954; Saxén and Toivonen, 1962). According to these investigations (Fig. 11A) the "prechordal plate", or more likely only the rostral portion of the chordamesoderm in that plate, acts as an archencephalic inductor, whereas that part of the chordamesoderm which develops into the notochord induces the deuteroencephalon and the spinal cord (deuteroencephalic or deuteroencephalospinal inductor). However, if the mesencephalon, which together with the prosencephalon and the rhombencephalon constitute the brain in the three-vesicle stage (Fig. 10B), is formed under the influence of the archencephalic or the deuteroencephalic inductor has been impossible to establish. Before these experiments were done Kingsbury (1920; 1922; 1924; 1930) and Johnston (1911; 1923) claimed that the mesencephalon, with reservation for its ventral part, goes together with the prosencephalon and they divide the brain into an anterior "prechordal" or "premetameric" portion, comprising the prosencephalon and the mesencephalon, and a posterior "epichordal" or "metameric" portion represented by the rhombencephalon. A subdivision of the brain into two portions seems to be justified although—due mainly to the new conception regarding the forward extent of the notochord—in a way differing somewhat from that suggested by Kingsbury and Johnston.

Most appropriate seems to be to divide the brain into anterior or premedullary and posterior or medullary portions along a line running from the preoptic recess, which bounds the optic chiasma (chismatic ridge) anteriorly via the tip of the tuberculum posterius to the isthmus (Fig. 11D). The optic chiasma arises at the most anterior end of the neural plate (Fig. 11A; Johnston, 1909; Kingsbury, 1920; 1922; 1924; 1930; Kuhlenbeck, 1973); and to the anterior part of the embryo reach also the archenteron, the chordamesoderm and the metameric mesoderm. The optic chiasma is therefore a most important landmark in the vertebrate head. However, on each side of the rostral portion of the chordamesoderm the foremost two somites (terminal and premandibular) are formed; and it is the sclerotomes of these somites that produce the most anterior parts of the mesomesenchymatic axial skeleton. That this is so is confirmed by the fact that, as shown experimentally (Stone, 1926; Hörstadius and Sellman, 1946) or in other ways (Landacre, 1921; de Beer, 1947), the mesomesenchymatic ("endomesodermal") part of the neurocranium extends forwards to the level of the canal for the n. opticus (Fig. 35C). Accordingly also the exit of the n. opticus from the brain and the opticus canals give us important information as to the approximate position of the original anterior end of the embryo. However, in the adult these structures are found far behind the anterior end of the head. This implies that the head

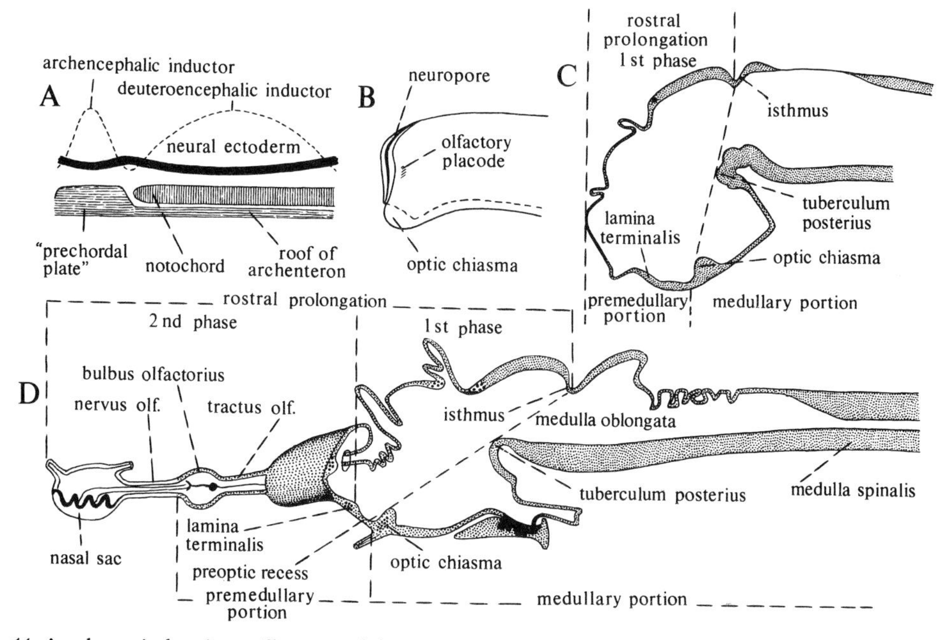

Fig. 11. A, schematic drawing to illustrate subdivision of neural system in archencephalic and deuteroencephalic parts. From Bergquist and Källén (1954). B–D, schematic drawings to illustrate extent of premedullary and medullary portions of brain and first and second phases of rostral prolongation. B, early stage, before rostral prolongation has started. Rostral end of embryo marked by optic chiasma. From Kingsbury and Adelmann (1924, after Johnston); C, D, mainly after Figs 10 C and 13.

in ontogeny has grown considerably forwards and this remarkable lengthening has influenced not only the anterior ectomesenchymatic part of the neurocranium, the brain and other associated soft parts, but also the hyobranchial skeleton, the subbranchial series, the hypobranchial musculature, etc (Jarvik, 1963, p. 29). What really happens in this growth forwards of the head is not easy to make out, and we can only say that a number of intricate developmental processes are involved. This complex of processes will collectively be referred to as the rostral prolongation (Fig. 11B–D).

THE PREMEDULLARY BRAIN PORTION

Subdivision. The premedullary portion of the brain, situated in front of the isthmus–chiasma line, seems essentially to be a product of this remarkable rostral prolongation. Two phases may be distinguished, one (Fig. 11C) dominated by the formation of the visual sense organs (the lateral eyes and the neuroepiphysial complex) and one (Fig. 11D) concerned mainly with the development of the telencephalon and the sense of smell.

Visual Sense Organs

The eye-forming area, when first discernible in the early embryo, lies in the most anterior part of the neural plate (Jacobson, 1959). In connection with the formation of the neural tube the walls of that tube expand laterally at the same time as they grow forwards (Kingsbury and Adelmann, 1924). Very soon a pair of lateral saclike protrusions of the brain wall are formed. Each of these protrusions or optic vesicles is connected with the brain proper by an optic stalk. Due to the development of the lens from an ectodermal lens placode the outer wall of the optic vesicle is pressed inwards and the vesicle is transformed into a double-layered optic cup (Fig. 15A). The two layers of the cup form the retina, which as is of importance to note, is a laterally displaced part of the brain wall; and from surrounding mesenchyme cells the sclera and the choroid are built up. The sclera often forms a cartilaginous scleral capsule (Figs 8, 10, 58, 286, Volume 1) outside of which there may be an exoskeletal sclerotic ring composed of a varying number of sclerotic plates. Neurites (axons) of the ganglionic cells in the retina grow inwards through the optic stalk. In the floor of the diencephalon, or below that floor, most or all of these fibres cross those coming from the other side of the head in the optic chiasma (Fig. 12) and continue in the wall of the diencephalon to the contralateral part of the tectum opticum where they end. This long strand of fibres formed by retinal neurities constitutes a brain tract, but generally the portion running from the retina to the chiasma, together with the surrounding optic stalk, is called the nervus opticus or nerve II, whereas the term optic tract is reserved for the portion running in the wall of the brain proper.

The tectum opticum belongs clearly to the visual system as does also the neuroepiphysial complex in the roof of the diencephalon. The diencephalic roof (epithalamus) is mainly a thin, vascularized choroid plexus, the tela choroidea anterior, which continues forwards into the telencephalon. The boundary between the dienceph-

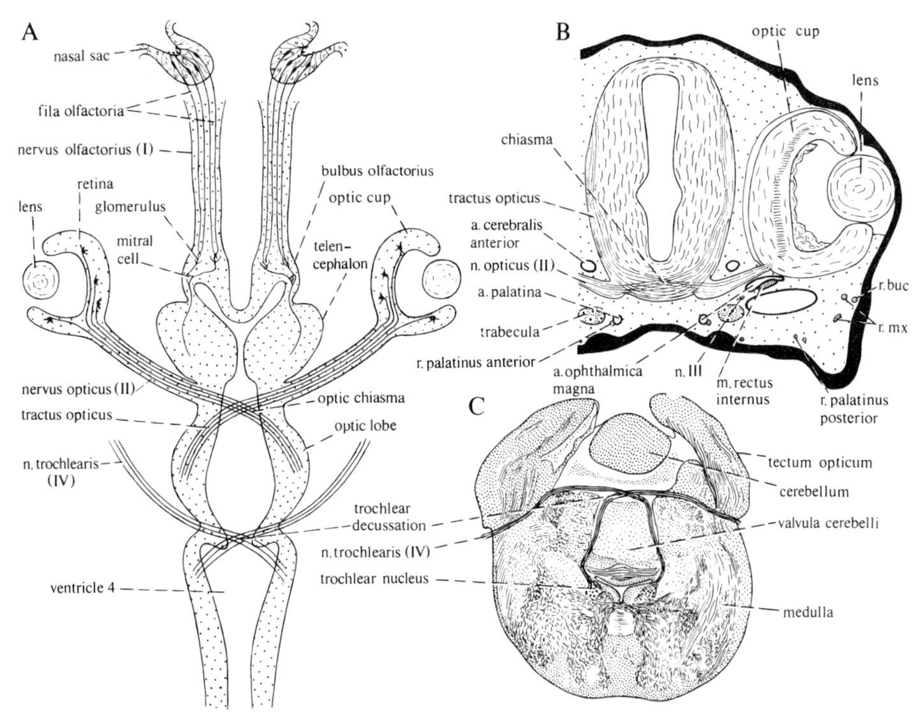

Fig. 12. A, diagram of brain to show composition of bulbus olfactorius and position of optic and trochlear decussations. B, *Amia calva*, embryo 8 mm. Transverse section in region of optic chiasma. C, *Anguilla anguilla*, adult. Composite transverse section to show trochlear decussation. From Kappers *et al.* (1960; after van der Horst).
r.buc, r. buccalis lateralis; r.mx, r. maxillaris.

alon and the telencephalon is marked by an invaginated part of the tela, the velum transversum (Fig. 13). The tela generally shows three saclike protrusions. The foremost of these sacs, the paraphysis (Figs 121, 122; Figs 60, 252, Volume 1), is formed by an evagination of the anterior part of the velar fold and belongs accordingly to the telencephalic part of the tela. In the adult the paraphysis is generally a digitate vesicle connected with the wall of the velum by a short paraphysial stalk. Behind the paraphysis follows the second protrusion, the dorsal sac. This sac which sometimes (*Amia, Lepisosteus*) may be large (Figs 60, 252, Volume 1), is probably, like other parts of the tela (Netsky and Shuangshoti, 1970), developed for the exchange of cerebrospinal fluid.* The third (posterior) saclike protrusion, the neuroepiphysis (Kappers, 1965; Oksche, 1965; Eakin, 1973; Bjerring, 1975), in contrast, is a light-sensitive structure (photoreceptor). This structure, which bears relations to the paired habenular nucleus, is often divided into two organs, parapineal and pineal, both of which may include an eyelike terminal vesicle (Figs 13, 14). These two organs, which generally differ in size, are sometimes asymmetric in position; and the one habenular nucleus is often larger than its fellow. In the lamprey *Geotria* (Dendy, 1907; Eddy, 1972) the pineal organ lies to the right and behind the parapineal. Since the pineal is large and is in nervous

* cf. *Polypterus*, Volume 1, p. 297.

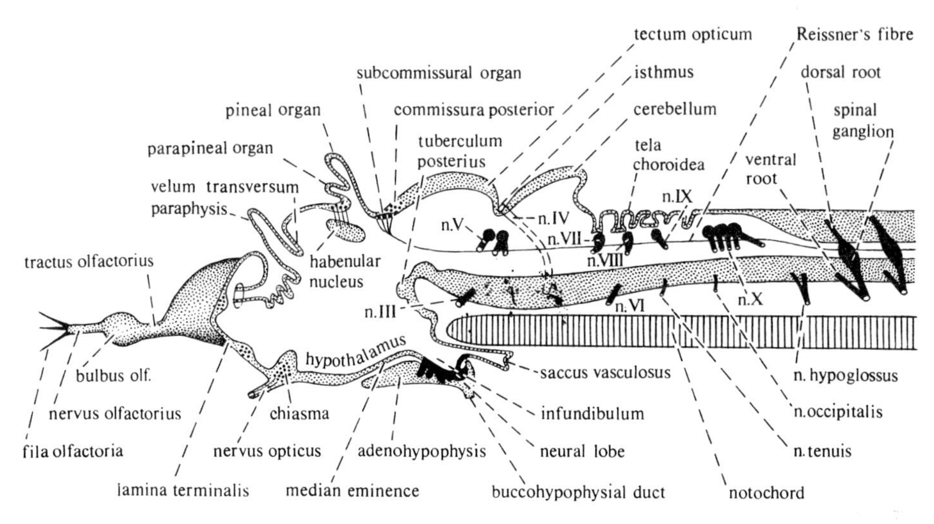

Fig. 13. Diagram of vertebrate brain in median section. Partly after Bütschli (1921) and Plate (1922).

connection with the large right habenular nucleus, whereas the parapineal is small and connected with the small left habenular nucleus Dendy suggested that vertebrates originally possessed a pair of dorsal eyes. In support of this view it has been pointed out (Edinger, 1956; see also Eakin, 1973) that arthrodires* often show asymmetric paired "pineal" pits or foramina. Moreover, the material in the neural plate destined to form the neuroepiphysis is paired (Jacobson, 1959; Eakin, 1973). However, when the neuroepiphysis appears in ontogeny it is generally a median evagination of the diencephalic roof (Bjerring, 1975). This evagination (Fig. 14) may develop into the undivided neuroepiphysis characteristic of elasmobranchs, holocephalans and dipnoans and present also in *Polypterus, Latimeria* and *Acipenser.* However in many forms the original evagination divides into anterior parapineal and posterior pineal organ. In certain forms, e.g. in *Lampetra*, osteolepiforms and anurans, both organs retain their median position in the adult, whereas in others, e.g. in porolepiforms, hynobiid urodeles (Figs 121, 122; Jarvik, 1972; Bjerring, 1977), *Amia* (Fig. 60B, Volume 1), *Lepisosteus* and teleosts (U. Holmgren, 1965) the parapineal organ may be displaced (usually to the left) and may eventually be located behind the pineal organ as in certain teleosts. Since the two organs in embryonic stages of *Geotria* are also median in position it is evident that the asymmetry of the neuroepiphysial organs is secondary. Another problem concerns the interpretation of the median foramen in the skull roof of fossil forms. The so called frontal organ in recent anurans develops from the rostral end of the neuroepiphysial evagination (Kappers, 1965) and is accordingly a parapineal organ and not a "pineal terminal vesicle" or pineal organ (Oksche, 1965; Jarvik, 1967a). The so called pineal opening in the skull roof of osteolepiforms also almost certainly lodged the parapineal organ and no doubt this organ occupied the foramen in

* Volume 1, p. 367.

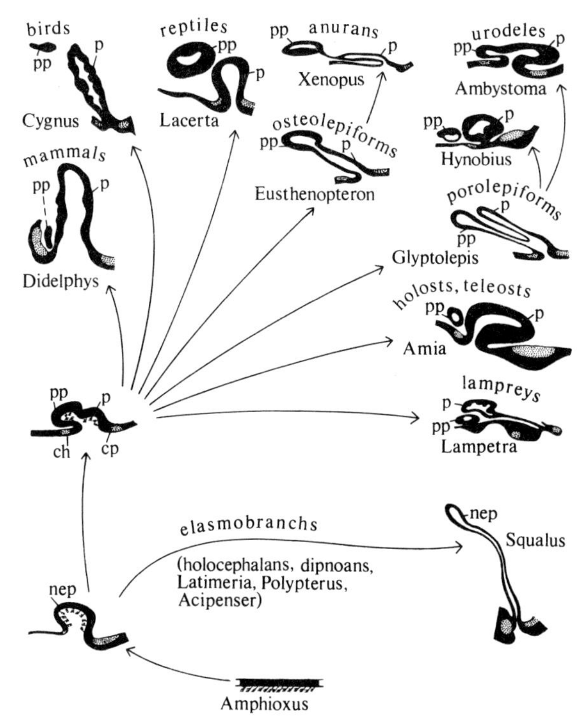

Fig. 14. Diagrammatic representations to illustrate evolution of neuro-epiphysial complex in vertebrates from a simple median area of photo-receptors as found in *Amphioxus*, From Bjerring (1975; simplified). ch,cp, habenular and posterior commissures; nep, neuro-epiphysis; p, pineal organ; pp, parapineal organ.

the skull roof in fossil tetrapods, as it does in recent anurans and reptiles. The fact that the parapineal foramen in the skull roof in osteolepiforms, anurans and fossil batrachomorphs lies far forwards, between the frontals, whereas in recent reptiles and fossil reptilomorphs it usually is found between the parietals, is due to displacement backwards of the brain within the cranial cavity and the development of flexures in the posterior part of the brain stem in the reptilomorphs (Fig. 124; Jarvik, 1967a). In palaeoniscids the foramen between the frontals most likely lodged the terminal vesicle of the pineal organ. This is probably the case in the recent teleost *Abramis farenus* which is remarkable in having a pineal foramen in the exoskeletal skull roof (Jarvik, 1967a, p. 186, plate 3b), but the possibility cannot be excluded that there was an undivided neuroepiphysis in palaeoniscids as in *Acipenser*.

Telencephalon and Olfactory Organs

The second phase in the rostral prolongation (Fig. 11D) results in a lengthening of the telencephalon and a displacement forwards of the olfactory organs.

The portion of the anterior wall of the prosencephalon situated between the processus neuroporicus and the optic chiasma, the lamina terminalis (Figs 10B, 11C, D, 13), is retained in the adult brain and is therefore an important landmark as well.

Early in ontogeny the telencephalon grows forwards on each side of the lamina terminalis forming two bulges with a common epithelial roof which is the anterior part of the tela choroidea anterior. In most vertebrates the thick lateral walls of these bulges are bent inwards and downwards (inversion) and the bulges are transformed into two tubes, the cerebral hemispheres. In *Amia* and other actinopterygians, in contrast, the walls of the bulges become thickened and are ultimately bent outwards and downwards in their dorsal parts (eversion; Figs 225–227, Volume 1). But no matter whether there is inversion or eversion the telencephalon in vertebrates extends far forwards in front of the lamina terminalis. The bulbus olfactorius (Figs 11–13), characterized by its content of glomeruli and mitral cells,* arises in ontogeny as an anterior paired evagination of the telencephalon. In the adult it often appears as a swelling which—the urodeles and porolepiforms excepted†—forms the anterior part of the telencephalon. However, in many vertebrates the bulbus is separated from the main part of the brain by a more or less long tractus olfactorius which transmits the neurites of the mitral cells. This may be due to a shifting backwards of the brain within the cranial cavity, as in reptilomorphs (Fig. 124B) and *Latimeria*, but seems in most cases to be a consequence of the displacement forwards of the olfactory organ.

The olfactory placodes which arise on each side of the processus neuroporicus soon deepen into olfactory pouches situated close anterolateral or lateral to the telencephalon. In connection with the transformation of the olfactory pouches into nasal sacs they move forwards and may in the adult be situated far in front of the brain. This remarkable change in position may cause a lengthening of either the tractus olfactorius (Stensiö, 1963a) as, for example, in elasmobranchs, holocephalans, *Neoceratodus* and certain teleosts (*Cyprinus, Gadus* and others) or of the n. olfactorius as in *Protopterus, Amia, Lepisosteus* and many teleosts (Figs 49, 252, Volume 1).

The fact that in certain teleosts (and *Neoceratodus* among dipnoans) it is the tractus, whereas in other teleosts (and *Protopterus*) it is the nervus olfactorius that is lengthened, is hard to explain. It is also doubtful if the brain really is retracted from the ethmoidal region in ontogeny leaving a space (e.g. cavum precerebrale in sharks; cavum internasale in urodeles) between the nasal cavities. It is more likely that this phenomenon is due to a growth forwards of the rostral part of the skull as a whole. That such a growth may occur independently of the nasal organs is evidenced by the conditions in *Acipenser* and several other vertebrates in which the nasal sacs are situated far behind the tip of the rostrum (this in contrast to other long-snouted forms, such as *Lepisosteus*, in which the nasal organs are terminal in position). Of interest in this connection are also the temnospondylous stegocephalians, many of which are characterized by a very strong rostral prolongation, a condition which has caused considerable confusion in the terminology of the dermal bones of the skull roof‡ (Jarvik, 1967a). In them (Fig. 123) there is differential growth and in forms (e.g. *Lyrocephalus*) in which preorbital and postorbital zones of intensive growth may be distinguished, both the nervus and the bulbus olfactorius have been much lengthened, whereas in others (e.g. *Benthosuchus*) in which the predominant growth has occurred in the preorbital zone only the n. olfactorius has been influenced.

* Volume 1, p. 84; † pp. 205–206; ‡ p. 206.

The nasal sacs and their incurrent and excurrent external openings have also undergone other changes in position. The cephalaspidomorphs are unique in these regards (Fig. 352, Volume 1). In several early fishes (e.g. rhenanid arthrodires, antiarchs) the nasal sacs had migrated to the dorsal side of the head, whereas in ichthyostegids (Fig. 171, Volume 1) the anterior and in the porolepiforms (Fig. 132) the posterior external nostril had moved in the opposite direction, towards the margin of the mouth. A more remarkable displacement of the posterior (excurrent) nostril had occurred even in the oldest known dipnoans in which it, like in holocephalans, lies in the roof of the mouth cavity (Figs 305, 310, Volume 1). This opening (the pseudochoana) must not be confused with the internal nasal opening or choana characteristic of tetrapods and their piscine ancestors; the osteolepiforms and porolepiforms. The origin of the choana has been much discussed and several suggestions have been made (Hinsberg, 1901; Fahrenholz, 1925; Matthes, 1934; Schmalhausen, 1968; Medvedeva, 1964; 1975; Bertmar, 1966; 1966a; 1969; Bjerring, 1972; 1977). Since separate anterior and posterior external nasal openings and tubes were present in the porolepiforms together with the choana it is evident that the choana has arisen independently of these openings or tubes; and also the view (Schmalhausen, Medvedeva) that a portion of the infraorbital sensory canal is involved is refuted by the conditions in the porolepiforms (Fig. 317, Volume 1; Jarvik, 1972). The choana may be a secondary formation as is probably the internal nasal opening in certain aberrant teleosts (Atz, 1952; 1952a; Jakubowsky, 1975), but it may also be a primitive feature and the possibility that it has arisen independently in porolepiforms and osteolepiforms cannot be excluded. In the ontogeny of urodeles (Bertmar) the nasal sac presents a posterior pouch (choanal process) which meets an anterior pouch or process of the gut. When the wall separating these pouches disappears, a communication, the choana, is formed between the cavities of the nasal sac and the gut. Consequently the choana, in this case at any rate, is formed in the same way as a gill slit (branchiothyrium, Fig. 7) and for this and other reasons (cf. Marshall, 1879) it has been suggested (Bjerring, 1972;1977) that the nasal sac and the choanal tube represent a pre-prespiracular gill slit (Figs 22–24; Fig. 135F, Volume 1). This view is supported by the fact that the olfactory epithelium in teleosts contains a special type of cells similar to the chloride cells of the pseudobranch (Bertmar, 1972a). According to some writers (e.g. Hinsberg, 1901) the choana in anurans arises in the same way as in urodeles although the choanal process is a solid rod and the gut process is short or perhaps non-existent. Fahrenholz (1925), in contrast, claims that the communication between the nasal sac and gut is established by the closure of a naso-buccal groove, which is the mode of formation of the choana said to be characteristic of the amniotes (Matthes, 1934). If there is a difference in this respect between urodeles and other extant tetrapods, which is still doubtful (cf. Medvedeva, 1975), this would be in accordance with the fact that the urodeles in many other respects stand apart and support the view that the tetrapods are diphyletic in origin. However, the origin of the nasobuccal groove is still obscure and further researches are necessary before the old and much debated problem of the origin of the choana is solved. Of interest in this connection are possibly the peculiar lateral diverticles of the preoral endoderm in *Lepisosteus* (Lindahl, 1944, fig. 5). Also the fact that the most anterior neuromasts of the snout in urodeles and porolepiforms (Fig. 130), in contrast to osteolepiforms and anurans, are innervated by a buccalis branch which runs forwards medial to the choana, may well turn out to have some bearing upon this problem. Other intricate and still unsolved problems concern the origin of the hypophysis.

Hypophysis

Among the basic structures of vertebrates the hypophysis or pituitary gland holds an outstanding position. This important structure, common to all vertebrates (for references and review see Wingstrand, 1966), is a double formation composed of two organs, the neurohypophysis and the adenohypophysis, which secondarily but obviously were early in vertebrate phylogeny have become associated with each other. The neuroepiphysis is situated in the diencephalic floor (hypothalamus), behind the optic chiasma, and thus in the medullary portion of the brain. It includes, in the adult (Fig. 59, Volume 1), the median eminence, the neural lobe and sometimes (elasmobranchs, actinopterygians) the saccus vasculosus. In ontogeny it is discernible quite early as an evagination, the infundibulum or saccus infundibuli, of the floor of the archencephalon. This evagination is found close in front of the transverse commissure between the premandibular somites and sometimes its ventral part, at any rate, which occupies the neurohypophysial fenestra (*Squalus*; Fig. 6D) is divided into lateral halves by the intervening rostral portion of the chordamesoderm (Bjerring, 1971), a condition which possibly means that it is paired in origin.

With respect to the origin of the adenohypophysis, which certainly is homologous in all vertebrates, we have to consider two different views.

(a) Derivation from Rathke's pouch. Partly at least as a consequence of the cephalic flexure the skin ectoderm forms a more or less deep pocket below the anterior part of the brain. This pocket, the ectodermal gut or stomodeum, is separated from the endodermal gut (archenteron) by the oral membrane (or plate) which is a double structure composed of ectodermal and endodermal epithelia (Figs 15–17). Close in front of that membrane there is in certain groups (elasmobranchs, amniotes) a pouch, Rathke's pouch or pocket, in the roof of the ectodermal gut, and a similar pouch, Seessel's pouch, is found behind the oral membrane in the roof of the endodermal gut (these two pouches are possibly vestiges of the foremost branchiothyrium according to Bjerring, 1973; 1977).

According to current views (Fig. 15) it is Rathke's pouch that is the primordium of the adenohypophysis; and also the epithelial hypophysial stalk (buccohypophysial duct) which in embryos and sometimes in the adult (e.g. *Polypterus, Elops, Eusthenopteron*) connects the adenohypophysis with the mouth cavity piercing the parasphenoid (Fig. 196, Volume 1), is considered to be ectodermal.

(b) Derivation from the area of the processus neuroporicus. The neuropore of the early embryonic brain (Fig. 9) is surrounded by so called lips which are the anterior ends of the paired neural fold. When these lips fuse in connection with the closure of the neuropore a process, processus neuroporicus (von Kupffer), is formed (Fig. 10B). This process is inseparably connected with an anterior median thickening of the

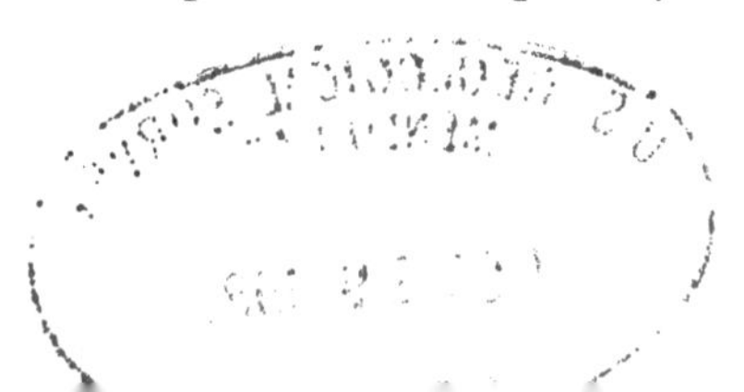

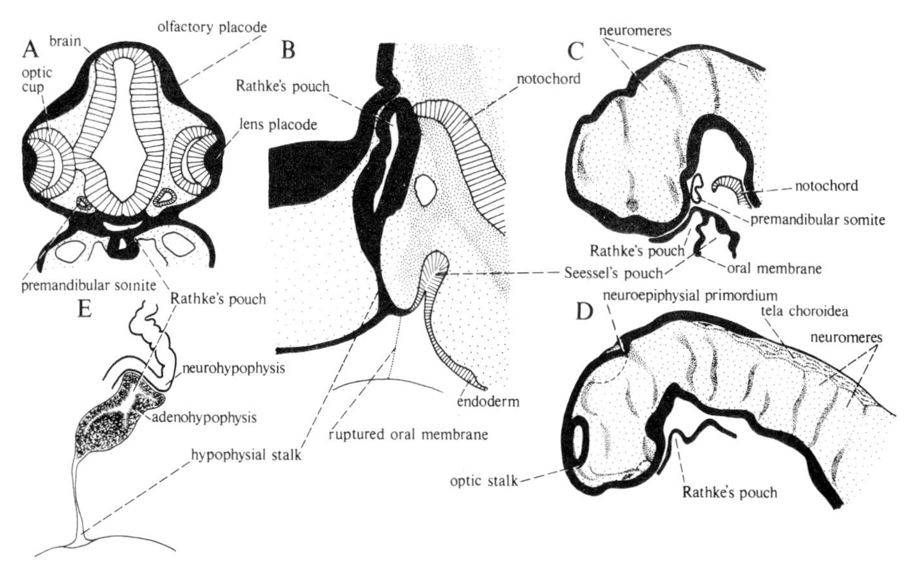

Fig. 15. Representations of embryos of various vertebrates to illustrate Rathke's and Seessel's pouches. A, shark, *Scyliorhinus canicula*, embryo 7 mm. Transverse section. B, bird, *Riparia riparia*, embryo 7·8 mm. Median section, C,D, graphical reconstructions of brain tube in lateral aspect with neuromeres. Rathke's and (in C only) Seessel's pouches in median section. C, *Squalus acanthias*, embryo 9·5 mm and, D, *Homo*, embryo, 3·5 mm. E. *Homo*, embryo (age group 19). Hypophysis in median section. A,C,D, from Bergquist (1952); B, from Wingstrand (1951); E, from Streeter (1951).

ectoderm which is the "unpaired olfactory placode" of von Kupffer. As is well established (von Kupffer, 1893; 1905; de Beer, 1923; Holmgren, 1931; Lindahl, 1944) the adenohypophysis in *Acipenser, Amia*, and *Lepisosteus* arises in this mass of cells (Fig. 16). The primordium grows backwards below the brain towards the infundibulum, but is separated from the stomodeum by the endodermal preoral gut in which the adhesive organ develops. The posterior part expands and forms the adenohypophysis in the adult. The ectodermal epithelial stalk, which in the embryo connects the adenohypophysis with the area of the processus neuroporicus, becomes lengthened, partly at least due to the rostral prolongation of the brain. Later this ectodermal stalk disappears and is replaced by a secondary endodermal stalk (Lindahl; subhypophysial cells, Holmgren) which (in embryos of *Lepisosteus*) pierces the parasphenoid. Upon studies of very early stages of *Squalus* and *Scyllium* Lindahl (1944) came to the conclusions that the adenohypophysis also in sharks—in which Rathke's and Seessel's pouches are present—is derived from the area of the processus neuroporicus and that in them, too, endodermal material is involved. Moreover a transient secondary endodermal stalk occurs in embryos of urodeles (*Hynobius*, Bjerring 1973; cf. birds, Wingstrand, 1951, fig. 58). Obviously the current view that the hypophysial stalk (buccohypophysial duct) is ectodermal is open to doubt; and also the view that the adenohypophysis is a derivative of stomodeal ectoderm (Rathke's pocket) needs confirmation (cf. *Rana*, Fig. 26C, D).

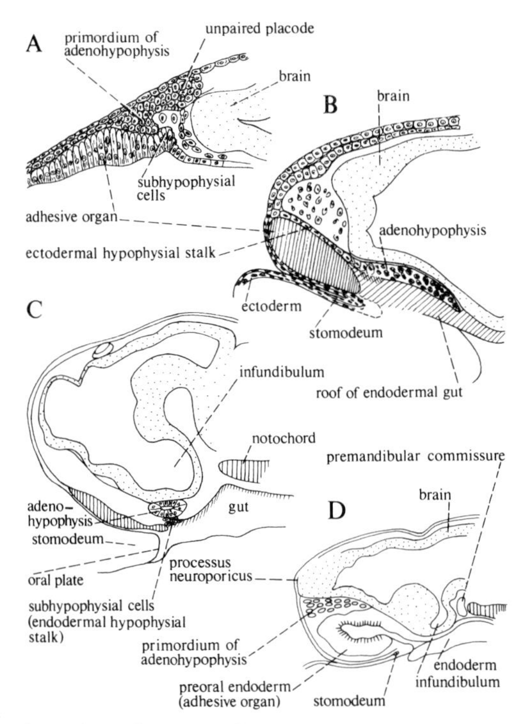

Fig. 16. Development of adenohypophysis from area of processus neuroporicus. A–C, *Acipenser ruthenus*, three stages in development of adenohypophysis. From Holmgren (1931). D, *Lepisosteus osseus*, embryo, 4·0 mm. Median section. From Lindahl (1944).

It is not easy to say from which kind of cells in the area of the processus neuroporicus the adenohypophysis really originates. Most likely it is derived from placodal ectoderm (Knouff, 1935), but neural crest ectoderm may be involved and it may originate from the processus neuroporicus itself and be formed by neurectoderm. If the latter alternative is true, this brings us over to another intricate and still unsolved problem. Since the processus neuroporicus is formed by the fusion of the paired lips of the neural folds it is possible that the adenohypophysis is paired in origin as it is claimed to be in urodeles (Fig. 17) and several other vertebrates (Kingsley and Thyng,

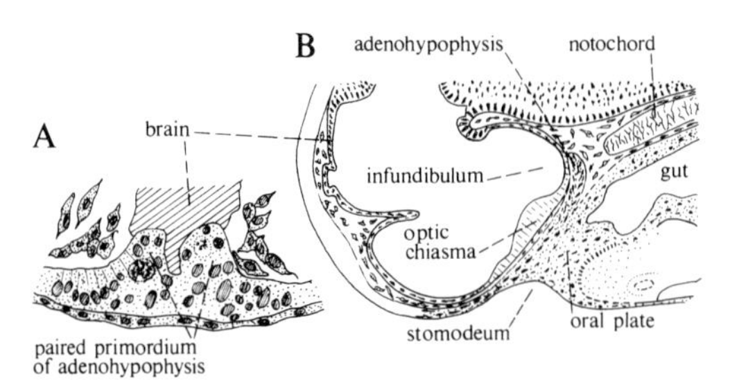

Fig. 17. Ambystoma. A, transverse section of embryo, 4·0 mm showing paired primordium of adenohypophysis. B, median section of later embryo. From Kingsley and Thyng (1904).

1904). This view is supported by the fact that the buccohypophysial duct in porolepiforms (Jarvik, 1972) and certain arthrodires (Stensiö, 1963; 1969) was paired. The neurohypophysis shows indications of being paired in early shark embryos* and since paired nervous tracts connect it with paired brain nuclei (van der Kamer, 1965, fig. 1), and it is vascularized by paired vessels, we cannot quite exclude the possibility that the neurohypophysis, and the hypophysis as a whole, is paired in origin.

MEDULLARY BRAIN PORTION, SPINAL CORD AND METAMERIC NERVES

Spinal Cord, Spinal Ganglia and Nerves

Characteristic of the spinal cord or medulla spinalis is that in each metamere of the trunk it gives off a dorsal and a ventral nerve root on each side. The dorsal root bears a spinal or root ganglion. This ganglion which contains afferent sensory nerve cells is

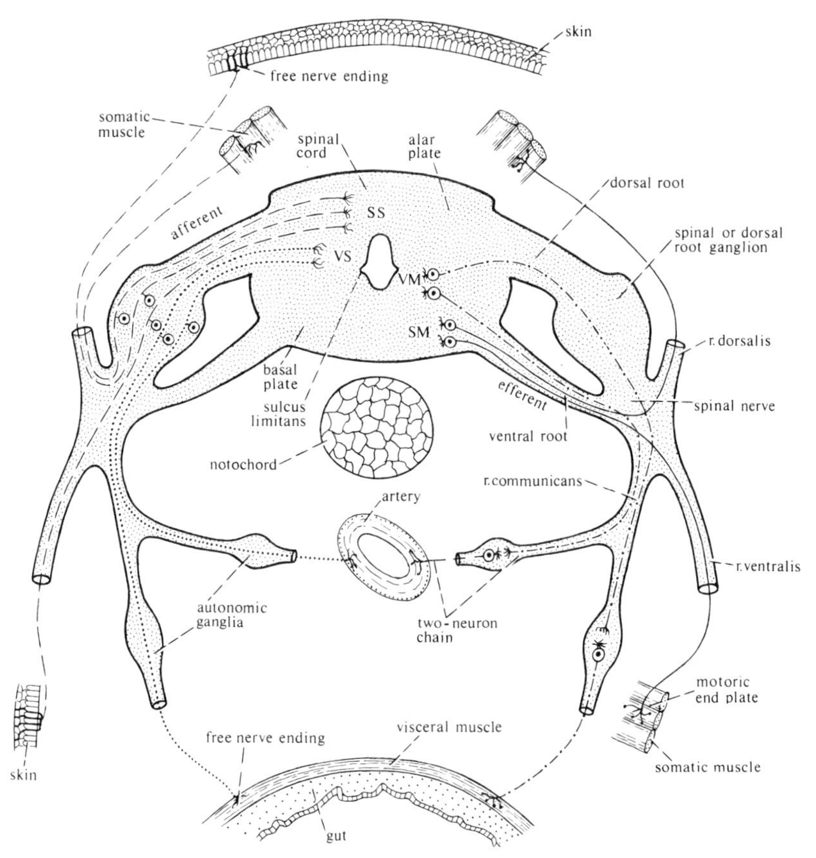

Fig. 18. Diagram of origin and peripheral distribution of systems of nerve components in region of spinal cord. SS, somatic sensory; VS, visceral sensory; VM, visceral motor; SM, somatic motor.

* p. 26.

derived from neural crest ectomesenchyme. Distal to the spinal ganglion the dorsal root, in all extant vertebrates except petromyzontids, joins the ventral root forming the spinal nerve. The spinal nerve and its roots contain numerous nerve fibres which are processes, neurites (axons), of special nerve cells, neurons. With regard to their function (Herrick, 1899; Johnston, 1905) we may distinguish between four kinds of nerve components (Fig. 18): somatic sensory (SS), visceral sensory (VS), visceral motor (VM), and somatic motor (SM). The neurons are by proximal neurites connected with longitudinal regions in the spinal cord known as the columns of His-Herrick (Fig. 10E). There are two dorsal sensory columns (SS, VS) and two ventral motor columns (VM, SM), separated by a groove, the sulcus limitans, generally discernible only in larval stages. The sensory or afferent neurons are bipolar cells with the cell body situated in the spinal ganglion and one proximal and one distal process or neurit which via the dorsal root conduct impulses to the spinal cord either (SS) from receptors in the skin and muscles or (VS) in the viscera (Fig. 18). The cell bodies of the efferent or motor neurons (VM, SM), in contrast, are located in the wall of the spinal cord. These cells have a short, branching proximal neurit (dendrit) from the visceral or somatic motor columns and a long distal neurit (axon) destined to motor end plates in the muscles. The somatic motor neurites go directly via the ventral roots to the muscles they supply. The visceral motor neurons, on the other hand, together with second order (postganglionic) neurons form a two-neuron chain which innervates smooth muscles in the gut, blood vessels, etc. The synaptic contacts between the distal neurites of the preganglionic neurons and the proximal neurites (dendrites) of the postganglionic neurons occur in special autonomic (sympathetic or parasympathetic) ganglia. As now seems to be well established (Balinsky, 1970) also these ganglia are formed by neural crest ectomesenchyme.

Medulla Oblongata, Cranial Ganglia and Nerves

The term medulla is sometimes used as a synonym of myelencephalon (von Kupffer, 1905; Plate, 1922), but most often also the part of the metencephalon situated ventral to the cerebellum is included (Kappers *et al.*, 1960). In this book the term medulla is used in a slightly widened sense, including also the ventral part (tegmentum) of the mesencephalon. According to Bergquist and Källén (1954) and Senn (1970) the somatic motor column (SM) extends forwards about to the tuberculum posterior, whereas the somatic sensory column (SS) has its anterior pole in the area of the isthmus (Fig. 10E). It may therefore be justified to define the medulla as that posterior part of the brain, exclusive the cerebellum, which runs forwards to the oblique line running from the isthmus to the tuberculum posterius (Fig. 11D). Given this definition the medulla will be that part of the brain from which all the metameric cranial nerves emerge. A remarkable fact is, however, that the two visceral columns (VS, VM) anteriorly extend to the optic chiasma (Bergquist and Källén, 1954; cf. Kingsbury, 1922), that is to the anterior limit of the medullary portion of the brain. Since the medulla is a widened anterior continuation of the spinal cord this may perhaps be taken to mean that the spinal cord originally extended forward to the anterior end of the animal marked by the optic chiasma. This is supported by the fact

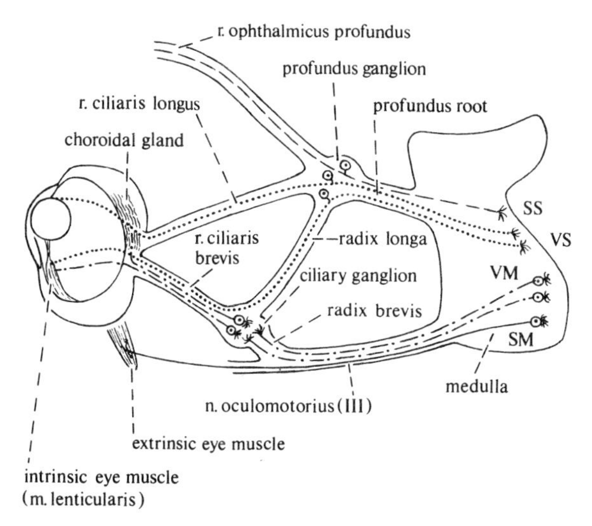

Fig. 19. Premandibular metamere. Diagram of origin and distribution of systems of nerve components in area of n. oculomotorius and r. ophthalmicus profundus (profundi). Based mainly on data in *Amia* presented by Allis (1897) and Norris (1925).

that a typical spinal nerve has been retained as far forward in the head as in the premandibular metamere (Fig. 19).

Metameric Cranial Nerves

Fig. 19 is a diagram of nerves emerging from the brain in the premandibular metamere. If we compare this diagram with the diagrammatic representation of a spinal nerve in Fig. 18 we will find a striking agreement. In the premandibular metamere, too, there is thus a dorsal and a ventral root. The dorsal root (the profundus root) exactly as in the trunk transmits proximal neurites of somatic sensory (SS) and visceral sensory (VS) neurons. Moreover the cell bodies of these neurons are contained in a ganglion (the profundus ganglion) which corresponds to and is serially homologous with the spinal (root) ganglia in the trunk. The r. ophthalmicus profundus may then be compared with the r. dorsalis of the spinal nerve. Furthermore there is a ventral root (n. oculomotorius) which like the ventral root of the spinal nerve contains distal neurites of somatic motor (SM) neurons together with distal neurites of preganglionic visceral motor (VM) neurons. In addition there is also in the premandibular metamere an autonomic (parasympathetic) ganglion (ciliary ganglion). This spinal nerve portion is present also in other cranial metameres but is more or less deficient. An important fact is, however, that in the head there are no less than three additional categories of nerves (Figs 10E, 20): sSS, sVS, sVM. These nerves supply three different organ systems: the acoustico-lateralis system, the taste bud system and the visceral arch musculature.

The acoustico-lateralis and the taste bud systems have several characters in common. The sensory organs of these systems, the neuromasts and the taste buds, are similiar in structure. Both are bulbuslike structures with sensory and supporting cells.

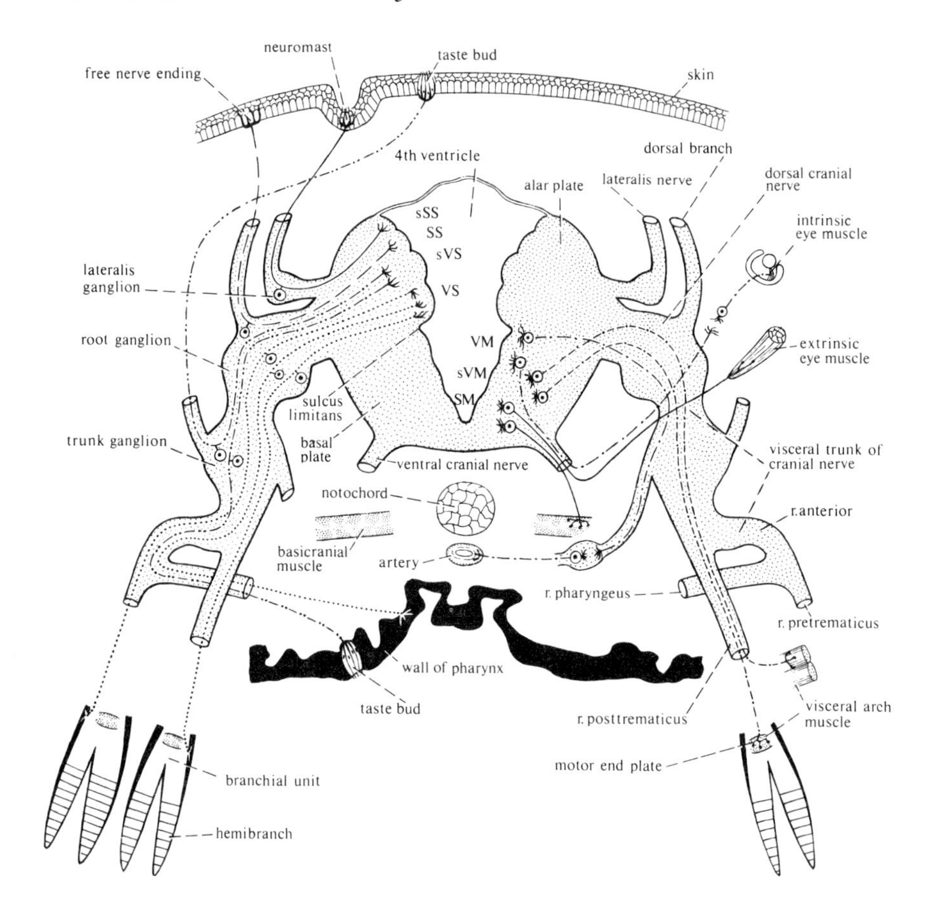

Fig. 20. Diagram of origin and peripheral distribution of systems of nerve components in middle part of medulla oblongata.

sSS, special somatic sensory (acoustico-lateralis area); SS, somatic sensory; sVS, special visceral sensory; VS, visceral sensory; sVM, special visceral motor, VM, visceral motor; SM, somatic motor.

The main difference is that the sensory cells of the taste buds are higher than in the neuromasts, extending through the whole depth of the epidermis. The taste buds, in contrast to neuromasts, may occur in the mucous membranes of the oral cavity, but although innervated exclusively by cranial nerves both kind of organs may be distributed in the skin all over the body. An interesting fact, well shown in larval stages of *Amia* (Fig. 50A, Volume 1), is that the external taste buds of the head may be arranged in lines suggestive of and running along the lateral lines; and new taste buds are, like new neuromasts, produced by budding from the peripherial cells of previously formed organs. Moreover, the ganglia and nerves of both systems are derived from ectodermal placodes. Two series of placodes are generally distinguished, one epibranchial related to taste buds and one dorsolateral belonging with the acoustico-lateralis system. The epibranchial placodes arise at the dorsal ends of the gill slits and accordingly they belong to the metameric system of the head. These placodes develop into trunk ganglia (Figs 20, 22), so called because they are borne by the visceral trunks of the cranial nerves. The trunk ganglia contain special visceral sensory neurons (sVS) which innervate taste buds and are connected with a special visceral sensory column (sVS) in the brain. The dorsolateral placodes, in contrast, form lateralis ganglia (Figs 20, 22) which contain the cell bodies of special somatic sensory neurons (sSS). These neurons are connected with a special somatic sensory area, the acoustico-lateralis area (sSS) in the brain and supply the neuromasts in the sensory lines of the head and trunk, the spiracular sense organs and the sensory maculae in the ear.

In addition to these two special sensory columns (sVS, sSS) there is also a special visceral motor column (sVM) in the brain (Figs 10, 20). The neurons (sVM) emerging from this column innervate the visceral arch musculature. However, in contrast to the visceral motor neurons (VM) their distal neurites utilize the dorsal root, and moreover they pass directly to the motoric end plates in the muscles they innervate without the intercalation of an autonomic ganglion.

In the brain there are accordingly three special columns, two sensory (sSS, sVS) and one motor (sVM) in addition to the four columns of His-Herrick in the spinal cord. Moreover the dorsal and ventral roots, disregarding the connections via the autonomic ganglion, are independent (as in petromyzontids) and for this reason it is generally spoken of dorsal and ventral cranial nerves. Like the spinal nerves these nerves are metameric structures. This is certainly true also of the acoustico-lateralis nerves which are constituents of the dorsal cranial nerves, but which, mainly for practical reasons, will be treated together with the lateral line system, the ear and the cerebellum.

Due to the weak development of the somatic (myotomic) musculature in the head the ventral cranial nerves (Fig.23) are generally weaker developed than are the ventral spinal roots, and are in some metameres tiny transient structures (e.g. the n. tenius) or absent. The dorsal cranial nerves (Fig. 22), in contrast, are usually strong. Disregarding the lateralis ganglion and nerve a typical dorsal cranial nerve includes: (a) one or several roots emerging from the brain, (b) a root ganglion (c) a dorsal somatic sensory (SS) branch and (d) a mixed (VS, sVS, sVM and sometimes SS and VM) visceral trunk. The visceral trunk bears: (e) a trunk ganglion derived from an epibranchial placode, (f) a r. posttrematicus and (g) a r. anterior. In the area where the suprapharyngeal, and infrapharyngeal and epal elements of the visceral arch meet (Fig. 48) the r. anterior divides into: (h) a r. pretrematicus which passes to the branchial unit next in front and (i) a r. pharyngeus which runs forwards in the roof of the pharynx and generally anastomoses with the corresponding branch of the metamere next in front (Figs 22, 24; Figs 45, 46, Volume 1).

The cranial nerves are thus, like the spinal nerves, not strictly confined to the metameres to which they belong. Moreover the rostral prolongation, the development of the mouth and the sense organs, modifications of the musculature and skeletal elements, and other modifications have contributed to disturb the metameric order of the cranial nerves. These conditions have rendered the interpretation difficult and that the cranial nerves are metameric structures has been doubted by several students (e.g. Kingsbury and Adelmann, 1924; Starck 1963). However thanks to the work of

Balfour and many others it became gradually clear that the cranial nerves display a distinct metamerism and in 1930 Goodrich (see also de Beer, 1937) could present a concise but in certain respects incomplete account of the cranial nerves and the metamerism of the head as a whole. Later investigations (Stensiö, 1927;1963a; Holmgren, 1940–1943; Jarvik, 1954; 1960; 1972; Millot and Anthony, 1958; 1965; Bertmar, 1959; Regel, 1961; 1964; 1968; Bjerring, 1967; 1968; 1970; 1971; 1972; 1973; 1977) have in most regards confirmed the views of Goodrich and de Beer, but they have also added a considerable amount of new relevant data filling out the gaps in our knowledge. Uncertainty still persists, e.g. as regard the terminal metamere (which has been distinguished quite recently, Bjerring, 1973; 1977), and the origin and composition of the cranial ganglia (see Weston, 1970), but nevertheless it is now possible to present a list of both the dorsal and the ventral cranial nerves in each of the cranial metameres. An attempt will also be made to identify the root and trunk ganglia.

The first or terminal metamere. Dorsal nerve is probably the n. terminalis (Figs 21, 22 together with the n. olfactorius and parts of the r. maxillaris. Like the r. ophthalmicus profundi the n. terminalis (Brookover, 1910; McKibben, 1911; Johnston, 1913; Haller, 1934; Pearson, 1941; 1941a; Larsell, 1950) contains general sensory (SS) fibres and since it bears a ganglion of neural crest origin it most likely is the dorsal branch of the dorsal cranial nerve of the terminal metamere. The ganglion terminale must then be the root ganglion and thence it follows that the nerve which in the literature usually is called n. terminalis is the root of that nerve. However, the ganglion terminale also includes sympathetic ganglion cells (cf. the ciliary ganglion, Fig. 19) and most likely distal neurites (VM) of these postganglionic neurons form part of the true n. terminalis and supply muscles in blood vessels and glands in the olfactory area (Fig. 21).

Visceral trunk in the terminal metamere is that branch of the r. maxillaris (r. palatonasalis in *Eusthenopteron* and *Rana*, Fig. 130; anterior superior alveolar branch in *Homo*, Gray, 1973) which supplies the premaxillary teeth and adjoining soft tissues. This branch, the r. premaxillaris terminali (Fig. 22) which often anastomoses with the r. palatinus by its pharyngeal branch (Fig. 24); the prechoanal anastomosis in *Eus-*

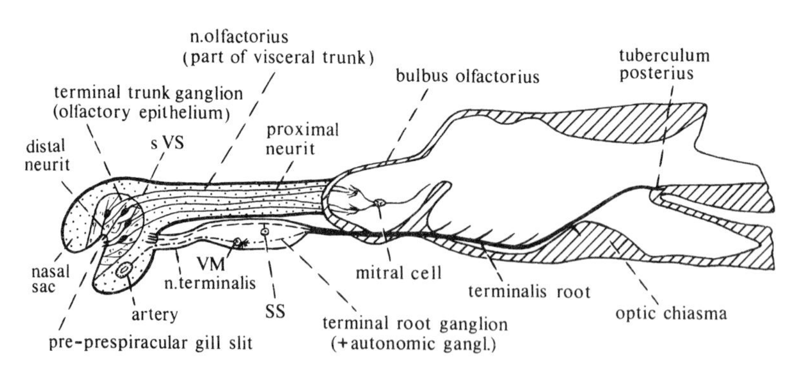

Fig. 21. Diagram of anterior part of brain in median section with right olfactory organ and nerve in longitudinal section to illustrate composition of n. terminalis. Note that the term n. terminalis is usually applied to the terminalis root. Brain and terminalis root of urodele (*Necturus*) from McKibben (1911).

thenopteron and *Rana*, Fig. 130) is somatic sensory (SS), but may, as probably in holocephalans, include motor fibres (sVM) to the most anterior preorbital muscles. It may be suggested that the olfactory epithelium is the trunk ganglion, and that the olfactory placode from which it is derived is the epibranchial placode of the terminal metamere (Figs 21, 22). Thence it follows that the n. olfactorius which carries proximal neurites of special visceral sensory neurons (sVS) and connects the terminal trunk ganglion with the brain (Fig. 21) is a part of the visceral trunk in the terminal metamere (cf. Fig. 20). However, because the trunk ganglion thus has migrated into the cavity of the nasal sac (= the pre-prespiracular gill slit) the sensory and motor fibres of the r. premaxillaris do not pass the trunk ganglion but have been gathered backwards and emerge from the brain close to the r. mandibularis trigemini (Fig. 22).

These suggestions need some comments. The olfactory cells are neurosensory cells which are a primitive kind of receptors related to ganglion cells (Kappers *et al.*, 1960; Torrey, 1971). The olfactory epithelium may be regarded as a ganglion in which the cell bodies have retained their superficial position. This peripheral ganglion is derived from the olfactory placode and since the olfactory cells are chemoreceptors as are the sensory cells in taste buds it is not unreasonable to regard the olfactory placode as an epibranchial placode. If true this means that there is an epibranchial placode in each cranial metamere. In the literature (see e.g. Goodrich 1930, fig. 741; Haller, 1934, fig. 497; Balinsky, 1970, fig. 292) the epibranchial placodes ("Kiemenspaltenorgane", Froriep, 1891), from the hyoid metamere backwards, are generally depicted as crescent-like bodies close posterodorsal to the gill slits. However, the epibranchial placodes have nothing to do with the gill slits. They are associated with the root ganglia and this is true also of the presumed epibranchial placode (the olfactory placode) in the terminal metamere, which is intimately connected with the neural crests cells that form the ganglion terminale (Johnston, 1913).

Ventral nerve in the terminal metamere (Fig. 23) is that part of the n. oculomotorius which innervates the m. obliquus inferior, a derivative of the terminal somite ("Platt's vesicle").

The second or premandibular metamere. Dorsal nerve (Fig. 22) is the n. profundus ("trigeminus I") which includes (Fig. 48; Jarvik, 1954; 1972) a dorsal branch (r. ophthalmicus profundi = "r. ophth. profundus" or "V_1") and a visceral trunk (r. maxillaris profundi = "r. max. trigemini" or "V_2"). Root ganglion is probably that part of the profundus ganglion which is of neural crest origin, whereas the trunk ganglion may be represented by the cells derived from the profundus (ophthalmicus) placode (Fig. 26) which most likely is an epibranchial placode (Haller, 1934; Holmgren, 1940; cf. Goodrich, 1930). The r. maxillaris profundi which like the corresponding branch in the terminal metamere often anastomoses with the r. palatinus (the postchoanal anastomosis in *Eusthenopteron* and *Rana*, Fig. 130) is generally regarded as a somatic sensory nerve but at any rate in *Amia* (Allis, Norris) it also contains fibres (sVS) to taste buds; and in several forms (cyclostomes, holocephalans, dipnoans; Haller, 1934; Holmgren, 1942) special visceral motor fibres (sVM) supplying the preorbital muscles (derived from the premandibular visceral tube) are distributed by the r. maxillaris. In *Amia* (Fig. 48, Volume 1) and many other fishes the preorbital muscles are innervated by separate branches which have been assigned to the r. mandibularis V. In cephalaspids the r. maxillaris profundi was a complete branchial nerve.

Ventral nerve (Fig. 23) in the 2nd metamere is that part of the n. oculomotorius (n. III) which supplies the superior, inferior and internal rectus muscles and (if present) basibranchial muscle 2 (Fig. 65; Bjerring 1977).

The third or mandibular metamere. Dorsal nerve (Fig. 22) is the n. trigeminus proper (n. V, "trigeminus II") which includes a dorsal sensory (SS, sVS) branch (r. ophthalmicus superficialis V) and a mixed (SS, sVS, sVM) visceral trunk (r. mandibularis V or "V_3"). Root ganglion is the trigeminus ganglion (g. gasseri or semilunare). However, like the profundus ganglion this ganglion is of dual origin including also a placodal component. The nature of the trigeminus placode (Fig. 26) has been much discussed (Goodrich, 1930; Balinsky, 1970; Weston, 1970), but most likely it is an epibranchial placode as suggested by Holmgren (1940). The ganglionic material derived from this placode may therefore be regarded as a vestigial trunk ganglion.

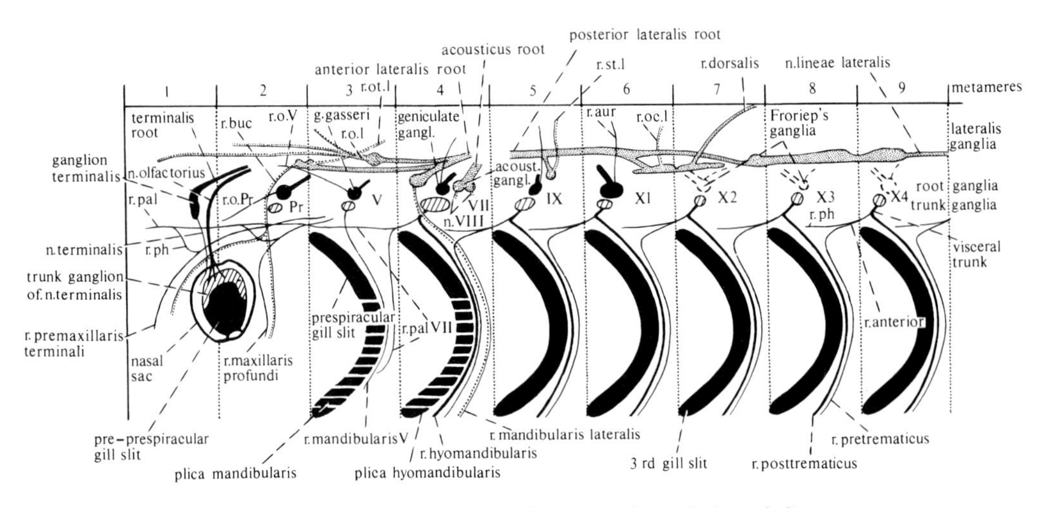

Fig. 22. Partly hypothetical interpretation of dorsal cranial nerves and ganglia in cephalic metameres. Pr,V,VII,IX,X1,X2,X3,X4, trunk ganglia of n. profundus, n. trigeminus, n. facialis, n. glossopharyngeus and portions of n. vagus; n. VIII, acoustic nerves; r.aur, r. auricularis; r.buc, r. buccalis lateralis branching into r. buccalis profundi and r. buccalis terminali; r.oc.l, r. occipitalis lateralis X1; r.o.l, r. ophthalmicus lateralis trigemini; r.o.Pr, r. ophthalmicus profundi; r.ot.l, r. oticus lateralis trigemini; r.oV, r. ophthalmicus trigemini; r.pal, terminal part of r. palatinus; r.palVII, r. palatinus facialis; r.ph, pharyngeal branches of visceral trunk of fourth vagus portion and of visceral trunk of n. terminalis (the latter forms with the r. palatinus the prechoanal anastomosis); r.st.l, r. supratemporalis lateralis IX.

The ventral nerve in the third metamere (Fig. 23) is the n. trochlearis (n. IV) together with the nerve (n. rarus; Bjerring, 1972; 1977) that innervates the large subcranial muscle (basicranial muscle 3) in "crossopterygians" (see Jarvik, 1972, pp. 239–241).

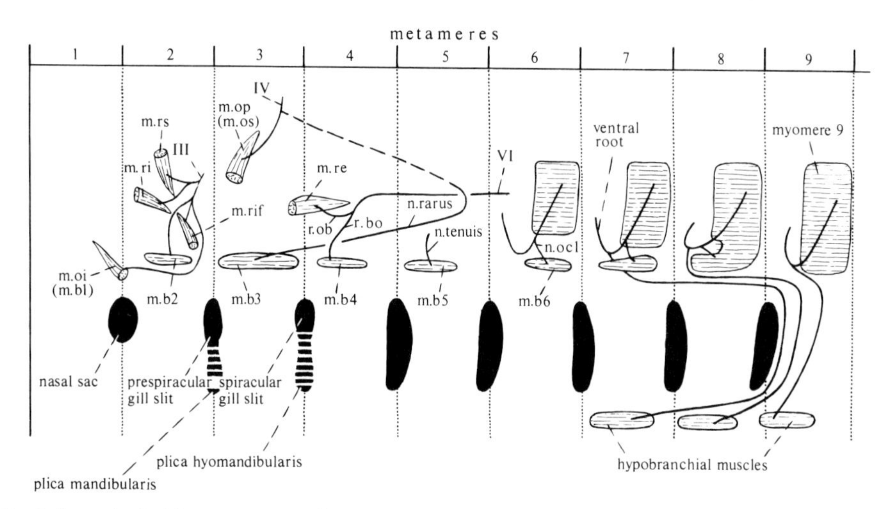

Fig. 23. Partly hypothetical interpretation of ventral cranial nerves and somatic muscles in cranial metameres. m.b1–m.b6, basicranial muscles 1–6 (m.b3 = subcranial muscle); m.oi (m.bl), m. obliquus inferior, probably representing basicranial muscle in terminal metamere (Bjerring, 1977); m.op (m.os), m. obliquus posterior in cephalaspidomorphs (m. obliquus superior in gnathostomes); m.re, m. rectus externus; m.ri, m. rectus internus; m.rif, m. rectus inferior; m.rs, m. rectus superior; n.ocl, n. occipitalis 1; r.bo, r. basioticus of n. abducens; r.ob, r. orbitalis of n. abducens; III, n. oculomotorius; IV, n. trochlearis; VI, n. abducens.

The fourth or hyoidean metamere. Dorsal nerve is n. facialis (n. VII). This nerve is represented mainly by its visceral trunk (r. palatinus VII and the posttrematic r. hyomandibularis), and the large geniculate ganglion which is the trunk ganglion. In certain forms, e.g. in *Amia* (Fig, 48, Volume 1) there is also a separate somatic sensory ganglion, which may be the root ganglion. Dorsal branches with somatic sensory (general cutaneous) fibres to the overlying part of the skin have been recorded in certain forms (van der Horst, 1928) but are still imperfectly known.

Remarks. The neuromasts in the main anterior part of the head in fish and amphibians are innervated by lateralis nerves which emerge from lateralis ganglia associated with the geniculate ganglion (Fig. 48, Volume 1). For this reason these nerves are most often regarded as branches of the n. facialis (e.g. r. buccalis and r. ophthalmicus superficialis VII). However, it is more likely that these nerves have retained their original metameric distribution and that instead it is the ganglia and the roots which have moved backwards into the facialis area. This displacement backwards, which may be compared with the gathering together forwards of nerves and ganglia in the vagus area (Figs 45, 46, Volume 1) is certainly not confined to the lateralis components. The visceral trunks of the second and third metamere and parts of the visceral trunk of the first metamere have obviously become brought together into a common maxillo-mandibular trunk which has been referred to the n. trigeminus; and because fibres to taste buds are distributed by nerves (r. maxillaris profundi and r. ophthalmicus superficialis V) belonging to the second and third metameres the geniculate ganglion may very well be a complex structure including most of the ganglionic cell material which originally was produced by the epibranchial placodes in the second and third metameres. What possibly remains of this cell material may be retained in the vestigial premandibular and mandibular trunk ganglia (cf. Froriep's ganglia).

The ventral nerve in the fourth metamere (Fig. 23) is the n. abducens (n. VI).

The fifth cephalic or first branchial metamere. Dorsal nerve (Figs 22, 24; Figs 45–48, Volume 1) is the n. glossopharyngeus (n. IX). Like the n. facialis this nerve is

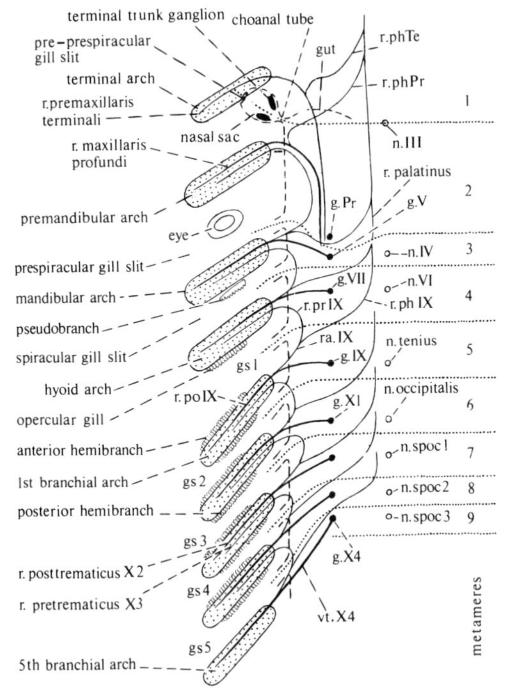

Fig. 24. Diagram of visceral arches with hemibranchs, and visceral trunks of cranial nerves with trunk ganglia (g.Pr, g.V, etc.), posttrematic, anterior, pretrematic and pharyngeal branches in cranial metameres. Ventral cranial nerves marked with o.

gs1–gs5, posthyoidean gill slits 1–5; n.spoc1, n.spoc2, n.spoc3, spino-occipital nerves; ra.IX, r. anterior of n. glossopharyngeus; r.phPr, pharyngeal branch of n. profundus (postchoanal anastomosis in *Eusthenopteron* and *Rana*); r.phTe, pharyngeal branch in terminal metamere (prechoanal anastomosis in *Eusthenopteron* and *Rana*); r.phIX, r.poIX, r.prIX, pharyngeal, posttrematic and pretrematic branches of n. glossopharyngeus; vt.X4, visceral trunk of fourth vagus portion.

represented mainly by its visceral trunk which bears a well developed trunk ganglion (g. petrosum). In mammals, birds and sometimes in reptiles there is a root ganglion (g. superius), whereas disregarding scattered ganglion cells found, e.g. in *Polyodon* (Norris, 1925), no such ganglion seems to have been observed in fishes and amphibians. Dorsal cutaneous branches may be present but are incompletely known.

The ventral nerve is the n. tenuis (Fig. 23). This tiny, transient nerve (predicted by Goodrich, 1930, fig. 240) has been found only in larval stages of *Hynobius* (Bjerring, 1970).

The sixth to ninth cephalic or second to fifth branchial metameres. The dorsal nerves of these metameres (Fig. 22; Figs 45–48, Volume 1) are included in the n. vagus (n. X). In most forms there is a strong root ganglion (g. jugulare) situated within the sixth metamere. Behind that follows in embryos of sharks and mammals three vestigial ganglia (Froriep's ganglia) belonging to the seventh, eighth and ninth metameres. These transient ganglia are regarded as spill-overs left behind when most of the ganglion cells of these metameres have become gathered forwards into the root ganglion of metamere 6 (Goodrich, 1930). The dorsal cutaneous (SS) branches are generally collected into the r. auricularis. The visceral trunks have also become gathered forwards and in fishes (sharks, Norris and Hughes, 1920; *Amia*, Fig. 45, Volume 1) their trunk ganglia lie in a row close behind each other. In tetrapods there is generally a single trunk ganglion (g. nodosum).

The ventral nerves in metameres 6–9 (Fig. 23) have retained much of their original position. In metamere 6 the ventral nerve (n. occipitalis) is a tiny transient structure recorded in sharks (*Scyllium*, Goodrich, 1918) and hynobiid urodeles (Fox, 1959; Regel, 1968; Bjerring, 1970). The ventral nerves in metameres 7–9 are in fishes (*Amia*, Figs 45, 46, Volume 1) and amphibians represented by spino-occipital nerves or ventral roots of spinal nerves. In amniotes they form the n. hypoglossus which supplies the hypobranchial muscles (in fish and amphibians the nerve to these muscles is generally called n. hypobranchialis, Fig, 48, Volume 1).

Acoustico-lateralis System and Cerebellum

Comprehension. The acoustico-lateralis or neuromast system is lacking in *Amphioxus* but was present in the earliest known vertebrates.* In cyclostomes, fishes and aquatic amphibians (Fig. 20) the system includes (a) the acoustico-lateralis area (sSS) of origin in the medulla oblongata, (b) nerve roots emerging from this area, (c) lateralis (and acoustic) ganglia and nerves, (d) sensory canals or pit lines with neuromasts and (e) the otic vesicle with the macula communis and its derivatives. The otic vesicle which develops into the membranous labyrinth may be regarded as an excessively developed neuromast. Other more or less modified neuromasts form the spiracular sense organ in elasmobranchs, dipnoans, ganoids and osteolepiforms; the ampullae of Lorenzini in elasmobranchs, holocephalans and certain teleosts (*Photosus*, Lekander 1949); the vesicles of Savi in the shark family Torpedidae; the nerve sacs in sturgeons; and possibly the groups of special sensory organs in osteolepiforms and porolepiforms. Also the cerebellum may be regarded as a part of the acoustico-lateralis system.† In amniotes the lateralis nerves and the lateral lines with their sensory organs have disappeared. However, according to one theory (Mauer, 1895), disputed by several writers (de Meijere, 1931; Gabe, 1967) but not yet definitely refuted, the supporting cells of the neuromasts are retained in mammals forming the primordia of the hairs.

Cerebellum

The cerebellum (Fig. 25) is an important coordination centre for sensory impulses and

* Volume 1, p. 491; † See Volume 1, p. 392 for organs of Fahrenholz in dipnoans and polypterids.

has for this reason attracted considerable attention during the last decades (Larsell, 1967; Nieuwenhuys, 1967; Llinás, 1969; Kuhlenbeck, 1975). The cerebellum includes two main components, the lateral auricles (flocculi), and the median corpus cerebelli (Figs 55, 56, Volume 1). The auricle which arises early in ontogeny and probably is the most primitive part of the cerebellum is formed by the anterior part of the acoustico-lateralis area (sSS), whereas the corpus is a derivative of the somatic sensory (SS) column (Fig. 10D). In cyclostomes and amphibians the corpus is small and forms a

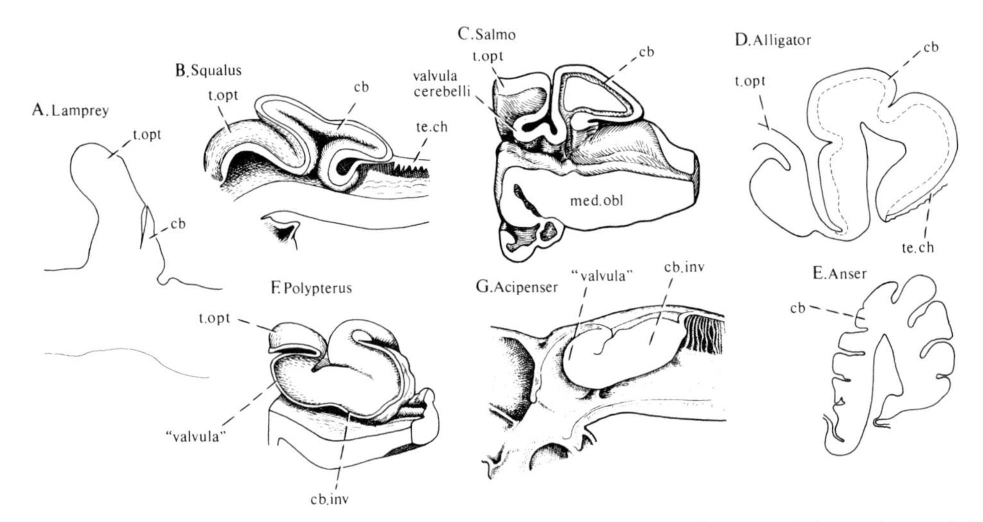

Fig. 25. Cerebellum in median longitudinal section of various vertebrates to illustrate solid type characteristic of cyclostomes (A) and amphibians, evaginated type in elasmobranchs (B), actinopterygians (C), reptiles (D), birds (E) and mammals, and invaginated type in polypterids (F) and acipenseriforms (G). After Saito, Sterzi, Schaper, Larsell, van der Horst and Johnston.

cb. solid or evaginated cerebellum; cb,inv, invaginated type of cerebellum; md.obl, medulla oblongata; te,ch, tela choroidea; t.opt, optic tectum.

solid plate (Fig. 25), whereas in fishes it usually is a more spectacular vesicular structure. In polypterids (Fig. 25F; Fig. 228, Volume 1; Senn, 1976) and *Acipenser* (Fig. 25G) this vesicle is formed by an invagination of the roof of the brain, whereas in most fishes as well as in amniotes it is mainly or wholly an evagination. In actinopterygians the anterior part of the corpus is invaginated into the third ventricle forming the valvula cerebelli, but if this structure is homologous with the so called valvula in polypterids is doubtful.*

Placodes

The dorsolateral placodes certainly play an important role in the development of the lateral line system of the head. However, among the partly transient epidermal thickenings on the dorsolateral side of the head in the vertebrate embryo it is not easy to identify and define these placodes (Landacre, 1910; 1912; 1916; 1921; Landacre and

* Volume 1, p. 295.

Conger, 1913; Stone, 1922; Knouff, 1927; 1935; Goodrich, 1930; Holmgren, 1940; Devillers, 1958; Balinsky, 1970). As far as I can see we have to distinguish two different kinds of placodes: the true dorsolateral placodes which form the lateralis ganglia and nerves, and the lateralis placodes which develop into the lateral lines and the neuromasts. With this distinction in mind let us turn to the lamprey as described by Fisk (1954; 1957; also Damas, 1951; Hagelin, 1974).

According to Fisk the caudal dorsolateral or ear (acoustic) placode in early ammocoete stages of the lamprey (Fig. 27A) includes three different primordia: (1) the primordium of the otic vesicle; (2) the primordium of the acoustic (vestibular) ganglion; (3) the primordium of the preauditory lateralis (neuromast) ganglion. As in other vertebrates (Streeter, 1907; Campenhout, 1935) the acoustic ganglion in the lamprey develops from a thickening in the bottom of the otic vesicle, close to the root ganglion of the n. facialis which is derived from the neural crest. It may be suggested that only the acoustic and the preauditory lateralis ganglia are derivatives of dorsolateral placodes whereas the otic vesicle (like other neuromast organs) is developed from a lateralis placode, the otic placode proper; and most likely it is the expanding otic vesicle that has pressed the dorsolateral placode, which forms the acoustic ganglion, inwards.

Also in the pre- and postauditory areas lateralis placodes arise close to dorsolateral placodes, a condition which explains why it usually is difficult to recognize the dorsolateral placodes. Let us therefore turn to their derivatives, the lateralis ganglia, which may give us an idea also of the number and original distribution of the dorsolateral placodes.

Lateralis Ganglia and Nerves

In elasmobranchs and holocephalans (Figs 29, 30) the preauditory lateralis root bears three separate ganglia, one continued by the r. buccalis lateralis, one by the r. ophthalmicus lateralis and one by the r. mandibularis lateralis (externus). Also in actinopterygians (Fig. 50C, Volume 1) and amphibians (Fig. 26E) the r. mandibularis lateralis emerges from a separate ganglion and no doubt this ganglion and the nerve belong to the fourth (hyoid) metamere and are to be regarded as parts of the n. facialis. Since the r. buccalis accompanies the r. maxillaris profundi and that part of the r. maxillaris (r. premaxillaris) which has been ascribed to the first (terminal) metamere it is reasonable to assume that also the r. buccalis and the buccalis ganglion have a corresponding composition and contain parts of the n. profundus in the second metamere and of the n. terminalis in the first metamere (Fig. 22). However, the buccalis ganglion is more complex. In *Latimeria* (Fig. 32; Millot and Anthony, 1965) the r. buccalis gives off branches which innervate the jugal canal. This is most remarkable since the jugal (or jugalo-oral) sensory canal in all other extant fishes is supplied by branches of the r. mandibularis lateralis. Millot and Anthony claim that *Latimeria* in this respect is primitive and that the r. buccalis as a whole belongs to the n. trigeminus. This may be true of the buccalis branches in *Latimeria* which innervate the jugal sensory canal and consequently also of the corresponding branches of the r. mandibularis in other fishes. Also the r. oticus lateralis which emerges from the

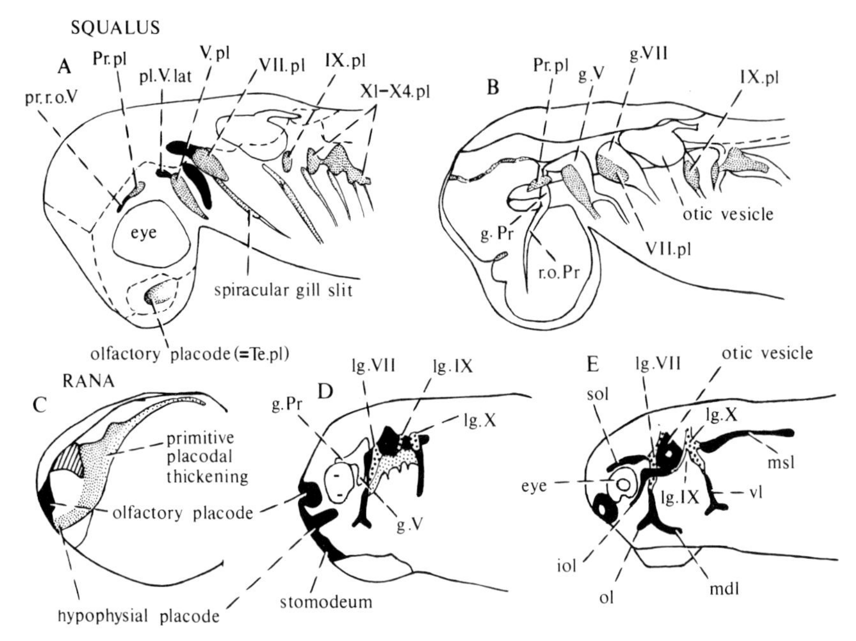

Fig. 26. A, B, *Squalus acanthias*, embryo 12 mm, Restorations of epibranchial placodes (stippled), lateralis placodes (black), and cranial nerves. From Holmgren (1940). C, D, E, graphic reconstructions of three embryonic stages of *Rana pipiens*. From Knouff (1935).

Pr.pl, Te.pl, V.pl, VII.pl, IX.pl, X1–X4.pl, epibranchial placodes of n. profundus, n. terminalis, n. trigeminus, n. facialis, n. glossopharyngenus, and vagus portions 1–4.

g.Pr, g.V, g.VII, profundus, trigeminus (gasserian) and facialis (geniculate) ganglia; iol, infraorbital line placode; lg.VII, lg.IX, lg.X, lateralis ganglia; mdl, msl, ol, placodes of mandibular, main sensory, and oral lines, pl.V.lat, placode which according to Holmgren forms a transient nerve, (although no neuromasts have been recorded this nerve is said to be a lateralis nerve and it is with reference to this nerve that Holmgren (1942a), introduced the term profundus line); pr.r.o. V, thickening which according to Holmgren (1942a, p. 24) forms the r. ophthalmicus superficialis trigemini but which he also refers to as a sensory line component; r.o.Pr, r. ophthalmicus profundus; sol, vl, placodes of supraorbital and ventral sensory lines.

buccalis ganglion or from a separate part of the ganglion (*Squalus*, Norris and Hughes, 1920) and at any rate the posterior part of the r. ophthalmicus lateralis may be regarded as parts of the n. trigeminus (Fig. 22).

Like other ganglia or parts of ganglia also the lateralis ganglia have obviously been gathered backwards in the preauditory region. A corresponding displacement, but forwards, is met with in the postauditory region.

The postauditory lateralis root generally bears two ganglia, one small which pertains to the n. glossopharyngeus and gives off a dorsal branch (r. supratemporalis lateralis), and one large in the vagal area (Fig. 22). However in *Amia* (Fig. 45, Volume 1) a small anterior portion of the large ganglion is almost independent and no doubt this portion and the emerging nerve (the r. occipitalis lateralis) belong to the sixth metamere (Fig. 32; Fig. 46, Volume 1) and are parts of the first vagus portion (X 1). Similar conditions are found also in *Squalus* (Fig. 29A, Landacre, 1916; Norris and Hughes, 1920) but in this form also the posterior part of the large lateralis ganglion is incompletely divided into an anterior portion which gives off the r. dorsalis and a larger posterior portion which is continued by the main lateralis nerve of the trunk (the lineae lateralis). This condition (Fig. 22) indicates that the anterior portion and the r. dorsalis pertain to the seventh metamere and are parts of the second vagus portion, whereas the posterior ganglion portion and the lineae lateralis belong to the eighth and ninth metameres.

Membranous Labyrinth, Spiracular Sense Organs and Lateral Lines

Membranous labyrinth. One of the most important derivatives of the lateralis or neuromast placodes is no doubt the membranous labyrinth, present in all vertebrates.

The membranous labyrinth arises from the otic placode proper* which is only part of a more extensive placodal area including also dorsolateral placodal components forming lateralis and acoustic ganglia. The placodal thickening of the epidermis is by invagination converted into an otic vesicle (Fig. 27). This vesicle which contains a large neuromast, the macula communis, is soon pinched off from the epidermis and

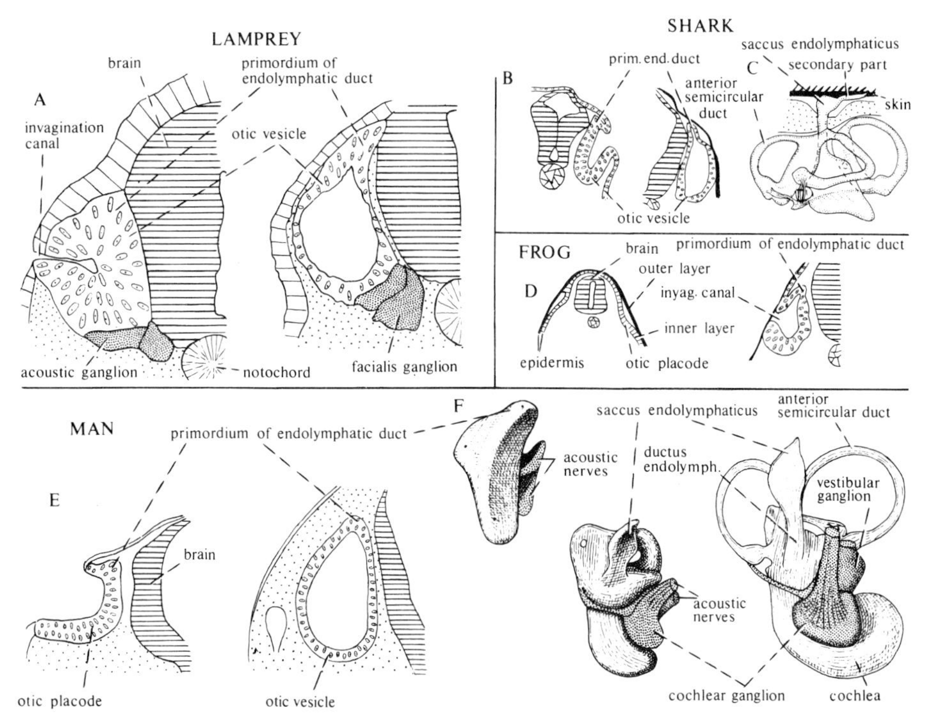

Fig. 27. Otic vesicle and membranous labyrinth. Note similarities in position of primordium of endolymphatic duct. A, transverse sections of early embryos of *Lampetra fluviatilis*. Drawings after photographs in Hagelin (1974). B, *Squalus acanthias*. Two stages in ontogenetic development of otic vesicle. After Quiring. C, *Chlamydoselachus anguineus*, membranous labyrinth of left side with overlying parts of endoskeletal skull roof and skin in lateral aspect. After Goodey. D, *Rana temporaria*, two stages in early ontogenetic development of otic vesicle. After Villy. B–D, from Jarvik (1975). E, F, *Homo sapiens*. E, two early and, F, three later stages in ontogenetic development of otic vesicle. From Streeter (1945; 1907).

* p. 41.

differentiates into the compartments characteristic of the membranous labyrinth in the adult (Retzius, 1881; 1881a; 1884; de Burlet, 1934; Cordier and Dalcq, 1954; Werner, 1960). Of particular interest is the ductus endolymphaticus.

The ductus endolymphaticus, present in all vertebrates, is a slender tube ascending from the medial side of the sacculus. In certain forms (*Amia*, Fig. 54, Volume 1; *Lepisosteus*, and most teleosts) it is uniform in thickness and terminates within the cavity of the otic capsule, but usually it extends into the cranial cavity and expands into a vesicle, the endolymphatic sac (Fig. 27). This sac which is filled with a fluid (endolymph) often containing calcareous material is in certain forms (*Protopterus*, urodeles, anurans and some reptiles) large; and in anurans and probably in osteolepiforms (Jarvik, 1975) it extends far backwards and forms paired calcareous sacs along the vertebral column (Figs 117, 122). In extant elasmobranchs (Fig. 27C; Fig. 259, Volume 1) and holocephalans the duct has an external opening as it obviously had also in cephalaspids (Fig. 357, Volume 1), arthrodires (Figs 280, 283, 286, 294, Volume 1) and some Devonian dipnoans (Miles, 1977).

The studies of the ontogenetic development of the endolymphatic duct has led to many controversies. Forms (dipnoans, urodeles, anurans) with a double epidermis excepted (Fig. 27D), the otic vesicle opens while the invagination proceeds to the exterior via a short invagination canal (Fig. 27A). According to current views (de Burlet, Werner and others) this canal persists in elasmobranchs and forms the ductus endolymphaticus in the adult, and this is according to Thornhill (1972) true also of the lamprey. However, as is established in various vertebrates and well shown in the lamprey (Fig. 27A; Fisk, 1954; Hagelin, 1974) the endolymphatic duct arises by evagination of the dorsal wall of the otic vesicle independently of the transitory invagination canal. In elasmobranchs the external opening of the duct is formed at its distal end, far from the site of the invagination canal (Fig. 27B, C); and as demonstrated by Ranzi (1962) the part of the duct situated distal to the saccus endolymphaticus and which opens outwards is a secondary formation.

The fact that the endolymphatic duct arises early in ontogeny—while the invagination process proceeds—as a characteristic dorsal process of the wall of the otic vesicle is a most remarkable basic feature common to all vertebrates, the lamprey as well as man (Fig. 27). Another peculiar structure also belonging to the lateralis system is the spiracular sense organ in gnathostomes.

Spiracular sense organs. The spiracular organ (Allis, 1889; Norris and Hughes, 1920a; Goodrich, 1930) is known in several extant fishes (elasmobranchs, holocephalans, dipnoans, *Acipenser, Lepisosteus, Amia*) and was present also in palaeoniscids and osteolepiforms (*Eusthenopteron*). It consists of neuromasts innervated by branches of the r. oticus lateralis and is situated in a diverticle of the spiracular gill tube (Fig. 52, Volume 1) or in a separate vesicle. In actinopterygians (Figs 17–21, 249, 251, Volume 1), osteolepiforms (Figs 86, 88, 92, Volume 1), and *Neoceratodus* (Fig. 313, Volume 1) the diverticle lies in a canal, the spiracular canal, in the neurocranium. This canal is bounded laterally by a skeletal bar which probably is a derivative of mandibular gill rays, as is also the peculiar lid of the dorsal opening in *Neoceratodus*. The ontogenetic development of the spiracular organ is imperfectly known but probably (see Allis,

1889; *Eusthenopteron**) it is formed by the invagination of a lateralis placode within the third metamere in about the same way as the otic vesicle is formed within the fourth metamere.

Sensory lines. The lateralis placodes which form the sensory canals and pit-lines are disposed in several groups (Landacre, 1910; 1916;1921; Landacre and Conger, 1913; Ruud, 1920; Stone, 1922; Knouff, 1935; Holmgren, 1940; Pehrson, 1949; Holmgren and Pehrson, 1949; Lekander, 1949; Devillers, 1958; Medvedeva, 1961; Schmalhausen, 1968). The placodes of the supra- and infraorbital canals may form continuous lines but often there are several separate placodes. In sharks, teleosts and urodeles there are thus anterior and posterior supraorbital placodes (Figs 28, 29), and in the infraorbital line the ethmoidal part may be formed by a separate placode. The significance of these conditions will be discussed below. The sensory lines innervated by the n. glossopharyngeus and n. vagus are derivatives of a series of postauditory placodes (Fig. 29). A remarkable fact, demonstrated by Harrison in a classical paper (1903), is that cell material of the hindmost of these placodes migrates backwards to form the neuromasts of the main sensory line of the trunk and tail. The oral–jugal–mandibular sensory lines are derivatives of a large independent placode usually shaped as an inverted V or U (Fig. 34C, D). Other separate placodes form gular lines, but neuromasts may also arise directly in the epidermis without previous placode formation (Lekander, 1949; Devillers, 1958).

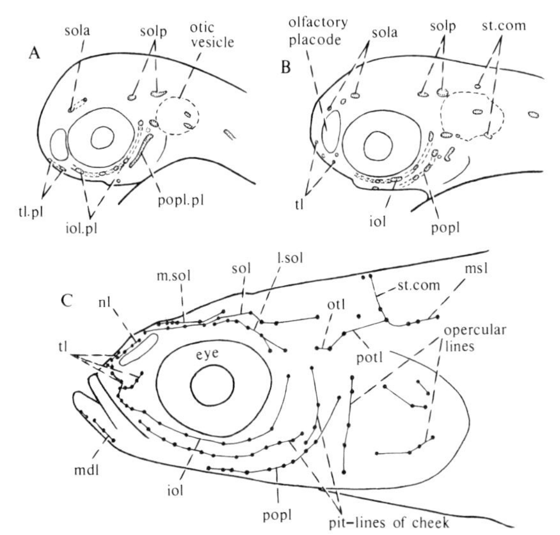

Fig. 28. Leuciscus rutilus. Three stages (141 hours, 6·6 mm, 14 mm) in development of lateralis placodes and sensory lines of head. From Lekander (1949).

iol, iol.pl, infraorbital sensory line and placode (premandibular arch line); l.sol, lateral supraorbital line; mdl, mandibular line (hyoidean arch line); m.sol, medial supraorbital pit-line; msl, main sensory line; nl, nasal pit-line; otl, otic line; popl, popl.pl, preopercular line and placode (mandibular arch line); potl, postotic line; sol, supraorbital line (canal); sola, solp, anterior and posterior supraorbital line placodes; st.com, supratemporal commissural line; tl, tl.pl, terminal arch line (including rostral commissure and antorbital line) and placode.

* Volume 1, p. 119.

Since 1889 when Allis published his masterly account on *Amia*, the lateral line system in extant cyclostomes, fishes and aquatic amphibians has been treated by a great number of writers and from various aspects (for review and references see Devillers, 1958; see also Disler, 1960; Medvedeva, 1961; Branson and Moore, 1962; Dijkgraaf, 1963; Schmalhausen, 1964; 1968; Reno, 1966; Nybelin, 1956; 1967; 1979; Tester and Nelson, 1967; Shelton, 1970; Nelson, 1972). In fossil forms the bony canals housing the membranous sensory canals and the grooves for pit-lines are often well preserved and it has been possible to make out not only the structure and disposition in several forms (Figs 127, 141, 144, 239, 240, 292, 298, 322, 333, 348, 378, Volume 1) but sometimes also to infer as to the number of neuromasts and their innervation.

In spite of the many researches devoted to the lateral line system several intricate problems remain to be solved. One such problem concerns its origin. Gross (1956) and Denison (1966a) have claimed that it is derived from the pore canal system of the early vertebrates, a system which according to Gross is retained in extant dipnoans and polypterids as the organs of Fahrenholz (1929). However, the view that these sensory organs belong with the lateral line system has been disputed (Pfeiffer, 1968) and we have to admit that the significance of both the organs of Fahrenholz and the pore canal system is unknown; nor have the attempts (Säve-Söderbergh, 1941; Holmgren, 1942a; Holmgren and Pehrson, 1949) to derive the pattern of lateral lines in gnathostomes from the conditions in heterostracans or the interpretations (Stensiö, 1947; Poplin, 1973) based on Pehrson's binary primordia (1940) led to any convincing results. According to Devillers (1958) attempts to demonstrate primary relations between the metamerism of the head and the distribution of the sensory lines have also failed.*

The reasons for these and other difficulties are several. The heterostracans are certainly no primitive predecessors of gnathostomes, what Säve-Söderbergh believed, and probably had it been more profitable to start from a system of longitudinal and transverse sensory lines as found in the trunk of e.g. *Protopterus* (Fig. 31; Jarvik 1968) than from the modified system in heterostracans (Fig. 36B, Volume 1). In gnathostomes the rostral prolongation, the development of gill slits and gill covers, and the changes in direction of the visceral arches have contributed much to modify the sensory line system of the head. Moreover fusions of bones have caused considerable displacements of the sensory lines (Jarvik, 1947; 1948; 1967a; 1972; Ørvig, 1975; cf. Graham-Smith, 1978; 1978a). As is well known the neuromasts of the sensory canals may induce the formation of dermal bones and as is most likely (Ørvig, 1972) these lateral sensory components have in phylogeny arisen independently of the membranous components which form the main parts of the dermal bones and scales. The neuromasts of pit-lines are, in contrast, said to lack inducing capacity (Pehrson, 1940; cf. Stensiö, 1947, p. 101) but nevertheless the pit-lines are influenced to the same extent as the sensory canals at the fusion of bones. This condition is hard to explain as is also the remarkable migration of the pit-lines in the ontogeny of *Protopterus*.†

In fishes the relations to sensory canals and pit-lines have frequently been used to settle bone homologies, but when using this instrument it is of course necessary to be

* However, see pp. 48–55; † Volume 1, p. 417.

sure that the utilized sensory lines are homologous. In extant fishes the homologization of sensory lines is based mainly on the innervation. The discovery of the aberrant buccalis innervation in *Latimeria* (Fig. 32) has, however, shaken our confidence in this method and there are also other problems: for example, the supraorbital canal which is innervated by r. ophthalmicus lateralis runs in plagiostomes (including dipnoans) lateral to the nostrils, in many actinopterygians (e.g. palaeoniscids and sturgeons) between the anterior and posterior nostrils, and in porolepiforms and osteolepiforms, medial to these openings.* It may then be questioned if the anterior parts of the supraorbital canal really are homologous in these fishes; and are the anterior nasals in actinopterygians and "rhipidistids" homologous (Jarvik, 1942)?

Other problems concern the commissures in the snout (Jarvik, 1942; Pehrson, 1947; Gardiner, 1963; Nybelin, 1967; Wenz, 1967; Poplin, 1973), the double sensory lines and the supernumerary lines. Sensory canals are often accompanied by superficial pit-lines. The ontogenetic development of such double lines has been described in certain teleosts (Pehrson, 1944; Lekander, 1949) and it has been shown that the lateralis placode forms a row of neuromasts, all similar in size and structure and innervated by twigs of the same nerve. Later in ontogeny some of the organs grow bigger and become enclosed in sensory canals whereas most organs remain outside the canal and form the neuromasts of the pit-line. In *Leuciscus* (Fig. 28) the supraorbital canal and the lateral supraorbital pit-line are formed in this way. The medial supraorbital pit-line, in contrast, arises independently late in ontogeny and without any signs of previous placode formation. Also in other groups supernumerary sensory lines or areas of neuromasts may occur (arthrodires, Stensiö, 1969; Ørvig, 1971; dipnoans, Fig. 327, Volume 1; extant amphibians, Schmalhausen, 1955). Sometimes (most arthrodires, post-Devonian stegocephalians) the bony sensory canals are replaced by open grooves or the canals are continued by such grooves or by pit-lines. Moreover sensory canals may be replaced by pit-lines as has happened in the phylogeny of the snout in teleosts (Nybelin, 1967). Conversely, pit-lines may be developed as canals as is the case in one specimen of *Holoptychius*, in which the squamosal and quadratojugal pit-lines are represented by a canal (Jarvik, 1948; 1972). Also the evaluation of the ampullae of Lorenzini and other aberrant lateralis-innervated organs which most often occur close to sensory lines (as do the lines of taste buds) is difficult.

The modifications of the sensory lines have resulted in the different patterns of arrangement which are characteristic of the various groups. However, at any rate in gnathostomes the differences are on the whole rather small and as a matter of fact most of the sensory canals and pit-lines which have been described in *Amia*† infraorbital canal, supraorbital canal, middle pit-line, etc) can be recognized also in other groups. This agreement must be taken to mean that we are concerned with a basic pattern common to all gnathostomes. As we have seen the lateralis ganglia are probably metameric structures which like other components of the dorsal cranial nerves have been gathered together; backwards in front of, and forwards behind the otic vesicle. These displacements have influenced also the proximal parts of the lateralis nerves but most likely the distal parts of these nerves have retained much of their original

* Volume 1, p. 369; † Volume 1, p. 76.

metameric distribution; and if so this must be true also of the sensory lines which they innervate. It now remains to see to which extent it is possible to trace this supposed metamerism of the sensory lines of the head.

Metameric Disposition and Visceral Arch Lines

The trunk of aquatic lower gnathostomes presents sensory lines, transverse as well as longitudinal. To the transverse system belong the accessory pit-lines of the lateral line scales (*Amia*, Fig. 5, Volume 1; Allis, 1889; osteolepids, Fig. 137, Volume 1; Jarvik, 1948; *Neoceratodus*, Fig. 336, Volume 1). In teleosts numerous transverse lines may occur and also *Protopterus* is well equipped with such lines (Fig. 31). However, since the ontogenetic development and innervation of these lines is unknown they are of little use for the interpretation of the head lines. The longitudinal lines are usually three in number (main, dorsal and ventral) as are the longitudinal lines which according to Allis (1934) form parts of the lateral line system of the head (cf. Poplin, 1973).

The ventral longitudinal trunk line represented by a canal (*Holoptychius*), a continuous pit-line (*Protopterus*, Fig. 331, Volume 1) or a series of pit-lines (fossil dipnoans, Ørvig, 1969a; osteolepids, Jarvik, 1948; Jessen, 1973; palaeoniscids, Westoll, 1944) often bends medially in its anterior part (as does the ventrolateral fin fold in this area) and may be joined to its antimere by a transverse commissure, as sometimes in *Protopterus*. The umbilical line in arthrodires and sharks (Miles, 1965a; Tester and Nelson, 1967; Stensiö, 1969) is possibly an anterior part of this line but if it reaches the head is uncertain (the gular pit-lines are situated on the hyoidean gill cover). The dorsal longitudinal line which may be represented either by a series of pit-lines (*Amia*, Fig. 50, Volume 1; osteolepids, Fig. 137, Volume 1) ending in the skin outside the supratemporal bone or by a canal or groove joining the main sensory canal in that bone (certain actinopterygians and arthrodires, Fig. 290, Volume 1; Jarvik, 1944a). Whether this line originally continued forwards to the head is difficult to say, but as possible anterior parts we may consider the pit-line of the lateral extrascapular, the posterior oblique parietal pit-line in osteolepiforms and porolepiforms, and the supraorbital canal (Figs 121, 187, Volume 1). Judging from its position the pit-line of the lateral extrascapular may be homologous with the posterior pit-line in *Amia* (Fig. 50, Volume 1) which is innervated by a dorsal lateralis branch of the first vagus portion in metamere 6. The posterior oblique parietal pit-line in "rhipidistids" is continuous with the transverse parietal pit-line. The latter forms together with the supratemporal pit-line a transverse line which most likely corresponds to the middle pit-line in *Amia* which is innervated by a lateralis branch of the n. glossopharyngeus. This may be true also of the posterior oblique line which then is to be referred to the fifth metamere. The supraorbital canal posteriorly joins the infraorbital canal or continues backwards independently of that canal (e.g. *Dipnorhynchus*, Fig. 330A, Volume 1; palaeoniscids, Fig. 240, Volume 1), sometimes (*Chimaera*, Fig. 30; Fig. 298, Volume 1) to the supratemporal commissure. No explanation can be given to these variations. Of interest is, however, that the supraorbital canal often arises from two placodal primordia (Figs 28, 29). Moreover the neuromasts in *Chlamydoselachus* (Allis, 1923, p. 196) "are separated, by the manner of their innervation, into two large groups, one

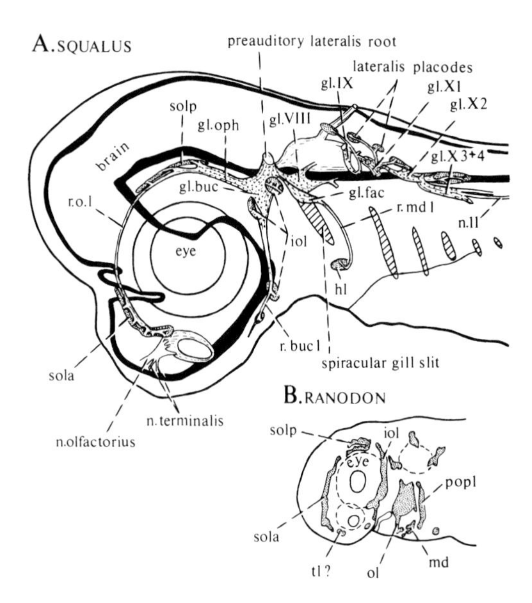

Fig. 29. A, Squalus acanthias. Lateralis placodes, ganglia and nerves of embryo, 22 mm. From Landacre (1916). B, *Ranodon sibiricus.* Lateralis placodes of embryo, 15mm. From Medvedeva (1961).

gl.buc, gl.fac, gl.oph, gl.VIII, gl.IX, gl.X1, gl.X2, gl.X3+4, lateralis ganglia of r. buccalis, n. facialis, r. ophthalmicus, n. acousticus, n. glossopharyngeus and vagus portions; hl, iol, md, placodes of hyomandibular, infraorbital and mandibular sensory lines; n.ll, n. linae lateralis; ol, popl, placodes of oral and preopercular sensory lines; r.bucl, r. buccalis lateralis; r.mdl, r. mandibularis lateralis; r.o.l, r. ophthalmicus lateralis; sola, solp, anterior and posterior placodes of supraorbital sensory line; tl?, placode probably of terminal sensory line.

lying dorsal to the orbit and the other anterior to it, these two groups apparently corresponding to the two primordia of these organs described by Landacre (1916) in embryos of *Squalus acanthias* and by Ruud (1920) in *Spinax niger*". These conditions suggest that the supraorbital canal is a complex structure, composed of a posterior portion belonging to the mandibular and an anterior portion to the premandibular and terminal metameres. This implies that the r. ophthalmicus lateralis also is complex including fibres belonging to the n. trigeminus, n. profundus and n. terminalis. The polymetameric nature of the supraorbital canal is further illustrated by the conditions in *Chimaera* (Fig. 30; Cole, 1896) in which a short part of the canal (the profundus canal of Cole) is innervated by fibres associated with the r. ophthalmicus profundus.

Of the three longitudinal trunk lines only the main sensory canal clearly passes to the head where it, together with the postotic and infraorbital canals, forms a continuous canal extending to the tip of the snout (Figs 30–33; Figs 50, 122, 144, 223, 235, Volume 1). This continuous canal which possibly is a longitudinal canal pressed upwards to pass over the branchial region and downwards by the eye, is clearly innervated by metameric lateralis nerves (Fig. 50, Volume 1). In the cephalic part of the main sensory canal this longitudinal canal is thus innervated by a dorsal lateralis branch of the first vagus portion (metamere 6. Fig. 22). Next in front follows a portion, the postotic canal, innervated by a dorsal lateralis branch of n. glossopharyngeus (metamere 5) and in front of that the otic portion of the infraorbital

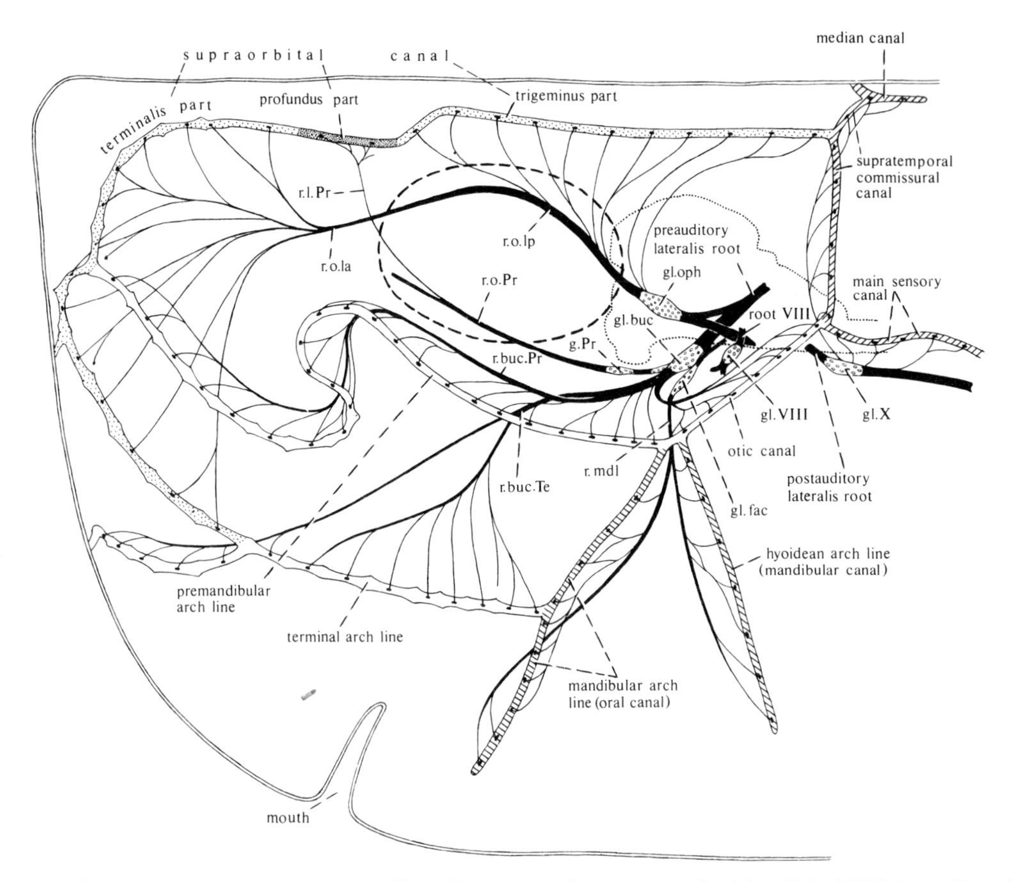

Fig. 30. Chimaera monstrosa. Interpretation of lateralis nerves and sensory canals. After Cole (1896) (ampullae of Lorenzini omitted).

gl.buc, ganglion of r. buccalis lateralis; gl.fac, lateralis ganglion of n. facialis; gl.oph, ganglion of r. ophthalmicus lateralis; gl.VIII, acousticus ganglion; gl.X, lateralis ganglion of n. vagus; g.Pr, profundus ganglion; r.buc.Pr, r. buccalis profundi; r.buc.Te, r. buccalis terminali; r.l.Pr, lateralis branch of n. profundus (innervating Cole's profundus canal); r.md.l, r. mandibularis lateralis; r.o.la, anterior part of r. ophthalmicus lateralis (probably belonging to terminal metamere); r.o.lp, posterior part of r. ophthalmicus lateralis (probably belonging to mandibular metamere); r.o.Pr, r. ophthalmicus profundi.

canal innervated by the r. oticus lateralis which may belong to the n. facialis (metamere 4) but more likely is a part of the n. trigeminus (metamere 3). So far the metameric innervation is indisputable but there are reasons to believe that the postorbital and the main part of the suborbital portions of the infraorbital canal are innervated by buccalis branches belonging to the n. profundus (metamere 2) and that the buccalis branches which supply the remaining anterior part of the infraorbital canal including the ethmoidal commissure are parts of the n. termianlis (metamere 1). The fact that—according to the interpretation now suggested—no part of the infraorbital canal is innervated by the n. facialis is easily accounted for if we assume that all the neuromasts derived from the lateralis placode in the fourth metamere (the otic placode

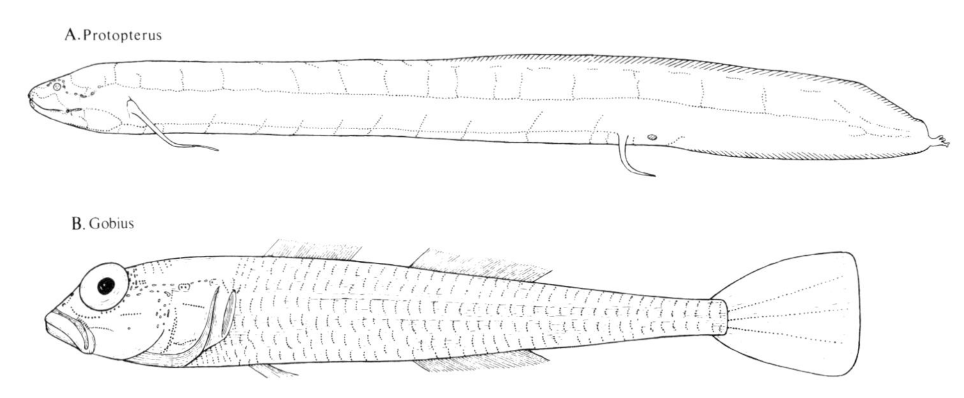

Fig. 31. Two fishes with well developed transverse trunk lines. A, *Protopterus dolloi*. From Jarvik (1968). B, *Gobius quadrimaculatus*. From Sanzo (1911).

proper) enter into the formation of the macula communis of the otic vesicle. Thence it follows that the n. acousticus (n. VIII) most likely is a dorsal lateralis nerve in the fourth metamere and a part of the n. facialis.

It is, however, by no means certain that the infraorbital canal is a part of an original longitudinal canal. The so called anterior dorsolateral canal (Poplin, 1973; sometimes called the profundus canal, a term introduced by Holmgren, 1942a, on doubtful premises, see Fig. 26) may be an anterior continuation of the otic part of the infraorbital canal, but as suggested by Stensiö (1947) it is also possible that the dorsolateral ("profundus") canal is homologous with the lateral supraorbital pit-line (Fig. 28) in teleosts which arises by duplication of the supraorbital canal* (Lekander, 1949). Be this as it may, as we now shall see the conditions in *Chlamydoselachus, Latimeria* and *Amia* suggest the following alternative interpretation of the infraorbital canal (Figs 32, 33). In these forms we may distinguish four metameric arched sensory lines on the lateral side of the head, one related to each of the four foremost (prootic) branchial units.

(1) Hyoidean arch line. This line is well developed in *Chlamydoselachus* (Fig. 32A), in which it is represented by the spiracular and gular lines of Allis (1923). The continuous line formed by these canals starts dorsally behind the spiracle, it runs outside the musculature of the hyoidean gill cover and it is innervated by the r. mandibularis lateralis (externus) of the n. facialis. It thus clearly belongs to the hyoidean branchial unit (cf. Jollie, 1962). In operculate fishes the dorsal, postspiracular part is usually lacking and the hyoidean arch line is represented by the subopercular and mandibular canals (*Latimeria*, Fig. 32B) or only by the mandibular canal (*Amia, Eusthenopteron*, Fig. 33C, D).

(2) Mandibular arch line. This is in *Latimeria* represented by the jugalo-preopercular canal which runs outside the epimandibular (pars pterygoquadrata) and is innervated by buccalis branches ascribed to the n. trigeminus (Fig. 32B). In other fishes (Fig. 50, Volume 1) these lateralis branches are, probably secondarily, associated with the r. mandibularis lateralis VII and most likely the mandibular arch line is formed (Fig. 33)

* p. 47.

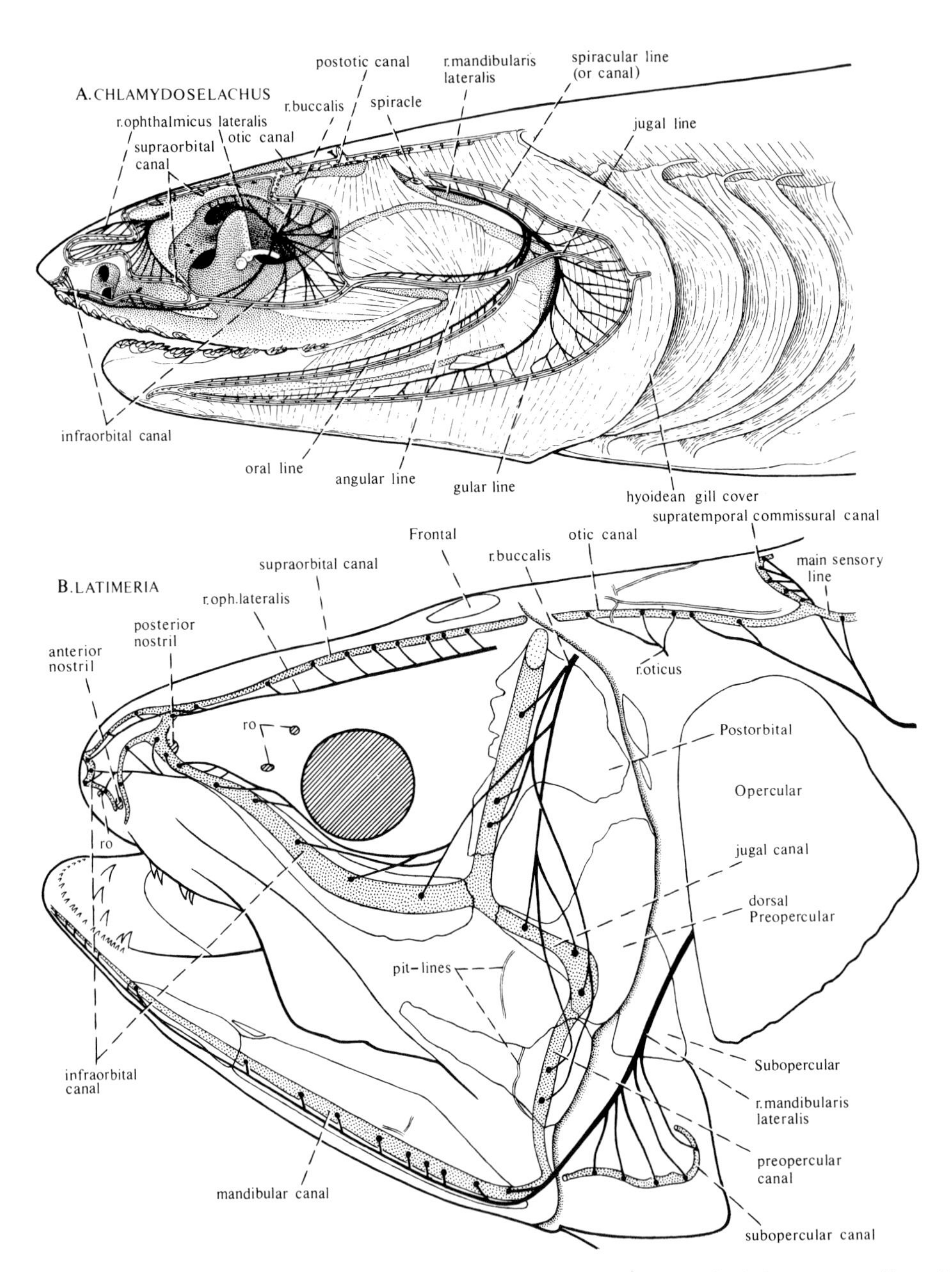

Fig. 32. Head in lateral aspect to show sensory lines and their innervation of, A, *Chlamydoselachus anguineus* (from Allis, 1923) and B, *Latimeria chalumnae* (from Millot and Anthony, 1965).
ro. openings of rostral organ.

in *Chlamydoselachus* by the angular-oral canal of Allis and in *Amia* by the horizontal cheek line and the preopercular canal (note that the dorsal part of the preopercular canal in *Amia* is a secondary formation; Holmgren and Pehrson, 1949). Hitherto it has been customary to regard the preopercular and mandibular canals as parts of a continuous preoperculo-mandibular canal. However, the conditions in *Chlamydoselachus, Latimeria* and *Amia* indicate that this is erroneous. The preopercular canal should be combined with the oral canal; and accordingly the mandibular arch line in *Eusthenopteron* is represented by the jugalo-preopercular–oral canals whereas the mandibular canal, as mentioned, most likely is a part of the hyoidean arch line (Fig. 33D).

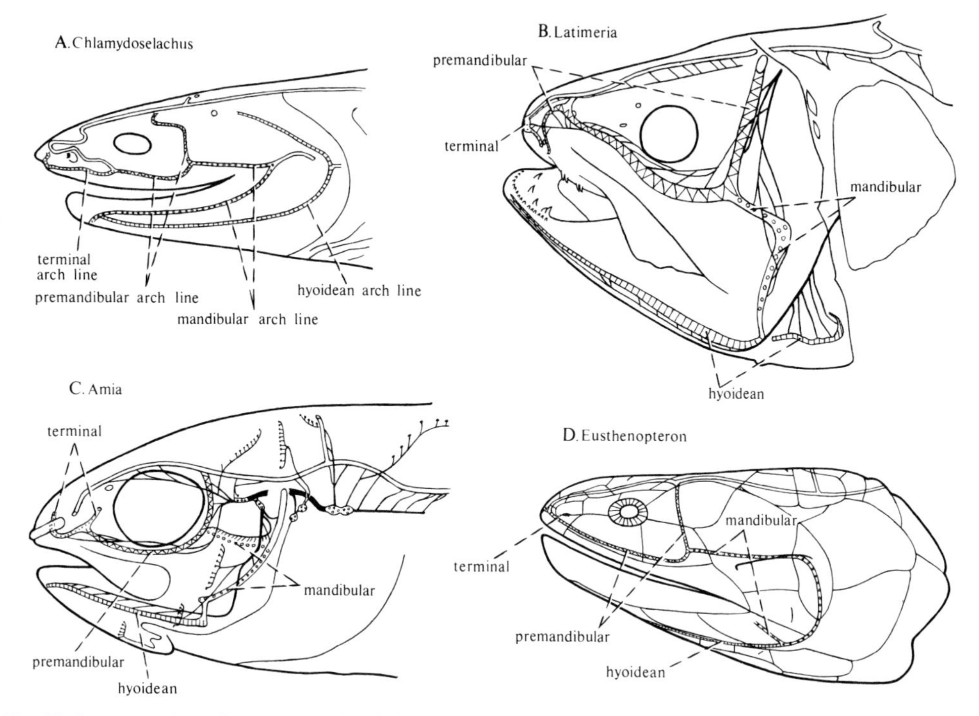

Fig. 33. Interpretation of sensory canals of cheek and lower jaw in terms of visceral arch lines in four fishes.

(3) Premandibular arch line. This line (Fig. 33) is in all forms represented by the suborbital portion of the infraorbital canal which runs along the epal element (pars autopalatina) of the premandibualr arch, and the postorbital portion of that canal. These portions are innervated (Fig. 32; Fig. 50, Volume 1) by branches of the r. buccalis which probably belong to the premandibular metamere and are parts of the n. profundus (Fig. 32).

(4) Terminal arch line. This line (Fig. 33) is represented by the anterior (ethmoidal) part of the infraorbital canal. This part is separate in *Latimeria* and in urodeles (Fig. 130; premaxillary line, Schmalhausen, 1964; 1968). In *Amia*, in which it lodges the foremost five infraorbital neuromasts (Fig. 50, Volume 1), and in other actinopterygians it

arises from a separate terminal placode (Fig. 28; anterior branch of infraorbital placode, Lekander, 1949) as it does in urodeles (Fig. 29B). In *Leuciscus* and other teleosts it is innervated by an upper rostral branch of the n. buccalis, whereas in *Latimeria, Amia, Polypterus* and sharks there is one twig to each neuromast.

These interpretations are, at the first sight, difficult to apply to dipnoans (Figs 329, 331, Volume 1) and other forms (arthrodires, Figs 292, 293, Volume 1; holocephalans, Fig. 30; Fig. 298B, Volume 1) in which the oral canal is a branch of the infraorbital canal. In order to get an explanation to this variation let us first turn to *Chlamydoselachus* (Fig. 34A). In this form Allis's jugal line sometimes extends backwards to join the dorsal end of the gular line at the point where this line meets the spiracular line; and at this point there is also a short posteriorly directed groove with one neuromast (the opercular line). In the 10 mm embryo of *Neoceratodus* (Fig. 34D; Pehrson 1949) the oral and mandibular placodes are joined by a "jugal loop" and where the latter joins the mandibular placode there is a posteriorly directed appendage which develops into a few transient opercular neuromasts. Moreover there is a small placode "y" between the "jugal loop" and the infraorbital placode. This small placode

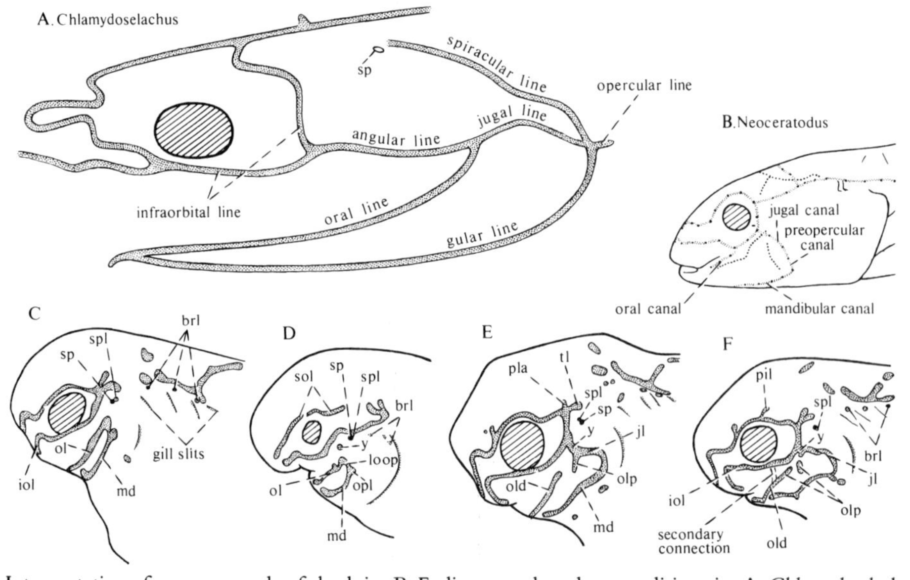

Fig. 34. Interpretation of sensory canals of cheek in, B–F, dipnoans, based on conditions in, A, *Chlamydoselachus*. B, *Neoceratodus forsteri*, head in lateral view. From Holmgren (1942a). C–F four stages in development of sensory lines in dipnoans. C, E, F, *Protopterus annectens*, 12·4, 13·6 and 14 mm. D, *Neoceratodus forsteri*, 10 mm. From Pehrson (1949).

brl, suprabranchial neuromasts of vestigial metameric branchial arch lines; iol, infraorbital line placode; jl, loop, jugal line placode (called "jugal loop" in early stages); md, ol, mandibular and oral line placodes; old, distal part of oral line placode forming oral canal in adult; olp, proximal part of oral line placode forming pit-lines of cheek; opl, opercular line placode; pil, pla, placodes of pineal and anterior pit-lines; sol, supraorbital line placodes; sp, spiracle; spl, placode of vestigial spiracular line; tl, temporal (otic) line placode; y, angular line placode.

forms a connection between the "loop" and the infraorbital placode and as soon as this connection is established (Fig. 34E, F) the similarities to the conditions in *Chlamydoselachus* are striking. As is readily seen the connecting canal ("y") must be homologous with Allis's angular canal in *Chlamydoselachus*, whereas the "jugal loop" corresponds to his jugal canal, and the opercular appendage to the short posterior opercular line. Moreover, it is evident that the oral placode (Fig. 34C, D) is homologous with the selachian oral canal and that the mandibular placode represents the gular canal. However, very soon the oral placode in both *Neoceratodus* and *Protopterus* (Fig. 34E, F) divides into proximal and distal portions. The proximal portion develops into the oral and mandibular pit-lines of the adult, whereas the distal portion which forms the oral canal joins the infraorbital canal. Accordingly the connection between the oral and infraorbital canals in dipnoans (and probably in arthrodires as well) is secondary; and it may be of interest to note that this connection has not yet materialized in Devonian dipnoans (*Dipterus*, Fig. 328, Volume 1; *Scaumenacia*, Fig. 333, Volume 1).

According to this interpretation, which differs considerably from those given by Stensiö (1947) and Holmgren (1953), the so called preoperculo-mandibular canal in dipnoans, using Allis's terminology, is an angular-jugal–gular canal. The equivalent to the preopercular canal in *Latimeria* and *Amia* must in *Chlamydoselachus* be represented by the proximal part of the oral canal, whereas in dipnoans it is to be searched for among the pit-lines which develop from the proximal portion of the oral placode. The spiracular line of *Chlamydoselachus* is in dipnoans probably represented by a transient neuromast placode which lies either close behind the spiracle (*Protopterus*), as does the dorsal end of the spiracular line in *Chlamydoselachus*, or (*Neoceratodus*) forms a lateral branch of the placode of the otic portion of the infraorbital canal (cf. the marginal canal in arthrodires). According to Pehrson (1949); also Holmgren and Pehrson, 1949) this transient spiracular neuromast is the foremost organ in a suprabranchial sensory line which contains three further neuromasts, one situated dorsal to each of the three foremost posthyoidean branchial units. Since the spiracular neuromast most likely is a vestige of the spiracular line which (in *Chlamydoselachus*) is a part of the hyoidean arch line it seems likely that the three other suprabranchial neuromasts are vestiges of the dorsal parts of three metameric branchial arch lines similar to those in adult lampreys (Jollie, 1962).

In view of what has now been set forth it is likely that each branchial unit originally was provided with a metameric sensory line as assumed by Goodrich (1930). These visceral arch lines, which in the foremost four branchial units have been more or less completely retained to form the essential parts of the lateral line system of the cheek and lower jaw, are obviously transverse lines.

Metameric transverse lines are certainly present on the dorsal side of the head as well, but because of secondary displacements they are not always easy to identify. However, to this category belong no doubt the supratemporal commissure which traverses the extrascapular series dorsal to cranial vertebra 6+7 (Fig. 61; innervation first vagus portion in metamere 6, Fig. 22) and the middle pit-line, related to the posterior parietal and the supratemporal, and situated dorsal to cranial vertebra 5+6 (innervation n. glossopharyngeus, metamere 5). Also the frontal pit-line in rhipidistids carried by the posterior frontal dorsal to CV2+3, is probably a transverse line as are also the transient pineal line in *Protopterus* and the so called profundus line in *Kujdanowiaspis* and a few other arthrodires (Stensiö, 1945; 1969; Denison, 1958; Ørvig, 1971; 1975; Poplin, 1973).

3 Somitic Derivatives

The distinct metamerism demonstrated above with respect to the cranial nerves and ganglia is a consequence of the metamerism displayed by the cephalic somites and their ventral continuations, the mesodermal visceral tubes (Fig. 9). The somites form important parts of the head and a remarkable fact is that not only the sclerotomes but also the myotomes and the dermatomes contribute to the formation of the endoskeleton.

DERMATOMES

The disintegration of the dermatomes begins remarkably late (note that the dermatomes persist in advanced embryonic stages of *Amia*, Fig. 13, Volume 1). Because the space between dermatome and skin ectoderm (the epidermis in the adult) is already filled by migratory neural crest cells when the proliferation of dermatomic material begins, it is difficult to decide to which extent this material contributes to the formation of the connective tissues (the dermis) of the skin. However, besides parts of the dermis the dermatomes probably give rise also to portions of the neurocranium (Bjerring, 1967). Judging from figures of early stages of sharks and teleosts (Bertmar, 1959) it is thus reasonable to assume that the anterolateral and posterolateral otic cartilages are derived from the dermatomes in the fourth and fifth metameres (Fig. 44). Possibly also parts of the scleral capsule are dermatomic in origin.*

MYOTOMES

Hypaxial Myotomic Derivatives

Myotomic cells are soon transformed into muscle fibrils to form metameric myomeres. In the trunk the myomeres are separated by intermetameric myosepta (myocommata; Figs 22, 66, Volume 1) and an horizontal septum divides the myotomic (somatic) musculature into dorsal or epaxial and ventral or hypaxial portions (Figs 13, 26, Volume 1). In the head myotomic material is present in all

* p. 11.

metameres, but has undergone considerable modifications (forming eye muscles and certain skeletal elements) and only in the posterior part are there distinct myomeres.

Hypobranchial Musculature

The hypaxial portions of some of these posterior cephalic myomeres have been retained to form the hypobranchial musculature (Figs 35, 36). This musculature (Edgeworth, 1935) includes two principal muscles, the geniobranchialis (Fig. 69, Volume 1) or geniohyoideus as it is called in tetrapods, and the rectus cervicis. However, in tetrapods in which the hypobranchial musculature forms the muscles of the tongue there are generally two additional muscles, the hyoglossus and the genioglossus (Fig. 116).

The number and numerical order of the myotomes that contribute to the hypobranchial musculature are subject to variation as is also the innervation (Edgeworth, 1935). It has, however, been established that the number of myotomes in certain groups at least (sharks, urodeles, mammals, Fig. 23; Goodrich, 1911; 1918; Edgeworth, 1926; 1928; 1935; Fox, 1954; Hazelton, 1970) is three, and most likely these myotomes belong to the seventh, eighth and ninth metameres (in *Amia*, Fig. 46, Volume 1, the composition of the n. hypobranchialis indicates that myotome 10 may be involved). The hypaxial portions of the myomeres formed by these myotomes grow forward ventral to the branchial skeleton, lose their connections with the epaxial parts and coalesce into a single muscle mass. In this mass the derivatives of the various myotomes are of course difficult to keep apart, but this does not necessarily mean that they have lost their individuality. At least as far as the osteolepiform–tetrapod stock is concerned it is likely (Figs 23, 35; Jarvik, 1963) that the rectus cervicis is formed by the hypaxial portion of myomere 9, the hyoglossus and the genioglossus by that of myomere 8, and the geniohyoideus by that of myomere 7.

Subbranchial Series

The hypobranchial muscles are inserted into a median ventral series of endoskeletal elements, the subbranchial series, first recognized in *Eusthenopteron* (Jarvik, 1954). In osteolepiforms (Fig. 35; Fig. 112, Volume 1) this series includes two elements, the sublingual rod and the urohyal. Vestiges of the sublingual rod and probably also of the urohyal are present also in anurans and amniotes (Fig. 115). In porolepiforms (Fig. 199, Volume 1) and most other teleostomes (cf., however, the prelingual cord in *Hepsetus*, Bertmar, 1959; the sublingual ossifications in cyprinids, Nelson, 1969; and the basihyal in *Lepisosteus*, Hammarberg, 1937) there is only one element, the urohyal. It has been shown (Jarvik, 1954; 1963; 1972) that this element is homologous with the so-called basibranchial 2 in urodeles (Figs 35C, 111, 112) which long was thought to belong to the visceral endoskeleton. However, as shown by experimental and other methods (Stone, 1926; Hörstadius, 1950; Jarvik, 1963; Weston, 1970) neither the urohyal ("basibranchial 2") nor the sublingual rod are of neural crest origin. Hence it follows that the subbranchial series must have another source than the visceral endoskeleton; and because of its intimate relations to the myotomic hypobranchial musculature it has been concluded (Jarvik, 1963) that it has arisen to form a support for that musculature and that it is derived from myotomic mesomesenchyme. Accordingly the subbranchial series most likely belongs to the myotomic skeleton* and together with the hypobranchial muscles and associated nerves and vessels it forms a unit, the subbranchial unit (Fig. 109; Fig. 11, Volume 1), different in origin and primarily separate from the overlying visceral structures.

Epaxial Myotomic Derivatives

The epaxial portions of the cephalic myomeres are retained in all metameres, but are partly much modified to form eye muscles or skeletal structures. A characteristic feature is that they, except most posteriorly, are divided into a dorsal or tectocranial and a ventral or basicranial series (Fig. 36; Bjerring, 1975; 1977).

Basicranial Series

Basicranial muscles are known with certainty in cephalic metameres 2–8, and are in all these metameres paired longitudinal structures situated close to the basis cranii (Figs 23, 36, 41; Bjerring, 1970; 1975; 1977; Jarvik, 1972). However, the basicranial muscles, which in some metameres are transient embryonic structures, may in ontogeny be replaced by (or possibly transformed into; Jarvik, 1972, p. 238) skeletal structures of myotomic origin and this has probably happened also in phylogeny. Because of these intimate relations between myotomic skeleton and muscles it is most convenient to treat the myotomic skeletal components in the basis cranii together with the basicranial muscles.

In the first metamere no longitudinal muscle is present but according to Bjerring basicranial muscle 1 is possibly represented by the m. obliquus inferior innervated by a branch of the n. oculomotorius (Fig. 23; Fig. 65, Volume 1; as to other eye muscles see the tectocranial series).

Basicranial muscle 2 (innervation n. oculomotorius) has been found only in embryonic stages of *Amia* and in the adult of *Scomber*† (Figs 23, 36, 42, 44; Fig. 65, Volume 1). In contrast to *Amia* the muscle in *Scomber* lacks anterior attachment, fading out anteriorly in the area of the eye bulb. As to its position and extent the muscle in *Scomber* reminds of, and is probably homologous with, the eye stalk in sharks which also is a derivative of myotome 2. To the myotomic skeleton in metamere 2 belongs probably also the basiorbital in petromyzontids (Fig. 365, Volume 1; Bjerring, 1977).

Basicranial muscle 3 is represented by the large subcranial muscle in *Latimeria* and other "crossopterygians" (Fig. 40; Figs 78, 190, Volume 1; innervation n. rarus‡). As suggested by Bjerring (1967) and later confirmed (Jarvik, 1972, p. 228) this muscle is probably homologous with the polar cartilage which forms the myotomic skeleton in metamere 3 (Fig. 44). The substitution of basicranial muscle 3 by a skeletal element (the polar cartilage) has probably contributed to the obliteration of the intracranial joint at the transition from fish to tetrapod.

* Volume 1, p. 43; † Volume 1, p. 94; ‡ Volume 1, p. 180.

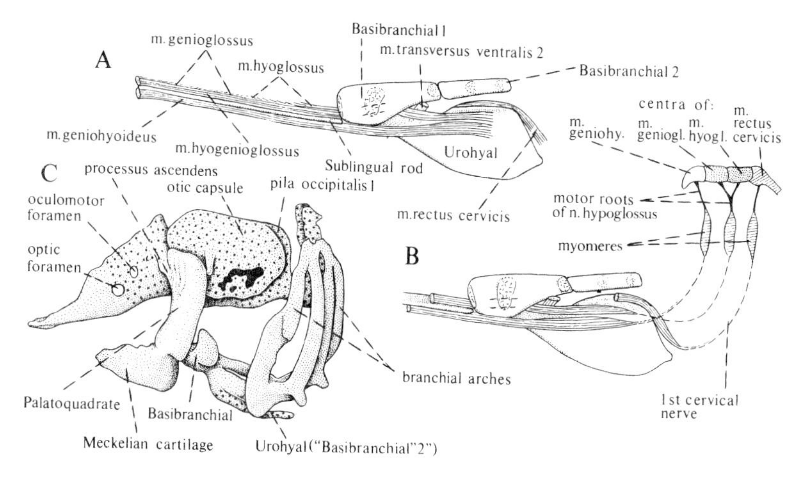

Fig. 35. A, B, *Eusthenopteron foordi* A, basibranchial and subbranchial series with hypobranchial muscles in lateral aspect. B, tentative diagram based on A to illustrate presumed relations of motor centra of n. hypoglossus and first cervical nerve (in hypoglossus nucleus), myomeres and hypobranchial musculature. From Jarvik (1963). C, *Ambystoma punctatum*. Neurocranium, visceral endoskeleton and subbranchial series (urohyal) of larva in lateral aspect. Parts developing from somitic mesomesoderm marked with coarse dots; parts developing from neural crest ectomesenchyme finely dotted. After Stone (1926).

Basicranial muscles 4 (innervated by r. basioticus of n. abducens) and 5 (innervated by n. tenuis, Fig. 23) have been described in embryonic stages of urodeles (Regel, 1961; 1964; 1968; Bjerring, 1970) and anurans (Smit, 1953; Regel and Epstein, 1972) but were probably retained in the adult in their piscine ancestors (Fig. 40; Jarvik, 1972). Basicranial muscle 5 is present also in embryos of *Lampetra* (Damas, 1944). As shown by Regel in urodeles basicranial muscles 4 and 5 ultimately become buried in the overlying cartilaginous components of the basis cranii (Fig. 36C; also Jarvik, 1972, fig. 98). These components, the lamina basiotica anterior in the fourth and the lamina basiotica mesotica in the fifth metamere, are also myotomic derivatives (Figs 42, 44; Holmgren, 1940; Bertmar, 1959) and constitute the myotomic skeleton of these metameres.

Basicranial muscle 6 (innervated by n. occipitalis; Fig. 23) is generally well developed but in most forms (*Amia*, Fig. 63, Volume 1; urodeles, Goodrich, 1911) it is an embryonic structure. In embryos of *Amia* its anterior end penetrates forwards into a deep pit in the posterior end of the lamina basiotica, a condition which indicates that this lamina includes also the lamina basiotica occipitalis which has been recognized in several teleostomes (and dipnoans) and represents the myotomic skeleton in metamere 6 (Bertmar, 1959). In *Polypterus* basicranial muscle 6 is retained in the adult but has an unusual position. It extends forwards ventral to the otic region but is separated from the neurocranium by the parasphenoid and is inserted on the ventral side of that dermal bone behind the complex processus ascendens posterior* (Figs 234, 237, Volume 1). The suggestion (Nelson, 1970) that this muscle is equivalent to the subcranial muscle in "crossopterygians" has been disputed by Bjerring (1973†).

* Volume 1, p. 307; † see also p. 261.

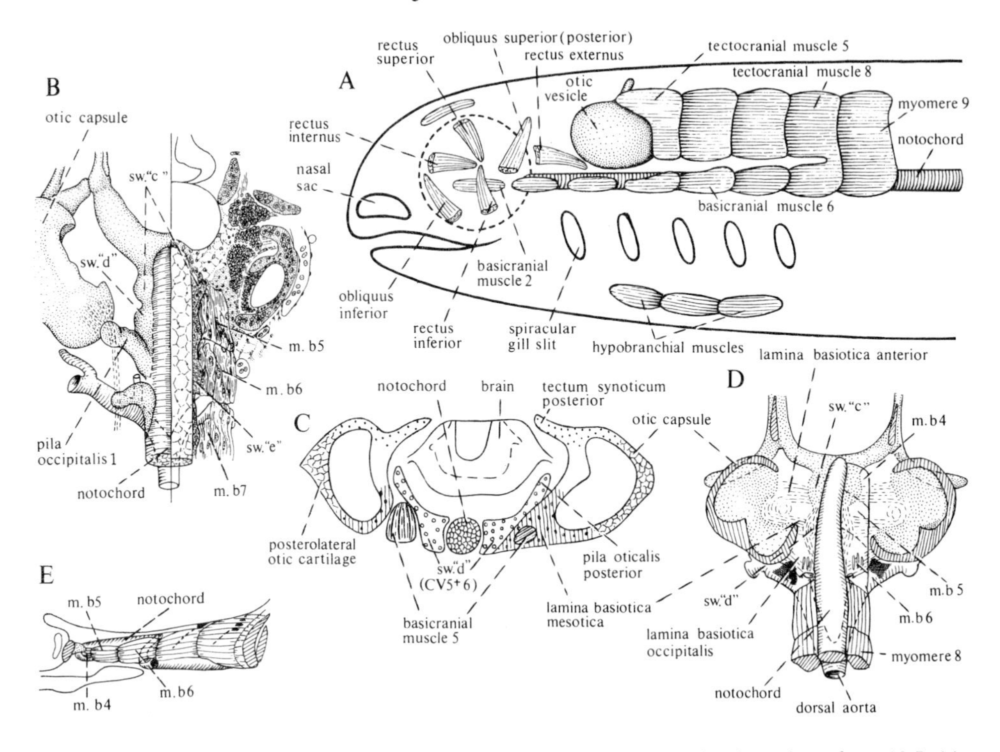

Fig. 36. A, diagram of cephalic myotomic muscles. From Bjerring (1977). B, *Hynobius keyserlingii*, larva 13·5–14 mm. Part of neurocranium in dorsal aspect to show position of basicranial muscles in relation to swellings. From Regel (1964). C, *Hynobius keyserlingii*, larva 14–17 mm. Transverse section of neurocranium through otic capsules and cranial vertebra 5+6. From Jarvik (1972, after Regel, 1964). D, *Pelobates fuscus*, larva. Posterior part of neurocranium with basicranial muscles in dorsal view. E, *Rana esculenta*, early larva. Basicranial muscles and notochord in lateral aspect. D, E, from Regel and Epstein (1972).

m.b4, m.b5, m.b6, basicranial muscles 4, 5, and 6; sw."c", sw."d", sw."e", swellings representing ventral vertebral arches of cranial vertebrae 4+5, 5+6, and 6+7.

Basicranial muscles 7 and 8 are generally preserved in the adult. In *Amia* (Fig. 66, Volume 1) these muscles are inserted on the bulla acustica. It seems likely that it is equivalents to one or both these muscles that form the m. opercularis in recent amphibians (Dunn, 1941; Monath, 1965) and possibly in osteolepiforms (Fig. 131, Volume 1). When this muscle is present in amphibians it is attached to a cartilaginous plate, the opercular plate ("operculum"), in the fenestra ovalis (Figs 90, 103). Most likely this element, the nature of which has been much debated (Swanepoel, 1970), has arisen to form a support of the m. opercularis and belongs to the myotomic skeleton.

The epibranchial muscles* in selachians (Fig. 263, Volume 1) and holocephalans are possibly also to be regarded as basicranial muscles. In *Scyllium* these muscles, according to Edgeworth (1935), arise as ventral outgrowths of myotomes X, Y and Z (outgrowths of the latter two also form hypobranchial muscles) and since the foremost of these myotomes most likely (see Goodrich, 1918) is myotome 6 (myotome 5

* Volume 1, p. 335.

of Goodrich) the epibranchial muscles may be modified basicranial muscles in metameres 6, 7 and 8 which secondarily have gained contact with the branchial skeleton (see acanthodians; Fig. 268, Volume 1).

Tectocranial Series

Separate or almost separate tectocranial muscles (Fig. 36; Bjerring, 1975; 1977) are certainly present in the occipital region in many forms but only rarely (*Scyllium*, Goodrich, 1918; *Amia* Fig. 12, Volume 1; *Ambystoma*, Goodrich, 1911) has their number and metameric position been safely established. In *Amia* tectocranial muscles 6, 7 and 8 are present, but a remarkable fact is that they change their areas of attachment from the fossa tectosynotica in larvae (Fig. 12, Volume 1) to the fossa bridgei in the adult (Fig. 49, Volume 1). Tectocranial muscle 5 seems to have been observed only in the lamprey (Damas, 1944). A portion of this muscle forms the cornealis muscle (Fig. 354D, Volume 1), whereas the main part grows forwards and develops into the m. supraocularis.*

No distinct tectocranial muscles have been found in metameres 1–4, but an attempt to identify also these muscles has been made by Bjerring. In his opinion the m. rectus externus (innervation n. abducens) which is a derivative of myotome 4 represents tectocranial muscle 4. Musculus obliquus superior (innervation n. trochlearis), a derivative of myotome 3, is assumed to be tectocranial muscle 3. In lampreys this muscle is found in the posterior part of the orbit (and hence called m. obliquus posterior), a position which may be primitive. Tectocranial muscle 2 (innervation n. oculomotorius) is assumed to have formed the three eye muscles (rectus internus, superior and inferior) originating from the second myotome. As tectocranial muscle 1, finally, Bjerring suggests the enigmatic striated muscle fibres found in the neuro-epiphysial area in many mammals (see, e.g. Dill, 1963).

SCLEROTOMES

Introduction. Ever since the scientific presentation (1807; 1820) of Goethe-Oken's vertebral theory of the skull (see Peyer, 1949; Singh-Roy, 1967) the question to which extent the vertebral column enters into the formation of the cranium has been the subject of much discussion (for review see de Beer, 1937). The external rings of dermal bones which the German poet J. W. Goethe thought he could distinguish in the mammalian skull cannot be true vertebrae.† The cranial vertebrae must, like those in the trunk, be formed by the somitic sclerotomes. It is generally agreed that the occipital region, at least partly, is made up of fused vertebrae, and as is well known the number of vertebrae plastered to the hind end of the neurocranium varies considerably and may be different even in closely related forms (Fig. 37). Of greater interest, however, is the question of how far forwards in the basis cranii the vertebral column

* Volume 1, p. 460; † cf., however, p. 104.

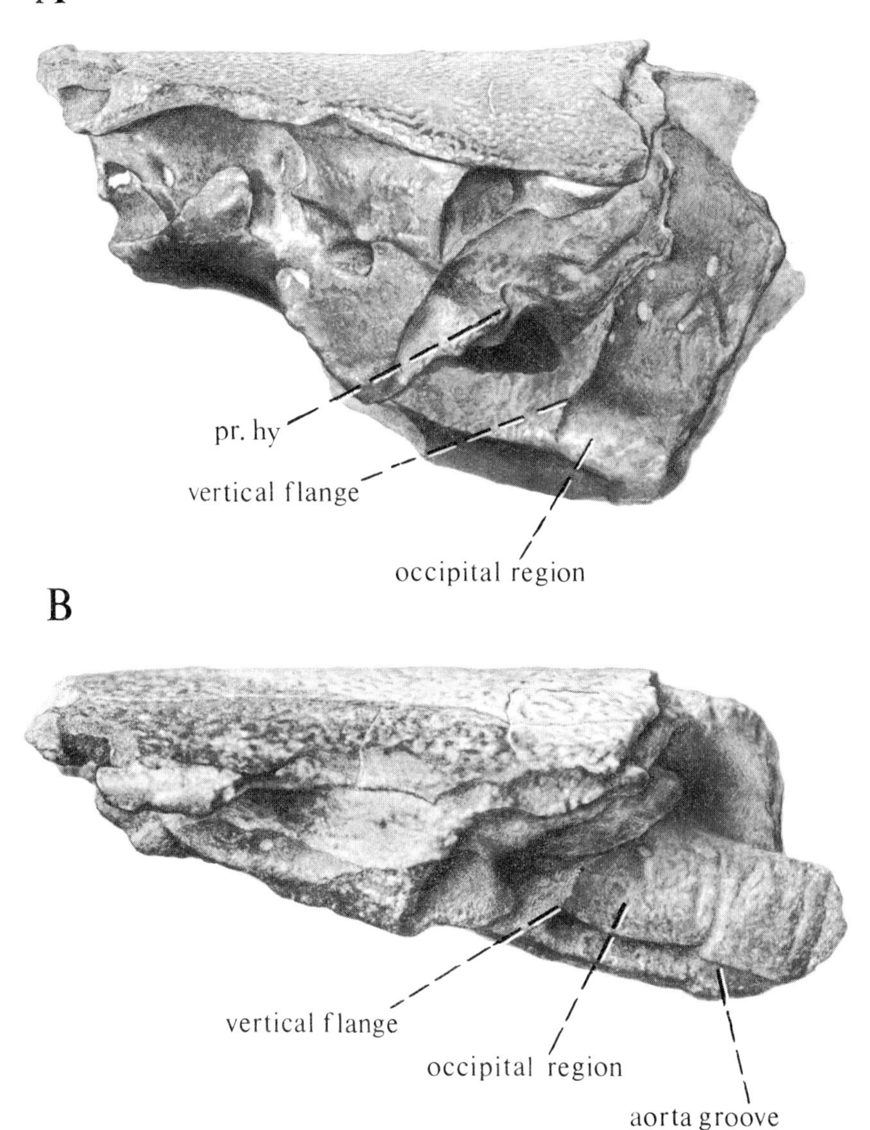

Fig. 37. Photographs of divisio cranialis posterior of two osteolepiforms in lateral aspect to show differences in length of occipital region. Parietal shield in both specimens made equal in length. A, *Eusthenopteron foordi*. B, small rhizodontid, Upper Devonian, East Greenland (this new form which in most respects is suggestive of *Eusthenopteron* will be described as *Spodichthys bütleri*).
pr.hy, process probably for attachment of m. adductor hyomandibulae.

may be traced. Gegenbaur (1872; 1887) suggested that it reaches as far as the notochord, that is to the hypophysial area (also Matveiev, 1925), whereas in the opinion of almost all other students it ends behind the otic capsules. A precise statement in this regard was made by de Beer (1937, p. 13). In his opinion "the evidence in favour of the formerly vertebral nature of the hinder part of the gnathostome chondrocranium begins in the fifth segment, and never farther forwards". However, the exploration of the fossil "crossopterygians", in particular *Eusthenopteron* (Jarvik, 1954), offered new possibilities to attack this intricate problem.

The occipital region in *Eusthenopteron* is, as seen in posterior view, very similar to the vertebrae following next behind and is no doubt composed of fused vertebrae. However, if we examine the posterior part of the ethmosphenoid we will find that this part, too, is very suggestive of a vertebra (Fig. 38). I therefore found it reasonable to assume (Jarvik, 1959; 1960) that the posterior part of the ethmosphenoid represents a vertebra and that the peculiar intracranial joint in the "crossopterygian" fishes is a persistent vertebral joint. Moreover, it was suggested that the various embryonic elements, which according to Holmgren (1940) are derivatives of the cephalic somites, and which are found in association with the cranial portion of the notochord and the overlying part of the brain, are arcual elements of cranial vertebrae.

These partly new ideas, implying that the basis cranii, forwards to the hypophysis, includes vertebrae of a long cranial portion of the vertebral column soon gained support by new discoveries. Independent cranial arcual elements were thus found in *Latimeria* (Millot and Anthony, 1958; 1965; Stensiö, 1963a) and fossil "crossopterygians" (Bjerring, 1967) and, moreover, metameric basicranial muscles and dis-

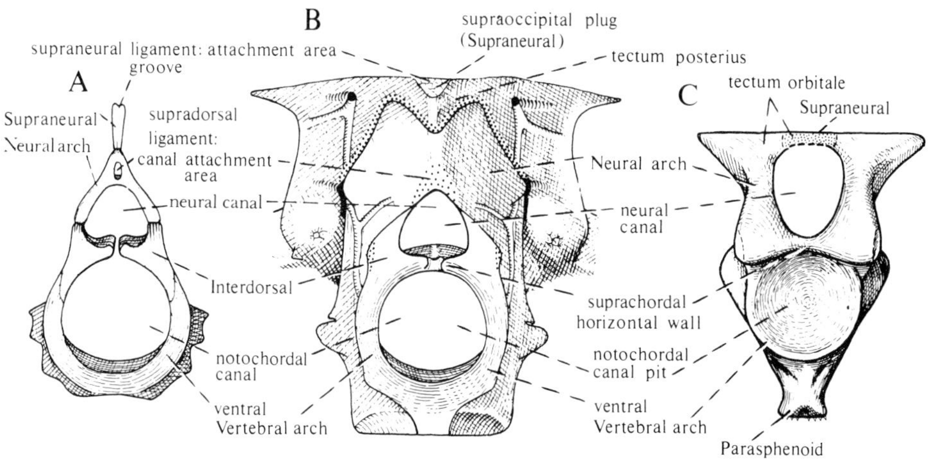

Fig. 38. Eusthenopteron foordi. A, vertebra, B, median parts of otoccipital and C, ethmosphenoid, in posterior aspects to demonstrate vertebralike structures in neurocranium.

tinct traces of segmentation of the sclerotomic derivatives were discovered in larval urodeles (Regel, 1961; 1964; 1968; Bjerring, 1970). On the basis of these and other new facts an attempt was made (Jarvik, 1972) to identify the cranial vertebrae and the intrametameric cranial vertebral joints in *Eusthenopteron* and in hynobiid urodeles. The important fact* that the notochord originally reached in front of the hypophysis was considered, but it was not realized that there is one metamere, the terminal,† in front of the premandibular metamere which up to that time was thought to be the

* p. 7; † p. 11.

foremost one. In the following brief exposé this new fact will be taken into account and the cranial vertebrae (CV) will be accordingly renumbered. However, before turning to the presumed cranial vertebrae it is necessary to consider the vertebral column.

Remarks on the Vertebral Column

There is reason to assume (Jarvik, 1965a) that the early vertebrates were swimming animals (Fig. 69) with metameric muscles derived from metameric somites. The skeletal elements in the vertebral column most likely are secondary formations arisen to form a support of the pre-existing metameric muscles (cf. the subbranchial series*). When in vertebrate phylogeny the condensation of the sclerotomic mesenchyme began and skeletal elements arose, the vertebral joints—in consequence of the undulating swimming movements of the body—necessarily must be placed opposite the middle of the metameric muscles (Fig. 39). This resulted in a primary subdivision or segmentation—not a "resegmentation" (= "remetamerism")—of the sclerotomic mesenchyme and this explains why the vertebrae usually are dimetameric structures derived from two half-scleromeres, one caudal from one metamere, and one cranial from the metamere following next behind. Because the vertebrae not only serve the purpose to give attachment to the metameric muscles, but also protect both the notochord and the overlying neural tube it is easy to understand why there arose one paired dorsal and one paired ventral element in each half-scleromere. Accordingly each vertebra will include four paired primary arcual elements, two anterior (basidor-

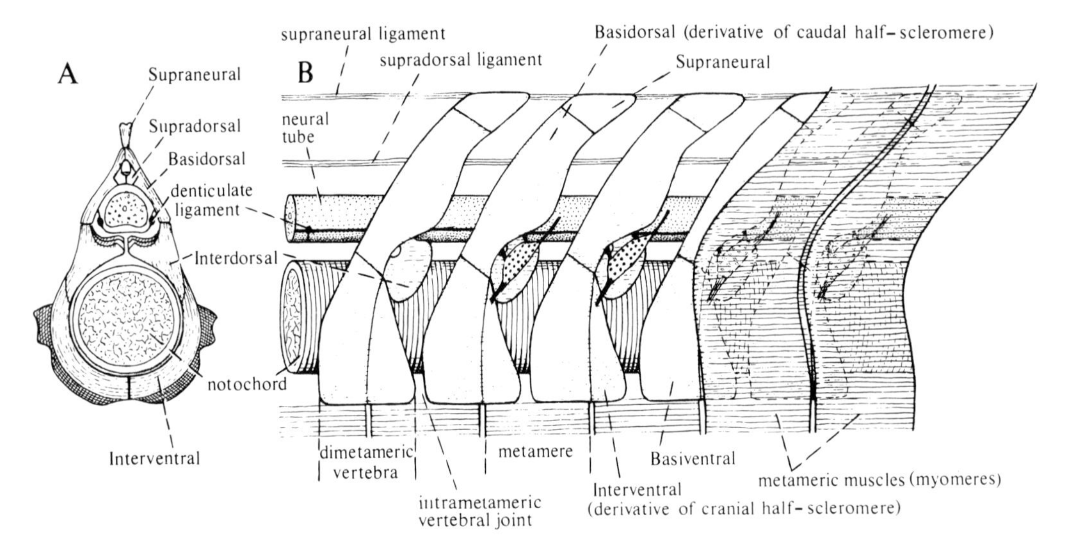

Fig. 39. Diagrams to illustrate compostion of dimetameric vertebrae and their relations to metameric myotomic musculature. Based mainly on conditions in *Eusthenopteron*. A, vertebrae with certain soft parts in posterior view. B, five vertebrae with spinal ganglia, ligaments, and metameric muscles in lateral view.

* pp. 57–58.

sal and basiventral) derived from the caudal half-scleromere of one metamere and two posterior (interdorsal and interventral) formed from the cranial half-scleromere of the metamere following next behind. The existence of these four paired primary components is documented by embryological studies of various vertebrates (Schauinsland, 1905) and there is no reason to reject Gadow's terms basidorsal, etc. for these components.*

These four components may fuse in various combinations and there are also many other variations† which have rendered the interpretation difficult and have caused lively debates (Schauinsland, 1905; Goodrich, 1930; Gadow, 1933; Williams, 1959; François, 1966; Schaeffer, 1967a; Shute, 1972). Metameric vertebrae separated by intermetameric joints are present in *Amia*, (Fig. 22A, Volume 1) but generally the vertebrae, in fishes (*Amia*, Fig. 22B, Volume 1) as well as in tetrapods, are dimetameric structures separated by intrametameric joints spanned by metameric muscle portions (myomeres, Fig. 39). These latter facts are of great importance for the interpretation of the axial part of the skull, as are also the following conditions. The basiventral often fuses with the interventral into a single element which I (1952), in order to avoid the special terminology used in tetrapods, called the ventral vertebral arch both in *Eusthenopteron* and *Ichthyostega* (Fig. 88). The basidorsal is generally provided with an ascending process, the neural arch, which forms part of the lateral wall of the neural canal. Also the interdorsal may share in that wall and is sometimes (urodeles, reptiles, birds; Smit 1953, p. 121) provided with an ascending process suggestive of the neural arch. In *Eusthenopteron* (Fig. 38A; Fig. 97C, Volume 1) both the basidorsal and the interdorsal present medial lamellae which separate the neural and notochordal canals and together with the lamellae of the other side form a median groove for the basilar artery. The horizontal, suprachordal wall thus formed (cf. the suprachordal cartilage in anurans Jarvik, 1972, fig. 24) and the basilar groove are continued forwards into the occipital region. The posterior end of the ethmosphenoid (Fig. 38C) shows an anterior part of that wall. However, the vertebral column includes also other elements than the arcuals. Sometimes there is a supradorsal, paired as, for example, in *Amia* (Fig. 23C, Volume 1) or unpaired as in sharks (Goodrich, 1930) and urodeles (Mookerjee, 1930). Moreover, *Eusthenopteron* (Fig. 38; Fig. 97, Volume 1) and certain other forms (Goodrich, 1930, p. 88) present a median series of supraneurals. In all vertebrates there is a strong longitudinal ligament. This ligament which runs close dorsal to the supradorsals has been termed the supradorsal ligament in order to distinguish it from another longitudinal ligament, the supraneural ligament, situated on top of the supraneurals (Fig. 131, Volume 1; Jarvik, 1975; ligamentum supraspinale in man‡). These two ligaments are attached anteriorly to the occiput. Of interest is also a third ligament associated with the vertebral column. This ligament, the ligamentum denticulatum, present in all vertebrates (Haller, 1934, p. 315, fig. 292), is a paired structure situated within the neural canal close to the spinal cord (Fig. 39; Fig. 13, Volume 1). Of particular interest are that this ligament gives off metameric lateral processes towards the inner sides of the neural arches and that the ligaments of both

* cf. p. 154; † p. 156; ‡ p. 267.

sides sometimes are connected by transverse metameric bands ventral to the spinal cord. This is of interest because Bjerring recently (1977) in the orbital region of *Amia* has discovered a transverse intracranial fibrous band which probably is a vestige of the dentate ligament. This may be true also of the ligamentum tenaculum oculi* (Fig. 44) known in many fishes (Harman, 1899; Holmgren, 1943), but if there are also other metameric vestiges of the dentate ligament in the head is unknown and further investigations are necessary.

Cephalic Sclerotomic Derivatives: Cranial Vertebrae

The cephalic sclerotomes produce skeletogenous material in the same way as those in the trunk. By studies, in the first place of sharks, it has also in many cases been possible to decide from which sclerotomes the various skeletal components in the axial part of the skull have arisen and to decide to which metamere they belong (Holmgren, 1940; 1943; Bertmar, 1959; Bjerring, 1968; 1971; 1973; 1977). As suggested previously† Jarvik, 1959; 1960) these skeletal components are probably arcual elements of cranial vertebrae, although it is now evident that the vertebral column extends farther forwards in the basis cranii (to the area of the optic chiasma) than there was reason to believe in 1959.

The purport of these statements is that the cranial vertebrae, like those in the trunk, are dimetameric structures formed by material from two half-scleromeres. These views have been confirmed in the first place by ontogenetic studies of urodeles and anurans (Regel, 1961; 1964; 1968; Regel and Epstein, 1972). As most distinctly seen in urodele embryos (Figs 36B, C, 49B) the parachordals show a series of swellings separated by constrictions and provided with dorsal processes situated along the brain. An important fact is also that the basicranial muscles run from the top of one swelling to the top of that next in front. These conditions, the relations to the notochord and the similarities to the vertebrae in the anterior part of the trunk (see Fox, 1954; 1959) can only be taken to mean (Jarvik, 1972) that the swellings represent the ventral vertebral arches and that the dorsal processes are the neural arches of cranial vertebrae (cf. Matveiev, 1925). Hence it follows that the constrictions spanned by the metameric basicranial muscle are vestigial intrametameric vertebral joints.

The subdivision of the parachordals in other recent vertebrates has been much discussed and generally two or three portions have been recognized (Bertmar, 1959). These portions (anterior, mesotic and occipital; Fig. 42) are generally considered to be metameric structures. However, as is well shown in *Amia* (Fig. 44; Fig. 62, Volume 1; Bertmar, 1959, Fig. 42B) the basiotic laminae which are metameric myotomic derivatives alternate with the parachordal portions, a condition which suggests that the parachordal portions in *Amia*, and probably in vertebrates in general, are dimetameric structures (ventral vertebral arches) as they are in urodeles and anurans.

These interpretations are further supported by the conditions in *Eusthenopteron, Glyptolepis, Latimeria* and other "crossopterygians" (Jarvik, 1954; 1972; Millot and Anthony, 1958; 1965; Bjerring, 1967; 1971; 1973). In porolepiforms and

* p. 72; † p. 62.

osteolepiforms the intrametameric joint persists within the third metamere and as in the otic region of urodeles and in the trunk this joint (the intracranial joint) is spanned by the basicranial muscle of the metamere (subcranial muscle or basicranial muscle 3). Behind this joint, which in *Ichthyostega* (Figs 171, 172, Volume 1) is represented by a transverse fissure, the fissura preoticalis, traces of intervertebral joints are found in several vertebrates (sharks, acanthodians, palaeoniscids, ichthyostegids). These joints are well shown in particular in *Eusthenopteron* in which they are represented by a series of fissurae in the cranial base (fissura oticalis anterior and posterior, fissura occipitalis lateralis), and distinct impressions indicate that also these joints were spanned by basicranial muscles (Fig. 40). Of interest in this connection is also that the ventral

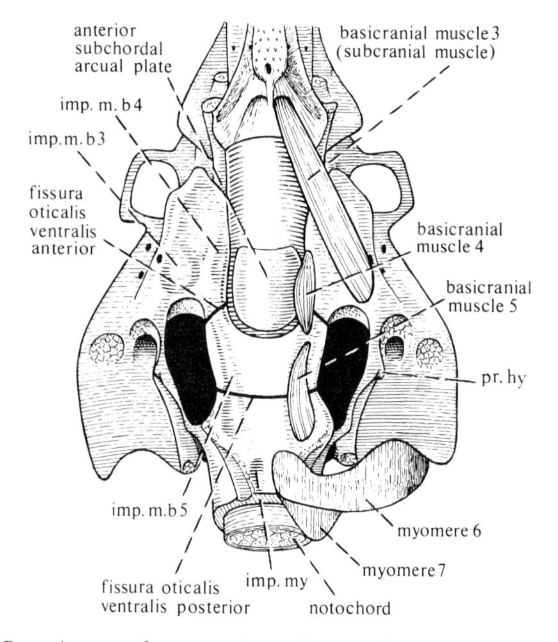

Fig. 40. Eusthenopteron foordi. Posterior part of neurocranium with parasphenoid in ventral aspect to show subdivisions in cranial base and impressions (imp) for attachment of basicranial muscles (m.b3, etc.) and myomeres (my). Basicranial muscles of left side tentatively restored.
pr.hy, process probably for attachment of m. adductor hyomandibulae.

vertebral arches may be independent forming subchordal (ventral) arcual plates (the suprachordal arcual plates are independent parts of the suprachordal wall and conceivably one such dimetameric plate was present in each vertebral segment forwards to the third metamere). Of interest in this connection is, finally, the conditions in anurans, which as is now well established are descendants of primitive osteolepiforms. In embryonic anurans (Fig. 36D) the basicranial muscles are well developed in the otic region (Smit, 1953; Regel and Epstein, 1972) and it is therefore not surprising that these muscles have been retained in the adult in their piscine ancestors (Fig. 40). But of greater interest is perhaps that contractions of the embryonic basicranial muscles in anurans caused a bending of the embryonic vertebral column in the area between the otic capsules (Regel and Epstein). This most remarkable condition that the vertebral column still is movable in the otic region, the presence of the intracranial joint and the structure of the posterior part of the ethmosphenoid in "crossopterygians", show that the vertebral column reaches as far forwards as to hypophysis; and obviously the sclerotomic material at least as far forwards as to the third metamere has undergone a primary segmentation into half-scleromeres forming dimetameric vertebrae, as in the trunk. However, the presence of basicranial muscles in the second (premandibular) and probably also in the first (terminal) metamere indicates that the conditions have been similar in the sclerotomes of the foremost two somites which are situated along the suppressed anterior part of the notochord. In other words it seems likely that the vertebral column extends as far forwards as the rostral portion of the chordamesoderm, the brain and the gut, that is to the area of the optic chiasma.

In accordance with the views advanced in 1972 an attempt will now be made to interpret the various components described by the embryologists and anatomists along the notochord (including the exchordals and the brain in the terms of arcual elements of dimetameric cranial vertebrae. These tentative and partly hypothetical interpretations are presented in diagrammatic representations (Figs 41–44, 61) and only a few remarks will be given here (see also Jarvik, 1972).

Basicranial muscle 6 in urodeles (Fig. 36B) and anurans (Fig. 36D, E) spans the constriction (vestigial intrametameric vertebral joint) between swellings d and e. Accordingly swelling e (occ3, Regel; occipital parachordal, Bertmar) must be the ventral vertebral arch of a vertebra composed of basiventral 6 (formed by the posterior half-scleromere of metamere 6) and interventral 7 (formed by the anterior half-scleromere of metamere 7; Fig. 41). This ventral vertebral arch has therefore been termed V.a 6+7 and together with its paired dorsal process, pila occipitalis 1 (first occipital arch), it forms cranial vertebra 6+7 (CV 6+7). Pila occipitalis 1 is most likely formed mainly by the basidorsal as are the neural arches (pila occipitalis 2, 3, etc) in the

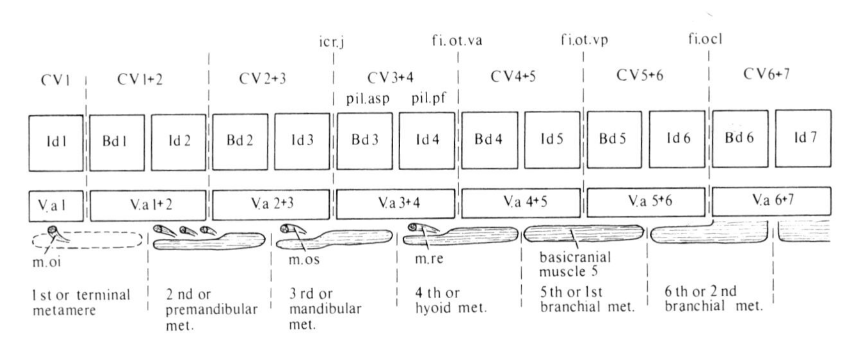

Fig 41. Key to interpretation of sclerotomic and myotomic (muscles only) derivatives of cranial somites.
Bd, basidorsals; CV, cranial vertebrae; Id, interdorsals; V.a, ventral vertebral arches; fi.ocl, fi.ot.va, fi.ot.vp, fissura oticalis lateralis, fissura oticalis ventralis anterior and posterior, icr.j, position of intracranial joint in osteolepiforms and porolepiforms; m.oi, m.os, m.re, inferior and superior oblique and external rectus eye muscles; pil.asp, pila antotica spuria; pil.pf, pila prefacialis.

various number of vertebrae which may be added to the neurocranium behind CV 6+7. Farther forwards, however, in the otic region, where the ascending processes (pilae) form continuous lateral walls of the cranial cavity also the interdorsals may be involved. In CV 3+4, including swelling b and the pila prefacialis, and in CV 4+5, including swelling c (ppocc 1, Regel; anterior parachordal, Bertmar) and the pila oticalis anterior (dorsal parachordal process of van Wijhe), the ventral vertebral arches (V.a 3+4, V.a 4+5) may be independent forming the anterior and posterior subchordal arcual plates in "crossopterygians". A remarkable fact is also that the pila antotica spuria (presumably basidorsal 3; Bjerring, 1973), situated between the profundus and trigeminus foramina, in osteolepiforms and porolepiforms forms part of CV 3+4, whereas in coelacanthiforms it lies in front of the intracranial joint which accordingly in this group is intermetameric in position (Fig. 42; Fig. 214, Volume 1).

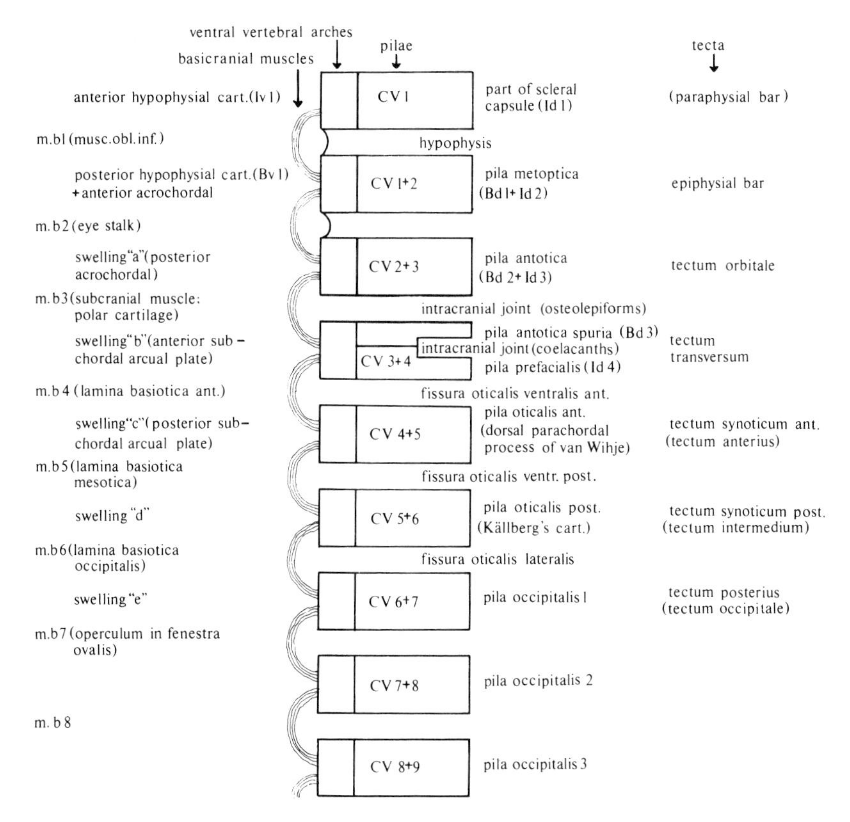

Fig. 42. Diagram of interpretations of (from the left): cephalic metameric myotomic derivatives (metameric basicranial muscles and skeletal elements); sclerotomic derivatives (dimetameric cranial vertebrae, CV, composed of ventral vertebral arches or swellings and neural arches or pilae), separated by intrametameric intervertebral joints (partly marked by cranial fissurae); and cranial tecta.

CV 2+3, which in osteolepiforms and porolepiforms forms the posterior part of the ethmosphenoid, includes swelling a (anterior acrochordal) and the pila antotica. The acrochordal tissue is according to Holmgren (1940) derived from the third (mandibular) somite, but it cannot be excluded that material from the premandibular commissure is involved, as suggested by van Wijhe (1922). The pila antotica (presumably

Bd2+Id3) may in its posterior part (Id3) be provided with a lateral process (the suprapterygoid process) which articulates with the suprapharyngomandibular (processus ascendens palatoquadrati). This is remarkable since also the presumed interdorsal in the fourth and fifth metameres are provided with lateral processes (the pre- and postotical processes; Fig. 44) which articulate with the suprapharyngeal elements of the hyoid and first branchial arches.

The most anterior of the presumed cranial vertebrae, CV1 and CV 1+2, are situated along the anterior suppressed portion of the notochord and are for this and other reasons difficult to interpret. However, the 1st or terminal somite (Platt's vesicle) produces sclerotomic material which forms "the first separate rudiment of the trabecula cranii" (Holmgren, 1940, p. 67). This rudiment must be a part of the posterior so-called somitic portion of the trabecle. In this region of the basis cranii there are sometimes (*Talpa*; Fig. 43) two paired cartilages, the hypophysial cartilages,

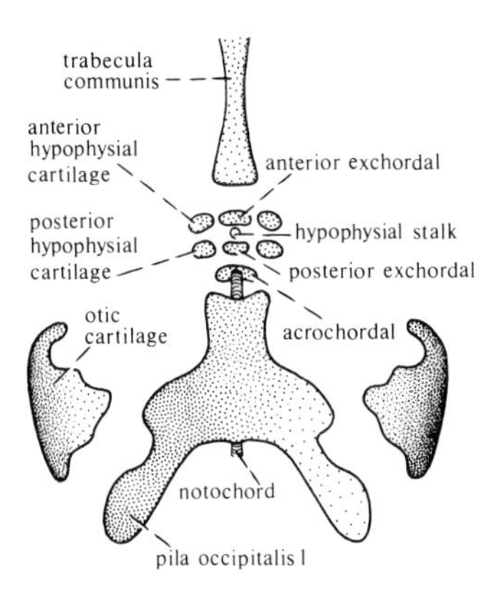

Fig. 43. Talpa europaea, embryo, 11 mm. Endocranial structures in dorsal view. From Bjerring (1971, after Noordenbos).

and it is reasonable to assume that these cartilages are derivatives of the terminal somite and represent the interventrals (Iv1) and the basiventrals (Bv1) of the terminal metamere (Figs 42, 44). Provided this be true the vertebral joint within the first metamere must be situated between the anterior and posterior pairs of hypophysial cartilages that is in the area of the buccohypophysial stalk as I for other reasons has suggested (1972, p. 247). Hence it follows that the anterior pair of hypophysial cartilages form the ventral vertebral arch of the foremost incomplete cranial vertebra (CV1), while the posterior pair together with material derived from the second (premandibular) somite forms that arch in CV 1+2. The two median cartilages, the exchordals (Bjerring, 1971), which in *Talpa* are found between the hypophysial cartilages are, like the corresponding structures in turtles (the taenia intratrabecularis) and porolepiforms (the median hypophysial crest) probably derived from the rostral portion of the chordamesoderm as suggested by Bjerring. The dorsal process of CV 1+2 is formed by the pila metoptica which at least mainly is a derivative of the premandibular somite (Holmgren, 1940). As regards the dorsal process of CV1,

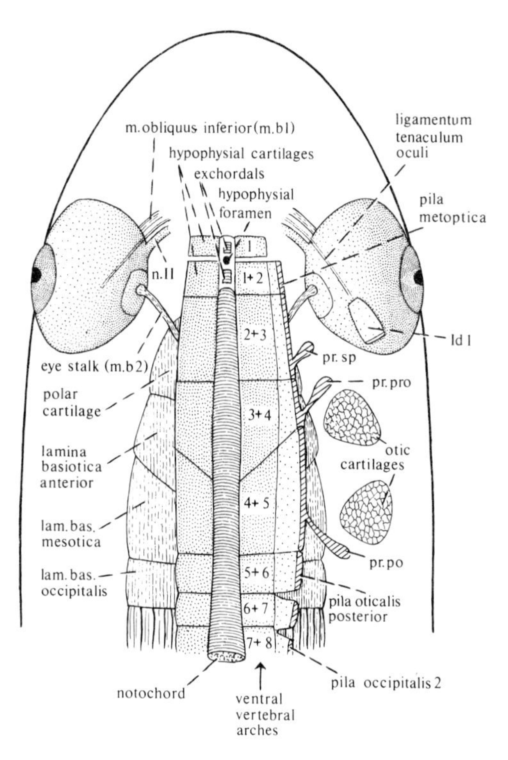

Fig. 44. Diagram of somitic derivatives (myotomic, sclerotomic, dermatomic) in cranial base of gnathostome (cf. Fig. 42). Myotomic muscles shown only in first (terminal) and seventh cranial metameres. Note that metameric myotomic derivatives (polar cartilage and basiotic laminae) alternate with sclerotomic derivatives which form dimetameric cranial vertebrae.

Id 1, part of scleral capsule, probably representing displaced neural arch (pila terminalis) of terminal metamere; n.II, nervus opticus; pr.po, postotical process (articulating with suprapharyngobranchial 1); pr.pro, preotical process (articulating with suprapharyngohyal); pr.sp, suprapterygoid process (articulating with suprapharyngomandibular); 1– 7+8, ventral vertebral arches of cranial vertebrae.

finally, there are several alternatives. It may also be a part of the orbital wall or it may be represented by independent cartilages (e.g. the cartilago hypochiasmatica in mammals, see de Beer, 1937). However, another possibility, suggested by Bjerring (1977), is that the posterodorsal portion of the sclera which according to Holmgren (1940, p. 87) is derived from the terminal somite ("Platt's vesicle") is the dorsal process in the terminal vertebra (Fig. 44). This process may very well have been displaced laterally in connection with the formation of the eye cup which is a part of the brain wall. This assumption is supported by the fact that the ligamentum tenaculum oculi, which possibly is the lateral process of the ligamentum denticulatum* in the terminal metamere, is attached to this area of the sclera, extending inwards close in front of the optic nerve to the interorbital septum.

In the trunk the neural arches in each pair generally fuse into a median neural spine wheras in the head they—due to the development of the brain—have been pressed to

* p. 66.

the sides (except most posteriorly). Characteristic is that the tips of the arches (pilae), which accordingly lie far apart, in each pair may be joined by a transverse bar or tectum. Examining the conditions in various vertebrates (Fig. 59) and considering the bars that have been described* (de Beer, 1937) we will find that there is one bar to each of the presumed cranial vertebrae from CV 1+2 backwards (Figs 42, 61). These bars are (Bjerring, 1975; 1977): the neuro-epiphysial bar, the tectum orbitale, the tectum transversum, the tectum synoticum anterior (= tectum anterius), the tectum synoticum posterior (= tectum intermedium) and the tectum occipitale (= tectum posterius; in *Amia*, Pehrson, 1922, there is also a paraphysial bar which possibly is connected with CV1). According to Holmgren (1940) these bars are ectomesenchymatic. However, it is also possible that they are somitic, derived from the cephalic sclerotomes, as suggested by Bjerring (1975; 1977). This is supported by the fact that the tectum posterius in *Eusthenopteron* includes the supraoccipital plug (Figs 88, 97, 131, Volume 1) which most likely belongs to the supraneural series (Jarvik, 1975); and since at least some of the bars arise from median and paired lateral primordia† (Figs 59, 61) Bjerring assumes that the median components are serially homologous with the supraneurals, whereas the paired lateral components represent cephalic supradorsals.

* pp. 100–105;† p. 102.

4 Visceral Endoskeleton and Musculature, Aortic Arches

Visceral Endoskeleton

Basibranchial Series

As a result of the formation of gill slits (Fig. 7) there arises a paired series of metameric double tubes which are the primordia of the branchial units.* Ectomesenchymatic cells derived from the neural crest migrate downwards within each embryonic branchial unit to form the endoskeletal visceral arches which accordingly are metameric structures as well. Ventrally, in the area of the undivided parasomitic mesoderm (which forms the heart, truncus arteriosus and ventral aorta), the separate ectomesenchymatic streams of both sides reach the ventromedian area dorsal to the conus arteriosus and the ventral aorta. In these area the ectomesenchyme cells form the median basibranchial series, which together with adjoining parts of visceral arches and associated soft parts constitute the basibranchial unit (Fig. 11, Volume 1); this unit is separated from the underlying subbranchial unit by the conus arteriosus and the ventral aorta (Fig. 109). Because of the metameric arrangement of the migrating ectomesenchyme it is likely that the basibranchial series, in contrast to what has recently been assumed (Nelson, 1969), was originally composed of metameric elements connecting the visceral arches of both sides. Independent such median elements have been found between the mandibular and hyoid arches (basimandibular, basihyal; de Beer, 1937, pp. 410, 418), but farther back there is considerable variation (Nelson, 1969) and only rarely (*Heptanchus*, Gegenbaur, 1872; Daniel, 1934) independent metameric basibranchials are discernible. Sometimes, as in one of the earliest known palaeoniscids, the basibranchials form a long median rod (Gardiner, 1973), a condition which has been taken to support Nelson's view that the segmentation of the basibranchial series is secondary. Another problem relates to the elements of the mesomesenchymatic (myotomic) subbranchial series which show a tendency to fuse

* pp. 12–13.

with elements (basibranchials, copulae) of the basibranchial series. Thus, for example, it is difficult to say if the anterior process of the basihyal found in many fishes and tetrapods is a part of the ectomesenchymatic visceral skeleton or if it belongs to the subbranchial series as does the similarly situated sublingual rod in *Eusthenopteron*.

Branchial Arches

An interesting fact is that the visceral arches in embryos of gnathostomes are transverse in position (*Amia*, Fig. 70, Volume 1; Edgeworth, 1935) as they are in cyclostomes. The characteristic oblique position in adult gnathostomes is therefore to be regarded as secondary and obviously the change in direction of the visceral arches has caused modifications also in other parts of the head, and in the anterior part of the trunk including the anterior part of the ventrolateral fin fold (Fig. 69; Jarvik, 1965a). Another remarkable fact is that parts of the anterior arches have moved forwards as a consequence of the rostral prolongation (Fig. 56).

The posthyoidean visceral arches or, as they are generally called, the branchial arches (usually five in number) may be divided into two parts, dorsal and ventral. The dorsal part includes three elements, infrapharyngeal, suprapharyngeal and epal (Figs 34, 110, 111, Volume 1). The infrapharyngeal and epal elements are situated in the roof of the pharynx and in teleostomes they form a row pointing backwards and bear teeth or dental plates. The suprapharyngobranchials, if present (see Nelson, 1968), are directed upwards and lack dentition. The ventral part comprises two tooth-bearing elements, ceratal and hypal, situated in the lateral and ventrolateral walls of the pharynx (Figs 33, 110–113, Volume 1). In teleostomes they form a row pointing forwards. A remarkable fact is that the infrapharyngeal and epal elements in elasmobranchs (Fig. 270, Volume 1) meet at acute angles as do also the hypal and ceratal elements. As viewed from the sides the branchial arches in elasmobranchs are sigma-shaped whereas in teleostomes they are V-shaped. No satisfactory explanation of this difference (or of the fact that the presumed epibranchials in *Neoceratodus* are directed forwards; Figs. 316, Volume 1) has been presented so far. Another remarkable difference is that there is only one series of gill (branchial) rays supporting the gill laminae in elasmobranchs, but two in teleostomes (Fig. 35, Volume 1). In dipnoans and holocephalans gill rays are lacking. A remarkable fact, too, is that the gill rakers in many elasmobranchs (Fig. 271; including acanthodians, Fig. 268, Volume 1), in dipnoans (Fig. 53E), and in holocephalans are supported by endoskeletal rods (Jarvik, 1977). Such rods or cores are unknown in teleostomes in which the gill rakers, as in *Amia* (Figs 35, 42, 43, Volume 1) and *Eusthenopteron*, are thickenings of the mucous membrane strengthened by dental plates.

Prootic Visceral Arches (Exclusive Terminal Arch)

Since the time of Cuvier, Rathke and other early anatomists it has been generally agreed that there are at least two prootic visceral arches, the hyoid arch including two principal elements the hyomandibula and the ceratohyal, and the mandibular arch including the palatoquadrate and the ceratomandibular or Meckelian element. Also it has been taken for granted that these arches are modified branchial arches. However, already Gegenbaur (1872) claimed the there are two further arches (represented by labial cartilages) in front of the mandibular arch, and later Jaekel (1899; 1906; 1925; 1927) and Sewertzoff (1916–1917; 1923; 1931) expressed similar views. Allis (1923; 1938) and de Beer (1937) admitted only one premandibular arch which in their opinion was incorporated in the neurocranium forming the trabecula. In 1927 Stensiö demonstrated that gill-bearing dorsal parts of the premandibular arch are present in cephalaspids and that these parts like the dorsal parts of all the other visceral arches have fused with the neurocranium. Holmgren (1943) was the first to claim that all the three dorsal elements of the branchial arches (epal, infrapharyngeal and suprapharyngeal) may be represented also in the prootic arches and he maintained that certain of these elements may participate in the neurocranium. However, Holmgren accepted only two prootic arches (hyoid and mandibular), as did Goodrich (1930).

Such was the situation, briefly in the early 1950s when the grinding series of *Eusthenopteron* was finished and the wax models were made. On the basis of this new material a complete description could for the first time be given not only of the visceral arches but also of the associated dental plates in a Devonian gnathostome fish and the intricate problem as to the prootic arches could be tackled from a new angle (Jarvik, 1954).

In *Eusthenopteron* (Fig. 111, Volume 1) the three dorsal elements are well shown in the first and second branchial arches. In the first branchial arch the infra- and suprapharyngobranchials have fused into a prominent L-shaped structure. The ventral sides of the infrapharyngeal and epal elements are covered with horizontal dental plates. At the lateral margin of the infrapharyngeal element the horizontal plates bend dorsally and are continued by ascending infrapharyngeal dental plates. These plates (Figs 45, 52; Fig. 110, Volume 1) occupy a triangular field in the medial wall of the first posthyoidean gill slit, in front of the ascending suprapharyngeal portion which lacks teeth. A similar triangular field is found also in the medial wall of the dorsal part of the second gill slit.

Proceeding forwards to the area inside the medial wall of the spiracular gill slit we will find that the neurocranium presents an L-shaped structure of about the same size as, and also in other respects similar to, that in the medial wall of the first gill slit (Fig. 46). The horizontal part of this structure is formed by the otical shelf and the ascending part by the lateral commissure, situated outside the jugular vein. On the ventral side of the otical shelf there is a large dental plate, the paraotic dental plate (Figs 45, 52; Figs 92, 110, Volume 1). At the lateral margin of the shelf this plate is continued by the spiracular dental plates which occupy a triangular field in the medial wall of the spiracular gill slit in front of the lateral commissure which lacks dentition on it outside.

These most striking resemblances can only mean that we are concerned with serial homologies. Accordingly the parotic dental plate must represent the horizontal and the spiracular dental plates the ascending infrapharyngeal dental plates of the hyoid arch. Moreover, the lateral commissure which is a visceral structure arising separately (Holmgren, 1943) must be the suprapharyngohyal, whereas the otical shelf, also visceral in origin, is the infrapharyngohyal (Fig. 46). Hence it follows that the hyomandibula which runs parallel with the epibranchials, at any rate in the main,* is the epihyal.

In *Eusthenopteron* the ascending infrapharyngohyal (spiracular) dental plates, which occupy a triangular area in the medial wall of the wide spiracular gill tube dorsally reach the ventral opening of the spiracular canal (Figs 92, 134, Volume 1). In *Amia* (Figs 17, 18, 41, 52, Volume 1) the spiracular tube which is narrow, is situated on the outside of a triangular process of the parasphenoid extending dorsally to the spiracular canal. This process, the processus ascendens posterior, which arises separately (Fig. 55B; Pehrson, 1940), is evidently homologous with the ascending infrapharyngohyal dental plates in *Eusthenopteron* and has arisen by fusion of such plates. That this is so is supported by the conditions in the palaeoniscids (Fig. 45B) in which the groove for the spiracular gill tube on the outside of the processus ascendens posterior sometimes is provided with denticles in its bottom. In front of this process the parasphenoid in palaeoniscids (and *Amia*, Figs 17B, 41, Volume 1) shows another process, the processus ascendens anterior, which supports the basipterygoid process. Such an ascending process is present also in osteolepiforms and porolepiforms (Fig. 45A) and on the outside of that process there is a groove decreasing in width upwards and sometimes (*Glyptolepis*) with denticles in its bottom. This groove, situated in front of the intracranial joint, is no doubt serially homologous with the very similar and sometimes toothed groove on the outside of the processus ascendens posterior in palaeoniscids. Because the latter groove housed the spiracular tube it is evident that the groove on the processus ascendens anterior in osteolepiforms and porolepiforms situated well in front of the space for the spiracular gill tube, must have been developed for another vestigial gill tube, namely the prespiracular gill tube (Fig, 134, Volume 1) which is present in embryos of their tetrapod descendants, the anurans and urodeles (Fig. 135, Volume 1). In palaeoniscids the prespiracular tube was lacking as it is in *Amia* and other extant actinopterygians and has most likely disappeared in connection with the reduction of the processus ascendens anterior and the basipterygoid process. That a prespiracular gill tube was originally present also in actinopterygians is evidenced by the fact that vestigial such tubes have been found in larval stages of certain forms.*

Because the processus ascendens posterior is derived from ascending infrapharyngohyal dental plates it is evident that its serial homologue, the processus ascendens anterior, must be formed by ascending infrapharyngeal dental plates of the mandibular arch, while the lateral parts of the parasphenoid from which the ascending process issues, and which (in *Eusthenopteron*, Fig. 45) on the outside show anastomosing ridges arising from fusion of teeth, are formed by the fusion of horizontal infrapharyngomandibular dental plates. Hence it follows that the suborbital ledge and the basipterygoid process (Figs 46, 47) which are covered by these parts of the parasphenoid must be the infrapharyngomandibular and homologous with that part of the trabecula cranii which has been shown to be ectomesenchymatic in origin (Fig. 35A; Stone, 1926; de Beer, 1947; Hörstadius, 1950) and which has about the same position and posterior extent. Behind the basipterygoid process follows the posterodorsal marginal part of the palatoquadrate or pars pterygoquadrata (Jarvik, 1972) which runs

* p. 81; †pp. 91, 260.

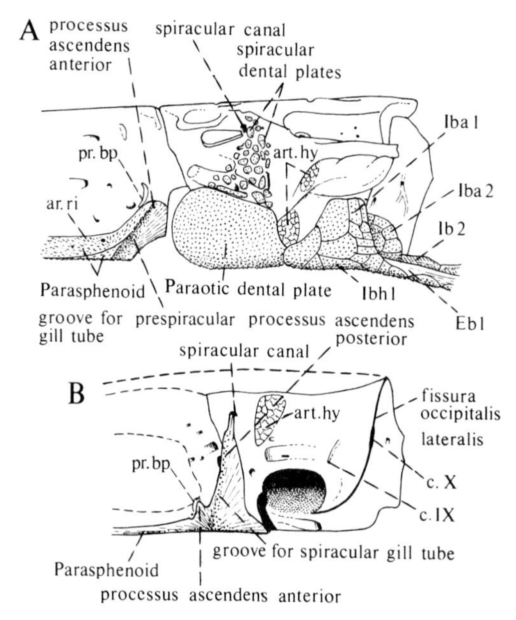

Fig. 45. Sketches to illustrate similar position of spiracular dental plates in *Eusthenopteron* (A) and processus ascendens posterior of parasphenoid in palaeoniscid (B) in relation to spiracular canal in neurocranium and spiracular gill tube. Note that the posterior ascending process dorsally reaches the spiracular canal and sometimes carries a toothed groove which lodged the spiracular gill tube. In *Eusthenopteron* the spiracular dental plates were situated in the medial wall of the wide spiracular gill tube (cf. Fig. 134, Volume 1) and partly within the spiracular canal. The toothed groove (adopted from porolepiforms) of the processus ascendens anterior lodged the vestigial prespiracular gill tube (lacking in palaeoniscids). From Jarvik (1954).

Eb 1, epibranchial 1; Iba 1, Ibh 1, ascending and horizontal dental plates of infrapharyngobranchial 1; Iba 2, ascending dental plates of infrapharyngobranchial 2; Ib 2, infrapharyngobranchial 2; ar.ri, area of ridges formed by fused teeth of horizontal infrapharyngomandibular dental plates; art.hy, articular areas for hyomandibula; c.IX, c.X, glossopharyngeus and vagus canals; pr.bp, basipterygoid process.

parallel with the epihyal and the epibranchials and evidently is in the main the epimandibular (Figs 46, 47). This part bears the strong processus ascendens palatoquadrati which like the suprapharyngohyal (lateral commissure) formed a bridge outside the jugular vein and no doubt is the suprapharyngomandibular. The paratemporal process and the independent prespiracular cartilage bones probably represent modified mandibular gill rays while the paratemporal articulation may be compared with the interarcual articulations in *Amia* (Fig. 33, Volume 1). Finally, it may be concluded that the basal articulation represents the original articulation between the epal and infrapharyngeal elements of the mandibular arch.

Since it has been shown that the three dorsal elements of the mandibular arch in *Eusthenopteron* are represented by the suborbital ledge with the basipterygoid process (infrapharyngomandibular = ectomesenchymatic portion of the trabecula), the processus ascendens palatoquadrati (suprapharyngomandibular) and the pars pterygoquadrata (epimandibular) it is evident that the pars autopalatina cannot possibly belong to the mandibular arch. Because, as is well shown on the model of *Eusthenopteron* (Figs 74, 113, Volume 1), the pars autopalatina runs parallel with the epal elements

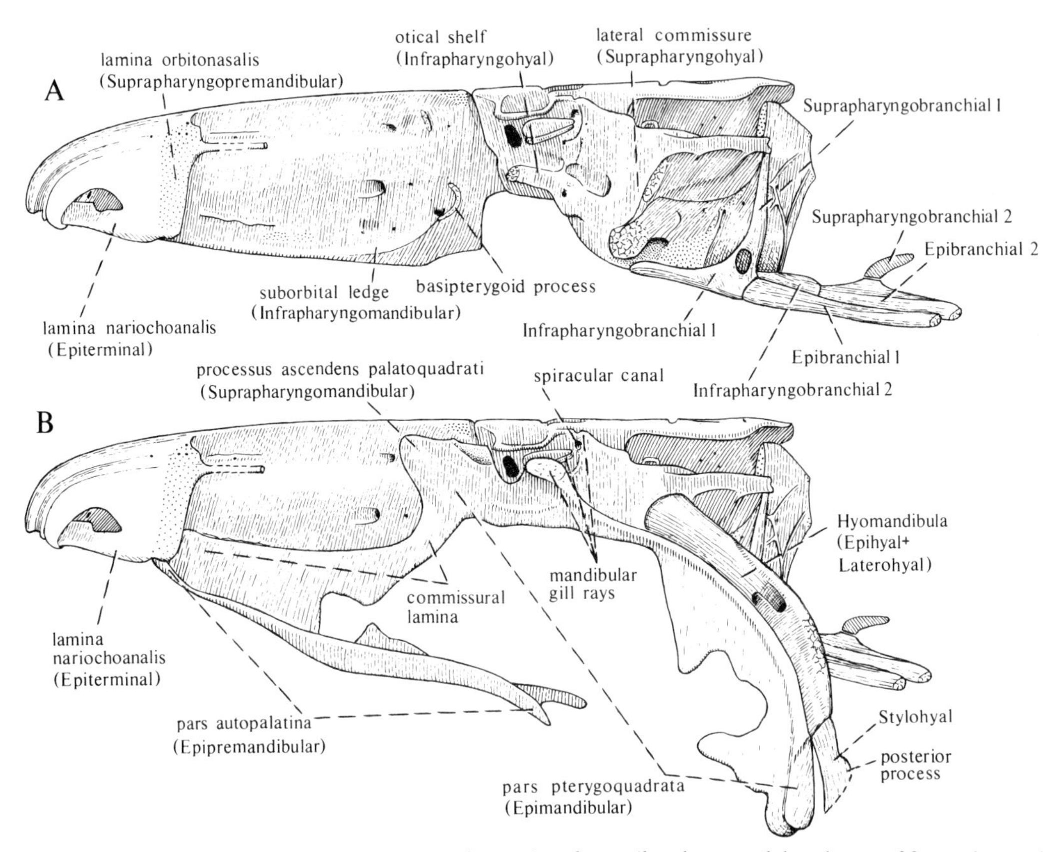

Fig. 46. A, *Eusthenopteron foordi*, neurocranium with prespiracular cartilage bones and dorsal parts of first and second branchial arches in lateral aspect. B, the same with addition of palatoquadrate and free dorsal part of hyoid arch. Interpretations of dorsal elements of anterior four visceral arches given in brackets. After models in wax on basis of grinding series 2.

in the mandibular, hyoid and branchial arches it is easy to see that it must be the epal element in the premandibular arch (Figs 46, 47). The epipremandibular thus identified is a barlike structure which on its ventral side bears two tooth-bearing bones (dermopalatine and ectopterygoid). These bones (Fig. 49) which must be modified horizontal epipremandibular dental plates are continued forward by a similar bone (the vomer) situated on the ventral side of the ethmoidal region. Supported by embryological evidence provided by Holmgren (1943) I (1954) concluded that the vomer represents modified infrapharyngopremandibular dental plates and that the infrapharyngopremandibular is formed by that ventral part of the ethmoidal region which carries the vomer. Accordingly the articulation between the articular head (processus apicalis) of the epipremandibular (pars autopalatina) and the ethmoidal region must be the original articulation between the epal and infrapharyngeal elements in the premandibular arch. However, because the epal element in the mandibular and in the second branchial arches articulates also with the suprapharyngeal element it was concluded that this articulation in the premandibular arch is rep-

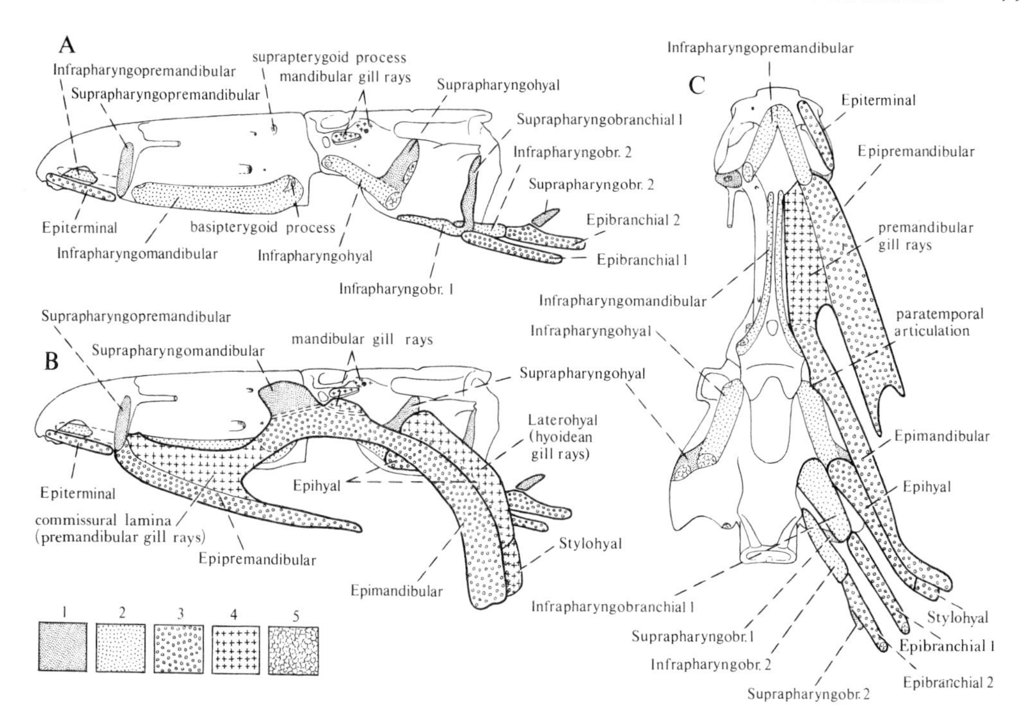

Fig. 47. Diagrammatic representations to show position of suprapharyngeal, infrapharyngeal and epal elements in terminal, premandibular, mandibular and hyoid arches in neurocranium and visceral endoskeleton of *Eusthenopteron*. A, B, lateral aspect (cf. Fig. 46), C, ventral aspect. Supra- and infrapharyngoterminals omitted. After Jarvik (1954).
1, suprapharyngeal elements; 2, infrapharyngeal elements; 3, epal elements; 4, skeletal parts formed by gill rays; 5, articular areas.

resented by the posterolateral ethmoidal articulation and that the suprapharyngopremandibular must be formed by the posterolateral part of the nasal capsule (lamina orbitonasalis), which in other fishes forms an ascending rod, arising independently and visceral in origin (Holmgren, 1943).

These studies (Jarvik, 1954) of a Devonian fish thus resulted in the discovery of well developed dorsal parts of the premandibular arch in gnathostomes (Figs 47, 48, 49A). Also the visceral trunk (the r. maxillaris) and the visceral musculature of the premandibular branchial unit (Fig. 51) were identified. Moreover it was shown that each of the three prootic visceral arches includes three dorsal elements (infrapharyngeal, suprapharyngeal and epal). These elements have retained much of their original position but the epal elements of the premandibular and mandibular arches together with a thin commissural lamina have fused to form the main part of the palatoquadrate, which thus is a complex dimetameric structure. In addition new interpretations of the parasphenoid and other dermal bones of the palate were given and the puzzling fact that the principal tooth-bearing bones (coronoids) of the ceratomandibular are opposed to the modified dental plates (vomer, dermopalatine and ectopterygoid) of the premandibular arch, that is the arch next in front, was explained.

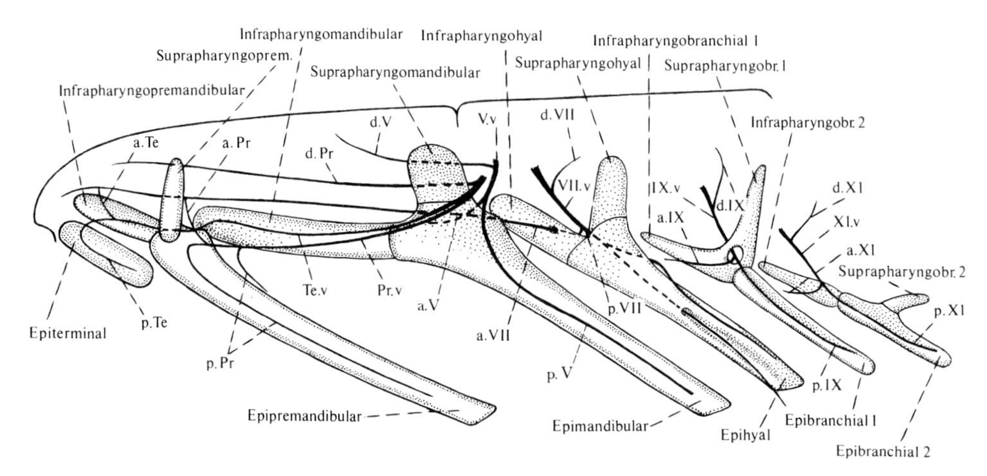

Fig. 48. Diagrammatic representation to show approximate position and course of main branches of dorsal cranial nerves in relation to dorsal parts of visceral arches in *Eusthenopteron* (cf. Fig. 22).

a, anterior (pharyngeal) branches of visceral trunks (pretrematic branches omitted); a.Pr = postchoanal anastomosis; a.Te = prechoanal anastomosis; a.V, hypothetical; a.VII = r. palatinus). d, dorsal branches (d.Pr = r. ophthalmicus profundi; d.V = r. ophthalmicus trigemini; d.VII, d.IX, incompletely known; d.XI = r. auricularis). p, posterior (posttrematic) branches of visceral trunks (p.Pr = r. maxillaris profundi; p.Te = r. premaxillaris terminali; p.V = r. mandibularis trigemini; p.VII = r. hyomandibularis facialis). Pr.v, Te.v, visceral trunks in premandibular and terminal metameres (= "r. maxillaris trigemini"); V.v, VII.v, IX.v, XI.v, visceral trunks of n. trigeminus, n. facialis, n. glossopharyngeus, and first portion of n. vagus.

These new views have been discussed and wholly or partly adopted by a great number of writers and have stimulated to further researches (de Beaumont, 1973; Bertmar, 1959; 1961; 1962; 1963; 1963a; 1966a; Bjerring, 1967; 1971; 1972; 1972b; 1973; 1977; Daget, 1964; Fox, 1959; 1963; 1963a; 1965; Frank and Smit, 1974; Lehman, 1966; Portmann, 1976; Regel, 1964; 1966; 1968; 1973; Stensiö, 1963; 1963a; 1969; Swanepoel, 1970; Vandebroek, 1969; Visser, 1972; van der Westhuizen, 1961). Critical voices have surely been heard (e.g. Nelson, 1969) and in particular the presence of a premandibular arch in gnathostomes as well as in cyclostomes has been doubted (Jollie, 1971; 1977; Miles, 1971; Schaeffer, 1973). However, already Dohrn and Zimmermann* described a premandibular visceral tube in sharks (Fig. 8). This tube is accompanied by a strand of neural crest ectomesenchyme which, as maintained by Zimmermann (1891) is the primordium of the premandibular arch. Moreover, the three dorsal elements of the premandibular arch have recently been indentified in embryonic stages of *Amia* (Bjerring, 1972b). Among other new discoveries may be mentioned that the processus ascendens palatoquadrati, interpreted as the suprapharyngomandibular, arises separately in urodele embryos (Regel 1968) and that vestigial prespiracular gill tubes (Fig. 135, Volume 1) have been discovered in embryos of both urodeles and anurans (Regel, 1966; 1973). Several new experiments (see Weston, 1970; Toerien, 1963; 1971) and other investigations (Bertmar, 1959) have, although, conflicting, confirmed that the anterior part of the trabecula is of neural crest origin. As regards the hyoid arch several embryological and experimental studies (e.g. Bertmar, Fox, Toerien, Swanepoel), have established that supra- and

* p. 13.

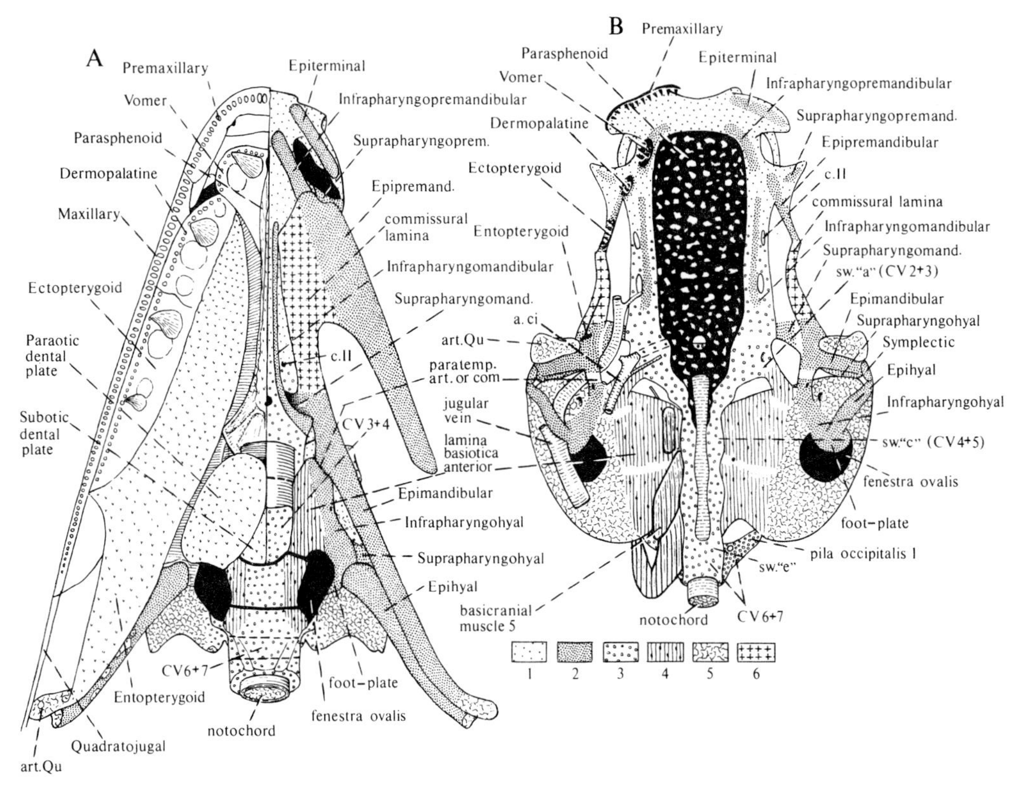

Fig. 49. Composition of neurocranium and dorsal parts of anterior visceral arches in, A *Eusthenopteron* and, B, primitive urodele. Dermal bones of right side omitted. Ventral (palatal) aspects. B, after Jarvik (1972; cf. Fig. 136).

a.ci, internal carotid; art.Qu, articular area of quadrate for lower jaw; c.II, canal for n. opticus; sw."a", sw."c", sw."e", swellings representing ventral vertebral arches of cranial vertebrae (CV) 2+3, 4+5 and 6+7. 1, general ectomesenchymatic skeletal parts; 2, visceral elements; 3, skeletal derivatives of sclerotomes; 4, skeletal elements derived from myotomes (myotomic endoskeleton); 5, skeletal elements derived from dermatomes; 6, skeletal elements derived from gill rays.

infrapharyngohyals are present and that they may be incorporated into the neurocranium as I claimed in 1954. The hyomandibula is in the main the epihyal but it includes often modified hyal gill rays (laterohyal, Bertmar) which form the opercular process and probably also the dorsal articular head in osteolepiforms and porolepiforms.

Composition of Palatoquadrate

As pointed out in 1954 the composition of the palatoquadrate may be somewhat different in the various groups of gnathostomes. On the basis of new evidence and what was previously known an attempt will now be made (see also Bjerring, 1977) to trace the fate of the dorsal elements of the premandibular, mandibular and hyoid arches in various vertebrates.*

* see p. 86 for terminal arch.

Cyclostomes. In cephalaspids the dorsal parts of the visceral endoskeleton have fused with the axial skeleton and form ridges interpreted as visceral arches in the roof of the oralobranchial chamber (Fig. 349, Volume 1). Since these ridges occupy the same position in the roof of the pharynx as do the infrapharyngeal and epal elements in gnathostomes it was suggested (Jarvik, 1954; 1964) that they include equivalents to these elements (Fig. 366, Volume 1). Also it was suggested that those parts of the visceral endoskeleton which are situated lateral to the jugular vein are suprapharyngeals. In support it was pointed out that the visceral nerve trunks in cephalaspids, exactly as they do in gnathostomes (Fig. 48), ran to the area where the three presumed dorsal visceral arch elements meet, dividing there into a strong r. post trematicus and a branch which ran from the distal end of the presumed infrapharyngeal elements in the proximal direction as does the r. anterior of the visceral trunk in gnathostomes. However, in contrast to gnathostomes the visceral arches in cephalaspids have retained their original transverse position in the adult. This implies that the medial portion of the dorsal part of the mandibular arch, which portion presumably corresponds to the infrapharyngomandibular or trabecular cranii in gnathostomes, is still a tranverse element. In view of these facts it is reasonable to assume that the infrapharyngomandibular (trabecula) in petromyzontids also is a transverse element and it has been suggested* (Fig. 365, Volume 1) that it is represented by one half of the transversal commissure which is ectomesenchymatic and is connected laterally with the parabuccal cell band, the posterior shank of which pertains to the mandibular arch. It was also suggested that the anterior shank of the parabuccal band, which is reduced early in ontogeny, is a part of the premandibular arch.

Osteolepiforms, porolepiforms and tetrapods. The palatoquadrate in porolepiforms (Jarvik, 1972) is, as in osteolepiforms, composed of three main components: the epipremandibular (pars autopalatina), the epimandibular (pars pterygoquadrata) and the suprapharyngomandibular (processus ascendens); and also in porolepiforms the remaining dorsal elements of the premandibular and mandibular arches partake in the neurocranium. In both groups the epal elements of the premandibular and mandibular arches are joined by a thin commissural lamina probably formed by premandibular gill rays. However, the paratemporal process and prespiracular cartilage bones of the osteolepiforms interpreted as derivatives of mandibular gill rays are lacking in porolepiforms and there are no distinct traces of mandibular gill rays in this group.†

As may be gathered from Figs 49 and 136 the dorsal elements of the prootic visceral arches in urodeles may be easily identified on the basis of the conditions in osteolepiforms and porolepiforms. In view of the similarities between these two ancestral fish it is not surprising that the anurans (Swanepoel, 1970, with references) in all essentials agree with the urodeles in these regards. However, in anurans that part (the commissura quadratocranialis anterior) of the epipremandibular which carries the dermopalatine (Fig. 134H) may, in contrast to urodeles, be incorporated in the postnasal wall, a condition which is due to differences between osteolepiforms and porolepiforms (Reinbach, 1939; Jarvik, 1942, pp. 522–524; 1972, p. 223). Following a

* Volume 1, p. 477; † see p. 85 for stylohyal in osteolepiforms and symplectic in porolepiforms.

general trend the basipterygoid process and the basal articulation have been reduced in both urodeles and anurans. Instead there is a paratemporal (hyobasal; Stephenson, Swanepoel) articulation or commissure between the epimandibular (pars pterygo-quadrata) and the infrapharyngohyal (otical shelf, "otohyoid ledge"). Moreover, the suprapharyngohyal (lateral commissure; processus dorsalis in anurans, van Eeden, 1951) has in both groups fused with the adjoining part of the pars pterygoquadrata forming the processus oticus which thus is a double formation (Jarvik, 1972, p. 217). The foot-plate of the columella auris is probably a detached posterior part of the infrapharyngohyal which has fused with the proximal end of the epihyal (Jarvik, 1972, p. 219; as to the cartilaginous opercular plate; operculum fenestrae ovalis, Westoll, 1943b, p. 400). The anulus tympanicus in anurans (Fig. 90C) is probably formed by modified mandibular gill rays (Swanepoel, 1970).

Disregarding the hyoid arch and the ear ossicles (Chapter 9) no analysis of the conditions in amniotes has been made and the fate of the dorsal elements of the mandibular and premandibular arches is still obscure.

Coelacanthiforms. The coelacanths are remarkable in several regards. In contrast to other gnathostomes the commissural lamina connecting the epipremandibular (pars autopalatina) with the epimandibular (pars pterygoquadrata) is lacking (Fig. 218, Volume 1). Moreover, the epimandibular is continued forwards below the lateral part of the ethmoidal region. This ethmoidal portion of the pars autopalatina (the auto-palatine ossification) is probably the infrapharyngopremandibular (Bjerring, 1977) and if so the dental plate on its ventral side (the predermopalatine, Jarvik, 1942) is the vomer.* The suprapharyngomandibular has, as in osteolepiforms, fused with the epimandibular forming the processus ascendens. The processus oticus is most likely formed by mandibular gill rays. As to the composition of the hyoid arch the coelacanths agree with osteolepiforms and also in coelacanths there is a stylohyal (together with a symplectic†). A remarkable fact is, however, that also the supra-pharyngeal element of the first branchial arch probably is incorporated in the neuro-cranium forming the parampular process (Bjerring, 1973).

Actinopterygii. Judging in the first place from the conditions in *Amia*, the actino-pterygians agree with osteolepiforms except in one remarkable regard. The supra-pharyngomandibular is incorporated in the neurocranium as the pila lateralis (ali-sphenoid pedicle, Figs 17, 18, Volume 1). This structure is retained, although more or less vestigial in palaeoniscids‡ and in teleosts (Rognes, 1973). In *Lepisosteus* and *Acipenser* this element seems to be lacking.

Dipnoi. The fact that the equivalents to the teleostome dermopalatine and ecto-pterygoid are carried by the so called trabecle (Fig. 317, Volume 1) has led to consider-able reinterpretations of the basis cranii and the palatoquadrate in dipnoans.§ In contrast to other gnathostomes the epipremandibular (pars autopalatina) has probably been incorporated in the cranial base forming a pseudotrabecle. The epipremandibular

* see p. 92 and Volume 1, p. 403; † Volume 1, p. 288; ‡ Volume 1, p. 319; § Volume 1, pp. 400–402.

is continued forwards by the infrapharyngopremandibular which together with its antimere forms a palatoquadrate commissure ("trabecle commissure") which carries the equivalents of the vomers but which, in contrast to sharks, forms part of the cranial base. However, if the epipremandibular has secondarily assumed the position of the trabecle the infrapharyngomandibular is to be sought elsewhere (Fig. 318, Volume 1). Most likely it is represented by the processus basalis which is an almost transverse plate (cf. cephalaspids) distally joined to the suprapharyngomandibular (processus ascendens) and the epimandibular (pars pterygoquadrata). In the hyoid arch the supra- and infrapharyngeal elements together form the so called lateral commissure which becomes incorporated in the neurocranium. In early dipnoans and in larval stages of *Neoceratodus* the epihyal and the laterohyal have fused forming the hyomandibula.* However, sometimes, as in the specimens shown in Fig. 313, Volume 1, the epihyal is independent and articulates with the laterohyal which like the opercular cartilages and probably also the suprabranchial roof† is formed by hyoidean gill rays.‡

Holocephali. The fate of the dorsal elements of the premandibular and mandibular arches is still obscure but an interesting fact is that the supra- and infrapharyngohyals have retained their independence.§ As in other plagiostomes the hyoidean gill rays are well developed.

Arthrodira. The dorsal parts of the three prootic arches in arthrodires (Fig. 284, Volume 1) have been interpreted by Stensiö (1963; 1969) on the basis of the conditions in *Eusthenopteron*. Several variations have been encountered in this large group, but most forms, at any rate, agree with *Eusthenopteron* although also the suprapharyngomandibular partakes in the neurocranium (as in *Amia*). The palatoquadrate (Fig. 288, Volume 1) is usually formed by the epipremandibular (pars autopalatina), the epimandibular (pars pterygoquadrata) and the commissural lamina. Mandibular gill rays may be added to the quadrate portion.

Elasmobranchs. Characteristic of sharks (Fig. 50) and other elasmobranchs (acanthodians, rays) is that the infrapharyngopremandibular is part of the palatoquadrate and that it together with its antimere forms a palatoquadrate commissure. Moreover the suprapharyngomandibular is lacking. The epal elements of the premandibular and mandibular arches which together with the infrapharyngopremandibular constitute the palatoquadrate are connected by the commissural lamina. Also the orbital process, present in sharks and acanthodians but not in rays, is conceivably formed by premandibular gill rays as are also the labial cartilages. Mandibular rays form the jugular bridges, the spiracular cartilages and the processus oticus externus. In the hyoid arch the supra- and infrapharyngeal elements have been incorporated in the neurocranium in rays, whereas in sharks and probably also in acanthodians the suprapharyngohyal has fused with the epihyal to form the hyomandibula. Hyoidean gill rays are well developed (as to the hamuloquadrate see next paragraph).

* Volume 1, p. 432; † Volume 1, p. 433; ‡ see p. 85 for hamuloquadrate; Volume 1, p. 387.

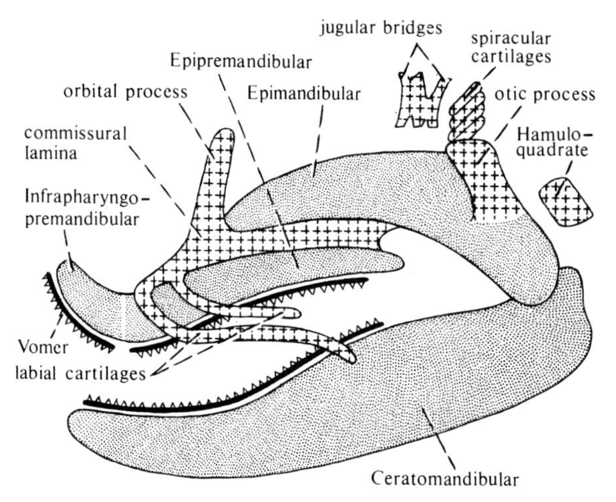

Fig. 50. Diagram to show composition of palatoquadrate in selachian. For explanation of symbols see Fig. 49: 2 and 6. From Bjerring (1977).

Stylohyal, Symplectic and Hamuloquadrate

The area behind the jaw joint presents two elements, the stylohyal and the symplectic, which have caused much controversies (Holmgren, 1943; Bertmar, 1959; 1961). The stylohyal or interhyal found, for example, in *Eusthenopteron* (Figs 109, 110, 123, Volume 1), *Latimeria* (Fig. 218, Volume 1), *Polypterus* (Fig. 234, Volume 1), actinopterygians (Figs 245, 246, Volume 1), reptiles, birds (Frank and Smit, 1974; 1976) and mammals (Figs 98, 100) connects the hyomandibula (epihyal) with the ceratohyal. It is probably a double formation formed by a part of the hyoid arch to which lateral hyoidean gill rays have been added. The symplectic which generally connects the ventral end of the hyomandibula, either with the quadrate portion (*Glyptolepis*, urodeles and certain actinopterygians; Fig. 91; Fig. 246B, Volume 1) or with the articular portion of the lower jaw (*Amia*, *Caturus*, *Latimeria*; Figs 39, 40, 218, 246A, D, Volume 1) arises as a process of the hyomandibula and is generally considered to be a detached part of that element (e.g. de Beer, Bertmar). However, according to Allis and Holmgren it is formed by mandibular gill rays and its origin is obscure. The element in *Neoceratodus* (Figs 313, 318, Volume 1) which Bertmar thought to be the stylohyal originates as a backgrowth of the quadrate portion (Fox, 1963a). Since a corresponding backgrowth (the processus posterior; Bertmar 1959, p. 226) is present in *Hepsetus* together with the symplectic and stylohyal, Bjerring (1977) has introduced the name "hamuloquadrate process" for this quadrate projection and the name "hamuloquadrate" for the part of the process in *Neoceratodus* that is independent. The ventral part of the hyomandibula in sharks formed by mandibular gill rays (Holmgren, 1940) is also termed "hamuloquadrate" by Bjerring (Fig. 50). To this category belong probably also the "accessory hyomandibula" in acanthodians and possibly the posterior quadrate process in *Polypterus* (Allis, 1922) which articulates with the anteroventral corner of the hyomandibula. The so called preopercular process (Saint-Seine, Nybelin, 1974) in leptolepids points backwards and is probably derived from hyoidean rays. Since both mandibular and hyoidean gill rays obviously are involved in the formation of the various structures found behind the jaw joint and since different parts of the series of gill rays may have been utilized it is by no means certain that the elements given the same name (e.g. the symplectic in *Amia* and *Leuciscus*, Fig. 246, Volume 1) are strictly homologous. Moreover, the fact that the mandibular and hyoid blastemas, formed by migrating neural crest cells, in this area lie close together (Fox, 1963a) and easily may be mixed up has contributed to make the interpretation of these structures difficult (cf. the origin of the malleus in mammals).

Terminal Visceral Arch

The terminal somite (Platt's vesicle) is continued by a tubelike structure, which most likely is the mesodermal visceral tube in the terminal metamere (Figs 6, 9). This tube is accompanied by migratory neural crest cells and accordingly the pre-requisites for the development of both the visceral arch and its musculature are at hand also in the terminal metamere. Unfortunately the fate of this embryonic material is unknown and Bjerring's attempt (1977) to identify the three dorsal elements in the terminal visceral arch and the musculature belonging to this arch in various groups is therefore hypothetical. As an example may be mentioned that the lamina nariochoanalis in *Eusthenopteron* is interpreted as the epiterminal (Figs 46–49), whereas the ventromedial part of the internasal wall and the septum nasi rising from it are considered to include the infra- and suprapharyngoterminals. Of interest is also Bjerring's suggestion that the so called vomer in dipnoans (Fig. 317, Volume 1), the glochinal, is formed by horizontal infrapharyngoterminal dental plates. If true this would explain the puzzling fact (Jarvik, 1967, p. 175) that bone MdX in the lower jaw of dipnoans is opposed to the "vomer" (glochinal) in the upper jaw. If the "labial cartilage" in *Neoceratodus* (Fig. 324, Volume 1) which carries bone MdX is the ceratopremandibular, which has been suggested as a possibility,* it is only natural that its dental plate (MdX) is opposed to a dental plate (the glochinal) belonging to the visceral arch next in front, that is the terminal arch (note that the coronoids carried by the ceratomandibular are opposed to the vomer, dermopalatine and ectopterygoid which are modified dental plates of the premandibular arch). Also the so called vomer in coelacanthids is, according to Bjerring, a glochinal.

Visceral Musculature and Aortic Arches

The metameric mesodermal visceral tubes soon separate from the somites above (at the somitic stalks) as well as from the parasomitic mesoderm below and are transformed into muscle plates which differentiate into the metameric visceral muscles (Figs 7, 9). Certainly many variations occur in vertebrates (Edgeworth, 1935) but as far as the gnathostomes are concerned the separation into individual muscles takes place in much the same way in all groups (see e.g. *Amia*; Figs 67, 68, Volume 1). The

* Volume 1, p. 431.

similarities between the various groups of gnathostomes are distinct in particular in the intermandibular division. In this area the mandibular muscle plate forms the musculature of the mandibular gill cover (Fig. 110; Fig. 70, Volume 1) and this musculature has been retained with surprisingly small changes and is well developed even in man (Fig. 118; Jarvik, 1963).

As to the premandibular visceral tube an important fact is that this tube, which is continuous with the premandibular somite in sharks (Figs 6, 9) and probably also in other forms (urodeles, Fig. 51A, B; Edgeworth, 1935), is situated as and joins the mandibular tube in the same way as the muscular process in *Acipenser*—which I in 1954 interpreted as the muscle primordium of the premandibular branchial unit—joins the mandibular muscle primordium (cf. Figs 9 and 51C). That a premandibular muscle primordium really exists has later been confirmed by studies of *Neoceratodus* embryos (Fox, 1963). Also in Amphibia (Fig. 51A, B) there are indications of this primordium. Thence it follows that the muscles which arise from this primordium (the preorbital muscle group), e.g. the m. preorbitalis (Luther, 1938; suborbitalis, Edgeworth) in elasmobranchs, the levator mandibulae anterior in *Neoceratodus* (Edgeworth, Fox), the adductor symphysialis anteriores in *Acipenser* (Sewertzoff; suborbitalis, Edgeworth), the m. palatomandibularis in *Amia* (Fig. 67, Volume 1) and *Lepisosteus*, and the m. levator bulbi in *Eusthenopteron* and anurans (Fig. 8, Volume 1; Luther, 1914; Jarvik, 1942) must pertain to the premandibular branchial unit, as assumed by Sewertzoff (1923; 1928), Fox (1963), and others. These muscles which generally originate from the epipremandibular (pars autopalatina; note that the m. levator mandibulae anterior in *Neoceratodus* originates from the "trabecula" (Edgeworth, 1935; Fox, 1963) which as shown above is a pseudotrabecle representing the epipremandibular) are innervated by branches belonging to the r. maxillaris profundi which is the visceral trunk in the premandibular metamere. No terminal muscle plate has been observed but since a terminal visceral tube exists it is possible that at least some labial muscles in holocephalans (Fig. 304, Volume 1) which are

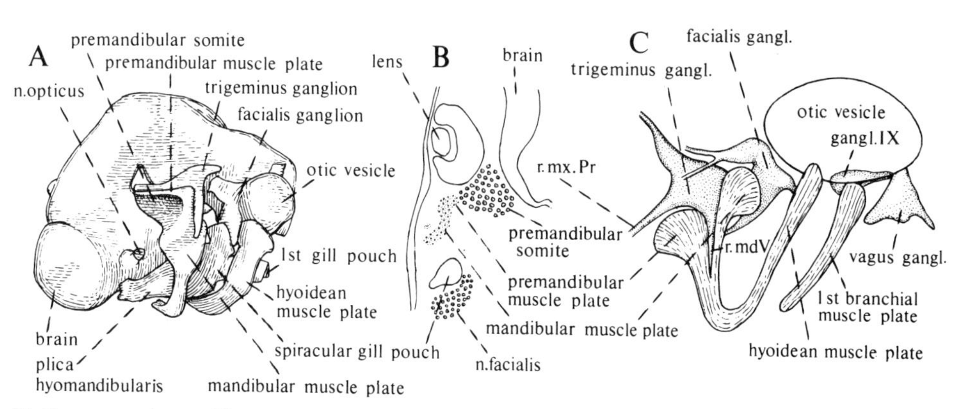

Fig. 51. Representations to illustrate premandibular muscle plate in, A, B, urodele (*Menopoma*, embryo, 15 mm; from Edgeworth, 1935) and, C, sturgeon (*Acipenser*, early larva; from Jarvik, 1954, after Sewertzoff).
r.md.V, r. mandibularis trigemini; r.mx.Pr, r. maxillaris profundi.

innervated by anterior branches of the r. maxillaris pertain to the terminal metamere.

The metameric aortic arches are well known from the mandibular metamere backwards (Figs 105, 106; Figs 61, 63, 64, Volume 1) whereas the identification of the premandibular aortic arch has offered great difficulties. Several writers (Dohrn, 1886; Allis, 1908; Holmgren, 1943; Bertmar, 1962; Fox, 1963b) assume that it is represented by the a. ophthalmica magna, but whereas Holmgren claims that this artery is an afferent vessel Allis, Bertmar and Fox are of the opinion that it is the efferent premandibular artery. A new interpretation has recently been suggested by Bjerring (1977). In his opinion the a. ophthalmica magna is an intermetameric artery (serially homologous with the intermetameric arteries in the trunk) which has been displaced laterally in connection with the formation of the eye cup (a part of the brain wall). The premandibular aortic arch is formed by the a. cerebralis anterior and as a consequence of his interpretation of Rathke's pocket* Bjerring suggests that the a. infundibularis is the terminal aortic arch.

* p. 26.

5 Exoskeleton

Early in vertebrate phylogeny—in the protovertebrates—there probably arose small toothed elements in the skin and in the mucous membranes of the oral cavity, the pharynx and the gill slits (Figs 56A, 69). These primary components (cf. the lepidomoria, Stensiö, 1961) soon fused into more or less complex units; in the head forming either tooth plates or external dermal bones arranged in characteristic patterns. It now remains to consider if and to which extent these patterns are influenced by the distinct metamerism displayed by the nerves, vessels, muscles and endoskeletal elements in the head. However, since a long median portion of the neurocranium is formed by cranial vertebrae we have also to examine to which extent the vertebral segmentation is reflected in the arrangement of the exoskeletal elements both dorsal and ventral to the cranial portion of the vertebral column. In other words has Goethe-Oken's vertebral theory p. 62, which according to current views (de Beer, 1937; Peyer, 1949) was definitely demolished more than a century ago, any justification today?

DERMAL BONES OF MOUTH CAVITY

The visceral arches are metameric structures. This is of course true also of the toothed plates carried by the visceral arch elements, regardless of whether these elements are independent or form part of the neurocranium. In *Eusthenopteron* (Fig. 52) the visceral arches and related dental plates show a regular disposition, a condition which is probably primitive and may be taken as a basis for the interpretation of the dermal bones of the mouth cavity.

Posthyoidean Branchial Units

The infrapharyngeal, epal, ceratal and hypal elements of the two foremost branchial arches in *Eusthenopteron* are well equipped with dental plates (Figs 52, 54). The epal plates are arranged in two rows or series (Fig. 53D), lateral and medial (ventral) whereas the cerato- and hypobranchials (Fig. 53B) present three series; lateral, dorsal and medial. The epal series are in both arches continued forwards by the horizontal infrapharyngeal plates. Moreover, there are ascending infrapharyngeal plates which occupy a triangular area in the medial walls of the first and second gill slits, in front of the suprapharyngobranchials (Figs 45A, 52).

Hyoidean Branchial Unit

The dental plates of the hyoidean branchial unit are arranged much as in the two foremost posthyoidean units. A remarkable fact is that *Eusthenopteron* is the only fish known so far in which the lateral series of ceratohyal dental plates is retained (Figs 110, 112, 132, Volume 1). Also of interest is that the horizontal infrapharyngohyal plates have fused into the paraotic dental plate which is situated in the anterior continuation of the epihyal (hyomandibular) plates (Fig. 52; Fig. 107, Volume 1) and is continued laterally by the ascending infrapharyngohyal plates (Fig. 45A; Figs 92, 110, Volume 1). These plates (the spiracular dental plates) occupy a triangular field in the median wall of the spiracular gill slit, in front of the suprapharyngohyal (lateral commissure). This implies that the conditions are exactly the same as in the two foremost posthyoidean units. The hyoidean triangular field dorsally reaches the spiracular canal, as does also the processus ascendens posterior of the parasphenoid in palaeoniscids (Fig. 45B) and other actinopterygians, which is also a triangular structure in the medial wall of the spiracular gill slit.

Mandibular Branchial Unit

A similar disposition of the dental plates is found also in the mandibular unit. The epimandibular (pars pterygoquadrata) forms part of the palatoquadrate and only some plates of the medial epal series (the palatoquadrate dental plates) are retained

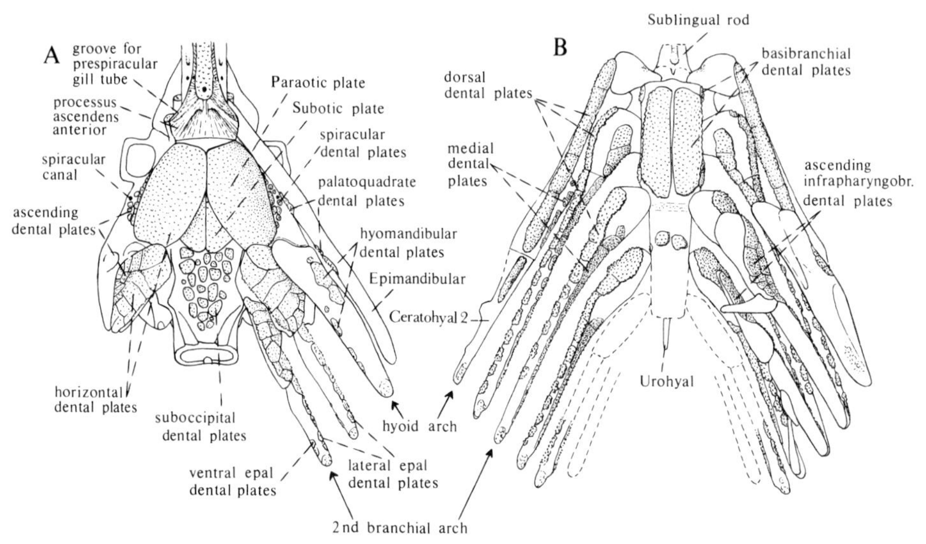

Fig. 52. Eusthenopteron foordi. A, part of neurocranium with dorsal parts of visceral arches of left side in ventral view to show upper oral dental plates. B, hyobranchial skeleton with lower oral plates and (on right side only) dorsal parts of visceral arches. Dorsal view.

(Fig. 52; Fig. 107, Volume 1). Also in the mandibular unit the ascending infrapharyngeal plates are situated in the medial wall of a gill slit (the prespiracular) and form, together with the horizontal infrapharyngeal plates, the lateral parts of the parasphenoid and its triangular ascending process (the processus ascendens anterior). This process which supports the basipterygoid process shows a groove (in porolepiforms sometimes toothed) for the vestigial prespiracular gill tube. As emphasized above* this groove is very similar to and serially homologous with the toothed groove for the spiracular gill tube of the processus ascendens posterior in palaeoniscids. Following a general trend the basipterygoid process and the processus ascendens anterior undergo a reduction in actinopterygians. They are, however, still discernible in *Amia*, in which the processus ascendens anterior (Fig. 17, Volume 1)—like the serially homologous ascending infrapharyngeal plates in the hyoidean and posthyoidean units in *Eusthenopteron*—rises in the angle between the infrapharyngeal (suborbital ledge) and the suprapharyngeal (pila lateralis) elements. In connection with these reductions also the prespiracular gill tube has been reduced and seems to have disappeared entirely in most actinopterygians (a transient prespiracular gill tube has been recorded in embryos of *Lepisosteus*, van Schrick, 1927; and *Acipenser*, Neumayer 1932; cf. also, Fig. 135, Volume 1). The dorsal ceratal plates in the mandibular unit have been modified to form the coronoids, while the lateral plates constitute the dentary. Moreover, the equivalents of the medial ceratal plates in the hyoidean and posthyoidean units have fused into the prearticular (Fig. 53A, B). The coronoids in osteolepiforms and porolepiforms (Figs 125, 188, Volume 1) are three in number but an anterior member of the series has been modified to form the parasymphysial dental plate. The total number is therefore four as in *Amia* (Fig. 40, Volume 1) and *Ichthyostega* (Fig. 174, Volume 1). In arthrodires (Fig. 280, Volume 1) the coronoids have fused into a mixicoronoid (inferognathal) which, however, probably includes also the prearticular (Stensiö, 1969, p. 418). Also in dipnoans (Fig. 325, Volume 1) there is a complex prearticular (bone MdI†) which bears a likewise complex tooth plate.

Premandibular Branchial Unit

The medial (ventral) epal dental plates in the premandibular branchial unit (Figs 53, 56‡) have been transformed forming two (*Eusthenopteron*) or three (*Amia*) tooth-bearing bones (one or two dermopalatines and the ectopterygoid). This series of bones is continued forwards by the vomer, arisen by fusion of horizontal infrapharyngo-premandibular plates. The ascending plates are probably (Bjerring, 1972b) represented by the rhinal bone in *Amia* (Fig. 16, Volume 1) and *Lepisosteus*. In larval stages this bone is situated in the angle between the infrapharyngeal (the ethmobasal) and suprapharyngeal (the lamina orbitonasalis) elements of the arch, and also as in other branchial units, it lies in the medial wall of a gill slit (N.B. if it is true that the walls of the nasal sac represent a pre-prespiracular gill tube (Figs 22, 24; Fig. 135, Volume 1)). In arthrodires (Fig. 280, Volume 1) there is a vomer (anterior superognathal) whereas

* p. 76; † Volume 1, pp. 410–412; ‡ see p. 93 for lateral series.

the dermopalatine and the ecto- and entopterygoids seem to be represented by a single bone, the palatino–pterygoid (posterior superognathal). In dipnoans (Figs 307, 323, Volume 1) the dermopalatine and ectopterygoid together with the vomer form the pterygoid and the palatine tooth plate. Entopterygoid is lacking. No ceratopremandidular has been found with certainty in gnathostomes, but may possibly be represented by the so-called "labial cartilage" in *Neoceratodus* (Fig. 324, Volume 1). If so the dental plates (MdX) carried by this cartilage, and bone MdX in fossil dipnoans, are more or less modified dorsal ceratopremandibular dental plates.

The commissural lamina, which probably is formed by premandibular gill rays, carries one (e.g. porolepiforms and osteolepiforms; Figs 107, 189, Volume 1) or two (*Polypterus*, palaeoniscids, *Amia*; Figs 41, 236, 245, Volume 1) tooth-bearing bones, the entopterygoid and dermometapterygoid. As shown long ago (Jarvik, 1937) the entopterygoid in *Eusthenopteron* is a double formation including a large bony plate and a dental plate. In this case we are conceivably concerned with delamination and it is to be assumed that the dental plate represents a second generation of fused dental plates. In porolepiforms some of these dental plates are independent and such plates are found also on the ventral side of the parasphenoid (Fig. 190A, Volume 1).

Terminal Branchial Unit

The so-called vomer in coelacanths and dipnoans is termed glochinal by Bjerring (1977) and is in his opinion derived from horizontal infrapharyngoterminal dental plates. Since, according to Starck (1967), the paired bone called vomer in *Eusthenopteron* and other gnathostomes is a true vomer homologous with that in mammals, it is possible that the prevomer (Broom) in mammals (the os paradoxum or dump-bell bone in *Ornithorhynchus*) is also a glochinal. Be this as it may (as to the intricate and still unsolved pre-vomer problem see also de Beer, 1937, p. 434), as we now shall see epiterminal dental plates are probably present as well.

Origin of Dental Arcades

Characteristic of teleostome fishes and tetrapods is that they have two dental arcades, outer and inner, in the upper and lower jaws (Figs 53, 54; Figs 124, 125, 171, 174, 188, 189, 236, Volume 1). The inner arcade, which is the only arcade present in plagiostomes, includes in the upper jaw the vomer, the dermopalatine, the ectopterygoid, the entopterygoid and sometimes a dermometapterygoid, and in the lower jaw the coronoids (including the parasymphysial dental plate) and the prearticular. As we have seen these various tooth-bearing bones have arisen by fusion of dental plates which in the upper jaw belong to the premandibular but in the lower jaw to the manidibular arch.

The outer dental arcade in teleostomes includes two tooth-bearing bones; the maxillary and the premaxillary, in the upper jaw and one, the dentary, in the lower jaw. Gaupp (1905) claimed that these bones arose in connection with labial cartilages, but no such cartilages are present in the proper position and the origin of the outer toothed jaw bones has hitherto been obscure. Nor has it been possible to explain why the outer dental arcade is lacking in plagiostomes. The conditions in *Eusthenopteron* suggest the following solutions to these problems.

Figure 53A, B presents sections of ceratobranchial 1 with associated dental plates and of the lower jaw in *Eusthenopteron*. As may be gathered from these sections the ceratal element (Meckelian bone) in the mandibular arch carries three series of toothed elements, exactly as does the ceratal element in the first branchial arch. Since the coronoids and the prearticular probably are derivatives of dental plates serially homologous with the dorsal and medial ceratobranchial plates, respectively, we are led to the conclusion that the dentary has arisen from lateral ceratomandibular dental plates serially homologous with the lateral ceratal dental plates in the branchial arches.

Turning to the upper jaw (Fig. 53C, D) it is also evident that the maxillary, which is situated on the lateral side of the epipremandibular (pars autopalatina) must be derived from lateral epipremandibular dental plates serially homologous with the lateral epal dental plates in the branchial arches. If the lamina nariochoanalis, as maintained by Bjerring, is the epal element in the terminal arch (Figs 49, 56), it may be justified to assume that the premaxillary is a derivative of lateral epiterminal dental plates.

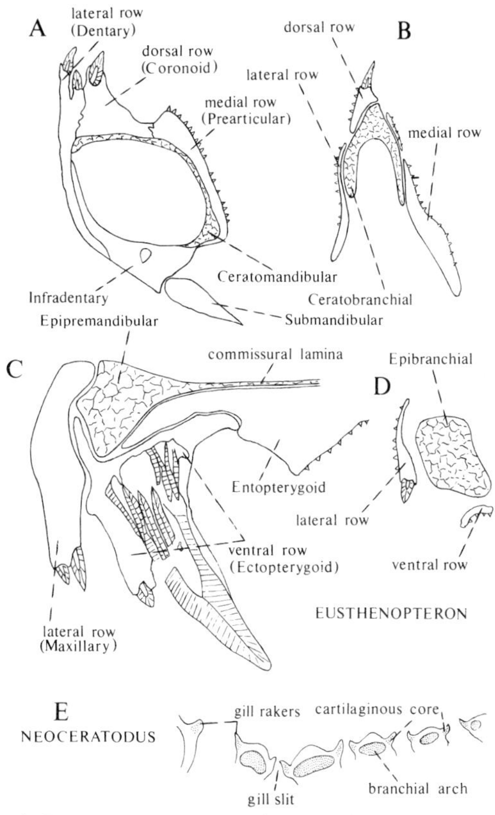

Fig. 53. A–D, *Eusthenopteron foordi*. Four transverse sections from grinding series 2 to explain origin of outer and inner dental arcades in teleostomes. A, lower jaw, B, D, first branchial arch, C, upper jaw. E, *Neoceratodus forsteri*. Sections of branchial arches to show endoskeletal cores of gill rakers. From Stadtmüller (1927). The presence of such cores probably prevented formation of outer dental arcade in plagiostomes (see also Fig. 271, Volume 1).

A remarkable difference between teleostomes and plagiostomes is that the gill rakers in the latter are supported by endoskeletal cores* (Jarvik, 1977). As exemplified by *Squalus* (fig. 271, Volume 1) and *Neoceratodus* (Fig. 53E) the anterior branchial arch carries only one series of such cores. This series is found in a position corresponding to that which in *Eusthenopteron* (Fig. 53B) is occupied by the lateral series of ceratal dental plates. It may therefore be tentatively suggested that it is the presence of such cartilaginous or (in acanthodians) ossified cores or rods that has prevented the formation of the outer dental arcade in plagiostomes.

In fishes like osteolepiforms and porolepiforms the main biting function is performed by the inner dental arcade which shows large tusks. However, proceeding to tetrapods we are witnessing a remarkable change from the inner to the outer dental arcade (Fig. 54). In some osteolepiforms the anterior teeth of the premaxillary and dentary may be enlarged. However, as early as in the late Devonian, in the oldest known tetrapods, the ichthyostegids, the outer arcade is the principal biting device and tusks are lacking in the inner arcade. Such tusks may be present in post-Devonian stegocephalians (Fig. 137) but if we turn to reptiles (Fig. 54C) we will find a gradual reduced or modified, a condition which has caused problems as to their identification Also the bones of the inner arcade which originally were tooth-bearing have been reduced or modified a condition which has caused problems as to their identification (as regards the pterygoid, the vomer and the prevomer see Stadtmüller, 1936a; de Beer, 1937; Starck, 1967).

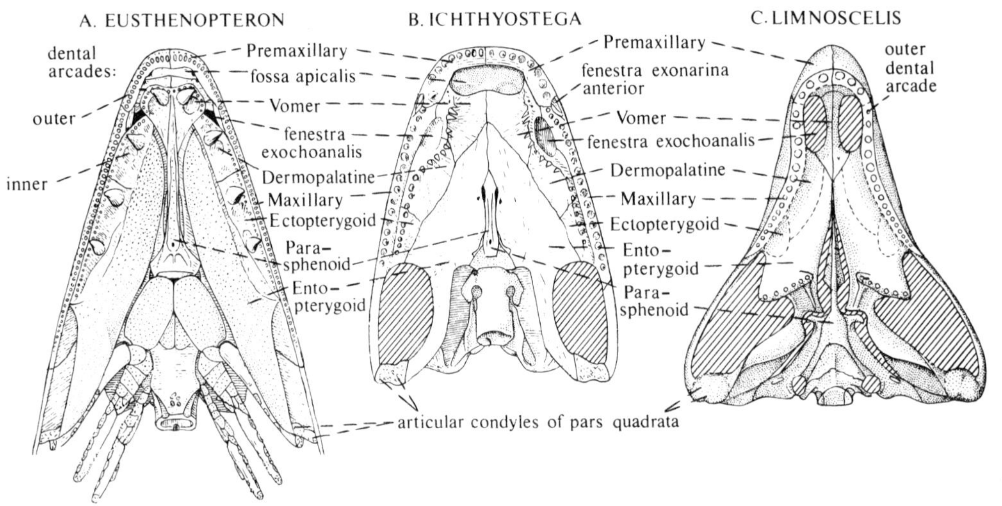

Fig. 54. Palatal aspects of, A, osteolepiform (*Eusthenopteron*), B, primitive tetrapod (*Ichthyostega*), and, C, reptile (*Limnoscelis*) to show change in dentition from inner to outer dental arcade in evolution of tetrapods.

* Volume 1, pp. 340–341.

Origin and Composition of Parasphenoid

The parasphenoid (Jarvik, 1954; 1972) usually consists of a median portion or stem (corpus) and paired lateral ascending processes. As is well shown, for instance, in *Acipenser* (Marinelli and Strenger, 1973), in which several vertebrae have been plastered on to the posterior part of the neurocranium, the stem posteriorly lies close underneath these vertebrae. This condition demonstrates the intimate relations between the median portion of the parasphenoid and the overlying vertbral column, and it may now be asked if not the vertebral segmentation in the basis cranii is reflected in the composition of the parasphenoid stem.

In *Eusthenopteron* (and other "crossopterygians") the parasphenoid stem is short extending posteriorly to the intracranial joint. Of great interest is that embryological studies (*Amia*, Pehrson, 1940; Holmgren, 1943; urodeles, Lebedkina, 1960; 1964; reptiles, de Beer, 1937; Pehrson, 1945) have established that there is a median prehypophysial parasphenoid primordium. That primordium (P1, Figs 55, 56) obviously arises close underneath that part of the basis cranii which is formed by cranial vertebra CV1 (Figs 42, 44, 61). However, it grows forwards and in the adult the parasphenoid may extend to the area underneath the anterior part of the ethmoidal region. This forward growth is probably a consequence of the rostral prolongation (Fig. 56B). Posteriorly the prehypophysial primordium ends at the hypophysial foramen which probably marks the position of the vertebral joint between CV1 and CV1+2. In urodeles there is a median posthypophysial primordium (P2+3) which probably develops into a part of the parasphenoid corresponding to that which in porolepiforms is situated behind the foramen for the buccohypophysial duct (Fig. 136; Jarvik, 1972, p. 256). In *Amia* (Fig. 55A) there is a long paired posthypophysial primordium. This primordium shows four thickenings separated by constrictions and it is tempting to assume that these four thickenings represent parasphenoid

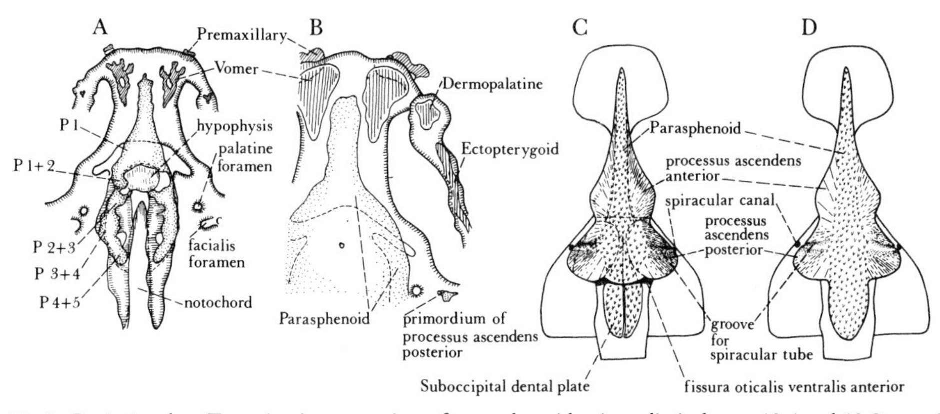

Fig. 55. A, B, *Amia calva*. Tentative interpretation of parasphenoid primordia in larvae 10·4 and 13·8 mm. From Pehrson (1940). C, palaeonisciform type of parasphenoid, formed by incorporation of equivalents to paraotic, subotic and spiracular dental plates in *Eusthenopteron*. D, holostean type; includes in addition suboccipital dental plates. From Jarvik (1954).

P 1–P 4+5, parasphenoid primordia in vertebral segments 1–4+5.

primordia developed underneath four cranial vertebrae (CV1+2, CV2+3, CV3+4 and CV4+5) and that the constrictions mark the position of former vertebral joints. This assumption is supported by the following conditions. The intracranial joint in *Eusthenopteron* (Figs 41, 42, 49) probably lies between CV2+3 and CV3+4. Behind that joint follows the paired subotical dental plate which is situated close below the vertebral column (Fig. 92, Volume 1) and posteriorly extends to the fissura oticalis ventralis anterior which is a vestige of the vertebral joint between CV3+4 and CV4+5. In palaeoniscids (Fig. 55C) the parasphenoid extends backwards to that fissura and no doubt its posterior median part includes equivalents to the subotical plates. However, this part carries the processus ascendens posterior and judging from the position of the primordium of that process in embryonic stages of *Amia* (Fig. 55B) it is evident that the process will fuse with the third thickening from in front (P3+4). Thence it follows that this thickening must include the equivalent of the subotical plate in *Eusthenopteron*. However, since it is broader than the other thickenings it probably also includes the equivalent to the paraotic plate. The posterior thickening in *Amia* (P4+5), finally, seems to be situated somewhat as that part of the basis cranii in *Eusthenopteron* which lies between the anterior and posterior otical fissurae and is formed by CV4+5 (Fig. 49A). Accordingly this paired thickening is probably the primordium of a separate segment of the parasphenoid developed in relation to CV4+5. In view of what has now been set forth, and considering also the presence of suboccipital dental plates in early palaeoniscids (*Moythomasia*) and in *Eusthenopteron* it seems likely that there has been one such segment underneath each of the cranial vertebrae. Since the parasphenoid, disregarding vestiges in certain forms (Starck, 1967), is lacking in mammals this segmentation has obviously nothing to do with the subdivisions in the basis cranii in mammals observed by Goethe and Oken.

To the complex parasphenoid thus formed paired metameric tooth-bearing elements belonging to the visceral exoskeleton have been added in most forms. In fishes the following six main types of parasphenoid may be distinguished (Jarvik, 1954).

The primitive or arthrodiran type. In this type (Figs 287, 296, Volume 1; Stensiö, 1969) characteristic of arthrodires only the median stem is present. This portion is short and judging from the shape of the bone in certain forms it has arisen by fusion of median pre- and posthypophysial primordia.

The crossopterygian type. Also in this type the stem is short extending backwards to the intracranial joint (Fig. 56D). To this portion infrapharyngomandibular dental plates have been added forming lateral parts of the parasphenoid and the processus ascendens anterior (Figs 45A, 56C). In osteolepiforms and porolepiforms this process lies in the medial wall of the prespiracular gill slit and presents a groove for the vestigial prespiracular gill tube (Fig. 134, Volume 1).

The palaeoniscid type. The intracranial joint has disappeared and the stem has been prolonged backwards by incorporation of an equivalent to the paired subotical dental plate in *Eusthenopteron* (Figs 49A, 55C). The processus ascendens anterior belonging to the mandibular metamere has undergone reduction and there are no traces of groove

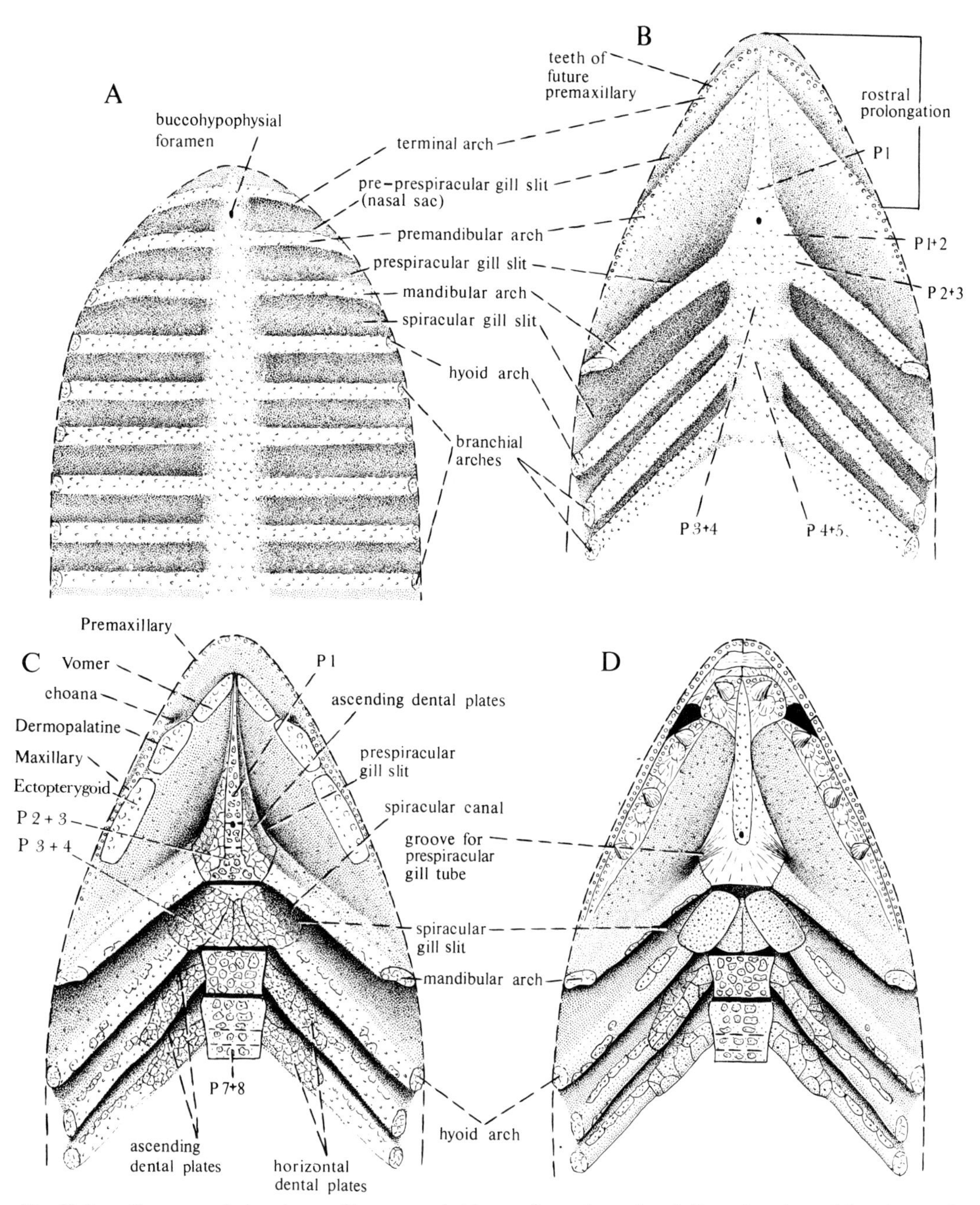

Fig. 56. Four diagrammatic drawings to illustrate probable transformations of roof of mouth cavity and dorsal parts of gill chambers in phylogeny of teleostome fishes. A, partly hypothetical primitive stage with transverse gill arches and numerous denticles in mucous lining of dorsal parts of visceral arches and walls of intervening gill slits. B, more advanced stage, in which rostral prolongation has occurred and gill arches have assumed an oblique position; presumed parasphenoid primordia underlying cranial vertebrae indicated. C, pre-osteolepiform stage; fusion of denticles into dermal bones has begun. D, osteolepiform stage (cf. *Eusthenopteron*, Fig. 54A; Fig. 124A, Volume 1).

for the prespiracular gill tube. Hyoidean infrapharyngeal dental plates corresponding to the paraotic and spircular dental plates in *Eushtenopteron* have fused with the posterior part of the stem. The spiracular (ascending infrapharyngohyal) dental plates form the processus ascendens posterior. This process, which extends upwards to the spiracular canal, is situated in the medial wall of the spiracular gill slit and shows a groove for the spiracular gill tube.

The holostean type. Holosteans and other advanced actinopterygians are as palaeoniscids but the stem has been prolonged backwards by incorporation of one or more parasphenoid segments (Fig. 55D). Since the stem often is bifurcate posteriorly it is to be assumed that these segments are paired in origin as is the suboccipital dental plate in *Moythomasia* (Fig. 55C).

The polypterid (brachiopterygian) type. The stem is as in advanced actinopterygians and is bifurcate posteriorly. The processus ascendens anterior is lacking and the ascending process of the bone is a complex structure (Fig. 237, Volume 1) formed by infrapharyngeal dental plates not only of the hyoid arch, but also of the first and second and probably the third branchial arches.

The dipnoan type. In dipnoans (Figs 307, 317, Volume 1) the parasphenoid is formed mainly by the stem which sometimes is bifurcate posteriorly. Ascending processes are lacking (as in arthrodires) but a lateral process situated on the ventral side the processus basalis, that is the infrapharyngomandibular, is probably formed by horizontal infrapharyngomandibular dental plates.

In urodeles and anurans the processus ascendens anterior has been reduced in connection with the reduction of the basipterygoid process. In anurans the processus ascendens posterior is generally well developed (Stadtmüller, 1936). In urodeles (Wiedershiem, 1877), in contrast, this process seems to be lacking, a condition which certainly is due to the fact that spiracular dental plates are lacking in their porolepiform ancestors (Jarvik, 1972). In post-Devonian stegocephalians (Romer, 1947) there is generally a processus ascendens anterior supporting the basipterygoid process and a large processus ascendens posterior.

EXTERNAL DERMAL BONES

Hyoidean Gill Cover

In teleostome fishes and dipnoans the metameric hyoidean gill cover was originally supported by a large number of elongated bony plates, the branchiostegal rays, carried by the hyoidean visceral arch (Fig. 58). Independent such elements may be retained, in particular in the middle part of the series (actinopterygians, porolepiforms) but generally all or most rays have fused to form the opercular and gular bones (Fig. 57; Jarvik, 1963). Branchiostegal rays are present also in acanthodians (Fig. 272, Volume 1; Jarvik, 1977) and many arthrodires possess an opercular bone (submarginal plate,

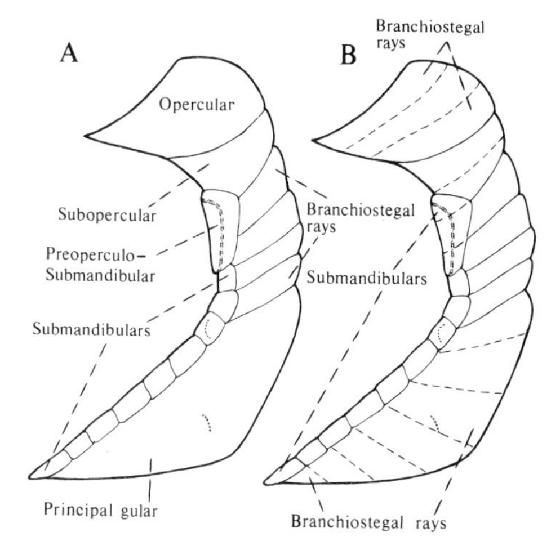

Fig. 57. A, submandibular series and dermal bones of operculogular membrane of porolepiform (*Holoptychius* sp.). B, probable composition of operculogular and submandibular series. From Jarvik (1963).

Figs 291–293, Volume 1). The vestigial subopercular in *Ichthyostega* excepted branchiostegal rays are absent in tetrapods.

Mandibular Gill Cover

A characteristic feature of porolepiforms and osteolepiforms is the presence of a series of submandibulars along the lower jaw (Jarvik, 1963). This series which has been indentified also in antiarchs (Stensiö, 1948), dipnoans (Jarvik, 1967) and possibly partakes in the gular plates in certain teleosts (Jessen, 1968b) has probably arisen to form a support of the mandibular gill cover. The submandibulars articulate with the infradentaries on the lateral side of the ceratomandibular and it may be tentatively suggested that the infradentaries represent modified proximal parts of the submandibulars to which secondarily sensory line components have been added (Fig. 58). Parts of the dermal bones of the cheek, related to the mandibular arch line (Fig. 33), may be tentatively assigned to the mandibular gill cover as well.

Premandibular and Terminal Gill Covers

The maxillary in *Eusthenopteron* formed by lateral epipremandibular dental plates (Fig. 53) is continued dorsally by the external dermal bones of the cheek. These bones which as, e.g. in *Lepisosteus* and certain coelacanths may be numerous (Holmgren and Stensiö, 1936), occupy the area between the epipremandibular (pars autopalatina) and the epimandibular (pars pterygoquadrata) and are situated outside the commissural lamina which probably is formed by fused premandibular gill rays. In view of these facts it may be justified to suggest that the external cheek bones to about the extent shown in Fig. 58 are formed by original but not necessarily raylike supporting elements of the premandibular gill cover and secondary lateral line components of the

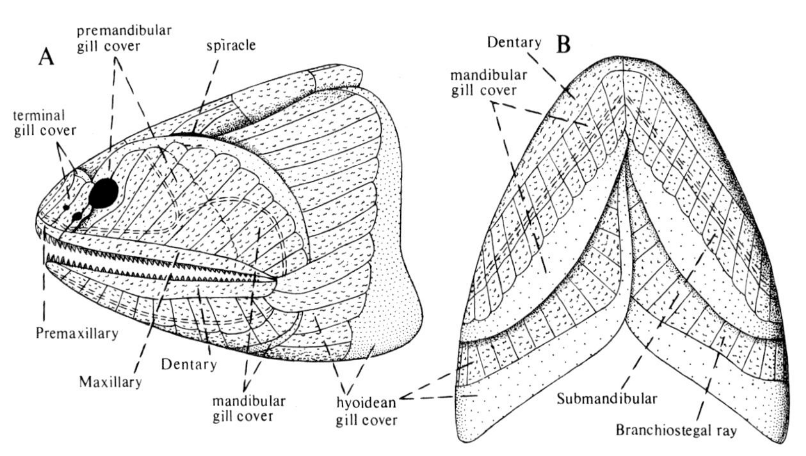

Fig. 58. Diagrammatic, partly hypothetical representations to illustrate presumed terminal and premandibular gill covers and composition of these and mandibular and hyoidean gill covers. Based partly on conditions in *Holoptychius* (cf. Fig. 57; Fig. 180, Volume 1). A, lateral, B, ventral aspects.

premandibular arch line (Fig. 33). Since the premaxillary probably is formed by epiterminal dental plates (Figs 49, 56) it seems likely that the dermal bones of the snout following dorsal to premaxillary represent supporting elements of an original terminal gill cover (Figs 58, 69).

Dermal Bones of the Cranial Roof

According to the tentative interpretations given above all the external dermal bones of the head hitherto considered are either supporting elements of gill covers, or, if we prefer to regard the bones of the outer dental arcade as external dermal bones, they are modified dental plates of visceral arches. This implies that they all belong to the visceral exoskeleton and are metameric structures. It now remains to consider the dermal bones situated close dorsal to the cranial portion of the vertebral column and to examine if these bones display a subdivision corresponding to the vertebral segmentation and to that encountered in the stem of the parasphenoid. As we have seen* the left and right dorsal processes or pilae are, in each cranial vertebra, connected by a transverse bridge or tectum (Fig. 61). Because the spaces between the tecta conceivably mark the position of original vertebral joints it may be justified to ask if not these conditions are reflected in the disposition of the overlying dermal bones.

A characteristic feature of teleostome fishes is the presence of a movable extrascapular series of bones. This series lies over the occipital region and is clearly related to the tectum posterius which is the bridge that interconnects the occipital pilae (Fig. 61). In *Eusthenopteron* the median extrascapular rests on the supraoccipital plug, which is a median part of the tectum posterius; and in the case the median extrascapular is retained in frogs (Fig. 126) it invests this tectum (Jarvik, 1975). In view of these facts it is of considerable interest that also in extant reptiles, birds and mammals (Figs 125D,

* p. 72.

126D) a dermal bone is associated with the tectum posterius (Stadtmüller, 1936a; de Beer, 1937; Starck, 1955; 1967). This bone, the interparietal (mammals) or postparietal (reptiles) is paired as are often the extrascapulars in fishes (Figs 37, 235, 240, Volume 1). The interparietal may even, as in man (Gray, 1973) arise from two paired centra, which is of interest because the median extrascapular in osteolepiforms probably is paired in origin (Fig. 145, Volume 1; Jarvik, 1948). This implies that the extrascapular series in osteolepiforms includes four elements, two on each side of the median line, as does the interparietal in man. Accordingly we can define the extrascapulars and their homologues (inter- or postparietals) in extant tetrapods as bones developed in relation to the tectum posterius; and possibly the fact that the extrascapular series generally is separate from the rest of the dermal skull roof is a consequence of the vertebral segmentation.

According to the definition given by Gaupp (1905) the parietal in extant gnathostomes belongs to the labyrinth region of the skull. More precisely it covers that part of the cranial roof which is situated between the otic capsules (Figs 126, 127; Figs 36, 120, Volume 1) and as indicated by Gaupp it is related to the synotic tecta. The term

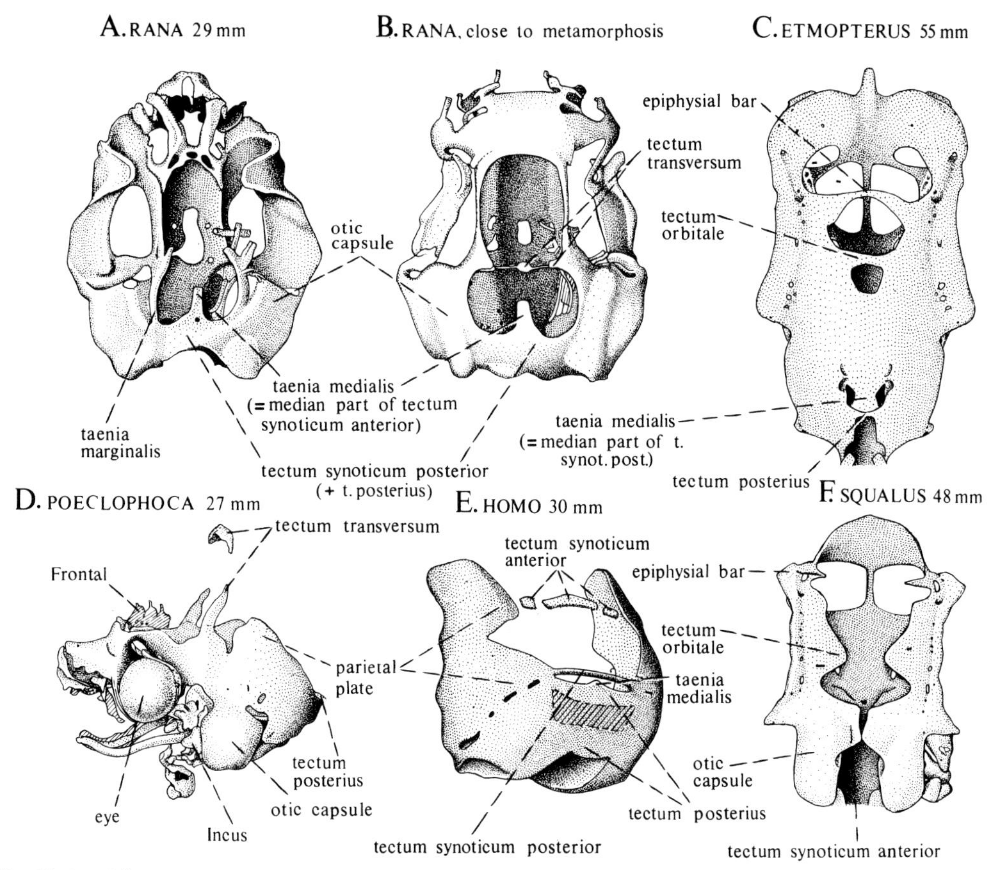

Fig. 59. Cranial tecta in various vertebrates. A, B, from Gaupp (1893); C, F, from Holmgren (1940); D, from Fawcett (1918); E, from Fawcett (1923).

synotic tectum was introduced by Gaupp (1893) in the frog (Fig. 59A, B). In this form the tectum synoticum is composed of a paired lateral primordium arising dorsomedial to the sinus superior of the membranous labyrinth, and a dorsal, median primordium. The lateral primordium is continued forwards by a bar, the taenia tecti marginalis, from which farther forwards another tectum, the tectum transversum, which also includes a dorsal median component, emerges. Moreover the median component of the tectum synoticum is continued forwards by a median process, the taenia tecti medialis. Also in elasmobranchs (Fig. 59C, F; Holmgren, 1940) the tectum synoticum arises from paired lateral and median dorsal primordia. However, in contrast to *Rana* the tectum synoticum in sharks and rays connects the anterior parts of the otic capsules. Moreover, the taenia tecti medialis is either a posterior process of the tectum synoticum (*Raja*), an anterior process of the tectum posterius (*Torpedo*), or joins these two tecta (*Etmopterus*, *Heterodontus*). These differences indicate that there are two synotic tecta, anterior and posterior. Of interest is also that elasmobranchs show two more bridges, the tectum orbitale and the epiphysial bar. Both these tecta arise from processes of the orbital cartilage, a structure which is independent in origin and obviously includes the lateral primordia of the two tecta. In view of these conditions and other available data (*Amia*, Pehrson, 1922; *Lepisosteus*, Hammarberg, 1937) it is evident that the endoskeletal cranial roof is composed of paired lateral and dorsal median series of components which originally formed segmental bridges (tecta), probably one in each vertebral segment (Fig. 61A). However, both the lateral and the dorsal components show a strong tendency to fuse into longitudinal bars (the taenia tecti marginalis and the taenia tecti medialis) and very often the bridges fuse with each other forming a more or less complete cranial roof. In mammals (and reptiles) the conditions are much the same as in fishes and amphibians. The lateral components form a long longitudinal bar or lamina. In mammals (Fig. 60) this lamina, which

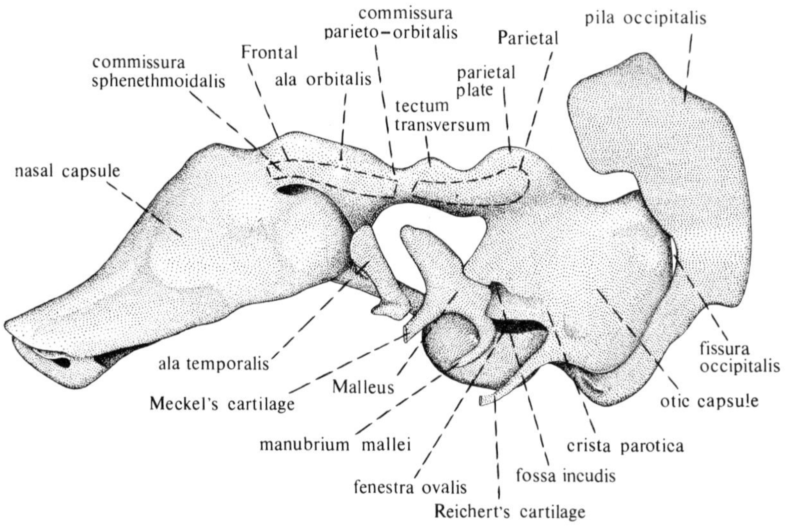

Fig. 60. Eremitalpa granti (Broom), embryo 28·5 mm. Chondrocranium in lateral aspect. Position of frontal and parietal indicated by interrupted lines. From Roux (1947).

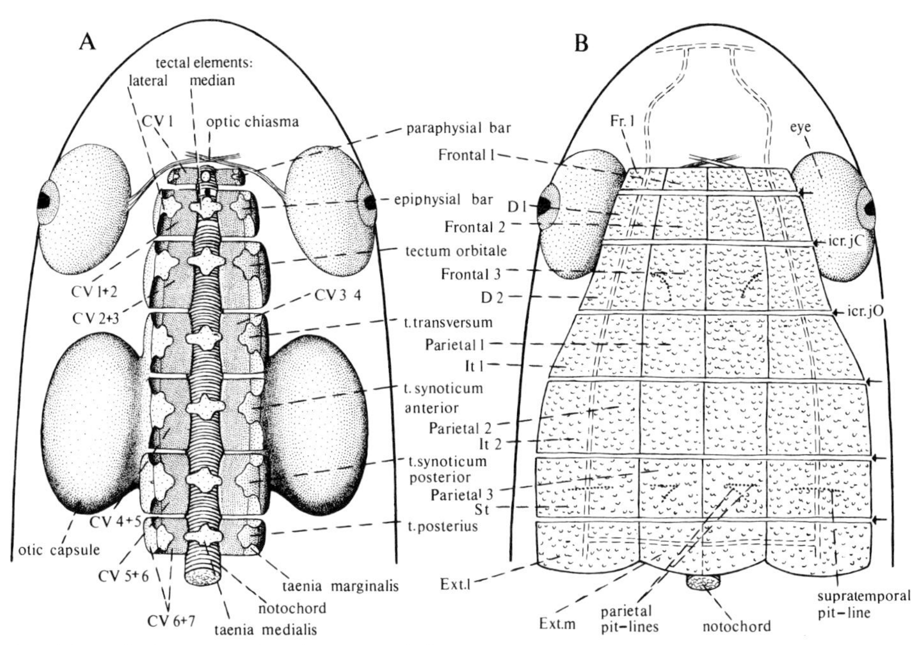

Fig. 61. Cranial vertebral segmentation. A, attempted diagrammatic reconstruction of cranial vertebrae (CV 1–CV 6+7) with tecta in dorsal aspect. B, tentative interpretation of dermal bones of cranial roof in relation to underlying cranial tecta. Arrows to the right indicate position of observed or hypothectical intervertebral joints as reflected in cranial roof.

D1, D2, anterior and posterior dermosphenotics; Ext. 1, Ext. m, lateral and median extrascapulars; Fr. 1, lateral frontal; It 1, It 2, anterior and posterior intertemporals; St, supratemporal; icr.jC, position of intracranial joint in coelacanthiforms; icr.jO, position of intracranial joint in osteolepifroms.

because of the large size of the brain is more or less vertical in position, includes two main parts; the lamina parietalis or parietal plate which surmounts the otic capsule, and the ala orbitalis which is situated in the orbitotemporal region. These two parts are often connected by a commissura parieto-orbitalis; and a commissura spheneth-moidalis joins the lamina orbitalis to the nasal capsule. The parietal plate may be provided with three dorsal processes. Two of these processes are well developed in embryos of *Homo* (Fig. 59E) in which each of these processes may be connected with its antimere by a cartilaginous bridge forming two separate synotic tecta (Fawcett, 1923). The anterior bridge includes a median dorsal component (sometimes paired or flanked by paired elements; de Beer, 1937). Moreover the tectum posterius presents an anterior median process (as in *Torpedo*). The third dorsal process in mammals extends from the anterior part of the parietal plate and represents the tectum transversum. This process is well shown in particular in seals (*Poecilophoca*, Fig. 59D) in which the tectum transversum includes a paired dorsal element (Fawcett, 1918; cf. the "pineal cartilages" in *Otomys*, Eloff, 1953). In view of these striking similarities to fishes and amphibians it is to be concluded that the parietal plate in mammals, when well developed, is formed by the lateral components of the anterior and posterior synotic tecta and the tectum transversum. Since the parietal arises on the lateral (external) side of the parietal plate (Fig. 60) this bone can be defined as an element developed in relation to the said three tecta (Fig. 61).

This is of great interest turning to the osteolepiforms and porolepiforms. In both these groups the anterior part of the otoccipital portion of the endoskeletal cranial roof connects the pilae facialis of CV3+4 (Figs 42, 61), and is consequently the tectum transversum (in *Glyptolepis*, Fig. 193, Volume 1, this tectum forms a transverse bar). The anterior parts of the parietals rest on this tectum, whereas the main posterior parts of these bones cover those parts of the cranial roof which are situated between the otic capsules and obviously are formed by the synotic tecta (Figs 120, 193, Volume 1). Accordingly the parietals in osteolepiforms and early porolepiforms (as to *Holoptychius* and urodeles, see Jarvik, 1972, p. 264) have exactly the same relations to the said three tecta as in mammals. This close agreement (see also Jarvik, 1967a; 1972) can only be taken to mean that the parietals in osteolepiforms and early porolepiforms are homologous with those in mammals and are true parietals (not postparietals as claimed by many students).

Lateral to the parietals in osteolepiforms follow two sensory canal bones, the intertemporal and the supratemporal. The intertemporal may be subdivided into two bones; anterior and posterior (Fig. 145B, Volume 1; Jarvik, 1948). In *Amia* (Pehrson, 1940) the other actinopterygians the corresponding space is occupied by a single bone which, however, arises from three primordia, two intertemporal and one supratemporal. Accordingly the lateral series in osteolepiforms and *Amia* includes three elements. Moreover the parietal in *Amia* arises from two primordia one situated medial to the supratemporal and one medial to the intertemporal portion. In osteolepiforms the parietal is often subdivided in a corresponding way, and in man (Fig. 125) the parietal arises from two primordia. Furthermore the parts of the parietal in osteolepiforms situated medial to the anterior half of the intertemporal are sometimes independent (Fig. 145B, Volume 1; Jarvik, 1948). Consequently we may distinguish three transverse series in the parietal shield of osteolepiforms and it is tempting to assume that the anterior of these series develops in relation to the tectum transversum, whereas the middle and the posterior series are related to the synotic tecta. It may also be tentatively suggested that the presence of median elements in the parietal area of osteolepiforms and other fishes may be conditioned by the presence of median elements in in the underlying cranial tecta.

If the three transverse series in the parietal shield in osteolepiforms are related to the tecta of three cranial vertebrae (CV3+4, CV4+5, CV5+6) the boundaries between these series must mark the position of original vertebral joints. That this is so is supported by the fact the cranial roof in osteolepiformes is divided into parietal and frontoethmoidal shields dorsal to the intracranial joint, that is the vertebral joint between CV2+3 and CV3+4. Thence it follows that the coronal suture between the parietals and frontals in the human skull (Fig, 125) is caused by that joint, too. Also it is evident that the lambdoid suture in man between the parietals and the complex interparietal (= extrascapulars) marks the position of another cranial vertebral joint (between CV5+6 and CV6+7). In other words Goethe-Oken's vertebral theory seems to have some justification as far as the cranial roof is concerned (cf. Peyer 1950, figs 16, 21).

The frontal may be divided into anterior and posterior frontals. In *Eusthenopteron* the posterior parts of the frontals rest on that part of the cranial roof which connects the antotic pilae of CV2+3, that is the tectum orbitale (Fig. 61). In coelacanths, (Figs 213, 214, Volume 1) in contrast, the posterior frontal has fused with the parietal at the same time as the underlying tectum orbitale has been incorporated in the otoccipital (Fig. 42). This shows that the posterior frontals are related to the tectum orbitale and thence it follows that the anterior frontals most likely are developed in relation to the epiphysial bar and in their anterior parts (frontal 1, Fig. 61B) possibly to the paraphysial bar. Moreover it is evident that the boundary between the frontoethmoidal and parietal shields in coelacanths marks the position of the vertebral joint between CV1+2 and Cv2+3.

II The Origin of Paired Extremities

6 The Origin of Girdles and Paired Fins

Historical review

The dual problem of the origin of the paired fins and the tetrapod limb has been vigorously debated for more than a century (Braus, 1904; Goodrich, 1930; Kälin, 1938; Nauck, 1938; Devillers, 1954a). In order to explain the origin of the paired fins C. Gegenbaur set forth his gill arch theory, but in spite of the endeavours of a great number of supporters this theory has been abandoned in favour of the rival fin fold theory. However, many of the terms introduced by Gegenbaur are still in use although sometimes in a different sense.

Gegenbaur (1865; 1870) based his conclusions in the first place on the conditions in sharks (Fig. 62). In the paired fins he distinguished three main portions; the metapterygium, mesopterygium and propterygium, each consisting of a large basal element (basale metapterygii, etc.) and a number of lateral rays or radials (Radien). The metapterygium which in his opinion could be recognized in the paired fins of all fish as well as in the tetrapod limb, was considered to be the most constant and therefore the most ancient part of the fin. Even in the original fin, the archipterygium, the basale metapterygii was developed and formed the proximal element of an important row of skeletal pieces, the stem or stem row (Stamm, Stammreihe), which was situated on the medial margin of the fin and carried a row of lateral radials. Thus a considerable morphological and phyletic importance was attributed to the basale metapterygii, while the basale mesopterygii and the basale propterygii were interpreted as secondary formations arisen by fusions of proximal portions of radials. According to this theory the archipterygium was unserial, consisting only of a jointed metapterygial stem and the lateral series of radials. However, later on, after the discovery of *Neoceratodus*, Gegenbaur changed his opinion and maintained (1872a) that the archipterygium was a biserial structure similar to the paired fins in *Neoceratodus* (Fig. 64A) and that the stem was represented by the principal axis of such a fin. Gegenbaur's famous gill arch theory (1872a; 1876) which soon became accepted and eagerly defended by a great number of students holds that the paired fins are visceral in origin and differ fundamentally from the median fins (Pinnae), the skeleton of which is regarded as axial. The endoskeletal girdles are interpreted as modified gill arches and

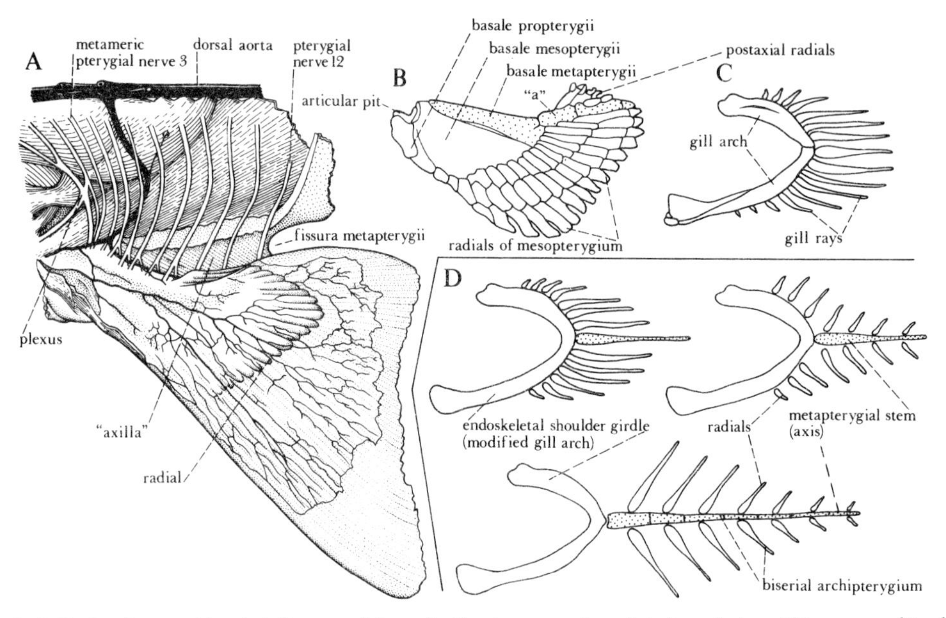

Fig. 62. A, B, *Squalus acanthias*. A, left pectoral fin and adjoining part of trunk in lateral view. Skin removed to show "axilla", traversed by metameric pterygial nerves. Note that the fan-shaped fin is externally long-based (eurybasal), but internally short-based (stenobasal). B, endoskeleton of pectoral fin. Presumed metapterygial stem (including element "a") dotted. From Jarvik (1965a, after E. Müller). C, *Heptanchus maculatus*. Ventral part of first branchial arch with gill rays. From Daniel (1934). D, diagrams illustrating Gegenbaur's gill arch theory.

the metapterygial stem is considered to be the central gill ray of the arch which has become enlarged and subdivided into several pieces (Fig. 62D). The radials are other gill rays which have shifted their bases of attachment from the visceral arch to the enlarged central ray.

As evidenced by embryological data the paired fins in fish are of the same nature as the median fins and are metameric structures arisen in longitudinal fin folds. This is the principal point in the fin fold, or metameric, theory of the origin of the paired fins advanced independently and almost simultaneously by Balfour (1876; 1881a) and Thacher (1876) and accepted and further developed by a great number of investigators.

The most common version of this theory advocates (Fig. 63; Goodrich, 1930; Devillers, 1954a) that the pectoral and pelvic fins were originally separate. The primitive paired fin was a long-based (eurybasal) ventrolateral fold stiffened by a transverse series of metameric endoskeletal rays. The rays were divided into proximal elements, basals, forming a longitudinal row in the body wall, and elongated distal elements, radials, situated within the fin. The longitudinal row of basals is generally referred to as the metapterygial stem or axis, and together with the lateral series of radials it forms a structure corresponding to Gegenbaur's uniserial archipterygium. The metapterygial stem is crossed in regular metameric order by the branches of the spinal nerves supplying the metameric radial muscles of the fin.

This primitive uniserial type of fin (Fig. 63A) is thought to have been modified in

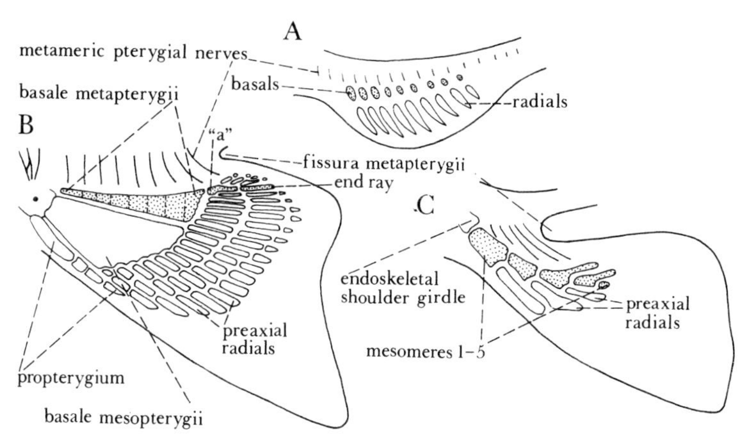

Fig. 63. Diagrams of pectoral fin to illustrate current views (Sewertzoff, Steiner, Holmgren, Westoll and others) as to "metapterygial stem" or "axis" and its freeing from body wall by posterior embayment (fissura metapterygii). A, hypothetical primitive condition, B, *Squalus* and C, *Eusthenopteron*. From Jarvik (1965a).

various ways. The endoskeletal girdles are inward extensions of the basals and in connection with a crowding together of the metameric elements, a process known as concentration and thought to be due to a differential growth between trunk and fin, the metameric basals have often fused with each other forming a large compound basale metapterygii. This has happened in modern sharks (Figs 62, 63B) which otherwise are believed to have preserved much of the primitive fin structure. According to current views the paired fin in sharks is thus still long-based and the metapterygial stem has retained its original position in the body wall and is crossed by the metameric nerves. However, in certain sharks (e.g. in *Squalus*) the posterior part of the metapterygial stem has become freed from the body wall by a slight embayment, the fissura (incisura) metapterygii, in the posterior margin of the fin and some radials have developed on its medial side. By the further deepening of the fissura metapterygii (Fig. 63C) the base of the fin became shortened from behind and the originally longitudinal metapterygial stem, composed of a row of metameric basals or mesomeres, became more and more freed from the body wall so as finally to form the principal axis of the short-based (stenobasal) fin known as the biserial archipterygium.

Common to the gill arch and metameric theories is the fundamental importance attributed to the metapterygial stem. However, among the supporters of both theories there is a considerable diversity of opinion as regards this stem, its position and the number of elements that enter into its formation.

Thus, in sharks several writers (Müller, 1909; Sewertzoff, 1926; Fig. 65B 1) with metapterygial stem mean the row of elements distinguished by Gegenbaur, although it has turned out to be difficult to decide which of the elements (for instance element "a"; Figs 62B, 63) following distal to the basale metapterygii that are true members of the stem. Other students (Moy-Thomas, 1936; Fig. 65A 1; Holmgren, 1939; 1952), in contrast, claim that the stem is represented by the three basalia together with an end ray. Of particular interest are the various interpretations of the metapterygial stem in the pectoral fins of *Neoceratodus, Eusthenopteron* and *Sauripterus*, that is those types of

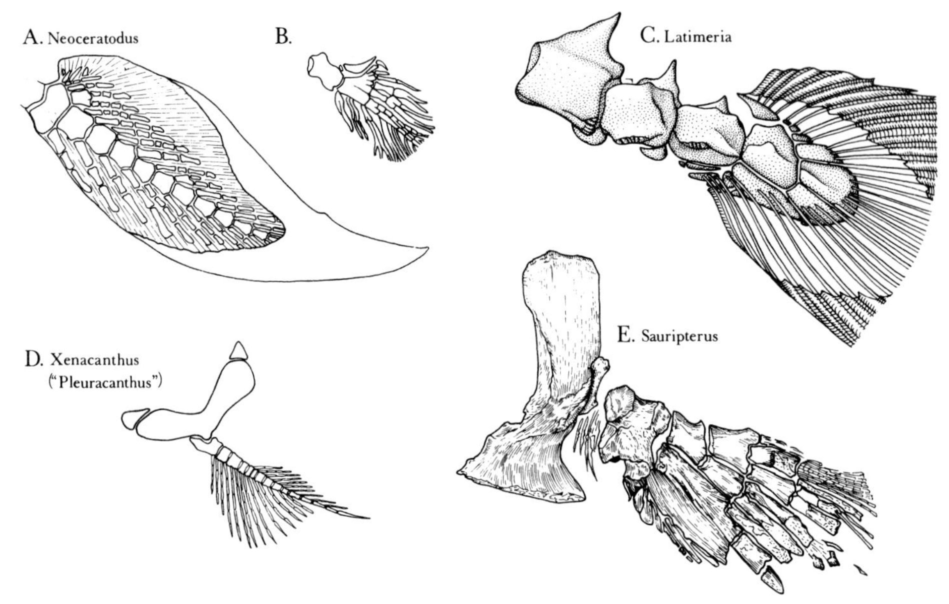

Fig. 64. Left pectoral fin of various fishes. A, B, *Neoceratodus forsteri*. Note subdivision of "metapterygial stem" in B. After Holmgren (1933) and Howes (1887). C, *Latimeria chalumnae*. From Millot and Anthony (1958). D, *Xenacanthus gaudryi*, Palaeozoic selachian. From Devillers (1954a, after Goodrich). E, *Sauripterus taylori*, Upper Devonian, USA. From Gregory (1935).

fin from which the tetrapod limb is generally derived, and in the biserial archipterygium of the Palaeozoic elasmobranch *Xenacanthus*.

In the paddle of *Neoceratodus* (Figs 64A, B, 65) the number of elements, so called mesomeres, of the principal axis is greater than the number of trunk metameres that may have contributed to the fin and several writers (Sewertzoff, 1926; Fig. 65B3; Steiner, 1935; Westoll, 1943a; Holmgren, 1952) therefore claim that a more or less long distal portion of the principal axis is a secondary formation and thus not metameric. As to the composition of the proximal portion, however, opinions differ.

Moy-Thomas (1936) thus maintains that the metapterygial stem in *Neoceratodus* as well as *Eusthenopteron* and tetrapods is represented only by the proximal "mesomere", which in his opinion has been formed by fusion of a series of once independent metameric elements (basals) situated in the body wall (Fig. 65A3). As regards *Xenacanthus*, in contrast (A2), he adopts the current view that the metameric metapterygial stem has become freed from the body wall and forms the principal axis. According to Sewertzoff (1926) only the proximal half of the axis in *Xenacanthus* (B2) represents the stem, whereas in *Neoceratodus* (B3) that stem should be formed by three metameric pieces (the first mesomere, the first postaxial radial and the part of the second "mesomere" situated between these two structures).

A similar interpretation of the metapterygial stem in *Neoceratodus* is given by Holmgren. Of fundamental importance for his extremity theories is the sharp distinction he makes (1933; 1939; 1949; 1952) between the "extremity stem" (not to be confused with metapterygial stem) or "archepodium" and the "secondary rays" or "neopodium". In embryonic stages the "secondary rays" are always distinguishable from the "extremity stem", in that they arise comparatively late and independently of the stem. Moreover, they may grow in both distal and proximal directions, whereas the elements of the "extremity stem" only grow in the distal direction. Holmgren categorically rejects all extremity theories disregarding these differences. In *Neoceratodus* (Fig. 65D 1) the "extremity stem" is said to be branched and consist of the metapterygial stem together with two branches. The metapterygial stem is three-jointed and is formed by the first "mesomere", the proximal part of the second "mesomere" and the postaxial portion of the second "mesomere" (the postaxial basal,

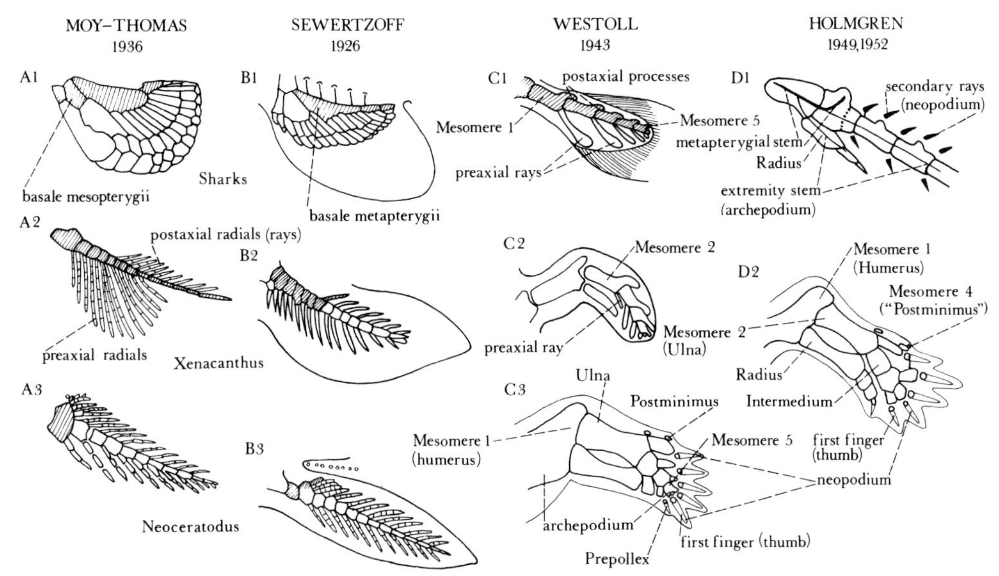

Fig. 65. Diagrams to illustrate various interpretations of metapterygial stem (shaded in A–C) in pectoral limb and (in C) presumed inturning of axis. Note that Holmgren also distinguishes an extremity stem and that both Westoll and Holmgren regard fingers in tetrapods as new formations (neopodium).

Holmgren, 1933). The two branches of the "extremity stem" are the first preaxial ray and the part of the principal axis following distal to the second "mesomere". "Secondary rays" in *Neoceratodus* are all radials except the first preaxial which forms part of the "extremity stem".

In the paired fins of *Eusthenopteron* all endoskeletal elements belong, according to Holmgren, to the "extremity stem". This stem consists in the pectoral fin of a four-jointed metapterygial (metameric) stem and three branches whereas in the pelvic, there is a three-jointed metapterygial stem and two branches (cf. Figs 81, 83). The so-called postaxial processes in *Eusthenopteron* are regarded as muscular processes, which implies that the paired fins in *Eusthenopteron* in Holmgren's opinion are uniserial, a view shared also by Steiner (1935; 1942). Gregory and Raven (1941) and Westoll (1943a), in contrast, interpret the osteolepiform paddles as biserial archipterygia with the metapterygial stem forming the principal axis. The number of metameric elements(mesomeres) in the stem are said to be five in the pectoral (Fig. 65C 1) and four in the pelvic of *Eusthenopteron* and four or possibly five in the pectoral of *Sauripterus* (cf. Fig. 64E).

Ever since Gegenbaur in 1865 made his first attempt to derive the tetrapod limb (cheiropterygium) from the paired fin (ichthyopterygium) in fish and claimed that the metapterygial stem is represented in tetrapods as well, it has been considered of fundamental importance to establish the position of the stem in the tetrapod limb. Gegenbaur first maintained (1865; 1870) that the stem in the tetrapod hand (Fig. 78) is represented by the humerus, the radius and the first digit, but later on (1876), after criticism by Huxley, he declared that the structure of the limb may be just as easily understood if it is assumed that the stem runs through the humerus, ulna and fifth digit. This uncertainty and arbitrariness displayed by Gegenbaur also stamps later attempts to derive the cheiropterygium from the ichthyopterygium and almost all possible alternatives, with regard to the position of the metapterygial stem in the tetrapod limb, have been tried. In two respects only there is general agreement: a metapterygial stem exists and it passes through the proximal element (humerus, femur) of the limb.

Hypothetical interpretations of the various types of fins or limbs founded on a hypothetical and arbitrarily chosen metapterygial stem are of course of no value. If we use the stem as a starting point it must be a necessary condition that it can be properly defined and that its position and extent can be determined without the slightest doubt. However, in view of the diversity of opinion as to this stem it is questionable if it is justified to distinguish such a structure and to attribute to it such a great morphological and phylogenetic importance as done by most students. And are really the principal axis of the biserial archipterygium or the somewhat similarly situated row of "mesomeres" in the paired fins of *Eusthenopteron* and *Sauripterus* true metameric structures which were once situated in the body wall? Furthermore it may be questioned if the digits and carpals (tarsals) are new acquisitions in tetrapods as claimed by Westoll (1943a) and for which he introduced the term neopodium, in contrast to the archepodium, that is those parts of the skeleton inherited from the ancestral fish (Fig. 65). And is it such a fundamental difference as claimed by Holmgren between the extremity stem (also called archepodium) and the "secondary rays" which Holmgren, adopting Westoll's term, although in a different sense, calls the neopodium? These were some of the questions I had to answer when I in the middle of the 1950s became interested in the problem of the origin of the tetrapod limb. In order to get a safe opinion as to the mysterious metapterygial stem I had to turn to sharks and other recent fishes and I devoted a long time to a perusal of the vast literature on the structure and ontogenetic development of the fins in fish and the tetrapod limb. These endeavours resulted in a new theory, or rather a new version of the fin fold theory, of the origin of the paired fins (Jarvik 1965a) and new interpretations (1964; 1965) of the tetrapod limb.

The Ventrolateral Fin Fold

According to Balfour (1876; 1878) the paired fins in elasmobranchs arise as "special developments of a continuous ridge on each side, precisely like the ridges of epiblast which form the rudiments of the unpaired fins" and he concluded "that the limbs are remnants of continuous lateral fins" Balfour's discovery was confirmed by Dohrn (1884a) who, in embryos of *Centrina*, described a strong fold connecting the pectoral and pelvic fins. However, other students (Rabl, Mollier) were unable to find any trace of the ectodermal thickening described by Balfour, except in the Rajiformes where secondary conditions were presumed to occur, and for a long time it was generally agreed that the pectoral and pelvic fins arise separately (Goodrich, 1930). This presumed discontinuity of the paired fins has frequently been used as an argument against the fin fold theory, and it is therefore of great interest that Ekman (1941), investigating *Squalus* and *Etmopterus*, could prove definitely that the paired fins in sharks develop from a continuous thickening of the ectoderm. This conclusion is strongly supported (1) by the presence of condensed mesenchyme derived from the somatopleure also inside that part of the thickening situated between the pectoral and pelvic fins which later disappears and (2) by the well known fact (Dohrn, 1884a; Mayer, 1885; Braus, 1899; Goodrich, 1906) that the myomeres in this area may produce a complete series of transient muscle buds (abortive buds; Fig. 66). In amniotes the primordia of the fore and hind legs are connected by a thickening, Wolff's crest, which soon disappears and also Balinsky's investigations on limb induction in urodele larvae, support the theory of an original ventrolateral fin (Jarvik, 1965a).

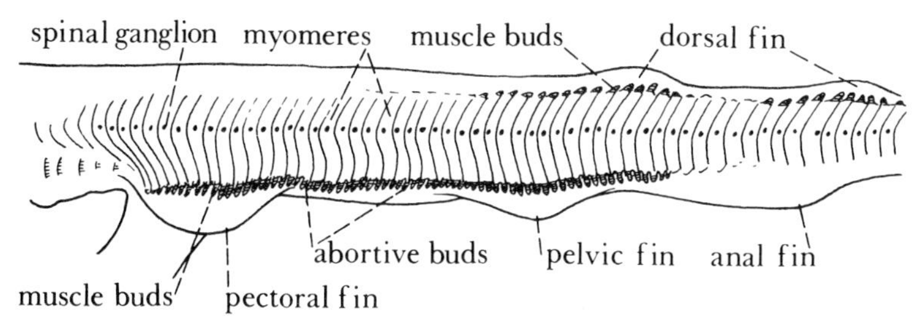

Fig. 66. Scyllium canicula, embryo 19 mm. Diagram to show spinal ganglia, myomeres and muscle buds in median and paired fins, and abortive buds between pectoral and pelvic fins. From Goodrich (1906). Note that each myomere produces two muscle buds in ventrolateral fin fold.

As shown by Balinsky (1933; 1935; 1937) the formation of accessory limbs may be induced, in urodele larvae, by the implantation of living tissue or pieces of inorganic matter within an elongated area (Extremitätenseitenfeld) of the flank, limited anteriorly and posteriorly by the normal fore and hind limbs. Balinsky paid special attention to the position of the induced limbs in relation to the trunk metameres, and to the time sequences in limb induction. His main results may be summarized as follows: accessory limbs may be induced on the flank in every metamere from those of the fore limb to those of the hind limb, but only during a strictly limited period of larval life. All induced limbs arise earlier than the normal hind limb, and earlier the more forwards the metamere of implantation is situated. In each metamere there is a critical phase of short duration during which induction is possible. This critical phase occurs, in the successive metameres, in a strict chronological order, appearing increasingly earlier the more forwards the metamere is situated. In each metamere the phase begins (and ceases) at a definite ontogenetic time, and the setting of the phase is not influenced by the time of inplantation.

All these conditions seem to have a simple explanation (Jarvik, 1965a) if we consider the fact that the myomeres arise in a strict chronological order from the front backwards and assume that the muscle material of the limbs like that of the paired fins is myotomic in origin being produced by the ventral ends of the downward-growing myomeres.

If it is true that the limb-muscle material also in tetrapods is myotomic in origin* it is readily seen that complete limbs with muscles can be formed by induction only when the inductor is so situated that it can influence, directly or indirectly, embryonic myomeres capable of producing limb muscle material. The period during which a growing myomere can give off myogenic limb material, whether by the formation of more or less distinct muscle buds or by migration of myogenic cells out of its free ventral end (Fig. 70A), is certainly of short duration. It is to be suspected that the potential zone of proliferation of myogenic material at the ventral end of the myomere moves rapidly downwards as the myomere grows in that direction. It seems likely therefore, that Balinsky's critical phase corresponds to the limited time in ontogeny when the myomere of the metamere of induction is in such an evloutionary stage that it can produce limb muscle material. Owing to the fact that the differentiation of the myomeres in vertebrates proceeds regularly in the posterior direction, the critical phases will of course occur later the farther back the metamere of induction is situated; and obviously the normal hind limb will be the last limb to receive its muscle material. It is also obvious that an early implantation in a metamere cannot have any inducing effect until the moment at which the differentiation of the myomere has proceeded so far that the production of limb-muscle material at its ventral end can begin. The "Extremitätenseitenfeld" in urodele larvae thus seems to represent an area of the flank within which the myomeres, during a short period, can give off myogenic limb material and which moves successively downwards and backwards as the growth and differentiation of the myomeres proceed.

Balinsky's experiments thus clearly indicate that the ventral ends of the growing myomeres are involved in the formation of the limb muscles and together with other evidence (Wiedersheim, 1892) they prove that these muscles are myotomic in origin. However, in these experiments the limb formation was initiated in an artificial way; but which is the natural inductor and where is it located? Since the paired fins in fish always arise in the continuous ventrolateral fold of the body wall formed by condensed mesoderm with a lining of thickened ectoderm, or in a corresponding position, and since the limb formation in tetrapods, at any rate in amniotes, is initiated in a similar thickening, Wolff's crest, it is quite evident that the natural inductor must be situated in this fold or crest, which accordingly is a potential ventrolateral fin. This view is strongly supported by the fact (Zwilling, 1961; Milaire, 1962; Balinsky, 1970) that the apical ectodermal thickening of the limb bud plays an active and important part in the formation of the tetrapod limb.

The presence of an ancestral ventrolateral crest or potential fin is also supported by the conditions in adult lower vertebrates. Many anaspids (Figs 344, 345, Volume 1)

* pp. 134–135.

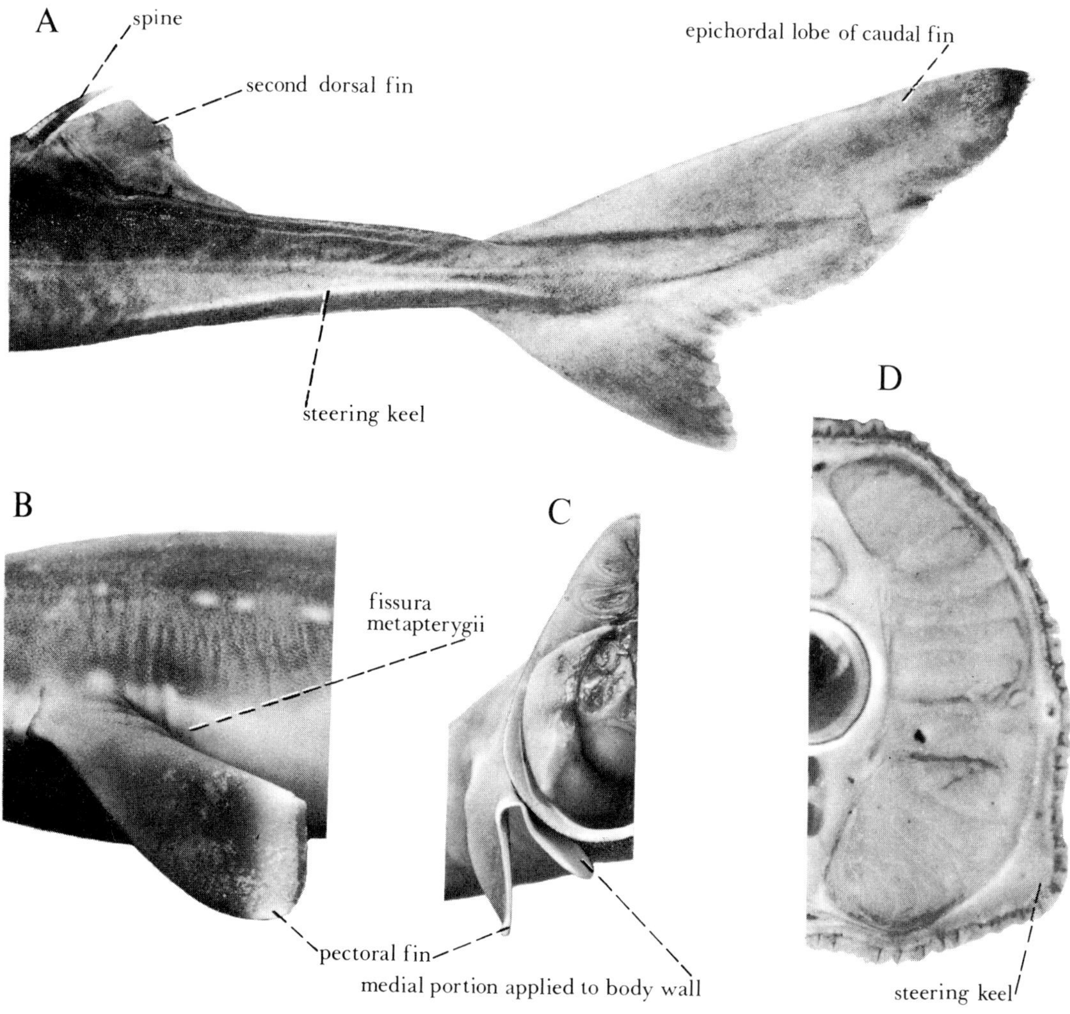

Fig. 67. Squalus acanthias. A, D, photographs of tail in lateral aspect and transverse section to show steering keel. B, C, photographs of left pectoral fin in lateral and posterior aspects to show medial, ventrally bent portion (similar conditions occur in *Acipenser* and *Eusthenopteron*). From Jarvik (1965a).

have long, probably movable, ventrolateral fins, and in fishes there is a strong variation in the rostrocaudal extent of both the paired and the median fins. Moreover many fishes show distinct ventrolateral crests which pass through the bases of the pectoral and pelvic fins. These crests may be marked by an angularity in the squamation as in osteolepids (Fig. 79; Figs 136, 137, Volume 1) and many palaeoniscids (Westoll, 1944; Jarvik, 1948; Jessen, 1968) but often (Fig. 67, Volume 1) there are bony ridges, ridge scales or, as in acanthodians, intermediate spines (often erroneously regarded as intermediate fins). A characteristic feature of the ventrolateral crest or fold is that it anteriorly bends medially (Figs 69D, 79) which is probably due to changes in the position of the gill arches. Posteriorly it runs lateral to the ventral median fin fold and the anal fin to disappear gradually on the side of the tail. The folds of both sides thus do not meet behind the anus to be continued by the ventral median fin fold as sometimes assumed (for instance by Säve-Söderbergh, 1951). The median fin fold is quite independent of the ventrolateral folds and continues far forwards in the front of the anus (Fig. 69; cf. also the preanal fin in myxinoids and heterostracans, Fig. 377, Volume 1).

Evolution of Girdles and Paired Fins

General Remarks

As is now well established (Jarvik, 1964) all the various main groups of Devonian and earlier vertebrates were highly specialized before their first appearance in the fossil record. It is therefore not likely that we will find truly primitive fins in the early vertebrates. The merosomactidial pectoral fin in arthrodires (Fig. 68), developed in early members of this diversified group, is highly specialized (Stensiö, 1959), and the peculiar monomesorhachic appendage of the Devonian antiarchs is among the most advanced types of pectoral fin encountered in vertebrates. Nor can we start from the

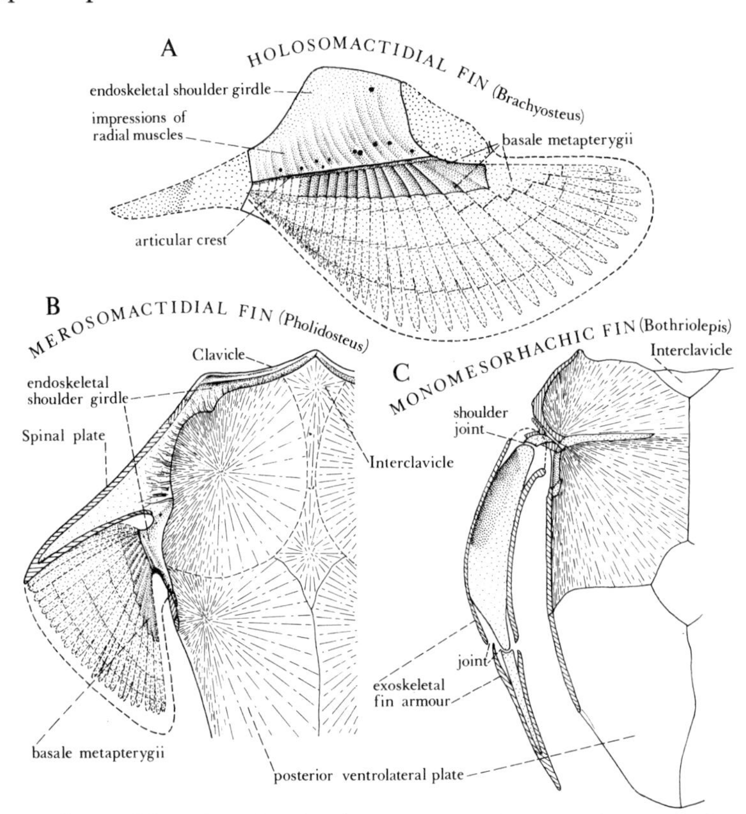

Fig. 68. Three types of pectoral fin in Devonian placoderms. A, most primitive and, C, most advanced type. From Jarvik (1959; 1960, after Stensiö, 1959).

early cyclostomes (ostracoderms) and, as is generally done, take it for granted that they are more primitive than the gnathostomes. Neither is it likely that the detached ichthyodorulites or other spines of early vertebrates are primitive. Like spines of extant fish they are no doubt highly modified, secondary formations; and we cannot, as done by several writers base a theory on the origin of fins solely on the fact that solid structures such as spines are often preserved in early vertebrates. Such a theory must be based on morphological and comparative anatomical data.

Because the cyclostomes are not the ancestors of the jawed vertebrates and the extant forms lack real paired fins (cf. *Neomyxine**), and also because the internal structure of the paired fins in ostracoderms is still unknown, the following account will be confined to the gnathostomes. In these the paired fins always articulate with special endoskeletal girdles, pectoral and pelvic, situated in the body wall, In most forms there is in addition, an exoskeletal shoulder girdle, but certain arthrodires excepted† there is no corresponding pelvic girdle, a condition which needs an explanation.

Influenced by the opinion that sharks are the most primitive gnathostomes it was for a long time generally assumed that a cartilaginous shoulder girdle of the type found in them (Figs 72, 75) represents a primitive condition (modified branchial arch, Gegenbaur; Fig. 62) and that the exoskeletal girdle was a later acquisition. However, in *Eusthenopteron* (Figs 100, 101, Volume 1) and other osteolepiforms the endoskeletal component is a fairly small tripodial structure, whereas the exoskeletal part is well developed and consists of five strong dermal bones on each side, and a ventral median interclavicle. In the tetrapod descendants of the osteolepiforms the evolution of the shoulder girdle is characterized by a gradual reduction of the exoskeleton and a corresponding progressive development of the endoskeletal part. This part first grows ventrally to form a strong coracoid plate, present even in the ichthyostegids (Figs 165, 168, Volume 1) and later a dorsal scapular blade. These conditions and the fact that the exoskeletal girdle, in the lower gnathostomes in general, is more strongly developed than the endoskeletal part indicates that it is rather the exoskeletal girdle that is primitive. However, a more likely alternative is that the endoskeletal and exoskeletal shoulder girdles have arisen independently of each other, the former, like the pelvic girdle, to support the fin, and the latter to fulfil special functions (Dohrn, 1884a; Howell, 1933; Gross, 1954).

In view of the facts presented above, as well as other data, it may be assumed that the girdles and paired fins in gnathostomes arose somewhat as follows (Jarvik, 1965a).

The fins are movable folds of the body wall. The musculature which is an essential part of a typical fin, consists of more or less modified radial muscles formed in ontogeny by migration of myogenic material out of the free dorsal and ventral ends of the growing embryonic myomeres (Fig. 70A). In fish the myogenic material generally forms muscle buds, and an important fact is that each myomere produces two dorsal and two ventral buds destined for the paired fins (Figs 66, 74). Since the radial muscles are derivatives of myomeres it is evident that in the gnathostome phylogeny myomeres must have been present before radial muscles and real movable fins came into existence (Fig. 70B). It is therefore to be concluded that the gnathostomes at an early

* Volume 1, p. 494; † Volume 1, Fig. 281.

phyletic, prognathostome, stage had a trunk with myotomic musculature. This prognathostome (Fig. 69A, B) obviously was a swimming animal and in order to stabilize its movements in the water it almost certainly had some steering keels formed by folds of the body wall, probably somewhat suggestive of the steering keel on the tail in *Squalus* (Fig. 67) and other sharks. Following what has been said above it may be assumed that there was at least a median (dorsal and ventral) and a paired ventrolateral crest or fold. The ventrolateral folds were independent of the ventral median fold and continued backwards to the tail where they gradually faded away. The head of the animal included an axial skeletal portion developed along the notochord (including the exchordal) and the brain, independent visceral arches, the foremost (terminal and premandibular) probably incomplete ventrally, and a terminal mouth. The visceral arches probably had a transverse position, as they have in the embryo, and in contrast to the procyclostomes they carried outwardly directed gills and were provided with independent gill covers.

Origin of Exoskeletal Shoulder Girdle

In my opinion the dermal fin rays (ceratotrichia, actinotrichia, lepidotrichia) are modified scale rows. They occur in several generations (Fig. 7, Volume 1) and most likely the first (oldest) generation arose at an early phyletic stage (Jarvik, 1959a; cf. Schaeffer, 1977). That this is so is supported by the facts that the actinotrichia, which probably represent the first generation in actinopterygians, arise very early in ontogeny, before the formation of muscle buds, and that they are represented in the preanal fin and other parts of the embryonic fin folds which later become reduced (Goodrich, 1930). Since modified scale rows thus conceivably were present in early phyletic stages, probably even before the muscularization of the fin folds, it seems likely that the early gnathostome ancestor had acquired an exoskeleton, and that the skin of the body, and probably also the mucous membranes of the oralobranchial cavity and the gill slits, contained small denticles or primary scales. However, already in this ancient creature the trunk musculature must have been interrupted anteriorly by the gill slits, and certainly at an early stage in phylogeny there was a need to support the trunk musculature at the transition between head and trunk. This need could be satisfied by consolidation of the small exoskeletal units in the skin, and probably the primary scales in the area of the future shoulder girdle soon fused into larger units, forming a primitive exoskeletal shoulder girdle (Fig. 69A). These fusions certainly took place in the same way as, and possibly about contemporaneously with, the formation of the dermal bones of the head. In the head the primary exoskeletal components have fused in more or less different ways in the various groups (Jarvik, 1948) and no doubt the same has happened in the shoulder girdle, a condition which explains the differences in the number and extant of the dermal bones of the shoulder girdle between the various groups. In view of the fact that the type of exoskeletal shoulder girdle characteristic of each group was established in the oldest known representatives, it is evident that this differentiation started very early in phylogeny, probably even in the prognathostome stage. However, the exoskeletal shoulder girdle has to serve other important purposes than to form an anterior anchorage for the trunk musculature and a support for the endoskeletal girdle (Jarvik, 1965a; see also Howell, 1933). It forms the posterior wall of the branchial cavity, it protects the heart and gives attachment for branchial muscles, the hypobranchial musculature, and often for fin muscles as well. It is evident that these manifold functions, which on the whole are fairly similar in the various groups, have strongly influenced the development of the individual dermal bones. This accounts for the fact that the exoskeletal shoulder girdle, although the number and extent of the individual dermal bones may be different, has on the whole a similar structure in the various groups (Jarvik, 1944a), and often presents an inwardly directed lamina situated in the posterior wall of the branchial cavity.

Since, like Howell, I am of the opinion that the exoskeletal shoulder girdle arose to form a support at the transition between head and trunk and accordingly was short, I cannot agree with Gross (1954) who maintains that the primary purpose of the exoskeletal shoulder girdle was to protect the trunk. This view rests on the fact that a long trunk armour is present in certain early groups of vertebrates, a condition which by Gross and others* was considered to be primitive. However, among the early gnathostomes a more or less long trunk armour is actually found only in some groups of arthrodires. Other arthrodires, and also acanthodians, dipnoans and true teleostomes (porolepiforms, osteolepiforms, struniiforms, coelacanthiforms, palaeoniscids) all have a short exoskeletal shoulder girdle. The presence of a long exoskeletal shoulder girdle (trunk armour) is certainly a secondary condition due to fusion of scales in the trunk (cf. cephalaspids†).

Origin of Endoskeletal Shoulder Girdle and Paired Fins

Whereas the exoskeletal shoulder girdle has thus probably arisen to fulfil special requirements at the transition between head and trunk, the endoskeletal shoulder girdle, like the pelvic girdle, certainly arose to form a support for the fin. On the basis of reliable embryological and morphological data it may be assumed that the course of events at the formation of the endoskeletal shoulder girdle and the paired fins was somewhat as follows (Fig. 70; Jarvik, 1965a).

The original ventrolateral crest of the prognathostome was a fold in the body wall, filled with somatopleuric mesomesenchyme (Fig. 70B). Radial muscles had not yet developed in this potential fin fold, but probably there were primitive scales in the skin, and seen in transverse sections it was therefore suggestive of the steering keel on the tail in *Squalus* (Fig. 67D). The crest certainly was a continuous structure extending forwards at least to the branchial region and backwards to the tail (Fig. 69A, B). However, at some early phyletic stage the muscularization began and the immovable crest was transformed into a movable ventrolateral fin. Stimulated probably by an interaction between the mesoderm and the ectoderm of the crest the ventral ends of the growing myomeres, when passing downwards inside the crest, began to produce migratory myogenic material which in the adult formed dimunitive radial muscles (Fig. 70C). In order to increase the flexibility of the ventrolateral fin, or for other

* Volume 1, pp. 354, 454; † Volume 1, p. 455.

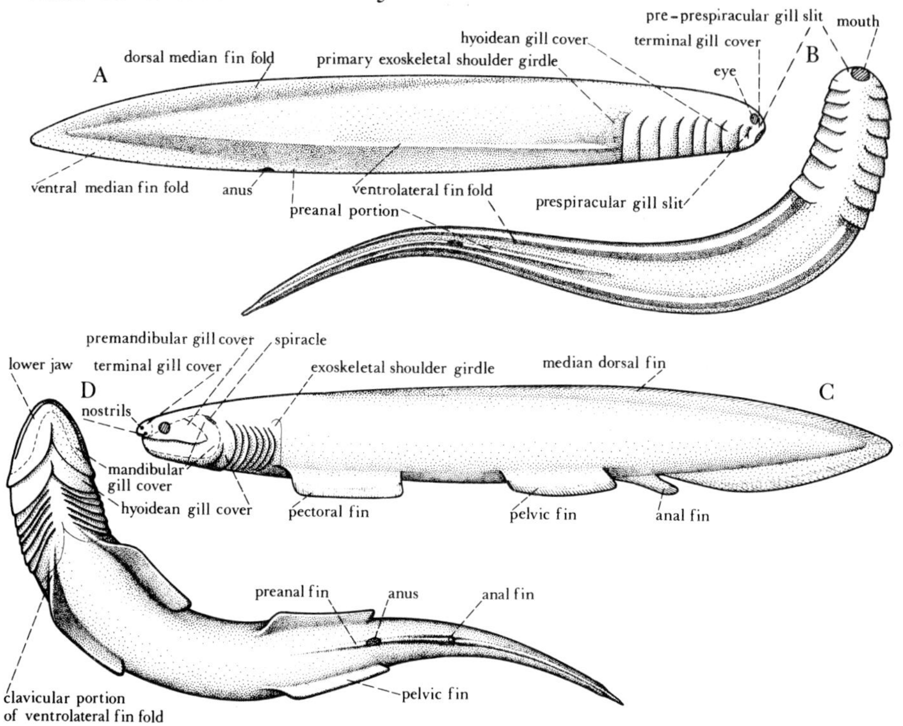

Fig. 69. Hypothetical representation of, A, B, pro-gnathostome and, C, D, primitive gnathostome in lateral and ventral aspects to show ventrolateral and median fin folds and change in direction of gill arches. From Jarvik (1965a, modified).

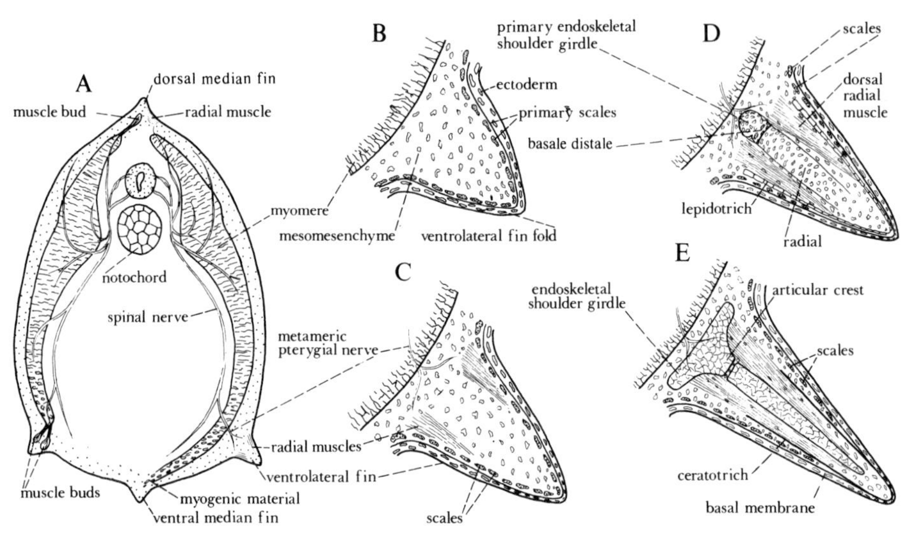

Fig. 70. A, diagrammatic representation to show relations between myomeres, muscle buds, spinal and pterygial nerves, and fin folds in embryo of gnathostome. On right side advanced stage; on left side somewhat earlier stage. B–E, four diagrammatic transverse sections through right ventrolateral crest or fin to demonstrate phyletic development of radial muscles, dermal fin rays and endoskeletal elements. Compilations after figures of embryonic stages by Balfour, Dohrn, Sewertzoff, Goodrich and others. From Jarvik (1965a).

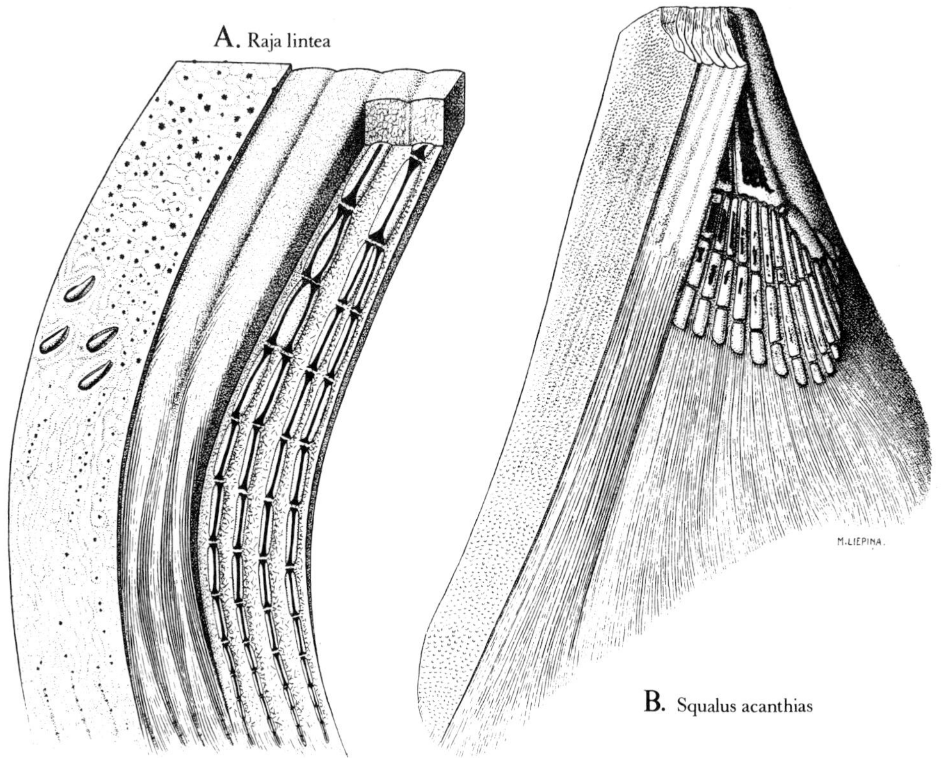

Fig. 71. Distal parts of, A, plesodic pectoral fin of ray (*Raja lintea*) and, B, aplesodic pectoral fin of shark (*Squalus acanthias*). Both figures show, from left to right, skin with placoid scales, radial muscles with ceratotrichia and endoskeletal fin supports (radials). Parts of endoskeletal elements formed by lime prismae drawn in black. From Jarvik (1959a).

reasons, there arose in each metamere two dorsal and two ventral radial muscles (Fig. 72A) innervated by twigs of a pterygial nerve which is a branch of the spinal nerve of the metamere. The radial muscles were originally confined to the base of the fin as they are in embryos, and each muscle, dorsal and ventral, took its origin in the connective tissue of the body wall and was inserted into the skin and the scales of the fin fold. The fin soon increased in breadth and the radial muscles grew out almost to the margin of the fold as they do in the embryonic fin in sharks. Contemporaneously with these changes the scale rows in the skin of the fin fold were transformed into lepidotrichia, which according to the principle of delamination (Jarvik, 1959a) became embedded in the skin in their proximal parts, whereas distally they still clung to the basal membrane. Moreover special supporting endoskeletal rods, primary radials, began to develop in the mesenchyme of the fin, one between each dorsal and ventral radial muscle (Fig. 70D, E). Accordingly there are two such radials in each metamere, a condition generally overlooked in the discussions of the origin of the tetrapod limb. As evidenced by the presence of abortive muscle buds in the middle part of the trunk and in the tail (Fig. 66), a middle and a posterior portion of the original fin fold became suppressed, probably partly at least, as a consequence of the undulatory swimming movements of the body, and independent pectoral and pelvic fins arose (Fig. 69C, D). These fins were plesodic (Fig. 71A; Stensiö, 1959) which is a primitive condition in vertebrates, and they certainly extended in the ventrolateral direction as they do in osteolepiforms (Jarvik, 1948) and many other adult fishes.

Due to the actions of the radial muscles the paired fins at this stage of evolution could move in dorsal and ventral directions bending along the body wall. This effected a subdivision of each primary radial into a proximal element (basal) situated in the body wall and a long distal element (radial) situated in the fin (Fig. 70D, E). The

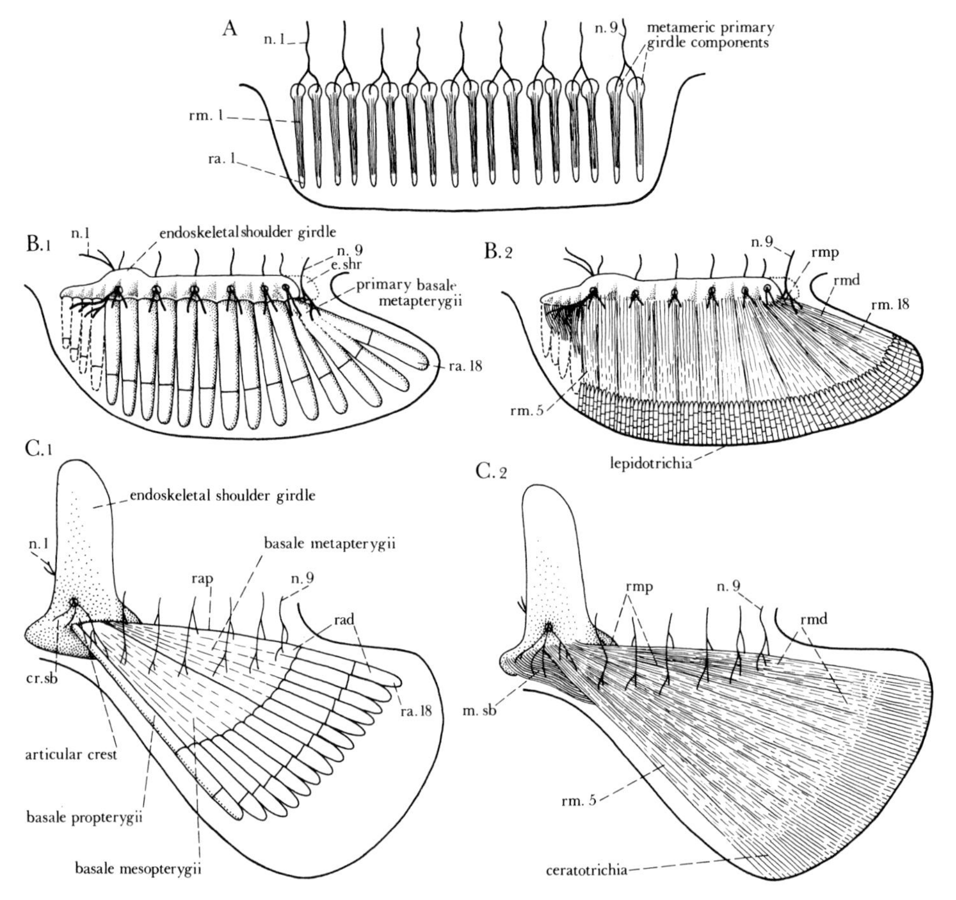

Fig. 72. Diagrams of left pectoral fin to illustrate main phyletic modifications of metameric nerves, radial muscles and endoskeletal elements in connection with shortening of shoulder joint from behind and change in direction of visceral arches. A, primary holosomactidial fin of primitive gnathostome derived from nine metameres. B.1, B.2, pachyosteomorph condition (cf. Fig. 68A) represented also in larval stages of *Acipenser* (Fig. 73) and sharks (Fig. 74). C.1, C.2, shark condition (cf. Figs 62, 63, 74–77). From Jarvik (1965a).

cr.sb, modified articular crest of anterior portion of shoulder girdle; e.shr, posterior reduced part of endoskeletal shoulder girdle; m.sb, modified radial muscles of anterior (subbranchial) portion of pectoral fin; n.1, n.9, metameric pterygial nerves; rad, rap, distal (primary) and proximal (secondary) portions of radials; ra.1, ra.18, radials (two in each metamere); rmd, rmp, distal (primary) and proximal (secondary) portions of radial muscles; rm.1 rm.5, rm.18, radial muscles (two dorsal and two ventral in each metamere).

proximal elements which have been observed by Sewertzoff (1926a) in embryonic pelvic fins of *Acipenser* (Fig. 73) sometimes at least divided into a basale proximale and a basale distale. The proximal basals represent the primary components of the endoskeletal shoulder girdle. These primary girdle components grew dorsally and ventrally in the body wall and fused with each other forming a primitive elongated girdle provided with a lateral articular crest formed by the distal basals. A shoulder girdle of this type, although in several respects more advanced, has been described by Stensiö (1959) in Devonian pachyosteomorph arthrodires (Fig. 72B1).

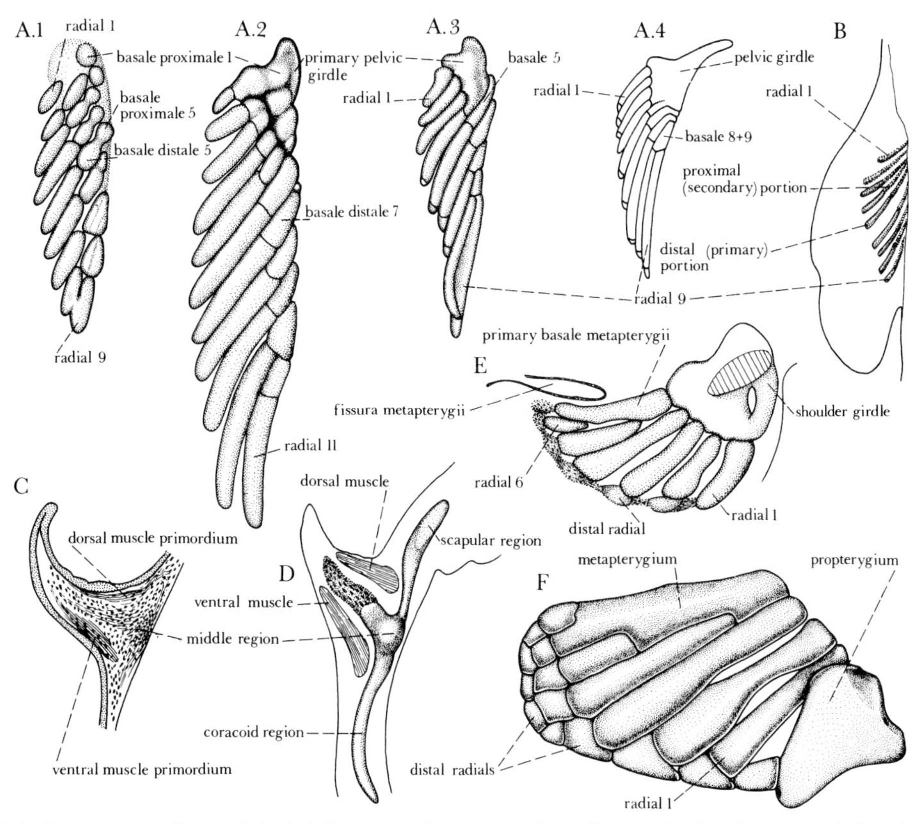

Fig. 73. A–E, *Acipenser ruthenus*. A.1–A.4, four stages in ontogenetic evolution of pelvic fin. A.1, early larval stage, A.4, adult. From Sewertzoff (1926a). B, primordium of endoskeleton of pelvic fin. From Sewertzoff (1934). C, transverse section and, D, reconstruction of endoskeletal shoulder girdle and pectoral fin of larvae. From Sewertzoff (1926). E, reconstruction of endoskeletal shoulder girdle and endoskeleton of pectoral fin of larva. From Sewertzoff (1926). F, *Acipenser sturio*, endoskeleton of right pectoral fin in dorsal view. From Jessen (1972).

The phyletic changes hitherto considered, which probably took place as early as in the prognathostome stage, are to a considerable extent recapitulated in the ontogenetic development of the paired fins in sharks and sturgeons (Figs 73, 74). They first resulted in the development of paired fins, pectoral and pelvic, with transverse metameric elements. Such a primitive fin, shown diagrammatically in Fig. 72A (primary

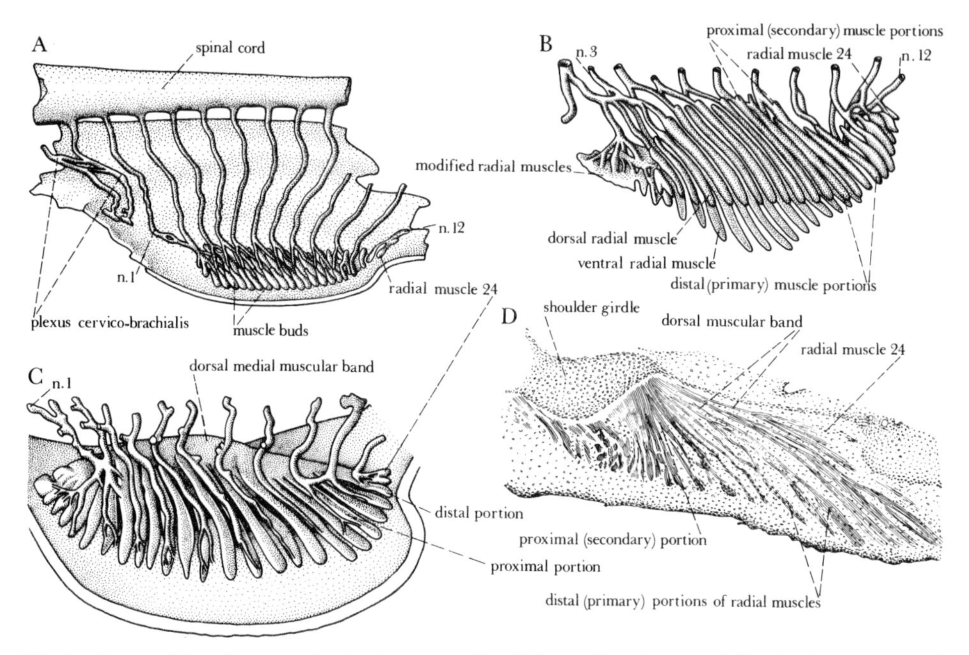

Fig. 74. Squalus acanthias. Ontogenetic development of radial muscles of pectoral fin. A–C, reconstructions of embryos 24, 27 and 30 mm. D, frontal section of embryo 38 mm. These figures explain why the pterygial nerves in the adult cross the "metapterygial stem" in metameric order (Figs 62A, 75D). Starting with a small bud at the entrance of the nerve, as shown in B, the proximal (secondary) portion of each radial muscle grows forwards towards the shoulder girdle, whereas the pterygial nerves retain their original position. Note that each metameric pterygial nerve supplies two dorsal and two ventral muscle buds and radial muscles. From Jarvik (1965a, after E. Müller, 1911).

holosomactidial fin, Stensiö, 1959) has hitherto not been found in any adult fish, (as to the pectoral fin in *Cladoselache* which has been thought to be primitive*). It is therefore of great interest that the pachyosteomorphs, which have retained a primitive type of shoulder girdle, also present a pectoral fin (Figs 68A, 72B) which in many respects is suggestive of such a primitive fin. However in them, as explained by Stensiö (1959; as regards *Acipenser* see also Fig. 73 and Sewertzoff, 1926a; 1934) the posterior radials have moved forwards in their proximal parts and have partly fused with each other, forming a primitive basale metapterygii. In connection with these changes the endoskeletal shoulder girdle and its articular crest were correspondingly shortened from behind. In the main middle part of the fin the primitive metameric disposition has been retained, and the metameric components (radials, radial muscles, pterygial nerves and vessels) were arranged much as they are in embryos of sharks (Figs 72, 74).

This shortening of the shoulder girdle and, more important, of the shoulder joint and the crowding together of the metameric elements from behind are the result of an important morphogenetic process leading ultimately to the formation of an internally short-based fin found for instance in sharks (Figs 62, 75–77), sturgeons (Fig. 73), and osteolepiforms (Fig. 72, Volume 1). Since there is nothing to indicate that the radial muscles and radials have moved forwards as a whole during this process it is obvious

* see Volume 1, p. 348.

that they must have undergone a successive forward lengthening in order to reach the shoulder joint. That this is so is strongly supported by embryological evidence (E. Müller, Sewertzoff).

As shown by Müller (1911) the pterygial nerves in early embryonic stages of *Squalus* enter the proximal ends of the pear-shaped muscle buds (Fig. 74). These buds are soon transformed into radial muscles which grow in the distal direction towards the margin of the fin fold. At a somewhat later stage a small processlike outgrowth appears at the proximal end of the muscle, close to the entrance of the pterygial nerve. This important secondary muscular portion (Fig. 74B) grows in the opposite, proximal, direction, but a most important fact is that it turns forwards towards the shoulder joint which it reaches in later stages. Accordingly we may in each radial muscle distinguish between a primary distal portion growing in the distal direction and a secondary proximal portion growing forwards to the shoulder joint (Figs 72, 74). The boundary between these two portions is marked by the point of entrance of the pterygial nerve. A most remarkable fact, too, is that the pterygial nerves apparently are not influenced by the secondary growth of the radial muscles, a condition which explains why they retain their original metameric position and in the adult cross the "metapterygial axis" in a regular metameric order (Figs 62A, 72, 74, 75).

As is readily seen, the hindmost radial muscle has the greatest distance to grow in order to reach the shoulder joint. Accordingly this muscle has a long proximal portion which runs forwards in an almost longitudinal direction. To this portion are added, successively, the proximal portions of the other radial muscles. In this way the muscles become packed close together, but they keep their individuality and, as may be easily ascertained by dissection of adult sharks (Figs 75, 76), they all continue to, or almost to the shoulder joint. The proximal portions of the caudal muscles, in the first

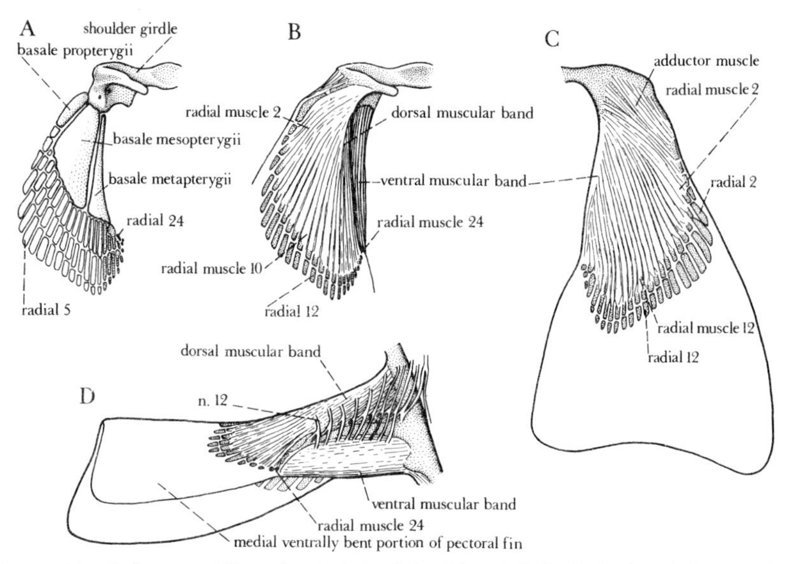

Fig. 75. Squalus acanthias. Left pectoral fin and endoskeletal shoulder girdle in, A, B, dorsal, C, ventral and, D, medial aspects to show skeleton, dorsal and ventral musculature, and (in D only) pterygial nerves. From Jarvik (1965a).

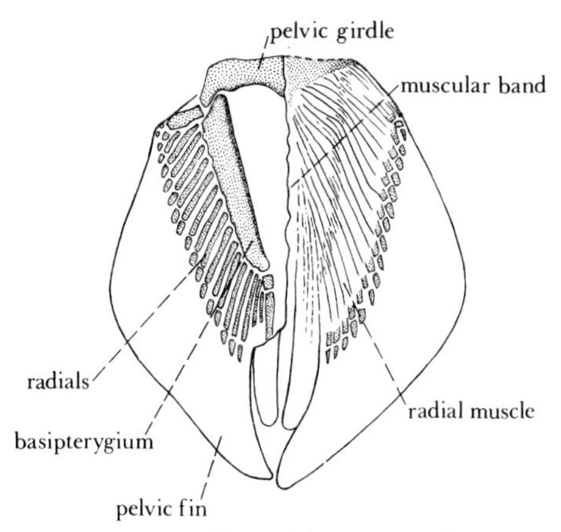

Fig. 76. Squalus acanthias. Pelvic girdles and fins in ventral view. From Jarvik (1965a).

instance those belonging to the metapterygial radials, form, on each side of the fin, a thick muscular band. This band together with the muscles which successively join it from the lateral side appears as a comblike structure. Although the conditions—due to the fact that the pectoral fin in *Squalus* (like in *Acipenser* and *Eusthenopteron*) is bent ventrally in its medial part (Figs 67, 75) and for other reasons—are a little more complicated than can be explained here, it is evident that the proximal portions of the metapterygial muscles ran forwards in the same direction as, and are intimately related to the basale metapterygii. This can only mean that the latter includes the original supporting portions of these secondary muscle portions. Accordingly it is to be assumed that the basale metapterygii in sharks is composed chiefly of the proximal portions of the metapterygial radials (Figs 72C1, 77) and that these portions have fused into a single piece, in the way indicated by the pachyosteomorphs and the ontogenetic development of the paired fin in *Acipenser*.

These conclusions are supported also by the great variations of the skeletal elements in the fins of *Squalus* and other sharks (Gegenbaur, 1865; Müller, 1909; E. G. White, 1937). Studies of these variations have revealed that the skeletal piece "a" (Figs 62B, 77A) which in *Squalus* and many other sharks follows distal to the basale metapterygii must have arisen by fusions of adjoining radials as indicated in Fig. 77A. Thence it follows that it cannot be a metameric basal or stem element as is generally assumed (Figs 62, 63). However, if element "a" is a product of fusion of portions of radials it is hard to imagine that the element following next in front of it, the basale metapterygii, could have been formed by fusions of elements of a longitudinal row of once independent metameric basals (Fig. 63B), that is in a quite different way. The only reasonable conclusion must be that the basale metapterygii, like the basale mesopterygii and other obviously compound skeletal element in the pectoral fin of sharks, is formed by the fusion of adjoining pieces of radials, the imaginary lines of fusion running longitudinally (Fig. 77A) and not transversely as generally assumed.

Of great interest is also the fact that the basale mesopterygii in *Carcharias* (Jarvik, 1965a, fig. 11A) and several other sharks (Fig. 65A1, B1) is much shorter than the basale metapterygii. In these sharks the basale metapterygii, like the basipterygium in the pelvic fin (Fig. 76) carries most of the radials, and together with them constitutes a comblike structure suggestive of the comblike structure formed by the radial muscles in *Squalus*. This variation, too, strongly suggests that the imaginary lines of fusion in the basale metapterygii (and in the basipterygium) ran longitudinally (Fig. 77A) and not transversely (Fig. 63B).

The development and variation of the skeleton in the pectoral fin of sharks are thus in accord with the evidence provided by the ontogenetic development and course of the radial muscles, and together they show that the basale metapterygii and other supposed stem elements are formed by adjoining pieces of radials which have fused longitudinally. Under these circumstances it is of course impossible to distinguish a metapterygial stem, and this concept which has played such a great role in the discussions of the origin and nature of the paired fins and of the tetrapod limb is to be regarded as a hypothetical construction without real significance.

As shown above, the radial muscles in *Squalus* have become lengthened forwards in connection with the shortening of the shoulder joint from behind and they include a secondary portion growing in the proximal direction. Since the proximal portions of the radials no doubt have arisen to form a support for the corresponding secondary muscle portions it is readily seen that the proximal radial portions, too, must be secondary formations and that they must have grown in the proximal direction (Figs 72C1, 77A). It is therefore of interest that the proximal portions of the radials in the pelvic fin of *Acipenser*, according to Sewertzoff, actually arise later than the distal portions and grow in the proximal direction (Fig. 73B). Since Sewertzoff (1926a) also states that the radials in *Acipenser* grow in both distal and proximal direction as do the "neopodial" rays in tetrapods (Fig. 65D; Holmgren, 1952) it is evident that there is a close agreement in the mode of growth between radial muscles and their supporting endoskeletal elements.

As we have seen the basale metapterygii in sharks, which is the principal element of the presumed metapterygial stem, is composed chiefly of proximal portions of radials. These portions have been crowded together in much the same way as the corresponding portions of the dorsal and ventral radial muscles and like them they are secondary formations which in phylogeny (as in ontogeny) have arisen by a lengthening forwards of the primary elements towards the shoulder joint (Figs 72, 74–77). The basale metapterygii is thus, in the main, a new formation and not the most ancient part of the fin as supposed by Gegenbaur. Nor is it composed of a row of metameric elements ("basals") originally situated in the body wall (Fig. 63B), as claimed by the partisans of the metameric theory, and obviously the statement, too, that in sharks it has retained its primitive position in the body wall, is inconsistent with available facts. It is true that the pectoral fin in sharks (Fig. 62A), as far as its external appearance is concerned is fairly long-based (eurybasal) as are the pectoral fins in *Acipenser* and *Eusthenopteron*. However, if the skin is removed it is clearly seen, that it is a short-based (stenobasal) fan-shaped structure in which all the metameric elements, with the exception of the metazonal nerves, converge towards the shoulder joint, which is short, again as in *Acipenser* and *Eusthenopteron*. The fan-shaped fin is separated from the musculature of the body wall by a large space filled with loose connective tissue. This almost empty space between fin and body, termed "Achselhöhle" (axilla) by Müller (1909), is traversed by the metazonal nerves in metameric order, a condition which has caused considerable confusion in the past, but which, as explained above, is due to the mode of growth of the radial muscles. The basale metapterygii lies in the medial margin of the fan-shaped fin. It has arisen by secondary modifications within the fin and neither this structure nor other elements included in the "metapterygial stem" have become freed from the body wall by a posterior embayment (incisura metapterygii; Steiner, Sewertzoff) as generally assumed (Fig. 63). The formation of this embayment is certainly due to the crowding together of the posterior metameric elements towards the shoulder joint (Fig. 72), through which these elements assume a more or less oblique position. The embayment is still moderate in sharks and has hardly modified the course of the most posterior pterygial nerves which run but slightly forwards in order to reach the fin (Figs 62B, 74). The same applies to sturgeons and conceivably to *Eusthenopteron* in which the pectoral fin is also externally long-based. In tetrapods (Fig. 80C), on the other hand, in which the base of the limb is short the posterior one or two pterygial nerves have been pushed more distinctly forwards and enter into the forma-

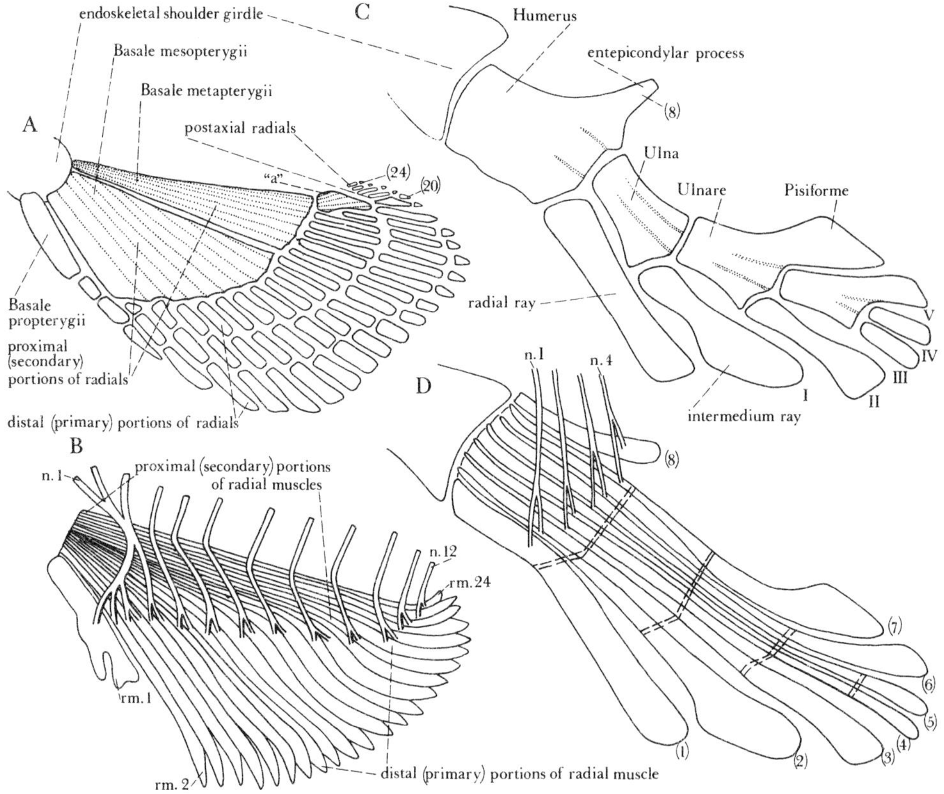

Fig. 77. Interpretations of pectoral fins in A, B, *Squalus* and, C, D, *Eusthenopteron*. Imaginary lines of fusion of proximal portions of radials indicated by dotted lines in A. From Jarvik (1965a).

n. 1, n. 4, n. 12, metameric pterygial nerves; rm. 1, rm. 2, rm. 24, radial muscles; (1)–(8), (20), (24), radials or rays; I–V, digits (fingers).

tion of the plexus cervico-brachialis. The so-called postaxial radials in *Squalus*, finally, are no new formations. As proved by the remarkable change in position of their radial muscles in ontogeny (Fig. 74) they represent the original most caudal radials of the fin.

Characteristic of *Squalus* and other sharks is that there are two dorsal and two ventral radial muscles and two radials in each metamere. In *Squalus* the pectoral fin is usually innervated by twelve metameric pterygial nerves and consequently the number of radials amounts to about 24 (Figs 62, 74, 77; Müller, 1909; 1911). However, each of the pterygial nerves supplies three or four radial muscles and another interesting fact is (Müller, 1911) that the dorsal (extensor) muscles have moved backwards in relation to the ventral (flexor) muscles (Fig. 74C). In *Acipenser* and other actinopterygians (Kryzanovsky, 1927; Jessen, 1972) the pterygial nerves are less numerous than in *Squalus*, generally about 4–6. However, also in them the radials are twice as many as the nerves, that is about 8–12, a condition which indicates that the pectoral fins in actinopterygians have arisen in the same way as in *Squalus*. In other groups of fishes the paired fins and their girdles have been modified in various ways, but as far as I can find, all these various types, too, have originated as outlined above; and obviously the many different types of more or less concentrated median fins in fishes have been formed according to the same fundamental principles. The median fins in amphibian larvae, which are induced by the trunk neural crest (Balinsky, 1970; Weston, 1970) lack muscles and skeletal elements (cf. however, *Palaeospondylus**). They are possibly to be regarded as larval adaptations (note also that *Amphioxus* presents a median fin fold in spite of the fact that neural crest is lacking).

Finally, the pectoral fin of *Eusthenopteron* will be briefly discussed. This fin, which has been found completely preserved only in one specimen (Fig. 102, Volume 1), is of particular interest since the tetrapod fore limb in the osteolepiform–eutetrapod stock, the Osteolepipoda, may be easily derived from this type of fin Jarvik, 1964; 1965). As is now well known the pectoral fin in *Eusthenopteron* is seven-rayed as is the tetrapod limb (Figs 77C, 81). However, the entepicondylar process of the humerus most likely represents an additional ray which increases the number of rays to eight. These rays no doubt represent radials and because usually two radials are formed in each metamere it is to be concluded that the musculature of the fin was derived from four myotomes and was innervated by four main spinal nerves, as in tetrapods (Sewertzoff, 1908; Goodrich, 1930; since branches of adjoining nerves contribute, Fig. 80C, the number of spinal nerves in the cervico-brachial plexus is generally five or six). However, in *Eusthenopteron* the radials have obviously become segmented and have fused in another fashion than in sharks and sturgeons. The proximal element, the 1st mesomere or humerus, probably includes the proximal segments of eight radials which obviously early in phylogeny have fused so intimately with each other that no boundaries between them are discernible in the ontogeny of tetrapods (cf. the basale metapterygii in sharks, Fig. 77, and the basipterygium in *Acipenser*, Fig. 73). Also the second (ulna), third (ulnare) and fourth mesomeres are complex structures. None of these four mesomeres, which generally have been considered to be "stem" elements, is thus a simple metameric structure originally situated in the body wall and there is no metapterygial stem either in fishes or in tetrapods.

* Volume 1, p. 218.

7 The Origin of Tetrapod Limbs

General Considerations

It is well known that the extremities in tetrapods are subject to strong variations. The differences are certainly great, e.g. between the wing of a bird and the foreleg of a horse, but also between the bird's wings and feet. The possibilities of variation, are, however, considerably restricted by the fact that the ontogenetic development of the extremities proceeds from a given number of metameric elements. According to a common opinion originating from Gegenbaur the extremity skeleton in tetrapods is constructed from a definite number of elements which are arranged, in the hand (manus) as well as in the foot (pes), according to an original common ground-plan (Steiner, 1921; 1934; 1942; 1965; Goodrich, 1930; Nauck, 1938; Schmidt-Ehrenberg, 1942; Cihák, 1972). Having regard to the great variations, especially in the carpus and tarsus, it has, however, proved difficult to recognize and differentiate the exact number of these canonical elements in the pentadactyle limb (Fig. 78), and the numerous attempts to homologize have led to most different results. But nevertheless much importance has been attributed to the primitive pentadactyle tetrapod limb and since in the opinion of many such a structure can have arisen only once the tetrapod limb has been the main argument in favour of the view that the tetrapods are monophyletic. These writers have, however, overlooked or ignored the following facts.

(1) The fore and hind legs of tetrapods have no doubt developed from the pectoral and pelvic fins of their piscine ancestors. An important fact is now that these two fins in fishes are practically always different as to the number of their skeletal elements and their arrangement (see for instance *Amia*, Figs 30, 32, Volume 1; *Eusthenopteron*, Figs 81, 83; *Squalus*, Figs 75, 76). This must be taken to mean that the fore limb and the hind limb in each tetrapod have arisen independently of each other from two different patterns of arrangement. In other words the pentadactyl limb has arisen at least twice. Disregard of this obvious fact has caused much confusion in the past.

(2) Practically all students (Zwick, Sewertzoff, Steiner, Holmgren) who have studied the ontogenetic development of the extremity skeleton in both urodeles and non-urodelan tetrapods, the eutetrapods, have strongly emphasized the great differences between these two tetrapod divisions. In 1908 Sewertzoff (Sewertzoff, 1908, p. 363) thus asked the question: "Wir müssten also annehmen dass die Extremitäten der

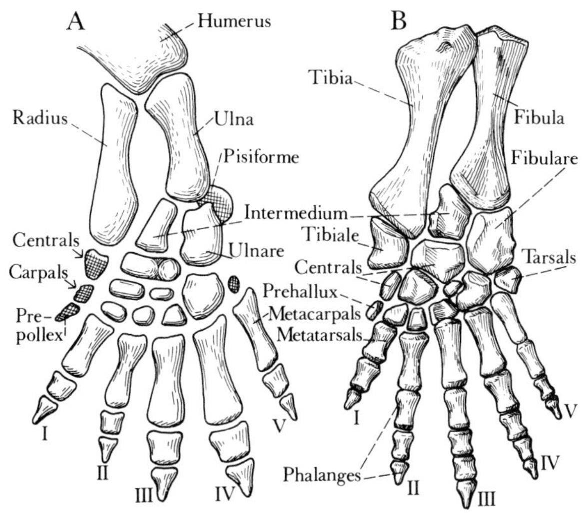

Fig. 78. "Canonical" elements in eutetrapod hand (A) and foot (B). After Steiner (1934; 1965) and Schaeffer (1941).

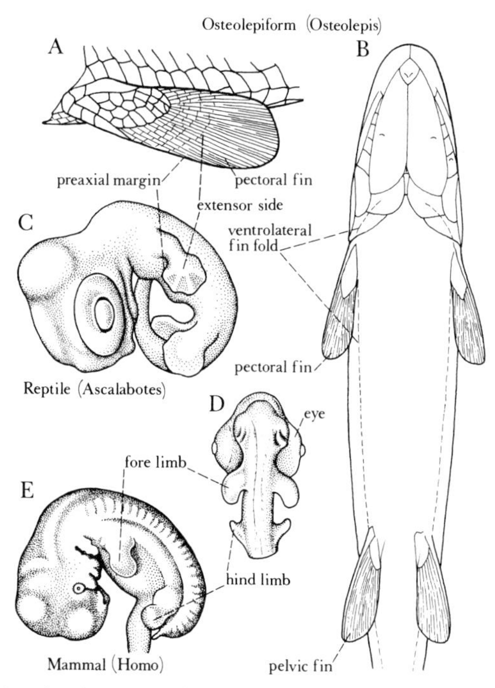

Fig. 79. Position of pectoral and pelvic appendages in Osteolepipoda (osteolepiform–eutetrapod stock). A, B, osteolepiform (*Osteolepis*). From Jarvik (1948). C, D, reptile (*Ascalabotes*), embryos in lateral and dorsal aspects. From Sewertzoff (1908). E, mammal (*Homo*, 10·5 mm, 31–34 days). From Keibel (1902, after His).

Reptilien und der Anura einerseits, die der Urodela andererseits phylogenetisch vollkommen unabhängig von einander entwickelt haben?". Steiner (1921) expressed the same opinion in the following way: "Die ontogenetische Entwicklung der Extremität der Urodelen steht in einem aufallenden Gegensatz zu allen übrigen Tetrapoden". Holmgren (1933) found these differences to be so great that he was forced to make what amounted to be rather startling conclusions for those days, namely that the tetrapods are diphyletic and that "the urodeles must be considered to have originated from fish ancestors, (crossopterygian or) dipnoan, with a short biserially arranged archipterygium". These unanimous results gained by prominent specialists must not be ignored, and as is quite evident the tetrapod limb cannot be used as proof of the monophyly of the tetrapods.

When discussing the origin of the tetrapod limbs we have thus to distinguish between the evolution of the fore limb and the evolution of the hind limb; and it is also necessary to treat the evolution of the limbs in the urodeles and the eutetrapods separately. Since the tetrapod limbs no doubt have developed step by step from the paired fins in the ancestral fishes and there can hardly be any fundamental or mysterious differences in the principles of construction of the paired appendages between fish and tetrapod: it is, moreover, necessary to consider the results obtained above (Chapter 6) as to the origin and nature of the paired fins. The following conditions deserve special attention.

(1) The conception of a metapterygial axis of great phylogenetic significance is due to the misinterpretation of the composition of the shark fin. Such an axis does not exist, and the proximal elements of the tetrapod limb (i.e. humerus, ulna, and so on) are not metameric elements originally situated in the body wall. Also, there is no reason to distinguish between archepodium (extremity stem) and neopodium (secondary rays) as done by Holmgren.

(2) The paired extremities of both fishes and tetrapods are composed of the following metameric elements: radial muscles, radials or rays which are secondary endoskeletal rods supporting the musculature, pterygial nerves (branches of spinal nerves) and blood vessels.

(3) Each myomere which takes part in the construction of the paired fins in fish gives rise, on each side of the animal, to two dorsal and two ventral radial muscles. Between each dorsal and ventral muscle a radial or ray is formed. Consequently there are in each metamere two pairs of radial muscles and two pairs of rays.

(4) The number of myomeres taking part in the construction of the paired fins varies considerably and can be different on both sides of the same individual (about 12 in *Squalus*, but only about four in *Acipenser*, other actinopterygians and, probably, in osteolepiforms). Because each pterygial nerve normally innervates three or more radial muscles on each side of the fin the number of pterygial nerves which supply the fin and which enter into the brachial plexus are usually more numerous than the myomeres.

In tetrapods two muscle primordia, one on each side of the limb bud, are usually discernible in early stages. Radial muscles are never formed and according to a widespread opinion (Streeter, 1949; Starck, 1955; Balinsky, 1970; Romer, 1970; Torrey, 1971) the extremity muscles are derived, not as in fishes from the myotomes,

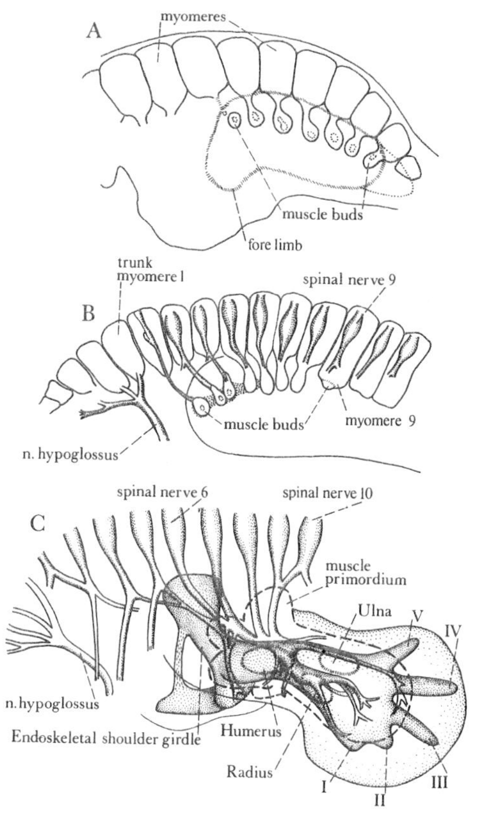

Fig. 80. Three stages in evolution of forelimb of reptile (*Ascalabotes fascicularis*) to show muscle buds in tetrapods. From Sewertzoff (1904; 1908).

but from migratory mesomesenchyme originating from the somatopleure. That it should be such a fundamental difference between fish and tetrapod is of course most unlikely and probably we are concerned with an abbreviated ontogenetic development. As we have seen* Balinsky's experiments indicate that the myogenic limb material in urodeles is produced by the ventral ends of the growing myomeres (see also Wiedersheim, 1892), and the observations by several early students (Field, van Bemmelen, Mollier, Paterson, Kolman; see Maurer, 1904) that ventral processes of the somites or migratory somitic mesenchyme enter the limb bud has been confirmed by Sewertzoff (1908) and more recently by Milaire (1957), Griffiths (1959) and Raynaud and Adrian (1976). According to Sewertzoff's accurate researches on the reptile *Ascalabotes* (Fig. 80) a distinct muscle bud originates at the ventral end of myomeres 2–9, each provided with a branch from the corresponding spinal nerve. During embryonic development the four anterior muscle buds atrophy together with their nerves, and the extremity musculature is formed only by the posterior four myomeres. From these myomeres cells migrate into the limb bud and form the

* pp. 115–116.

dorsolateral and dorsomedial muscle primordia. The branches of spinal nerves 6–9 belonging to these muscle buds make up the branchial plexus. However, a branch of spinal nerve 10 joins the plexus which thus in the adult receives branches of five spinal nerves (6–10). The conditions in this regard are thus the same as in fishes and since also in many other tetrapods, for instance in man, branches of five spinal nerves supply the musculature of the free extremity it seems likely that this musculature, like in *Ascalobotes*, is derived from four myomeres. Thence it follows that tetrapod limbs may ordinarily be constructed from eight radials and include the equivalents of eight dorsal and eight ventral radial muscles. In this respect the posterior extremity behaves like the anterior and there is no marked difference between the urodeles and the eutetrapods. A remarkable fact is, however, that also in tetrapods* the extensor muscles, judging from the innervation, have moved backwards in relation to the flexor muscles. This is distinct in urodeles (Francis, 1934) but seems also to be the case in man (Gray, 1973) in which, for example, the extensor muscle of the thumb (extensor pollicis longus) is innervated by cervical nerves 6–8, whereas the opposing flexor pollicis longus is supplied by cervical nerve 8 and thoracic nerve 1.

(5) The rays which in both fishes and tetrapods make up the skeleton of the free extremity, are usually divided into pieces or segments. As may be gathered from sharks, this dividing up takes place more or less irregularly, and not exactly in the same way even on the right and left sides of the same specimen. Therefore it is also not possible to homologize segments of rays with any great certainty either in fishes or tetrapods.

(6) Segments of rays can fuse very easily with segments of bordering rays to form more or less complex structures, a circumstance which helps to explain the great variation in the arrangement of the skeletal elements in the tetrapod carpus and tarsus. That side branches of rays are formed by budding, as assumed by Holmgren, is most unlikely.

(7) In the tetrapods the proximal segments of all the rays are fused into a single element, the humerus or femur. However, this is not a special characteristic of tetrapods. In many fishes a single proximal element is also present. Further common to fishes and tetrapods is that the first ray in the extremity is shorter and stronger than the one following immediately behind. This ray is the propterygium in elasmobranchs and several other fishes and the radial (or tibial) ray in tetrapods.

After these general remarks we may now turn to the Eutetrapoda which together with their osteolepiform ancestors constitute the osteolepiform–eutetrapod stock or the Osteolepipoda.

Girdles and Paired Limbs in the Osteolepipoda

Shoulder Girdle and Foreleg

As demonstrated above the endoskeletal shoulder girdle in the earliest known tetrapod, *Ichthyostega*, may be easily derived from that in the osteolepiform fish *Eus-*

* cf. p. 131.

thenopteron (Fig. 165, Volume 1). The main changes are that the glenoid fossa has become screw-shaped and that the girdle has grown downwards to form a large coracoid plate. This progressive development of the endoskeleton has been accompanied by a reduction of the ventral part of the cleithrum and adjoining parts of the clavicle and interclavicle. In post-Devonian eutetrapods the endoskeletal girdle has grown also in the dorsal direction forming a scapular blade which has led to a further reduction of the cleithrum. This element has been retained in extant anurans (Braus, 1919; de Villiers, 1922) but has disappeared in all living amniotes. The clavicle is found in most eutetrapods and is well developed in man, whereas the interclavicle is lacking in anurans and all mammals except monotremes.

The humerus of *Ichthyostega* (Fig. 166, Volume 1) is a complicated structure and not well suited for comparative purposes. However, as we have seen* practically all the structural details characteristic of the humerus of the Permian stegocephalian *Eryops* are exhibited by the first mesomere of the pectoral fin in *Eusthenopteron*, which although contained in the fin of a fish, was already a typical tetrapod humerus (Figs 103, 104, Volume 1). The endoskeletal elements in the pectoral fin of *Eusthenopteron* following distal to the humerus, too, agree surprisingly well with the corresponding elements in the eutetrapod hand and we can extend our comparisons to include also the embryonic hand of man (Fig. 81; Jarvik, 1965).

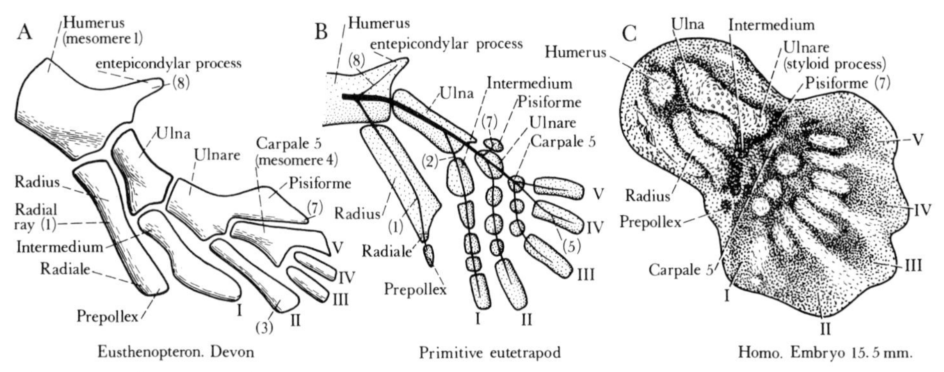

Fig. 81. Osteolepipoda (osteolepiform–eutetrapod stock). Diagrams to illustrate similarities in composition of foreleg and hand (manus) between, A, Devonian osteolepiform fish (*Eusthenopteron*), B, primitive eutetrapod (diagram illustrating Steiner's opinion; from Holmgren, 1952) and C, man (after Schmidt-Ehrenberg, 1942). From Jarvik (1965).

(1)–(8), rays or radials; I, first digit or finger (Pollex, thumb); II, digitus II (Index, forefinger); III, digitus III (Medius, middle finger); IV, digitus IV (Anularis, ring-finger); V, digitus V (Minimus, little finger).

As established by embryological investigations (e.g. Sewertzoff, Steiner, Holmgren) the eutetrapod hand is seven-rayed. An obvious but long overlooked fact is that the pectoral fin in *Eusthenopteron* is seven-rayed as well (Figs 77, 81). The first of these rays in osteolepiforms, the radial ray, is ventrolateral (preaxial) in position and consists of a single rod articulating with the humerus. In eutetrapods it has a corresponding position and extent and arises from a single elongated blastema, which,

* Volume 1, pp. 140–143.

however, later in ontogeny becomes separated into three elements (radius, radiale and prepollex). A characteristic feature of both osteolepiforms and eutetrapods is, moreover, that the remaining six rays radiate from the second mesomere or ulna. The second ray, the intermedian ray, forms the intermedian and the first finger (the thumb). This ray articulates directly with the ulna whereas the following five rays (3–7) radiate from the third mesomere or ulnare. Here again we find that the first of these rays (3) articulates directly with the mesomere (ulnare). This ray forms the second finger, whereas the third, fourth and fifth fingers, formed by rays 4–6, radiate from the fourth mesomere. The identification of the latter (carpale 5 in Fig. 81B) in the eutetrapod hand has led to controversy and several different interpretations have been suggested (Jarvik, 1965). However, one of the reasons to these differences of opinion is obviously that this element in the mammalian hand by Steiner and Schmidt-Ehrenberg has been misinterpreted as the ulnare. As shown by Kindahl and Holmgren (1949; 1952) and confirmed by Slabý (1958; 1967) and Čihák (1972) the ulnare in mammals has fused with the ulna and forms its styloid process (Fig. 81C). The seventh ray in *Eusthenopteron* is represented by the winglike process of the ulnare. In tetrapods it is formed by the pisiforme which develops in connection with the distal end of the ulnare. This is the case also in mammals in which the ulnare as mentioned forms the styloid process of the ulna. In addition to these seven rays there is probably also an eighth ray which forms the entepicondylar process in *Eusthenopteron* and the eutetrapods (medial epicondylus in man).

Pelvic Girdle and Hind Leg

The pelvic girdle (Fig. 82) in *Eusthenopteron* is a triradiate structure as it is in most eutetrapods (Wiedersheim, 1892; Nauck, 1938; Romer, 1970). This is true also of *Ichthyostega* (Fig. 162, Volume 1) if we disregard the posterior iliac process, which is formed by a postsacral rib (Jarvik, 1952). In *Ichthyostega*, too, it is a single ossification which is a primitive feature. Three divisions (pubic, ischiadic and iliac) may be distinguished, and on the external side there is a well developed articular fossa, the acetabulum, for the femur. Provided the orientation of the bone accepted above* is correct the iliac portion in *Eusthenopteron* is small which is in agreement with the ontogenetic development of the pelvis (Wiedersheim, 1892). During the evolution from the fish to a tetrapod like *Ichthyostega* the pelvic girdle has (like the pectoral girdle) grown downwards to meet its antimere in a median symphysis, whereas the iliac portion has grown upwards to gain contact with a sacral rib (Fig. 170, Volume 1). In connection with the ventral growth of the bone the obturator groove on the inner side (Fig. 106, Volume 1; cf. the sulcus obturatorius or obturator groove in man) has been transformed into an obturator canal. However, we can also compare the pelvic bone in *Eusthenopteron* directly with the pelvic girdle in a mammal like *Talpa* (Fig. 82) in which the symphysis is formed between the distal ends of the pubic portions, as was probably the case also in *Eusthenopteron* (Fig. 106, Volume 1). The subdivision of the pelvic bone (os coxae) into three bones meeting in the acetabulum, which is characteristic of the post-Devonian eutetrapods, is a secondary condition.

* Volume 1, p. 144.

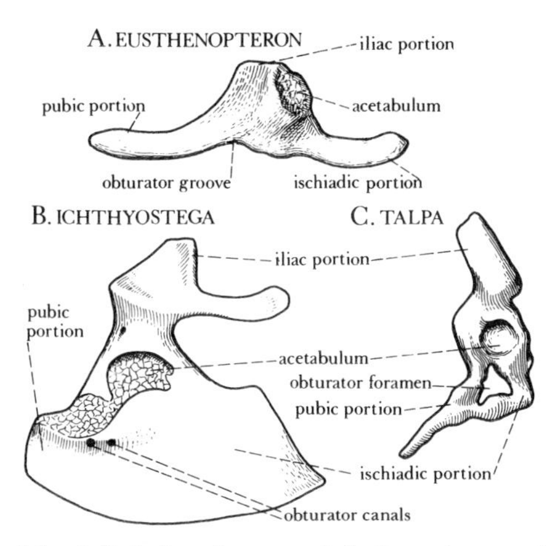

Fig. 82. Osteolepipoda. Left pelvic girdle in lateral aspect of, A, Devonian osteolepiform fish (*Eusthenopteron*), B, Devonian eutetrapod (*Ichthyostega*), and, C, embryonic extant mammal (*Talpa*, embryo, 13 mm. From Nauck, 1931).

Like the pectoral fin, the pelvic fin in *Eusthenopteron* and other osteolepiforms (Jarvik, 1948) is so situated that its preaxial (tibial) margin is ventrolateral in position (Fig. 79; Figs 71, 72, 136, 137, Volume 1). The endoskeletal elements (Fig. 83; Fig. 106, Volume 1) are arranged much as in the pectoral fins, but there are several important differences. In the pelvic fin there is thus only three mesomeres. Moreover the first mesomere or femur is much simpler in structure than the humerus and it lacks wing (entepicondyle). Another important difference is that the second mesomere, the fibula, in contrast to the ulna, is provided with a wing.

These differences are reflected in the eutetrapods, in which a smaller number of elements is present in the tarsus than in the carpus (Sewertzoff, 1908); and as we now shall see there is a most remarkable agreement between the pelvic fin in *Eusthenopteron*, the hind leg in *Ichthyostega* and the lower limb in the human embryo (Fig. 83).

The femur in *Eusthenopteron* (Fig. 106, Volume 1) is shorter than that in *Ichthyostega* (Figs 162–164, Volume 1) and lacks canals, but the tibial and fibular condyles are

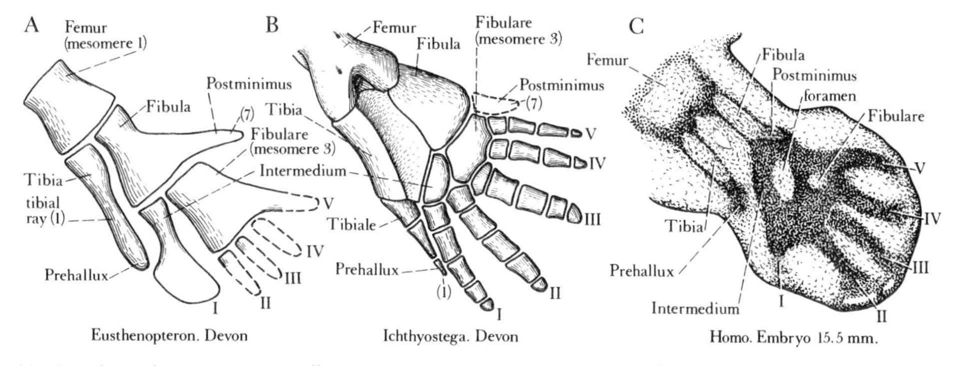

Fig. 83. Osteolepipoda. Diagrams to illustrate similarities in composition of hind leg and foot (pes) between, A, Devonian osteolepiform fish (*Eusthenopteron*), B, Devonian eutetrapod (*Ichthyostega*) and, C, early human embryo. From Jarvik (1965; C, after Schmidt–Ehrenberg, 1942).

I, digitus I (Hallux, great toe): II, III, IV, digits II–IV; V, digitus V (Minimus, little toe).

separated by a groove on the extensor side. On the flexor side there is also in *Eusthenopteron* a distinct ridge accompanied by a groove on the tibial side. The first ray, the tibial ray, is a long rod (*Eusthenopteron*) or a rodlike blastema (*Homo*) which articulates with the femur (Fig. 83). Like the radial ray it is in the adult divided into three or, as in *Ichthyostega*, four (tibia, tibiale, tarsale prehallucis and prehallux) pieces. The second ray, the intermedium ray, forms, like that in the manus, the intermedium and the first digit. The rays 3–6 which form digits II–V all radiate from the third mesomere or fibulare. The seventh ray is represented by the postminimus which is a branch of the second mesomere or fibula. The postminimus which probably was present in *Ichthyostega* and is well shown in the human embryo, must not be confused with the pisiforme in the manus, which is a branch of the third mesomere (ulnare).

Concluding remarks. As we have now seen the fundamental structures of both the girdles and limbs of the eutetrapods arose in their piscine ancestors and the agreement in the structure, in particular of the posterior limb, between *Eusthenopteron* and man (Fig. 83) is most striking indicating relationship. Since a typical tetrapod limb, in practically final stage, arose within the paired paddles of the osteolepiforms it is evident that the transformations at the transition from fish to tetrapod were inconsiderable, and may easily have occurred independently in the various lines of evolution that sprung from the osteolepiforms or related piscine ancestors. In order to understand the paired fins in those fishes which gave rise to tetrapods it would be best to forget about looking for the metapterygial axis both in the sense of Gegenbaur and of the partisans of the fin fold theory. The theory of Gregory and Raven (1941) and Westoll (1943a; Fig. 65C) that the digits (fingers and toes) are new formations must be rejected, as also Holmgren's ideas (1952; Fig. 65D) about the fundamental differences between the archepodium (extremity stem) and the neopodium (secondary rays). Morevoer, I think, the idea of a twisting or primary torsion of the humerus (Evans and Krahl, 1945) which arose in connection with the transformation of the pectoral paddle into the tetrapod limb can be abandoned as well.* The rotation of the humerus on its long axis in *Eryops* and other early tetrapods is brought about by the sliding in the screw-shaped glenoid fossa (Miner, 1925). Finally it is to be pointed out that there has been no inturning of the axis (Fig. 65C2) as assumed by Westoll and the view of a peculiar swinging and bending of the paired paddles advanced by Romer (1970, fig. 133) is certainly erroneous. As shown by Gregory and Raven and Westoll this view rests on a misinterpretation of the position of the osteolepiform paddles.

As is well shown in the specimen of *Eusthenopteron* in Fig. 71, Volume 1 and other material of osteolepiforms (Jarvik, 1948) the pectoral fin of the osteolepiforms, when in trailing position, was so situated that its radial ray was ventrolateral, its flexor side ventromedial and its extensor side dorsolateral. The pectoral fin was thus situated somewhat as the limb bud in the human embryo (see, e.g. Blechschmidt, 1963) and embryos of other eutetrapods (Fig. 79) and no doubt nerves, vessels and muscles were much as in the embryonic eutetrapod limb (Fig. 80; Sewertzoff, 1908). Restorations of the soft parts on the basis of urodeles (Miner, 1925; Andrews and Westoll, 1970) should be avoided.

* Volume 1, p. 140.

If we imagine that the pectoral fin in *Eusthenopteron* was spread out and lowered, the flexor side would touch the ground as it does in the tetrapod limb and by moving the fin slightly forwards and backwards the fish might conceivably have been capable of making short "steps". These movements cannot, however, have been large in *Eusthenopteron* and other osteolepiforms, where the pectoral fin is long-based externally and is anchored to the body by strong basal scutes (Fig. 79; Figs 71, 72, Volume 1). As a first phase towards the transformation into a tetrapod limb we must therefore assume a reduction of the basal scutes. This reduction was certainly accompanied by a reduction of the basal scales, dermal fin rays and posteroventral parts of the cleithrum. Contemporaneous with these retrogressive changes in the exoskeleton, the endoskeletal shoulder girdle and the pectoral fin underwent mainly the following changes. The endoskeletal shoulder girdle developed progressively in the ventral direction forming the coracoid plate, while the glenoid fossa became screw-shaped and, due to flattening of the girdle, assumed a posterior position. These conditions are found in *Ichthyostega*. In more advanced forms the endoskeletal girdle also grew in the dorsal direction forming a scapular blade and this process led to a more or less complete reduction of the cleithrum as well.* The main changes in the pectoral limb (Fig. 84) concern the

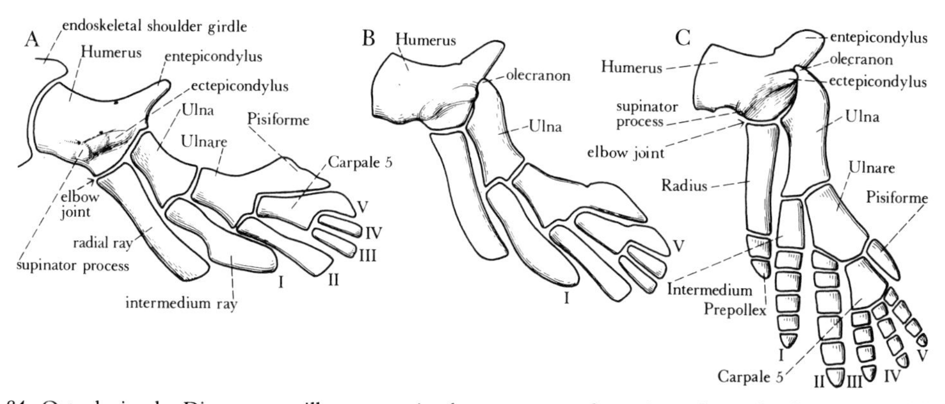

Fig. 84. Osteolepipoda. Diagrams to illustrate main changes at transformation of osteolepiform pectoral fin into eutetrapod limb. A, *Eusthenopteron* stage. B, intermediate stage; C, primitive tetrapod stage. From Jarvik (1964).

development of the elbow joint and the subdivision of the various rays into segments (carpals, phalanges). In the elbow joint, which is indicated already in *Eusthenopteron*, the position of the area of the humerus articulating with the radius has changed a little so as to be directed ventrally. Moreover, in order to facilitate the movements of the forearm, the humerus has become hollowed out in front of the radial condylus. However, perhaps the most important modification is the formation of the olecranon of the ulna which is well developed even in *Ichthyostega*. Regarding the segmentation of the rays it may be sufficient to refer to the well known fact that the digits in early embryonic stages of tetrapods (Figs 65D2, 81C, 83C, 86E) are represented by undivided blastemas obviously corresponding to the undivided rays of the fin in the ancestral fish. In this case we are concerned with a recapitulation of the phyletic development.

* p. 137.

Girdles and Paired Limbs in the Urodelomorpha

A noteworthy fact is that the paired limbs in larvae of primitive urodeles as to their external shape (Fig. 85) are suggestive of the pectoral fins of *Neoceratodus* (Fig. 336, Volume 1), but also of those in the porolepiforms which for other reasons are closely related to the urodeles and together with them form the porolepiform–tetrapod stock or the Urodelomorpha.

The endoskeletal shoulder girdle of a porolepiform (*Glyptolepis*) has been described above* and has been found to be a simple structure (Fig. 200, Volume 1). A remarkable fact is that it is provided with an articular condylus and that there is a corresponding articular fossa on the proximal element of the pectoral fin. The shoulder joint in porolepiforms is accordingly what Wiedersheim (1892) called a fish joint (Fig. 64) a condition which, it could perhaps be imagined, would debar them from ancestry to the urodeles, in which the articular (glenoid) fossa, as in other terrestrial tetrapods, is on the shoulder girdle and the articular condylus on the humerus. However, this would be a premature conclusion. In larval stages of urodeles (Fig. 85E, F) the endoskeletal shoulder girdle and the humerus form a continuous cartilaginous mass from which the articular condyle of the humerus is later tied off but which just as

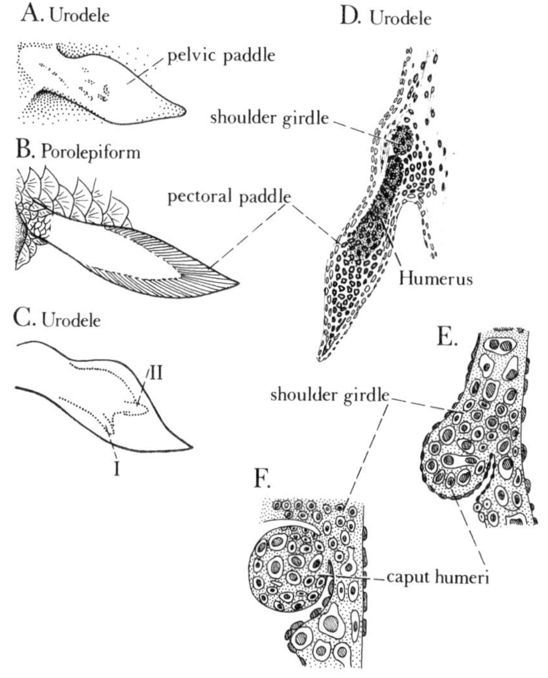

Fig. 85. Paired appendages in Urodelomorpha (porolepiform-urodele stock). A–C, paired paddles in porolepiform (*Holoptychius*) and urodele larvae (*Hynobius retardatus*, from Holmgren, 1933; 1949). Note that digits I and II are the first to appear in urodeles (cf. eutetrapods, Fig. 80 C). D–F, larval *Triturus helveticus*. Three sections to show evolution of shoulder joint. From Wiedersheim (1892).

* Volume 1, p. 269.

easily could develop into a fish joint (Wiedersheim, 1892). The endoskeletal shoulder girdle in urodeles (Engler, 1929) which is a rather simple mainly cartilaginous structure (in contrast to eutetrapods exoskeletal girdle is lacking) may very well be derived from that in porolepiforms. The endoskeleton of the pectoral fin in porolepiforms is still too incompletely known to form a basis for a discussion. Since the rays still are unknown it may very well be similar to that in *Neoceratodus* from which Holmgren derived the urodele limb but it may also be more like that in *Latimeria* (Fig. 64) or represent a type of its own.

It now remains to consider the urodele limbs which according to the embryologists (Sewertzoff, Steiner, Holmgren) differ from the eutetrapod limbs and it is most convenient to turn to the hind leg which, in contrast to the foreleg, carries five digits and is thus a pentadactyl limb.

The first ray in the urodele hind leg (Fig. 86A–C), the tibial ray, resembles that in the eutetrapods and becomes usually divided into four segments (tibia, tibiale, tarsale prehallucis and prehallux). A characteristic feature of the urodele foot (and hand) emphasized already by Zwick (1897) is that the first two digits are the first to appear in ontogeny (Fig. 85), and a most important character easily observed in embryos as well as in adult individuals is that they are carried by a common tarsal element, the tarsale (basale) commune. This character, by which the urodeles differ distinctly from the eutetrapods, is linked with a fact which is expressed as follows by Sewertzoff (1908, p. 165): "Wir kommen also zu dem wichtigen Schlusse, dass das Centrale des Tarsus von *Triton* sich von Anfang an im Zusammenhang mit dem Tarsale commune anlegt und

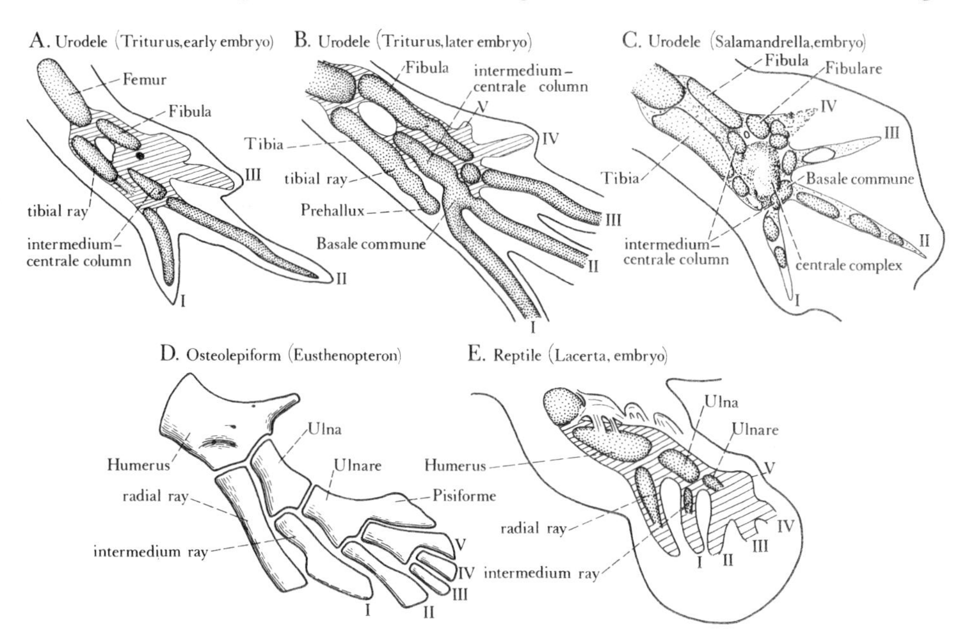

Fig. 86. A–C, Urodelomorpha. Three stages in development of urodele hind leg. A, B, *Triturus cristatus*. From Sewertzoff (1908). C, *Salamandrella*. From Schmalhausen (1910). D, E, Osteolepipoda. Diagrams to illustrate similarities in construction of pectoral limb between Devonian osteolepiform (*Eusthenopteron*) and embryo of reptile (*Lacerta*, from Steiner, 1935).

dessen proximale Fortsetzung bildet". If we add that the "centrale" is continued proximally by the so-called intermedium we can say (Jarvik, 1964; 1965) that the difference between urodeles and eutetrapods in this respect is that the intermedium ray in the former is branched and carries two digits while in the eutetrapods it is unbranched (as in osteolepiforms) and carries only one digit (Fig. 86D, E). In other words, in the urodele foot (and hand) the second and third rays have fused proximal to the digits into a compound intermedium–centrale column.

The intermedium–centrale column in the urodele tarsus (and carpus) includes four elements (Figs 86, 87); the intermedium which is a complex structure and therefore not homologous with the eutetrapod intermedium, the centrale proximale, the centrale distale and the basale commune. However, on the fibular (postaxial) side of the two centrals there are sometimes in the urodele tarsus (but not in the carpus; Holmgren, 1949) one or two accessory or supernumerary centrals (Holmgren; postcentrals, Westoll) which have caused much controversy (see Holmgren, 1952). Since Holmgren's view that they are side-branches formed by budding for principal reasons is unacceptable, I have tentatively suggested (1965) that they belong to the fourth ray which has been pulled towards the intermedium–centrale column in the form of an arch and often partly fuse with the centrals of that column (Fig. 87). The fourth ray,

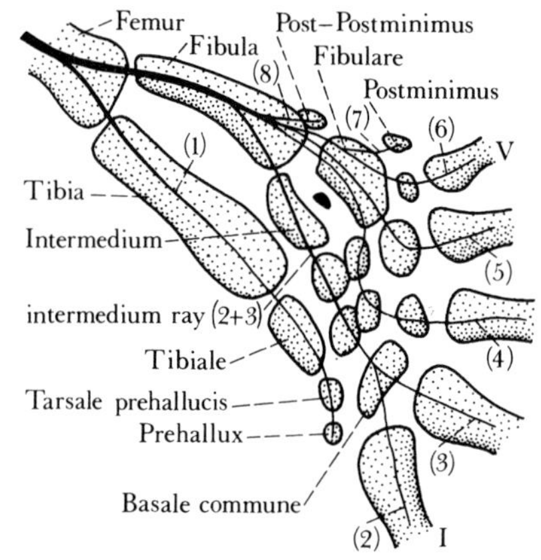

Fig. 87. Interpretation of composition of hind limb in urodele. From Jarvik (1965). (1)–(8), rays 1–8; I, V, digits I and V.

which forms the third digit, bends in its proximal part towards the fibular side and shares in the formation of the fibulare. In general the fibulare also includes segments of the fifth and sixth rays, whose distal parts give rise to the fourth and fifth digits. Also a segment of the seventh ray seems to be included in the fibulare. The distal segment of the seventh ray forms the so-called postminimus which thus, like the pisiforme of the eutetrapod hand, is a branch of the third "mesomere". The so-called post-postminimus (found only in *Cryptobranchus*) which possibly represents the eighth ray in the urodele tarsus, is comparable to the postminimus in the eutetrapod tarsus, which is also a branch of the second "mesomere" (fibula).

On the assumption that the views advanced in this section are correct the differences in the embryonic development of the extremities between urodeles and eutetrapods are in the first place due to a partial fusion of the second and third rays in urodeles. This resulted in a composite intermedium–centrale column which is completed by the basale commune and carries the first and second digits. In the osteolepiforms the second ray (intermedium ray) is free both in the pectoral and pelvic fins from which the anterior and posterior extremities in the eutetrapods have doubtless evolved, while segments of the third ray even in the fish stage (Figs 77C, D, 81, 83) have fused with the future ulnare (fibulare). Hence it follows that the urodele limbs cannot be derived from the basic patterns of the osteolepiform paired fins. Although the opinions stated above differ in important points from Holmgren's theories with regard to the origins of the tetrapod limbs, they still lead to agreement with Holmgren in the same main results; namely, that the urodeles were evolved independently from ancestors among fish other than those from which the eutetrapods have developed and that the tetrapods, in consequence, are of diphyletic origin.

III The Origin of the Tetrapods

Introduction

The lively discussions about the origin of the tetrapods* have resulted in unanimity in one important respect: the progenitors of the tetrapods are to be sought among the Devonian "rhipidistid crossopterygians". However, the "rhipidistids" include two distinct and anatomically very different groups: the Osteolepiformes and the Porolepiformes. In both groups aberrant forms have been described but on the whole the variations are inconsiderable and intermediate forms are unknown. Our knowledge of the internal structure is still incomplete, but we are in the fortunate position that two species, one osteolepiform (*Eusthenopteron foordi*) and one porolepiform (*Glyptolepis groenlandica*), are well known in this regard. Consequently it is upon these two species that we in the first place have to rely; and it is the detailed knowledge of the structure of these two forms gained with the aid of Sollas's grinding method and mechanical preparation of well preserved fossil material that forms the chief palaeozoological basis of the discussions about the origin and evolution of the tetrapods.

As shown in a series of papers since 1942 the osteolepiforms in numerous respects agree with anurans and other eutetrapods (ichthyostegids, most other early tetrapods, reptiles, birds and mammals), whereas the porolepiforms show many of the specializations characteristic of the urodeles. The only reliable conclusion from these data must be that the eutetrapods are descendants of osteolepiforms or osteolepiformlike ancestors, whereas the urodeles have evolved from primitive porolepiforms. This implies that the tetrapods are diphyletic and for practical reasons we may distinguish between osteolepiform–eutetrapod and porolepiform–urodele stocks or, in classification (Fig. 140), between Osteolepipoda and Urodelomorpha.

Such a distinction was necessary in the previous chapter in which it was shown that the urodeles in the ontogenetic development and structure of the paired limbs differ so much from the eutetrapods that an origin from different fish groups was to be postulated. Unfortunately the paired fins in porolepiforms are still too incompletely known to make a comparison with the urodele limbs profitable. As to the eutetrapods, in contrast, it could be shown that not only the fore and hind limbs but also the girdles could be easily derived from the conditions in *Eusthenopteron*. Of particular interest is the remarkable agreement found to exist between the patterns of arrangement both in

* p. 219; Volume 1, pp. 426–428.

the pectoral and pelvic appendages between *Eusthenopteron* and human embryos (Figs 81, 83) a condition which shows that also the mammals are related to *Eusthenopteron* and descendants of osteolepiforms. This relationship is demonstrated also by similarities in the cranial anatomy, but before turning to the head the vertebral column deserves some attention.

8 The Vertebral Column

When, in 1952, I presented the first description of the vertebral column in *Ichthyostega* and established that the vertebrae in this early tetrapod are similar to those in *Eusthenopteron* I had to choose between Gadow's terms used in fishes and the special terminology applied to the vertebral components in the post-Devonian early tetrapods. The two dorsal elements in *Eusthenopteron* and *Ichthyostega* (Fig. 88A, B) could

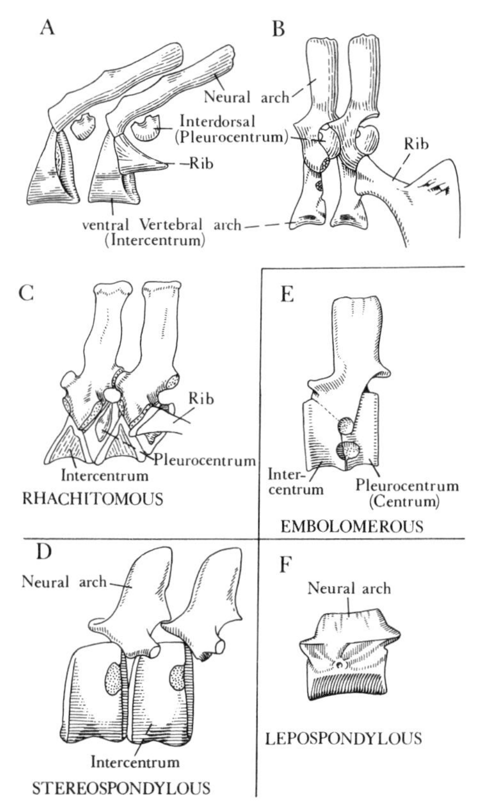

Fig. 88. Types of vertebrae in *Eusthenopteron* and early eutetrapods. A–C, two rhachitomous vertebrae in lateral aspect of, A, *Eusthenopteron*, B, *Ichthyostega* and, C, *Eryops* (C, from Moulton, 1974). D–F, diagrams from Lehman (1955, after Williston and others).

easily be identified as Gadow's basidorsal (neural arch) and interdorsal, and since the large ventral vertebral element in *Eusthenopteron* (and as has been established later, also in *Ichthyostega*) shows a groove for the intermetameric artery I found it reasonable to assume that it is a complex structure including the basi- and interventrals. This compound element was called the ventral vertebral arch.

In the post-Devonian early tetrapods the vertebrae are subject to strong variations and the attempts to interpret the various types have led to lively debates and considerable disagreement. Since the time of Cope and Gaudry (see Nilsson, 1943; Romer, 1947; 1966; 1970; Devillers, 1954; Lehman, 1955; 1959a; Williams, 1959; Panchen, 1977) it has been customary to distinguish four principal types of vertebrae (Fig. 88C–F) and to base the classification of the early tetrapods upon these types. In three of these types (embolomerous, rhachitomous and stereospondylous) the vertebrae are composed of two or most often three separate elements, whereas in the fourth, or lepospondylous, type the neural arch and the body ("centrum") form a single unit. The three elements in the rhachitomous vertebra are termed the neural arch, the pleurocentrum and the intercentrum (or hypocentrum), and most students use these terms also in *Eusthenopteron* and *Ichthyostega*. In *Eryops* (Moulton, 1974) the pleurocentrum may fuse with its antimere forming a horizontal wall between the neural and notochordal canals. This is reminiscent of the conditions in *Eusthenopteron* (Fig. 89; Fig. 97, Volume 1) and it seems likely that the rhachitomous pleurocentrum is homologous with the interdorsal in *Eusthenopteron*. However, according to Romer and others (Williams, Torrey) the pleurocentrum in one line of evolution, leading from the rhachitomes, via the embolomeres to the reptiles, develops progressively in the ventral direction, eventually to form the "centrum" in reptiles. According to this view the interdorsal in *Eusthenopteron*, the pleurocentrum in rhachitomes, and the definitive centrum in amniotes should thus be homologous. We certainly do not solve any problems by applying different names to structures claimed to be homologous, and Romer's view is also for other reasons unacceptable. The embolomeres and other reptilomorphs are no descendants of any stegocephalians classified as rhachitomes (Jarvik, 1967a; 1968a). Moreover, embolomerous vertebrae occur in several unrelated groups (embolomeres, certain microsaurs, Panchen, 1977; and the presumed brachyopid *Tupilakosaurus*, Nielsen, 1955), and the nature of this type of vertebra is still obscure (Schauinsland's suggestion that we are concerned with diplospondyly, as in the tail of *Amia*, has not been refuted). Furthermore the main part (the primary centrum) of the amniote centrum (Fig. 89) is most likely homologous with the ventral vertebral arch in *Eusthenopteron*.* In another presumed evolutionary line embodied in Romer's classification of the Labyrinthodontia (1947) and said to lead from rhachitomes to stereospondyls it should be the intercentum (ventral vertebral arch) that has developed progressively to form the centrum in stereospondyls. This may be true, but it may be noted that most of the Triassic stereospondyls cannot be descendants of any well known Palaeozoic rhachitomes† (Jarvik, 1942).

During the last few decades descriptions of new types of vertebrae (see Panchen, 1977; Holmes and Carroll, 1977) have contributed to obliterate the differences between the various types of vertebrae; and as has become increasingly clear it is not

* p. 84; † p. 223.

possible to base a classification of the early tetrapods on vertebral structures. This is true also of the so-called lepospondyls which because of some similarities in the structure of the vertebrae once were thought to include the ancestors of the urodeles and were classified together with them (e.g. by Dechaseaux, 1955). Our knowledge of the vertebrae in the early tetrapods is still incomplete in many respects and we have to admit that we still cannot interpret safely the various types that have been described. Several of the phyletic lines that have been distinguished are no more than a grouping of the fossils after their geologic age and it is at present impossible to follow the gradual transformation of the vertebrae in the phylogeny of the tetrapods. A more profitable way to an understanding of the vertebrae in extant tetrapods is by means of ontogenetic data and direct comparisons with *Eusthenopteron*.

In 1952 I found it reasonable to adopt Gadow's terminology of the arcualia. However, a few years later Williams (1959) rejected these terms, and moreover he claimed that "resegmentation", that is union of "half-sclerotomes" or "sclerotomites" belonging to two adjoining "segments", is primitive for tetrapods. These views have been partly or wholly accepted by a great number of students (Schaeffer, 1967; Chase, 1965; Romer, 1970; Andrews and Westoll, 1970; Wake, 1970; Y. L. Werner, 1971; Moulton, 1974; Panchen, 1977; Winchester and Bellairs, 1977) and only a few critical voices have been heard (e.g. Lehman, 1968; Verbout, 1976; Borchwarth, 1977). Let us first consider the terms sclerotome ("sclerotomite" is a small sclerotome), segment and "resegmentation".

The somites of the early embryo are usually called segments and it is often spoken of the segmentation of the head. However, since this kind of subdivision or segmentation may easily be confused with the segmentation of the vertebral column I have suggested to use the term metamerism for the subdivision into somites* (Jarvik, 1972). The sclerotome is a part of the wall of the metameric somite which produces skeletogenous mesomesenchyme. This migratory cell material soon assumes a position along the notochord and the neural tube, eventually to give rise to the vertebrae and connecting membranes. These displaced cells derived from the sclerotomes may for some time retain their metameric disposition and form metameric masses or blocks. In analogy with the myomeres† which are metameric derivatives of the myotomes, these blocks are most conveniently termed scleromeres (Schauinsland, 1905; as to the term scleromere see also Sensenig, 1949, p. 36). When vertebrae arose in the phylogeny of the vertbrates‡ (Jarvik, 1972, p. 235) they must for functional reasons alternate with the metameric myomeres. This caused a subdivision of the metameric scleromeres into half-scleromeres (= "half-sclerotomes" or "sclerotomites"), in amniotes separated by a fissure or sclerocoel. Two such half-scleromeres, one caudal of one metamere and one cranial belonging to the metamere following next behind provide the material for the individual diametameric vertebrae (Fig. 39). In this case we are obviously concerned with a primary subdivision or segmentation of the scleromeric material, not a remetamerism which must be the purport of the term "resegmentation". This primary subdivision occurs both in fishes and tetrapods and the fact that the scleromere-halves in amniotes are separated by a more or less distinct sclerocoel is of little interest. Of greater importance is that condensations of

* p. 8; † see p. 56; ‡ see p. 64.

the scleromeric mesenchyme may form skeletal elements, termed (Goodrich, 1930) basals in the caudal (posterior) and interbasals in the cranial half-scleromere of the metamere; and because of the dual function to support the neural tube and the notochord both basals and interbasals may be divided into dorsal and ventral elements. The dorsal elements, the basidorsal and the interdorsal, form the wall of the neural canal but an important fact, well shown in *Amia* (Fig. 13, Volume 1) and *Eusthenopteron* (Fig. 89; Fig. 97, Volume 1) is that both elements, but in particular the interdorsal, extend downwards and form part of the wall of the notochordal canal. Moreover, both dorsal arcual elements in *Eusthenopteron* are provided with medial laminae which together with their antimeres form the horizontal wall which separates the notochordal and neural canals. In the embryo of *Amia* (Fig. 13, Volume 1) this wall is represented by the skeletogenous layer of dense mesenchyme in the corresponding position. This skeletogenous layer of mesenchyme which laterally merges into the cartilaginous dorsal arcualia is found also outside the spinal cord and the notochord and is here, too, continuous with the arcualia (dorsal and ventral) forming extensions of these elements. Also the centrum arises in this layer. In the adult it is retained as a membrane which connects the vertebrae and forms the three longitudinal tubes described above.*

The four paired elements, the basidorsal, the interdorsal, the basiventral and the interventral, that is Gadow's arcualia, may, as in *Amia* and many other fishes, appear as separate skeletal pieces. However, because they arise in the continuous skeletogenous layer fusions and other modifications may easily occur and when we use Gadow's terms also in tetrapods this, in the first place, implies a mapping of the vertebra-forming material. Thus, for instance, the term "interdorsal" informs us that we are concerned with material derived from the dorsal part of the anterior half-scleromere of the metamere (cf. Fig. 39).

The neural arch in tetrapods (Schauinsland, 1905; Goodrich, 1930; Sensenig, 1949; Smit, 1953; Schmalhausen, 1968; Y. L. Werner, 1971; Shute, 1972; Wake and Lawson, 1973; Winchester and Bellairs, 1977) is formed mainly from the dorsal part of the caudal half-scleromere but also material from the anterior half-scleromere of the succeeding metamere is involved. Winchester and Bellairs, who like almost all students since 1959 claim that there is no evidence to support the concept of arcualia, express this well-known fact by stating "that the neural arches are formed mainly from the postsclerotomites, although it is probable that the very narrow portion of the presclerotomite which lies immediately anterior to the nerve ganglion contributes to the posterior border of the arch" (Winchester and Bellairs, 1959, p. 501). I can see no advantage in using this long circumlocution when it could be said in a much simpler and more precise way by stating that the interdorsal has fused with the basidorsal. Obviously the abandonment of Gadow's terms was a little too rash.

Also the so called primary centra in extant reptiles (Fig. 89C) as well as the centra in modern amphibians are complex structures being formed by the ventral parts of caudal and cranial half-scleromeres. Accordingly we can, using Gadow's terminology, say that they include the basiventral and interventral and correspond to the ventral vertebral arch in *Eusthenopteron*. The view that they are derivatives of the

* Volume 1, p. 39.

interdorsal (pleurocentrum) is disproved by the ontogenetic development. An interesting condition in reptiles is, however, that the ventral (proximal) part of the compound neural arch, the "secondary centre" (Y. L. Werner, 1971), separates in the view of several students (Goette, Schauinsland, Werner) form the main part of the arch at the neurocentral suture and fuses with the primary centrum to form the definitive centrum. This view (disputed by Winchester and Bellairs) is supported by the conditions in *Eusthenopteron*. Judging from microphotographs published by Werner, Winchester and Bellairs and others, the "secondary centrum" in reptiles is provided with a medial lamina, which almost meets its antimere in the median line. If this be true it is readily seen that the "secondary centrum" in reptiles includes equivalents to those parts of the basidorsal and interdorsal in *Eusthenopteron* (Fig. 89) which form the horizontal wall between the neural and notochordal canals and which in the skull may be represented by the dorsal arcual plates. We are thus led to the

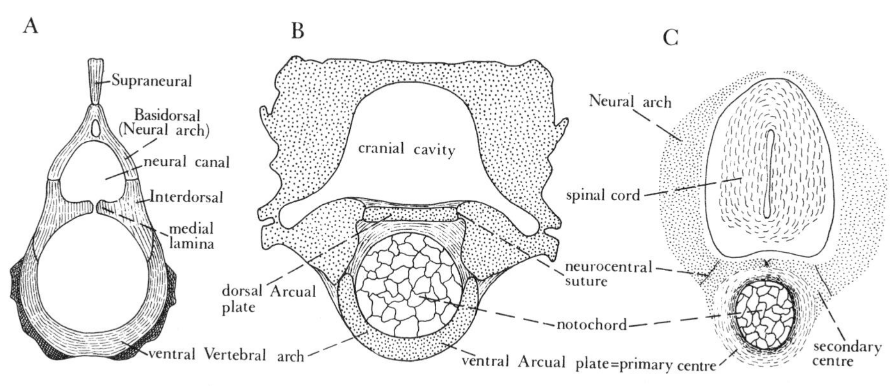

Fig. 89. Interpretation of amniote vertebra after *Eusthenopteron* model. A, B, vertebra in posterior aspect and transverse section of neurocranium of *Eusthenopteron*. C, diagrammatic transverse section of embryonic vertebra of reptile. Based on photographs published by Werner (1971) and Winchester and Bellairs (1977).

conclusion that the "secondary centra" in reptiles at least partly are serially homologous with the dorsal arcual plates in *Eusthenopteron*; and thence it follows that the boundaries between these plates and the cranial pilae (neural arches) are probably neurocentral sutures. Because dorsal arcual plates are present also in porolepiforms, coelacanthiforms and palaeoniscids, and similar structures are known also in other groups as well (e.g. the horizontal wall between the neural and notochordal canals in *Amia*, Fig. 13, Volume 1; the suprachordal cartilages and epichordal commissures in anurans, Schauinsland, 1905; Smit, 1953; Jarvik, 1972; the suprachordal connecting pieces in mammals, Sensenig, 1949) it seems likely that we are concerned with an ancient feature in the construction of the vertebrae.

The main difference between the vertebrae in *Eusthenopteron* and extant reptiles thus seems to be that the dorsal arcual plates ("secondary centra") in *Eusthenopteron* are independent or, as in the vertebral column, form part of the dorsal arcualia, whereas they in extant reptiles and conceivably in amniotes in general have fused with the "primary centra" (ventral vertebral arches) to form the definitive centra. However, in many respects the vertebral column is much more specialized in amniotes and extant tetrapods in general than in *Eusthenopteron*. The interdorsal has fused with the basidorsal to form a complex neural arch, the notochord is much reduced, the spaces between the half-scleromeres are often discernible as narrow sclerocoels, and the intervertebral joints are more elaborated. However, pre- and postzygapophyses are indicated in *Eusthenopteron* (Jarvik, 1952) and tissues corresponding to the intervertebral perichordal thickenings were certainly present as well. The dimetameric intervertebral skeletal elements (chevrons, hypapophyses, Werner, 1971) in amniotes are most likely secondary formations.

Concluding Remarks

The vertebral column is among the structures which may be easily modified (Jarvik, 1960; 1964;) and the variations encountered in the vertebrates are considerable. These variations are made possible by the presence of the skeletogenous layer (skeletoblastische Schicht, Schauinsland). This layer is well shown in *Amia* (Fig. 13, Volume 1) in which it in the adult is represented not only by the cartilaginous and ossified parts (including the centra) of the vertebrae, which arise in this layer, but also of the dense membrane which connects the vertebrae and forms three longitudinal tubes surrounding the supradorsal ligament, the spinal cord and the notochord, respectively. The tube surrounding the notochord is often referred to as the perichordal skeletogenous layer (skeletoblastische Schicht, Schauinsland, 1905; perichorda, Shute, 1972) but the dorsal and intervertebral parts of the skeletogenous layer, well shown in embryos of *Amia* (Fig. 13, Volume 1), are often disregarded. The representations of the vertebral column in the literature, not only in *Amia* but also in other vertebrates are, therefore, often incomplete (cf. the representations of ontogenetic stages of the neurocranium*). When, for example, the roots of the spinal nerves in *Amia* (Fig. 22, Volume 1; Goodrich, 1930; Schaeffer, 1967; Shute, 1972) are shown to pass out dorsal to the interdorsal ("intercalary") this is misleading. In fact the roots emerge through foramina in the dense membrane which is a dorsal extension of, and thus a membranous part of the dorsal arcualia (Fig. 13, Volume 1). Early in vertebrate phylogeny, when cartilaginous or ossified vertebrae arose, this necessitated a subdivision (primary segmentation) of the metameric scleromeres into half-scleromeres and this in turn led to the formation of dimetameric vertebral segments including caudal and cranial half-scleromeres belonging to two adjacent metameres. This regrouping of the metameric material into vertebral segments ("resegmentation") is certainly a phenomenon common to all vertebrates and not characteristic only of tetrapods, as claimed by Williams and others. However, the dispersal of the sclerotomic material may, as, for example, in urodeles (Wake and Lawson, 1973), occur so rapidly that neither scleromeres nor sclerocoels are discernible in ontogeny. In these cases we may speak of an abbreviated ontogenetic development.

In each dimetameric vertebral segment distinct skeletal elements, arcualia, may as in *Amia* and other fishes arise in the skeletogenous layer. The centra which originate in the perichordal part of that layer develop in connection with the bases of the arcualia and are therefore to be regarded as extensions or parts of the arcual elements. The idea

* Volume 1, p. 25, Figs 17–20.

of chordal centra in elasmobranchs (see Goodrich, 1930; Remane, 1936) is according to Shute (1972) due to a false identification of the notochordal sheaths. In his opinion the centra in elasmobranchs are formed in the perichordal layer (perichorda) in fundamentally the same way as in other vertebrates. When, as is usually the case in extant tetrapods, no distinct arcualia are discernible in ontogeny this is probably due to an abbreviated ontogenetic development and therefore of little importance. However, also other skeletal elements than the arcualia may arise in the three longitudinal tubes formed by the skeletogenous layer. Such additional elements are the supradorsals and supraneurals, and most likely also the dimetameric intervertebral chevrons in amniotes belong to this category. In certain cases material belonging to adjacent vertebral segments may be incorporated, as at the formation of the tips of the prezygapophyses in man (Sensenig, 1949). Moreover vertebrae may easily fuse forming, for example, urostyles in the tail or synarcuals in the anterior part of the column, and other modifications may also occur.

The vertebrae no doubt arise in principally the same way in all vertebrates. However, due to specializations there are differences between the various vertebrate groups and as an introduction to the next chapter it may be of interest to quote the following statement from Schmalhausen (1964; 1968, p. 241): "Urodele Amphibia occupy an apparently isolated position among terrestrial vertebrates not only with respect to the origin of the centra, but also in peculiarities of the structure of the transverse processes and the position of the ribs". Although an ardent monophyletist Schmalhausen thus has to admit that the urodeles hold a unique position among the tetrapods as regards the vertebral column, as also in limb structure and many other respects. Wake (1970, p. 33) expresses similar opinions as Schmalhausen stating: "Nothing in the vertebral column of modern amphibians supports the concept of the Lissamphibia, and in fact the vertebral evidence offers no suggestion of relationship of the living amphibian orders".

9 The Middle Ear

Non-mammalian Tetrapods

A remarkable difference between urodeles and eutetrapods is that the tympanic membrane, tympanic cavity and eustachian tube are absent from urodeles. This has caused much speculation about the mode of hearing in urodeles and it has been suggested that sound waves reach the inner ear either from the ground via the lower jaw or the foreleg (Goodrich, 1930; Francis, 1934), or via the venous system and the endolymphatic sac (Schmalhausen, 1968). Another problem is whether the absence of tympanic membrane and cavity is primitive or secondary. In support of the latter view Schmalhausen claims that he has discovered a vestigial tympanic cavity in urodele larvae. However, the dorsal diverticulum described by him is obviously only a remnant of the spiracular gill pouch, and the view that the urodeles have evolved from forms with a tympanic membrane remains unproved.

Because of the absence of the tympanic membrane it is, however, surprising that the urodeles have an ear bone, the columella auris, which as in eutetrapods is provided with a foot-plate in the fenestra ovalis (vestibuli). Moreover, behind the foot-plate there may be a cartilaginous plate, the "operculum" (operculum fenestrae vestibuli, Westoll, 1943b) and a special opercularis muscle, as in anurans. These similarities have been used as arguments for the view that the extant Amphibia (the "Lissamphibia") are monophyletic (see Jarvik, 1965; 1972, p. 209). Let us first consider the opercular plate and the m. opercularis.

The opercular plate which arises in the membrane closing the posterior part of the fenestra ovalis is joined to the dorsal part of the endoskeletal shoulder girdle by the m. opercularis (Kingsbury and Reed, 1908; Reed, 1920; Goodrich, 1930; Barry, 1956; Swanepoel, 1970). However, according to Dunn (1941) and Monath (1965) the m. opercularis is not homologous in urodeles and anurans, and most likely not even in the various urodeles. Since the opercular plate serves as the area of insertion of the opercularis muscle and occurs only when this muscle is present it is reasonable to assume that it is a secondary formation (referred to the myotomic skeleton*). Most likely it has arisen independently as a fixing device for those different parts of the axial musculature from which the m. opercularis has developed (cf. *Amia*,† *Eusthenopteron*‡). This is supported by the fact that the m. opercularis and the opercular plate are

* p. 61; † Volume 1, p. 92; ‡ Volume 1, p. 188.

lacking in hynobiids which are considered to be the most primitive urodeles and which according to Schmalhausen (1968) display the most primitive conditions among the urodeles with regard to the development of the columella auris.

In osteolepiforms and porolepiforms there is a large fenestra ovalis (vestibuli, Jarvik, 1954). This opening may have been closed by cartilage but it is also possible that it was occluded by a membrane and that a portion of the axial musculature extended forwards to be inserted into this membrane (Fig. 90A, B; Fig. 131, Volume 1). This implies that the prerequisites for the formation of the opercular plate and the opercularis muscle were probably present in the piscine ancestors of extant amphibians.

As is well shown in *Eusthenopteron* the anterior part of the fenestra ovalis is occupied by a tonguelike lamina which extends backwards from the margin of the fenestra and which most likely is the posterior part of the infrapharyngohyal (Figs 90, 100A). On the outside of this lamina lies the articular area for the ventral head of the hyomandibula. The ventral head is continued distally by the shaft of the hyomandibula, which in

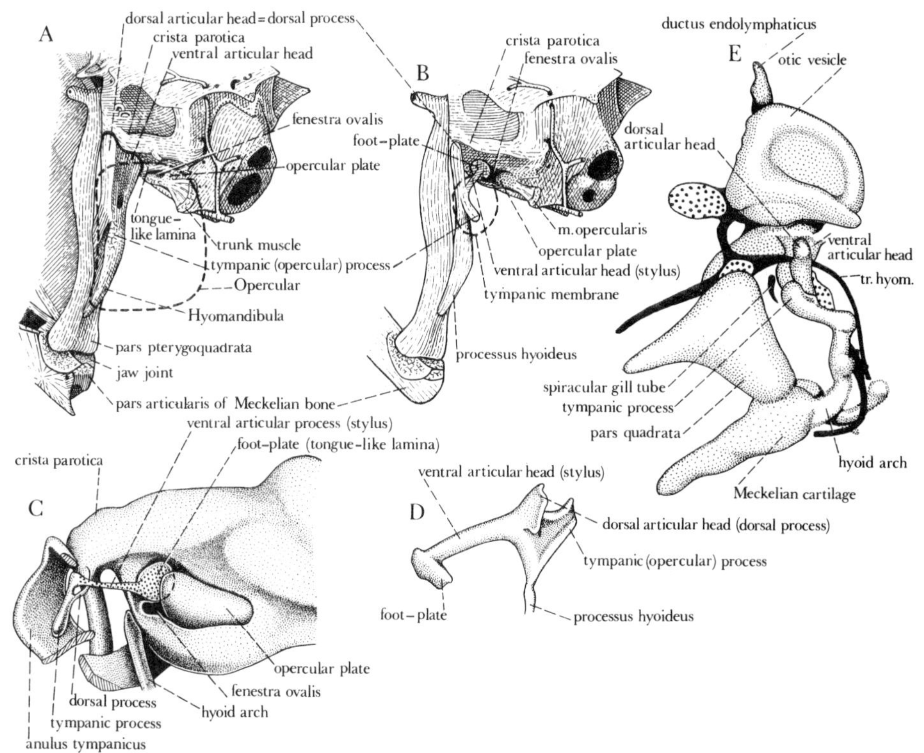

Fig. 90. Interpretations of columella auris in non-mammalian eutetrapods. A, hyomandibula, posterior parts of neurocranium, palatoquadrate, and lower jaw of *Eusthenopteron* in posterolateral view. Position of opercular bone marked with interrupted line. B, same as in A with hyomandibula modified into columella and opercular bone replaced by tympanic membrane. C, columella auris and adjoining parts of neurocranium of anuran (*Rana*) in posterior view. From Stadtmüller (1936, mainly after Gaupp). D, E, columella auris of reptile (*Crocodilus niloticus*). From Frank and Smit (1974). D, columella of embryo, stage 5, in posterior aspect. E, mandibular and hyoid arches of embryo, stage 4, in lateral aspect. Note connection between hyoid arch and Meckelian cartilage of mandibular arch (cf. mammals, Fig. 98).

its middle part is provided with a process, the opercular process, received by a pit on the inner side of the opercular bone. If we now assume that the opercular bone becomes reduced and replaced by a tympanic membrane sound waves from this membrane will be conducted to the inner ear via the processus opercularis, the proximal part of the shaft, and the ventral articular head of the hyomandibula. If we further assume that the ventral articular head fuses with the tonguelike lamina in the fenestra ovalis it is readily seen that a structure will arise suggestive of a columella auris of the type found in anurans, stegocephalians and reptiles (Fig. 90; Versluys, 1898; 1936; Goodrich, 1930; Stadtmüller, 1936; de Beer, 1937; Westoll, 1943b; C. F. Werner, 1960; Olson, 1966; Swanepoel, 1970; Frank and Smit, 1974; 1976; Shishkin, 1975; Lombard and Bolt, 1979*). The tonguelike lamina will then form the foot-plate which either retains its connection with the main part of the infrapharyngohyal (anurans) or is freed from the margin of the fenestra ovalis (amniotes). The ventral articular head and the part of the epihyal portion of the hyomandibula following distal to it will give rise to the main part (shaft, stylus) of the columella, whereas the laterohyal component will form both the dorsal process which is a derivative of the dorsal articular head of the osteolepiform hyomandibula, and the tympanic process which is a slightly modified opercular process. In *Eusthenopteron* the dorsal head of the hyomandibula articulates with the suprapharyngohyal (lateral commissure). In anurans the suprapharyngohyal probably shares in the formation of the otic process† and the dorsal process of the columella has fused with the crista parotica. In reptiles, in contrast, there may be an independent element between the dorsal process and the crista parotica. It is possible that this element, the intercalary of Versluys, represents the suprapharyngohyal.‡

In porolepiforms the hyomandibula in several regards reminds of that of osteolepiforms (Jarvik, 1972) and it is therefore not surprising that the columella in primitive urodeles (hynobiids, Schmalhausen, 1968) includes a foot-plate derived from the posterior part of the infrapharyngohyal and, as in anurans, continuous with the margin of the fenestra ovalis, and a dorsal process formed by the dorsal articular head of the hyomandibula (Fig. 91). Tympanic membrane is lacking in urodeles and accordingly there is no tympanic process. The presence of a pit on the inner side of the opercular bone indicates that the hyomandibula in porolepiforms, as in many other

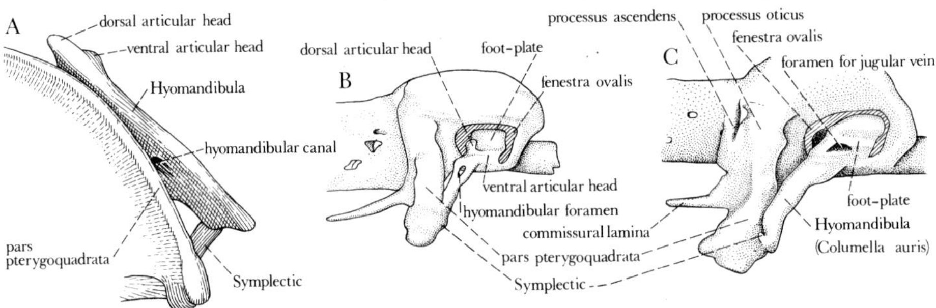

Fig. 91. Hyomandibula and columella auris in Urodelomorpha (porolepiform–urodele stock). A, *Glyptolepis groenlandica*. Posterior part of palatoquadrate with hyomandibula and symplectic in lateral aspect. From Jarvik (1972). B, C, *Hynobius keyserlingii*. Posterior part of chondrocranium of larvae, 27 and 34 mm, in lateral aspect. From Schmalhausen (1968).

* see p. 172 for mammals; † p. 83; ‡ cf. mammals, p. 171.

teleostomes (*Amia, Eusthenopteron*, etc.), is provided with an opercular process (Jarvik, 1972). This implies that the prerequisites for the formation of a tympanic membrane were present in porolepiforms as well. It is, however, possible that it is the special construction of the hyoid arch and the hyoid gill cover in the area behind the jaw joint in porolepiforms that has paved the way for the mode of hearing in urodeles. An important fact is that the ventral end of the hyomandibula in porolepiforms and urodeles, in contrast to osteolepiforms and eutetrapods, is joined to the pars quadrata by a sympletic (Jarvik, 1972). In porolepiforms this element, situated inside the preoperculo-submandibular (Fig. 91; Fig. 186, Volume 1), is independent as it is in early larvae of primitive urodeles (hynobiids, Schmalhausen, 1968). In adult hynobiids the symplectic has fused both with the pars quadrata and the ventral end of the columella (hyomandibula). In this way a route for the sound waves, from the ground via the lower jaw, the pars quadrata, the symplectic and the columella with the foot-plate, to the inner ear has been created. It may very well be so, that a tympanic membrane never arose in the porolepiform-urodele stock. Be this as it may, it is evident that the construction of the columella and in particular the presence of the symplectic furnishes us with another strong proof of the views that urodeles are closely related to porolepiforms and that the Amphibia are diphyletic.

In advanced urodeles the columella has usually been more or less reduced (Kingsbury and Reed, 1908) and also in eutetrapods reductions have occurred (Barry, 1963). However, in many extant forms, in particular in reptiles, the columella may be an elaborate structure, a condition which has resulted in considerable terminological confusion in the literature (Frank and Smit, 1974) and has worried those students (e.g. van der Klaauw, 1923) who think that the rather simple stapes in mammals is to be derived from the reptilian columella. However, as will be explained in the next part the mammalian ear ossicles may be interpreted from the conditions in *Eusthenopteron* without considering the specializations in the reptiles.

The Mammalian Ear Ossicles: A New Theory

Few problems in vertebrate anatomy have been more thoroughly penetrated than the origin of the three ear ossicles (malleus, incus, stapes) and of the squamoso-dentary jaw joint in mammals. The solutions to these problems advanced long ago (1837) by C. Reichert and further elaborated by E. Gaupp are well known even to laymen and are now generally accepted (Gaupp, 1913; Goodrich, 1930; Stadtmüller, 1936a; C. F. Werner, 1960; Devillers, 1961; Frick and Starck, 1963; Hopson, 1966; Starck, 1967; F. Müller, 1968; 1969; Lehman, 1973; Allin, 1975).

This theory, known as the Reichert-Gaupp theory, rests in the first place on the fact that the malleus in mammalian embryos is continuous with the Meckelian cartilage (Fig. 92). Moreover, it is clear that the original jaw joint between two endoskeletal elements of the mandibular arch, namely the quadrate portion of the pterygoquadrate and the articular portion of the ceratomandibular (Meckel's cartilage or bone), has at least partly been replaced by a new joint formed by two dermal bones, the squamosal and the dentary. Another fact is that the mammalian stapes, like the columella auris in

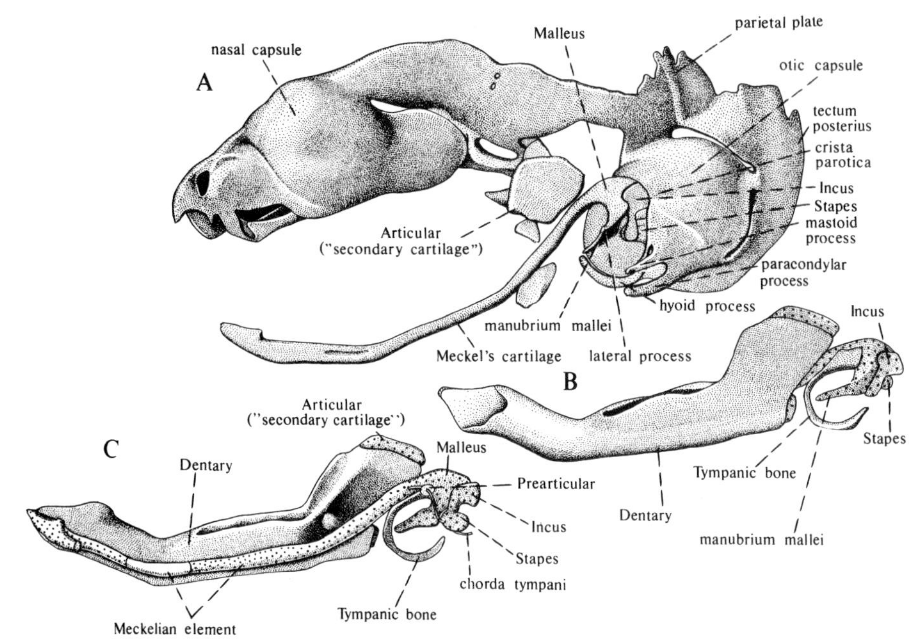

Fig. 92. Lepus cuniculus. A, neurocranium with Meckelian cartilage, parts of hyoid arch and "secondary cartilages" of embryo, 45, mm. Lateral aspect. B, C, lower jaw with primordia of ear ossicles in lateral and medial aspects. From Gaupp (1913).

reptiles, is chiefly a derivative of the piscine hyomandibula and belongs to the hyoid arch.

However, in spite of the prevailing unanimity it is evident that the Reichert-Gaupp theory embodies several peculiar statements which to many (see Fuchs, 1931) have been hard to accept. The following main objections may be raised.

(1) An essential part of the theory is that the quadrate and articular portions in the phylogeny of mammals have become loosened and have been "pushed in, so to speak, between the stapes and the tympanic membrane so as to form a chain of three firmly united elements connecting it to the fenestra ovalis" (Goodrich, 1930, p. 469). This postulated migration of the quadrate (incus) and the articular (malleus) into a pre-existing and functioning auditory apparatus can of course not have occurred before the new squamoso-dentary joint has completely overtaken the role of the old quadrate-articular joint, that is in an early mammalian stage of evolution. It is certainly difficult to imagine how such profound transformations could have happened in phylogeny and a weak point in the theory is that the postulated migration cannot be traced in ontogeny.

(2) A well established fact is namely that the squamoso-dentary joint in the ontogeny of mammals arises in front of the presumed quadrate-articular joint (that is the incus-malleus joint). Since it is difficult to understand how two jaw joints, one in front of the other, have functioned in the phylogeny of mammals, Rabl, Fuchs, and others have claimed that the new joint must have arisen outside the old one. This view has gained strong support by the palaeontological studies of mammallike reptiles and

early fossil mammals recently carried out by Crompton (1958; 1972), Kermack and Musset (1958), Romer, (1969; 1970a), and Kermack *et al.* (1973). As shown by these writers several early mammals have retained the old quadrate-articular joint inside the new squamoso-dentary joint (Fig. 93). It has also been shown (Crompton, 1972) that the formation of the squamoso-dentary joint was gradual in phylogeny. In early cynodonts the surangular (= infradentary 4 in *Eusthenopteron*), is in contact with the quadratojugal (as is the case also in *Eusthenopteron*), but soon there arose a contact also between the surangular and the squamosal. This contact developed into a real joint which formed a subsidiary squamoso-surangular joint outside the quadrate-articular joint. Later in phylogeny the dentary grew backwards and in connection with

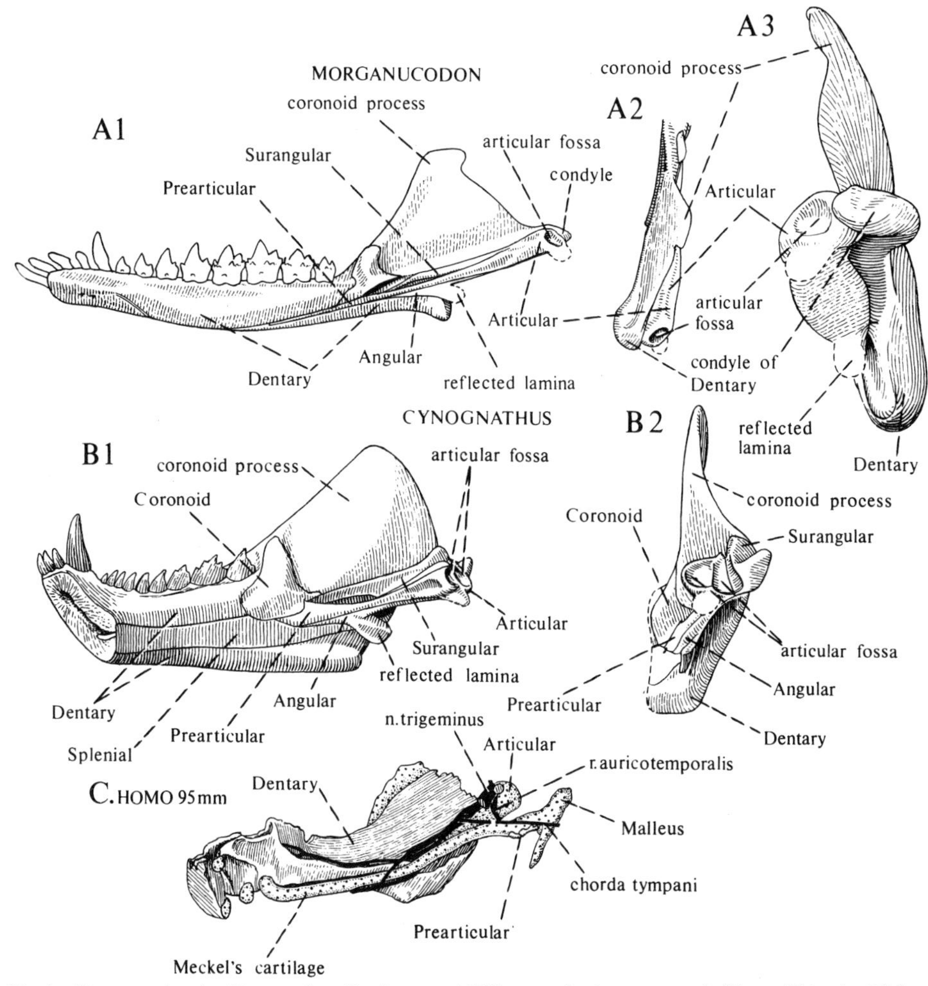

Fig. 93. A, *Morganucodon* (= *Eozostrodon*, Parrington, 1978), non-therian mammal, Upper Triassic, Wales, Great Britain, and Yunnan, China. A 1, right lower jaw in medial aspect. A 2, posterior part of left lower jaw in dorsal aspect. A 3, right lower jaw in posterior aspect. From Kermack *et al.* (1973). B, *Cynognathus*, mammallike reptile, Lower and Middle Triassic, Karroo, South Africa. Right lower jaw in, B 1, medial and, B 2, posterior aspects. From Kermack *et al.* (1973). C, right lower jaw in medial aspect of human embryo. From Low (1909).

reduction of the surangular the subsidiary joint was gradually replaced by the squamoso-dentary joint.

(3) In association with the squamoso-dentary joint several mammals including man (Figs 92, 93C, 94C) present a large mandibular cartilage with a distinct articular condyle and there are also corresponding cartilages in the articular fossa of the squamosal. According to Gaupp (1913) these cartilages are secondary formations which have arisen independently of the endocranium, a view which has been accepted by de Beer (1937), Starck (1967), and other partisans of the Reichert-Gaupp theory. These writers will thus have us to believe that the quadrate and articular first have been loosened and wandered backwards into the tympanic cavity and that then secondary articular cartilages have arisen in about the same place as that previously occupied by the quadrate–articular joint. That this has happened is most unlikely and according to Fuchs these and many other so called secondary cartilages are parts of the original endocranium, a view shared by Reinbach (1952) as regards the pterygoid cartilage. Since Fuchs also was unable to accept the view that the new joint has arisen in front of the old one he (1931) proposed a new theory (the Rabl-Fuchs theory). According to this theory the quadrate and the articular have split longitudinally. The medial parts of these elements have migrated backwards to form the incus and the malleus, whereas the lateral parts have become intimately associated with the squamosal and the dentary and form the so called secondary articular cartilages. This implies that the old jaw joint has been retained in mammals, but that it has been strengthened and more or less completely replaced by dermal bones. So far the theory is acceptable but the opinion that medial parts of the quadrate and the articular form the incus and malleus lacks support.

The view that the new joint arose lateral to the old one has recently been opposed by F. Müller (1968; 1969). On the basis of ontogenetic studies of the development of the jaw joints in mammals she claims that there is a correlation between the formation of the secondary jaw joint, lactation and fusion of lips in ontogeny (cf. Lillegraven, 1976). Since the secondary joint in mammals always arises in front of the incus-malleus joint (which is considered to be the primary quadrate-articular joint) the mammals cannot in her opinion be descendants of fossil forms in which the two joints, as described by Crompton, Kermack and other palaeontologists are found side by side. She therefore concludes that these forms are "rather products of mosaic evolution which lead into a blind alley" (1969, p. 715).

In view of this discrepancy between palaeontological and embryological data and considering the objections discussed above it may be justified to ask if the incus-malleus joint really is the old quadrate-articular joint as Müller and other supporters of the Reichert-Gaupp theory believe. In other words is this theory correct and which is the weight of the arguments in favour of the view that the malleus and the incus belong to the mandibular arch. Let us critically consider the following four arguments used by Gaupp (1913).

(1) The main argument is the fact that the malleus is continuous with the Meckelian cartilage (Fig. 92). The area of the jaw joint where this connection occurs in mammalian embryos is intricate in structure in the gnathostomes. In this area the spiracular gill slit is always closed, nerves and vessels pass from the hyoid to the mandibular arch

or vice versa, and the mandibular and hyoidean ectomesenchymatic streams are partly mixed in early ontogenetic stages.* According to Fuchs (1905) the incus and malleus in early stages of *Lepus* arise in a separate blastema, a condition which if true, must mean that the connection between the malleus and Meckel's cartilage is secondary. Moreover it has been shown (Lindahl, 1948) that the incus in early stages of *Procavia* is in blastemic connection with elements of the hyoid arch. If it is added that the hyoid arch in the crocodile embryo (see, e.g. Goodrich, 1930, fig. 483; Frank and Smit, 1974) is continuous with the Meckelian cartilage (Fig. 90E) it is evident that the statement that the malleus and incus are parts of the mandibular arch is unproved.

(2) The course of the chorda tympani. The discussions of this nerve (see also Goodrich, 1930; R. Fox, 1965; Lombard and Bolt, 1979) have mainly concerned the variations in its course in relation to the tympanic cavity and the processes of the columella auris in the non-mammalian tetrapods; and as far as I can see there is nothing in the course of the nerve that proves that the malleus and incus belong to the mandibular arch.† In the non-mammalian tetrapods the chorda tympani (r. mandibularis internus VII) runs downwards behind the quadrate, whereas in mammals (Figs 95, 98C, E, 99) it passes outside the presumed quadrate (incus) but inside the presumed articular (malleus).

(3) The position of a cutaneous branch of the r. mandibularis trigemini in front of the incus–malleus joint in mammals but behind the squamoso-dentary joint. This branch, the r. auricotemporalis (Figs 93C, 98C, E) runs in man (Gray, 1973) backwards inside the neck of the mandibula to supply the tympanic membrane and the adjoining parts of the skin. Also in other gnathostomes (sharks, Luther, 1909; *Amia, Lepisosteus*, Norris, 1925; urodeles, Francis, 1934; reptiles, Haller, 1934) cutaneous mandibularis branches pass to the area outside and behind the jaw joint and the position of the said branch is rather an argument against the view that the malleus and incus are homologous with the articular and quadrate.

(4) The fact that the m. tensor tympani which is a mandibular muscle is inserted into the malleus (Figs 95, 100B). Since mandibular muscles (innervated by the n. trigeminus) often (e.g. in *Amia*‡) are attached to elements of the hyoid arch, neither this argument proves that the malleus belongs to the mandibular arch.

Because the evidence in favour of the Reichert-Gaupp theory is unconvincing and the theory also in other respects is unsatisfactory let us turn to *Eusthenopteron*. As briefly discussed elsewhere (Jarvik, 1972, pp. 210–212) the structure of the hyoid arch and its relations to the hyoid gill cover in this form suggest new and simpler solutions to most of the problems concerning the ear ossicles and jaw joints in mammals.

In the serial sections of *Eusthenopteron* the part of the hyoid arch which is situated behind the jaw joint and includes the stylohyal and ceratohyal 2 is imperfectly preserved (Jarvik, 1954). However, it is well shown that the dorsal part of ceratohyal 2 has a lateral process (Fig. 96; Figs 109, 110, 123, Volume 1) which fits into a distinct pit on the inner side of the subopercular (Fig. 114A, Volume 1). This pit lies behind a prominent crescent-shaped flange of the subopercular which projects inwards from the anterior margin of the bone, encircles the jaw joint from behind, and separates the hyoid arch from the area of the joint. Moreover ceratohyal 2 is provided with a

* p. 86; † cf. p. 174; ‡ Volume 1, pp. 95–96.

tonguelike process (Fig. 96) which, although imperfectly preserved, most likely abuts against the inner side of the submandibulo-branchiostegal plate (Fig. 97A) in about the same way as the lateral crest of ceratohyal 1 supports submandibular 7.

In extant anurans, reptiles and birds it is, as we have seen,* probably the opercular bone of their osteolepiform ancestors that has been replaced by a tympanic membrane, and this applies certainly to the early tetrapods with a well developed otic notch as well (Fig. 94B). If we now assume that in other descendants of osteolepiforms or osteolepiformlike ancestors it is instead the subopercular or the submandibulo-branchiostegal plate or both these dermal bones that are replaced by a tympanic membrane (Fig. 94A) it is readily seen that conditions recalling those in mammals (Fig. 94C) will arise. The tympanic membrane replacing the said two bones will be situated behind and posteroventral to the jaw joint, that is about as the tympanic membrane in embryonic mammals, and a chain of three bones, (ceratohyal 2, stylohyal and hyomandibula) will connect it with the fenestra ovalis. As is also evident the tonguelike process of ceratohyal 2 is suggestive of the manubrium mallei and the

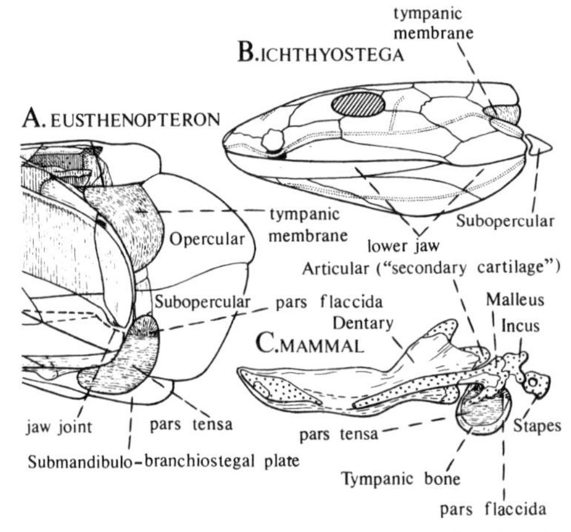

Fig. 94. Derivation of tympanic membrane in non-mammalian eutetrapods (B) and mammals (C) from conditions in *Eusthenopteron* (A).

contact between the tip of this process and the submandibulo-branchiostegal plate may be compared with the attachment of the tip of the manubrium to the inner side of the tympanic membrane at the umbo (Fig. 95B). However, the malleus in mammals has also a lateral process (Figs 95C, D, 96) which is attached to the tympanic membrane at the transition between the pars flaccida and the pars tensa. In *Eusthenopteron* ceratohyal 2 shows a similar lateral process which is received by the pit on the inner side of the subopercular. Moreover the malleus in mammals has an anterior process (processus folianus) strengthened by two small dermal bones, the prearticular ("gonial") and the ossiculum accessorium mallei (de Beer 1937, p. 441). In *Eusthenopteron* the anterior part of ceratohyal 2 lies close to the prearticular and carries a rather strong dental plate (Fig. 97A).

* p. 160.

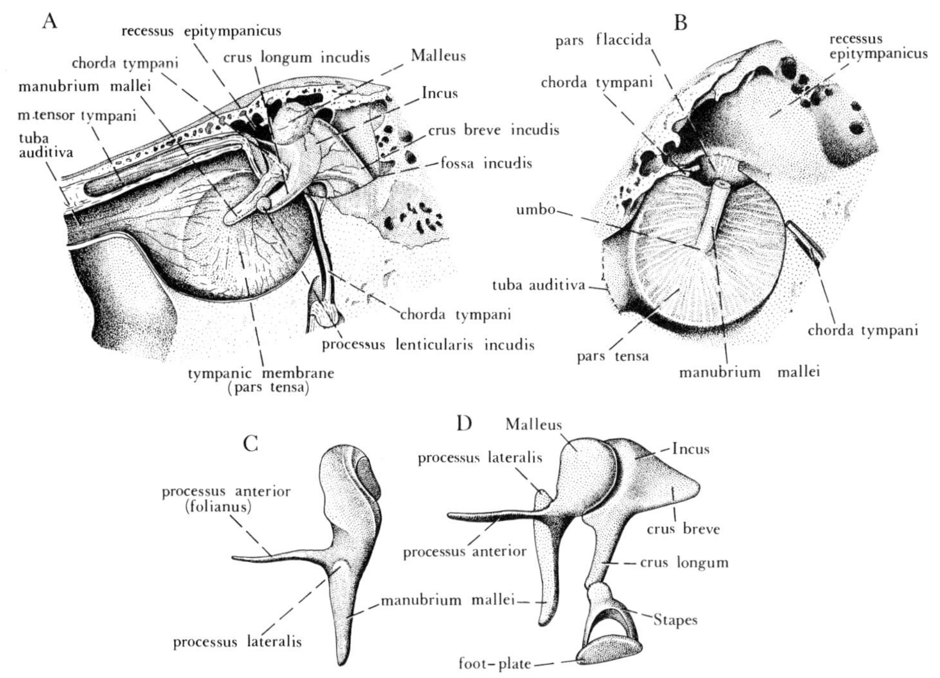

Fig. 95. Homo sapiens, ear ossicles. A, tympanic membrane, ear ossicles and adjoining structures of right side in medial view. B, right tympanic membrane in medial view to show pars flaccida. C, left malleus in lateral view. D, right ear ossicles in medial view. After Sobotta (1922).

In view of these striking similarities it may be justified to suggest (Fig. 96) that the malleus in mammals is homologous with ceratohyal 2 in osteolepiforms and belongs to the hyoid arch. It is then evident that the processes (manubrium mallei and lateral process) characteristic of the malleus and the dermal bones carried by its anterior process were developed already in their osteolepiform ancestors. Thence it follows that the mammalian incus most likely is derived from the osteolepiform stylohyal. Since the malleus–incus joint thus probably is a joint in the hyoid arch it is natural that this joint in the ontogeny of mammals arises behind the jaw joint, and we need not speculate about peculiar migrations of that joint backwards or how two joints, one situated behind the other, have functioned in phylogeny. Nor have we to explain how the malleus and incus were inserted between the stapes and the tympanic membrane. That membrane was, as soon as it arose in the phylogeny of mammals, in contact with the two processes of the malleus and it was formed in another area (behind the jaw joint, Fig. 94) and probably independently of that in anurans, reptiles and birds as claimed by Gaupp (1913).

The posterior articular part of the Meckelian bone in *Eusthenopteron* is besides by the canal for the chorda tympani (see below) pierced by a wide canal which from the bottom of the adductor fossa runs obliquely backwards and downwards to open on the ventral side of the lower jaw, ventral to the jaw joint. This oblique canal (Fig. 97B, C) is directed towards the area of the submandibulo-branchiostegal plate (Fig. 97A), where the main part of the mammalian tympanic membrane presumably was formed.

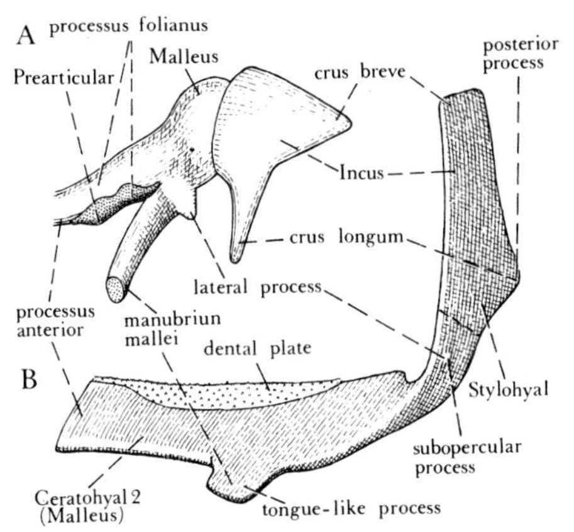

Fig. 96. Suggested homologies in malleus–incus complex between extant mammal and Devonian osteolepiform fish. A, *Homo sapiens*, embryo. Malleus and incus of left side in lateral aspect (partly after Low, 1909). B, *Eusthenopteron foordi*. Ceratohyal 2 and stylohyal of left side in lateral view. Restoration based on wax model after grinding series 2.

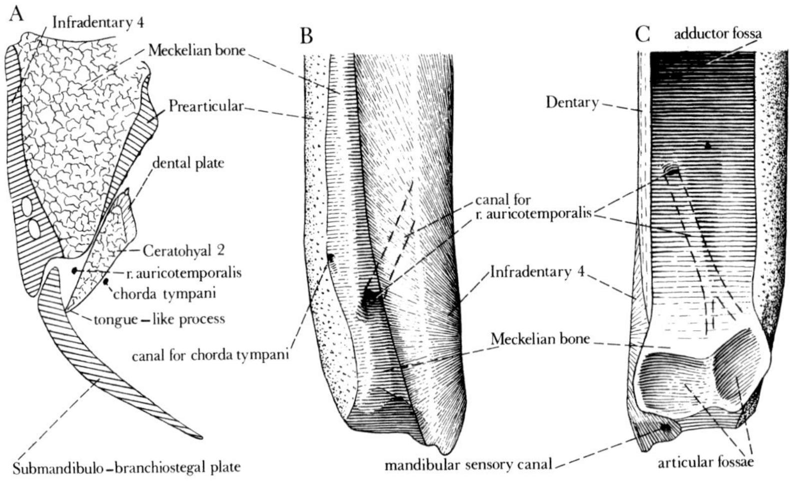

Fig. 97. Eusthenopteron foordi. A, restoration of transverse section of posterior part of lower jaw, ceratohyal 2, and submandibulo-branchiostegal plate. Mainly after section 402 of grinding series 2. B, C, posterior part of lower jaw in ventral and dorsal aspects. Course of canal for r. auricotemporalis indicated by interrupted lines.

It is therefore likely that this canal was developed for a trigeminus branch homologous with the r. auricotemporalis in mammals which passing outside the Meckelian cartilage (Figs 93C, 98C, E, 100), runs backwards to the area behind the jaw joint and supplies the tympanic membrane (Gray, 1973).

According to Fuchs (1905) the connection between the malleus and the Meckelian cartilage is secondary. In *Eusthenopteron* the anterior part of ceratohyal 2 lies close to the inner side of the Meckelian bone (Fig. 97A) and if a fusion occurs in this area conditions similar to those in embryonic mammals will arise. Ceratohyal 2, that is the future malleus, will be continued forwards by the anterior part of the Meckelian bone with the prearticular; and the dental plate of ceratohyal 2 will be situated as the splint of bone, the ossiculum accessorium mallie, associated with the dorsal side of the anterior process of the malleus. If we further assume that the dorsal and ventral walls of the wide canal for the r. auricotemporalis in connection with the formation of the squamoso-dentary joint are reduced (Fig. 98A, B), the articular portion will be separate and will be situated as, and be similar to the mandibular articular cartilage in man (Fig. 98C). Also as in man the r. auricotemporalis will run downwards and backwards between the articular portion and the malleus–Meckelian bar. The most likely conclusion must be that the mammalian mandibular articular cartilage is the piscine articular portion which, as claimed by Fuchs and many others, has been retained in mammals. Thence it follows that the cartilage in the articular fossa of the squamosal is probably a remnant of the quadrate portion of the piscine ancestors. In agreement with the results gained by the studies of the mammallike reptiles this implies that the original quadrate–articular joint, although more or less vestigial, persists inside the squamoso-dentary joint.

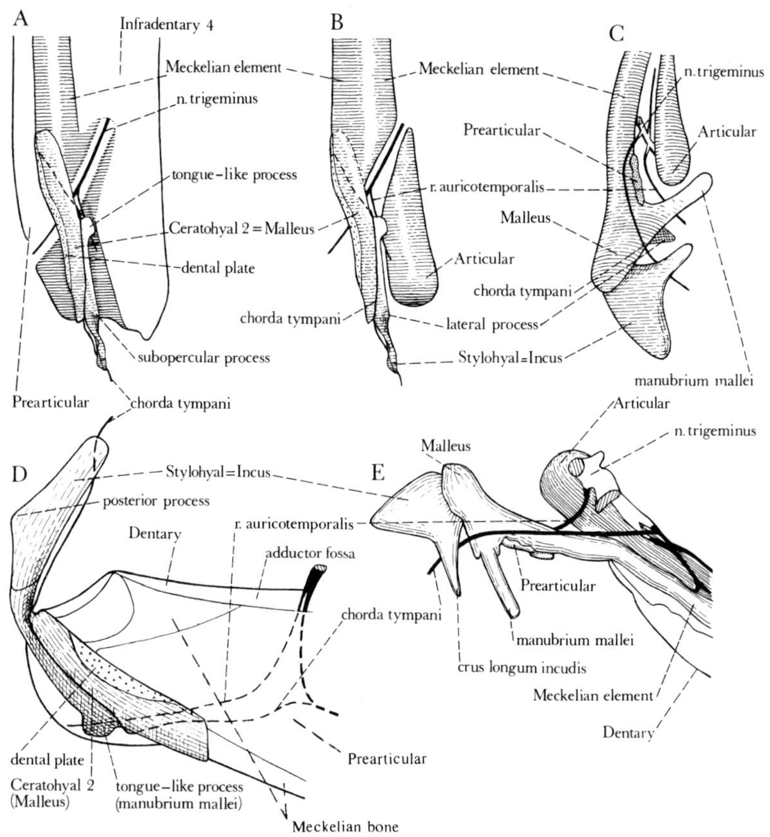

Fig. 98. Diagrams to illustrate derivation of mammalian ear ossicles (C, E) from *Eusthenopteron* (A, B, D). A, posterior part of lower jaw with ceratohyal 2 and stylohyal in ventral aspect. Skeletal parts of lower jaw dorsal and ventral to canal for r. auricotemporalis removed. R. auricotemporalis and chorda tympani restored. B, same as in A, with dermal bones of lower jaw removed to show independent articular portion of Meckelian element. Fusion between ceratohyal 2 and anterior part of Meckelian bone indicated. C, diagram of corresponding parts in mammal (*Homo*). D, left ceratohyal 2 and stylohyal with posterior part of lower jaw in *Eusthenopteron*. Medial view. R. auricotemporalis and chorda tympani restored. E, diagram of corresponding parts in mammal (*Homo*). C, E, after Low (1909).

As we have seen the prerequisites for the formation of the three mammalian ear ossicles were present in *Eusthenopteron* and this is one of the reasons why I long ago (1959; 1960) indicated that the mammals (and the theropsids as a whole) have evolved more directly from their osteolepiform ancestors than was generally assumed (Fig. 141). The mammallike reptiles illustrate how the lower jaw and the quadrate region have been gradually transformed in the phylogeny of mammals. Following a general trend the dermal bones have developed progressively to the cost of the endoskeleton. In connection with the formation of the secondary jaw joint the articular portions of the dentary and squamosal have increased in thickness and have partly replaced the articular and the quadrate situated inside these portions. As explained by Reinbach (1952) with respect to the pterygoid cartilage it is the intimate relations to the dermal bones that has modified the histological structure of the endoskeletal elements in such a way that they have been considered to be secondary formations.

The mammallike reptiles and the early fossil mammals certainly provide important informations regarding the evolution of the jaw joint. However, as to the ear ossicles in mammals, which according to the new theory advanced above all probably belong to the hyoid arch, the immediate mammalian predecessors are of little help. If we want to follow the evolution of these elements and adjoining structures from the conditions in *Eusthenopteron* to mammals we have therefore to rely mainly on the descriptions of the ontogenetic development of the auditory region. We are then confronted with a comprehensive literature (for review see Stadtmüller, 1936a; de Beer, 1937; Starck, 1967; Gasc, 1967) and it is not easy to find out the truth among the many often conflicting interpretations of the various structures in this complex region that have been proposed and the confusing terminology in particular with respect to the dorsal elements of the hyoid arch. However, in spite of these difficulties an attempt will now be made—starting from the conditions in *Eusthenopteron*—to unravel some of the essential features in the evolution of the mammalian ear ossicles and the auditory region as a whole.

The dorsal part of the hyoid arch in *Eusthenopteron* (Fig. 100A) includes three elements: (1) the suprapharyngohyal or lateral commissure, (2) the infrapharyngohyal which forms the otical shelf and the tonguelike lamina in the anterior part of the fenestra ovalis and (3) the hyomandibula. The latter includes two components separated by the hyomandibular canal, one dorsal (lateral) and one ventral (medial). The dorsal component, the laterohyal, is derived from hyoid gill rays and forms the dorsal articular head and the opercular process. The ventral component, the epihyal, comprises the ventral articular head and the distal part of the hyomandibula, which carries the groove for the chorda tympani. Ventral to the hyomandibula follow the stylohyal and ceratohyal 2. It seems likely that the lateral process and the tonguelike process of the latter are formed by hyoid gill rays. Probably also the posterior process of the stylohyal is derived from such rays as well (Fig. 47).

The n. facialis in *Eusthenopteron* (Fig. 100; Fig. 129, Volume 1; cf. Fig. 107, Volume 1) ran, accompanied by the jugular vein, backwards inside the lateral commissure. Reaching the inner side of the proximal part of the hyomandibula it divided into its two main branches, the r. hyoideus and the r. mandibularis internus or chorda tympani. The r. hyoideus continued on the inner side of the hyomandibula in the posterolateral direction, crossed the dorsal margin of the bone and, passing between the fossa for the insertion of the m. protractor hyomandibulare and the opercular process, continued downwards along the hyoid arch. The chorda tympani (Figs 97A, 98, 100A) traversed the hyomandibular canal and continued in the groove on the distal part of the epihyal portion to the external side of the stylohyal. However, farther ventrally, the stylohyal and ceratohyal 2 lie close to the inner side of the subopercular and close behind the crescent-shaped flange of that bone (Fig. 114A, Volume 1), and most likely the chorda tympani continued in the ventral direction in front of that flange and close behind the quadrate. The fact that ceratohyal 2 presents a notch in the proper place indicates that the nerve passing that notch reached the ventromedial side of ceratohyal 2. Thence it most likely continued forwards in the lateral wall of the first post-hyoidean gill slit, inside the base of the tonguelike process (the future manubrium mallei), crossed the ventrolateral margin of ceratohyal 2 and passed ventral to the posterior narrow or closed part of the plica hyomandibularis to enter the canal in the Meckelian bone (Fig. 97B; Fig. 125, Volume 1). Within that bone it gave off an anterior branch and entered the adductor fossa where it presumably anastomosed with a branch of the r. mandibularis trigemini as it does in mammals (Figs 93C, 98E) and other extant gnathostomes.

In mammals—due mainly to the strong development of the brain—the otic capsules with the stapes have been pressed downwards and have, moreover, rotated clockwise backwards and downwards (Werner, 1960). This has caused a compression of the space occupied by the free dorsal elements of the hyoid arch and the gently arched chain in *Eusthenopteron* (Fig. 100A) formed by the future ear ossicles has been bent at angles and the elements have been displaced in relation to each other and to adjoining structures (Fig. 100B). However, in spite of these and other modifications much of the original structural features and anatomical relations have been retained.

The suprapharyngohyal (lateral commissure) of *Eusthenopteron* seems in mammals to be represented by the intercalary of Fuchs (Fig. 99; if this element is identical with the intercalary of Dreyfuss or Broman's laterohyal is doubtful). The suprapharyngohyal (intercalary) which in *Eusthenopteron* (Fig. 100A) leans slightly backwards has in mammals (*Lepus*) assumed an almost horizontal position (Figs 99, 100B) but lies still outside the proximal part of the n. facialis and the jugular vein. Its anterior (originally ventral) end shows a pit which receives the tip of the crus breve incudis and is the future fossa incudis. The dorsal end of the suprapharyngohyal in *Eusthenopteron* carries the articular area for the dorsal articular head of the hyomandibula, that is the proximal part of the laterohyal. In embryonic mammals (Fig. 99; Lindahl, 1948) the proximal end of the equivalent to the osteolepiform hyomandibula is forked much as in *Eusthenopteron* and as in that form the dorsal prong (the dorsal articular head) is in contact with the originally dorsal (in mammals posterior) end of the suprapharyngohyal (intercalary). The latter also in mammals becomes incorporated into the neurocranium but fuses also with the laterohyal component of the hyomandibula which forms a descending rod known as Reichert's cartilage (Figs 99, 100B).

In mammals the distal and middle part of the epihyal portion of the hyomandibula probably form Spence's and Paauw's cartilages (see below). In *Eusthenopteron* the proximal part of the epihyal, i.e. the ventral articular head of the hyomandibula,

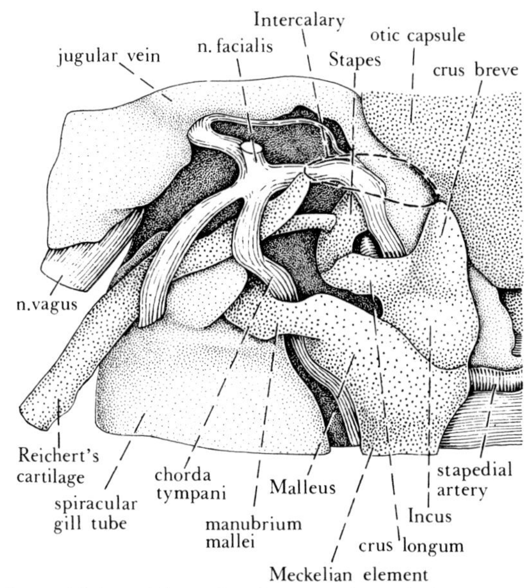

Fig. 99. Lepus cuniculus, embryo, 16 days. Primordia of ear ossicles and adjoining parts. Right side, lateral view. Position of intercalary indicated by interrupted line. From Fuchs (1905).

articulates with the external side of the tonguelike lamina in the fenestra ovalis which is formed by the posterior end of the infrapharyngohyal. In mammals the conditions are similar. The stapes (Figs 95D, 99, 100B) composed of a shaft, usually with two shanks surrounding the foramen for the stapedial artery, and a foot-plate in the fenestra ovalis, is a double formation (Lindahl, 1948). The shaft of the stapes is formed by the ventral prong (head) of the hyomandibula and represents the proximal part of the epihyal. The foot-plate is, as shown by Fuchs (1905) and confirmed by Lindahl, a part of the wall of the otic capsule, a part which, however, is obviously homologous with the tonguelike lamina in *Eusthenopteron* and represents the posterior part of the infrapharyngohyal (Fig. 100A). The fate of the main anterior part of the latter (the otical shelf) is obscure. A possibility is that it is represented by the skeletal piece which in early ontogenetic stages of *Lepus* (Fuchs, 1905) is continuous with the foot-plate; but also the tubotympanicum of Reinbach (1952) may be considered. Another structure is the tegmen tympani (see de Beer, 1937) which, however, perhaps more likely is a derivative of mandibular gill rays (cf. *Eusthenopteron**).

As a consequence of the changes in position and the rotation of the otic capsule the equivalents of the osteolepiform stylohyal and ceratohyal 2, i.e. the incus and the malleus, are in mammals found in front of the derivatives of the hyomandibula (Fig. 100B). The dorsal end of the crus breve incudis derived from the dorsal part of the stylohyal is received by a pit, the fossa inducis, formed at the anterior (originally ventral) end of the suprapharyngohyal (intercalary, Fuchs). The two long posteriorly directed processes of the malleus–incus complex (Figs 95A, 99, 100B), the manubrium mallei and the crus longum incudis, arise independently of the bodies of the malleus and incus (Kingsley, 1900; Fuchs, 1905) and sometimes (*Procavia*, Lindahl, 1948) so

* p. 82, Fig. 47.

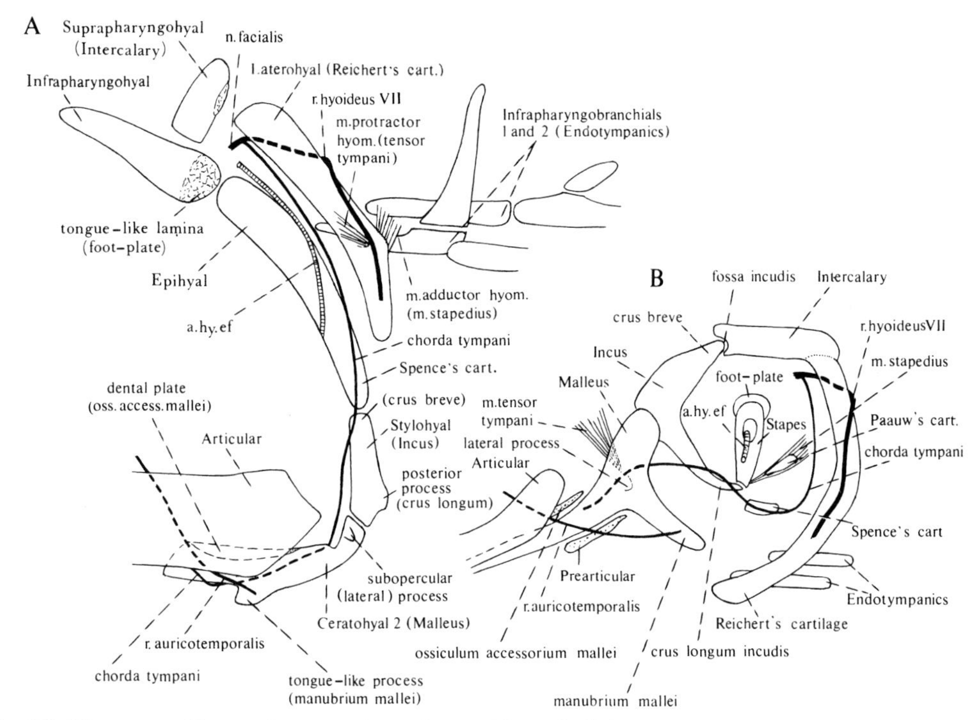

Fig. 100. Diagrams to illustrate interpretations of ear ossicles and adjoining nerves and muscles in mammals (B) on basis of conditions in *Eusthenopteron* (A). Left side, lateral view. Relevant skeletal parts shown independent in A. a.ef.hy, efferent hyoidean artery (proximal part of stapedial artery).

close that they appear to arise from a common blastema. This indicates that they most likely are of the same nature, and their independent origin suggests that they are formed by hyoid gill rays as are probably their presumed equivalents in *Eusthenopteron* (the ceratohyal tonguelike process and the posterior process of the stylohyal). The fact that the blastemic crus longum incudis sometimes (e.g. in *Procavia*) is continuous with the blastema forming the shaft of the stapes supports the view that the malleus–incus complex belongs to the hyoid arch. The fact that the embryonic malleus anteriorly is continuous with the Meckelian cartilage is probably due to secondary fusion.*

The ventral wall of the tympanic cavity which in many mammals forms a prominent bulla tympanica (auditory bulla, van der Klaauw, 1931) may include one or more endoskeletal ossifications (Fig. 101B), the endotympanics (or os bullae), described in several papers by van Kampen and van der Klaauw (1922) and more recently by Reinbach (1952), Frick (1954) and Jurgens (1963). The nature of these elements is still obscure. Frick and others (de Beer, Starck) think that they are secondary formations, whereas Reinbach claims that they belong to the hyoid arch. Another alternative suggested by the conditions in *Eusthenopteron* is that they are derivatives of the infrapharyngobranchials of the first and second branchial arches (Fig. 101A; cf. Jollie 1968, p. 276). It is also possible that the paracondylar process, which according to Reinbach includes material from the hyoid arch, or other processes in this area may be

* p. 168.

derivatives of the suprapharyngobranchials or other dorsal elements of the first and second branchial arches.

The n. facialis in mammals runs together with the jugular vein backwards in the space inside the suprapharyngohyal (intercalary) as it did in *Eusthenopteron* and exactly as in that form the main part of the nerve (the r. hyoideus) curved across the dorsal (lateral) margin of the laterohyal component (Reichert's cartilage) of the hyomandibula (Fig. 100). Also as in *Eusthenopteron* the chorda tympani was given off in the area of the proximal end of the laterohyal and as in *Manis* (van Kampen, 1905) but in contrast to most mammals it did not cross that element on the outside. Because the malleus and incus in mammals are found in front of the former hyomandibula the chorda tympani soon turns forwards (Figs 95, 100B) and enters the tympanic cavity. However, it obviously retained its relations to the distal part of the epihyal portion of the hyomandibula which was dragged forward to form the so called chorda support or Spence's cartilage (Fig. 100B; van der Klaauw, 1923). That this is so is evidenced by the fact that the nerve in some mammals (*Sus*, Bondy, 1907) runs in a groove on the

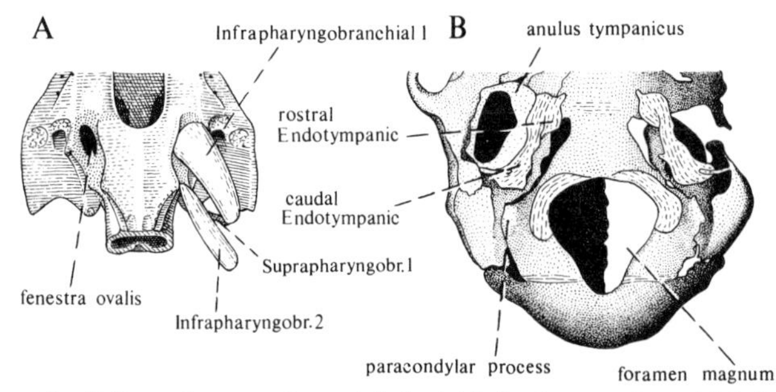

Fig. 101. A, *Eusthenopteron foordi*. Posterior part of otoccipital with infrapharyngobranchials 1 and 2 in ventral view. B, *Rousettus aegyptiacus*. Posterior part of neurocranium in ventral view to show position of endotympanics in mammals. From Jurgens (1963).

dorsolateral side of Spence's cartilage in much the same way as did the chorda tympani (r. mandibularis internus VII) in the groove on the distal part of the hyomandibula in *Eusthenopteron* (Fig. 107, Volume 1). Thence the chorda tympani in mammals (Figs 95A, 100B) passes outside the incus and continues forwards inside the base of the manubrium mallei, usually close dorsal to the insertion area of the m. tensor tympani and about on level with the lateral process of the malleus and the transition between the pars tensa and the pars flaccida of the tympanic membrane. The fact that the chorda tympani in mammals (Fig. 95A) thus runs outside (lateral to) the incus but inside the malleus is of great interest because the nerve in *Eusthenopteron* (Figs 98, 100A) probably had the same relations to the future incus (the stylohyal) and the future malleus (ceratohyal 2); and in *Eusthenopteron*, too, the nerve passed forwards on level with the lateral process of the future malleus.

These similar relations lead us to a brief consideration of the tympanic membrane and the tympanic cavity. If it is true that the tympanic membrane in mammals (as regards mammallike reptiles see Tatarinov, 1968; 1976) has developed in the area of the submandibulo-branchiostegal plate and the subopercular of their osteolepiform ancestors (Fig. 102) it is evident that the main part of the membrane (pars tensa) has arisen in the area of that plate and the main ventral part of the subopercular, whereas the pars flaccida has been formed in the part of the subopercular situated dorsal to the pit on the inside which receives the lateral process of ceratohyal 2. Since the spiracular gill slit in *Eusthenopteron* certainly was closed in the area of the future tympanic membrane and since the chorda tympani, inside the base of the future manubrium mallei, ran forwards in the lateral wall of the first post-hyoidean gill slit, it seems likely that this gill slit has contributed to the tympanic cavity. This is in contrast to the current view that the tympanic cavity is formed from the spiracular gill slit. However, the ontogenetic development of the tympanic cavity in mammals is a complicated process (e.g. Broman, 1899; Fuchs, 1905; van Kampen, 1905; Goodrich, 1930; Shute, 1956;) and without access to a series of models of relevant ontogenetic stages it is difficult to understand what really happens.

The stapedial muscle (Fig. 100) in mammals which is innervated by a branch of the n. facialis, was in *Eusthenopteron* probably represented by the m. adductor hyomandibulae inserted into the ridge on the posterodorsal side of the hyomandibula. This ridge (Fig. 107B, Volume 1) belongs to the laterohyal portion and of great interest is (Edgeworth, 1935) that the insertion of the muscle in the ontogeny of mammals shifts from the laterohyal ("ventrohyal", Edgeworth; "stylohyal", Goodrich) to the distal end of the short shaft of the stapes, formed by the proximal part of the epihyal. Paauw's cartilage (van Klaauw, 1923) found in the tendon of the stapedial muscle in many mammals may be a detached middle portion of the epihyal.

The m. tensor tympani (Figs 95A, 100B) which is innervated by the n. trigeminus and usually is inserted into the malleus, is a derivative of the posterior part of the masticatory muscle plate (m. levator mandibulae posterior, Edgeworth, 1935). In *Eusthenopteron* (Fig. 100; Fig. 129, Volume 1) this muscle was most likely represented by the presumed protractor hyomandibularis which also was innervated by the n. trigeminus and (like in *Polypterus*, Edgeworth) was a derivative of the posterior part of the masticatory muscle plate. This muscle was in *Eusthenopteron* inserted into the laterohyal component of the hyomandibula as is the tensor tympani (or pterygohyoideus) at least in one mammal (*Manis*, Edgeworth). As may be gathered from Fig. 99 the lateral side of the malleus lies close to the original insertion area of the muscle on the laterohyal (Reichert's cartilage). A shift of the insertion from the laterohyal to the malleus may have easily occurred within the tissues of the hyoidean branchial unit along the course of the chorda tympani.

As will be explained in the sequel* the stapedial artery may also be derived from the conditions in *Eusthenopteron*.

* pp. 181–183.

10 Endolymphatic Sac, Occiput and Associated Vessels

Endolymphatic sac

In most vertebrates the endolymphatic duct* extends into the cranial cavity and expands into an endolymphatic sac which sometimes (*Protopterus*, certain elasmobranchs, urodeles, anurans and certain reptiles; Retzius, 1881; 1884; de Burlet, 1934) may attain a large size. The conditions in anurans are of particular interest.

The endolymphatic sac in anurans (Fig. 102; Gaupp, Whiteside, Dempster, Birkmann, and others; see Jarvik, 1975) consists in the adult of two portions: the pars anterior which from the area of the sinus superior extends forwards, and the pars posterior projecting backwards. The pars posterior includes the processus ascendens posterior which meets its antimere dorsal to the medulla, and the pars spinalis. Behind the medulla the pars spinalis joins that of the other side and together the two partes form an externally unpaired spinal portion which dorsal to the spinal cord extends backwards through the neural canal of the vertebral column. Within that canal the spinal portion sends forth transverse diverticles which, accompanied by the dorsal roots of the spinal nerves, pass out from the neural canal through the intervertebral foramina. Outside the vertebral column the transverse diverticles expand forming calcareous sacs which more or less completely surround the spinal ganglia. In the presence of the spinal portion and calcareous sacs the anurans are unique among extant vertebrates (in urodeles, Fig. 102A, the endolymphatic sac also includes two portions, but the pars posterior is always short).

However, a saccus endolymphaticus of anuran type was probably present in *Eusthenopteron* (Fig. 102D) in which it was housed in the supraotic cavity† (Jarvik, 1975). As in anurans two partes may be distinguished: the pars anterior which extended forwards from the area close medial to the sinus superior towards the area of the facialis ganglion, and the pars posterior extending backwards. Close behind the sinus superior the pars posterior obviously met and probably joined its antimere forming a processus ascendens posterior, exactly as in anurans. Behind that commissure the two partes lying close together continued backwards dorsal to the medulla but were separated from that by a bony wall. Farther back they turned ventrally and

* p. 44; † Volume 1, p. 130.

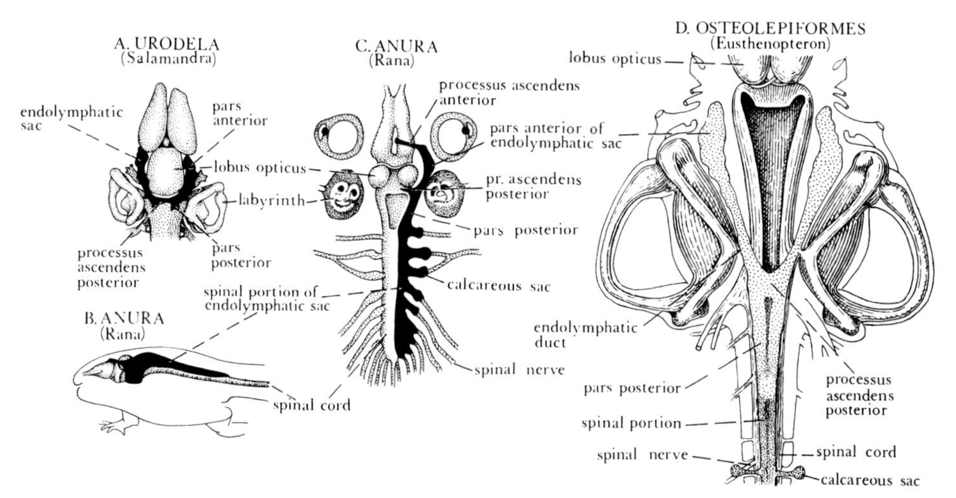

Fig. 102. Endolymphatic sac. A, brain, otic vesicles, and endolymphatic sacs of urodele (*Salamandra atra*) in dorsal view. From Jarvik (1975, after Dempster). B, C, brain, spinal cord, endolymphatic, and (in C only) calcareous sacs of anuran (*Rana temporaria*) in lateral and dorsal views. From Jarvik (1975, after Whiteside), D, restoration of brain, spinal cord, membranous labyrinths and endolymphatic and calcareous sacs of osteolepiform (*Eusthenopteron foordi*) in dorsal aspect.

entered the posterior part of the medullary portion of the cranial cavity. This condition indicates that the endolymphatic sac also in *Eusthenopteron* had a spinal portion extending backwards in the neural canal. How far backwards in the vertebral column this presumed spinal portion reached is impossible to say. Nor is it possible to decide if there were any calcareous sacs in *Eusthenopteron*. However, the openings for the dorsal roots of the spinal nerves are wide, allowing also the passage of transverse processes of the endolymphatic sac. Moreover, there was probably a strong a. vertebralis dorsalis, an artery which is characteristic of frogs and in them supplies the calcareous sacs. Since the special vein draining these structures was probably present, too,* it is reasonable to assume that calcareous sacs were developed, at any rate in the anterior part of the trunk in *Eusthenopteron* (Fig. 131, Volume 1).

Occiput

In the structure of the posterior part of the head there are also other similarities which support the view of a close relationship between osteolepiforms and anurans. Probably due to the flattening of the head in phylogeny the saccus endolymphaticus in anurans lies close above the brain and the skeletal wall which in *Eusthenopteron* separates the posterior division of the supraotic cavity from the cranial cavity has disappeared (cf. Jarvik, 1975, p. 206). Moreover, following a general trend, the dermal bones have usually sunk in to be deeply embedded in soft tissues. However, in some cases (*Pelobates*) they have retained their superficial position, are ornamented, and may form a roof of the fossa bridgei, which then opens backwards much as in *Euthenopteron* (Fig. 103). Also in anurans there are canals or grooves for the "occipital" artery (see

* p. 180.

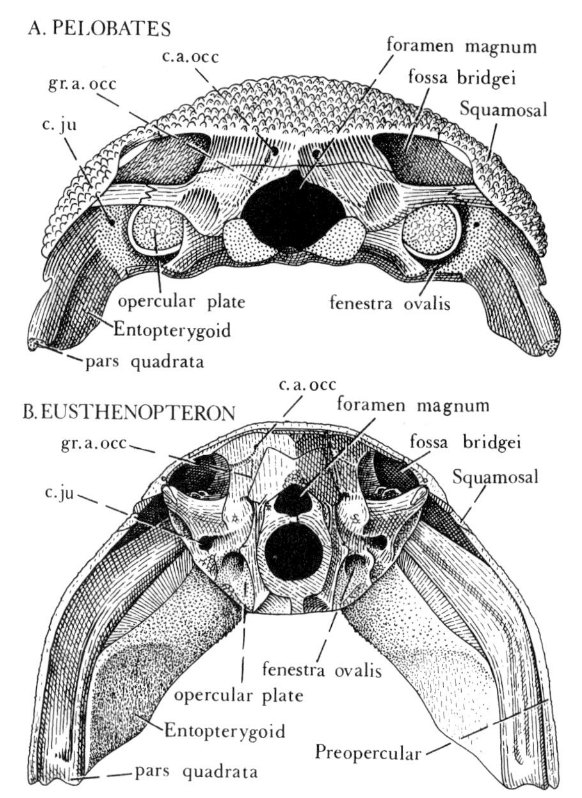

Fig. 103. Skulls of, A, anuran (*Pelobates cultripes*) and, B, osteolepiform (*Eusthenopteron foordi*) in posterior aspects. From Jarvik (1975).
c.a.occ, canal for "occipital" artery; c.ju, canal for jugular vein; gr.a.occ, groove for "occipital" artery.

below) and the tectum posterius may present a median element which corresponds to the supraoccipital plug in *Eusthenopteron* and belongs to the median series of elements of the cranial tecta (Jarvik, 1975, fig. 11). This median tectal element may also in anurans (in *Pelobates* larvae) be situated underneath an independent median extrascapular (Fig. 126B) and show an area of attachment for the supraneural ligament (in the case the dermal bones of the occiput have been thickened this area is found on the exoskeleton, Fig. 103A). Close dorsal to the foramen magnum there is also in anurans an attachment area for the supradorsal ligament (cf. Figs 86C, 97B, 131, Volume 1). The fenestra ovalis in *Eusthenopteron* contains an equivalent to the foot-plate of the columella and possibly there was also an opercular plate. Between these structures and the lateral commissure, which in anurans forms part of the processus oticus (p. 83), runs the jugular vein.

Intermetameric Vessels, Orbital, "Occipital" and Stapedial Arteries in the Osteolepipoda

A characteristic feature of vertebrates is that the dorsal aorta in the trunk on each side gives off a regular series of ascending intermetameric arteries (Fig. 107; Fig. 22, Volume 1). These vessels are often connected by anastomoses which form longitudinal arteries; and if so the ascending parts of one or more of the intermetameric arteries involved are usually reduced. A longitudinal vessel formed in this way is the dorsal vertebral artery in anurans (Figs 104, 107; Gaupp, 1896–1904). This paired artery which is a branch of the occipito-vertebral artery runs backwards outside the vertebral column, ventral to the oblique processes of the vertebrae, but dorsal to the calcareous sacs and the roots and ganglia of the spinal nerves (cf. urodeles;* Fig. 107D). It gives off dorsal and ventral branches to trunk muscles. However, certain twigs of the ventral branches supply the calcareous sacs, whereas other twigs enter the neural canal through the intervertebral foramina to join lateral twigs of the spinal portion of the basilar artery. Although a median vessel the basilar artery has, like the dorsal vertebral artery, arisen by fusion of longitudinal anastomoses.

In view of the distinct metamerism displayed in the head it is not surprising that the system of intermetameric and longitudinal vessels along the vertebral column continues forwards into the head region (Fig. 106A). As just mentioned the basilar artery is such a longitudinal vessel. However, anteriorly the basilar artery divides and merges into the posterior cerebral artery (Fig. 105A). This paired artery which is a dorsal branch of the internal carotid may be tentatively interpreted as a persisting intermetameric artery. The a. ophthamica magna is possibly also such a cranial intermetameric vessel (Bjerring, 1977), and probably the orbital artery belongs to this category as well.

The orbital artery is of particular interest. In *Amia* (Fig. 105A; Figs 61, 64, Volume 1; Bjerring, 1977) this artery, which arises late in ontogeny,† emerges from the lateral dorsal aorta at some distance in front of the origin of the efferent hyoid artery, and divides into three characteristic branches; supraorbital, infraorbital and mandibular. There are, however, considerable variations as to the orbital artery in vertebrates. In a 22 mm embryo of *Amia*, for example, the supraorbital branch on one side of the specimen emerges from the efferent hyoid artery. Of more interest, however, is that the three said branches in certain mammals (Fig. 106E) form the terminal part of the stapedial artery, whereas they in anurans (Figs 104A, 107A) are branches of the so called occipital artery. An attempt will now be made to explain these remarkable differences with the aid of the conditions in *Eusthenopteron*.

As evidenced by distinct grooves on the otoccipital an arteria occipito-vertebralis emerging from the lateral dorsal aorta was present also in *Eusthenopteron* (Figs 104, 107; Fig. 131, Volume 1). Exactly as in frogs this artery, which may be regarded as an enlarged intermetameric artery, divided dorsally into a posterior a. vertebralis dorsalis and an anterior so-called occipital artery. The existence of the dorsal vertebral artery is further supported by the fact that grooves for intermetameric arteries are lacking on the foremost vertebrae (Fig. 97A, Volume 1), these arteries being reduced in connection with the formation of the longitudinal vessel. Also it seems likely that ventral twigs of the dorsal vertebral artery in *Eusthenopteron*, too, supplied the spinal portion of the endolymphatic sac and, if present, the calcareous sacs. In this connection may be

* p. 184; † Volume 1, p. 89.

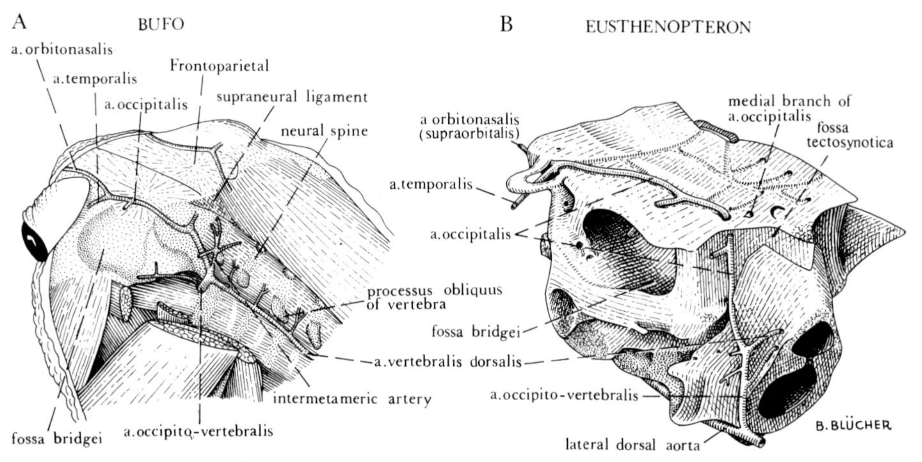

Fig. 104. A, *Bufo bufo.* Sketch of anterior part of body, dissected to show course of branches of occipito-vertebral artery. Posterodorsolateral view. B, *Eusthenopteron foordi.* Otoccipital in posterodorsolateral view. Occipito-vertebral artery and its main branches restored. From Jarvik (1975).

mentioned that these sacs in anurans are drained by a special vein, the v. vertebralis interna dorsalis (Gaupp). This vein anteriorly divides into a paired v. occipitalis vertebralis. The existence of an ascending paired canal (Fig. 87A, Volume 1) leading from the cranial cavity to the supraotic cavity indicates that a corresponding vein was present in *Eusthenopteron* also (Jarvik, 1975).

As in frogs the so called occipital artery in *Eusthenopteron* passed upwards in a groove on the posterior face of the otic region and after passing a short endoskeletal canal similar to that in *calyptocephalus* (Reinbach, 1939; Jarvik, 1975, fig. 8D) it reached the dorsal side of the otic region and ran forwards towards the orbit, exactly as it does in anurans. At the entrance into the orbit it gave off the supraorbital artery (r. orbitonasalis, Gaupp). Thence the remaining part of the artery (a. temporalis, Gaupp) obviously traversed the trigeminus notch in the palatoquadrate eventually to divide into infraorbital (maxillaris) and mandibular arteries (Fig. 131, Volume 1). In *Eusthenopteron* (Fig. 105), consequently, the supraorbital, infraorbital and mandibular arteries, which in *Amia* are the principal branches of the orbital artery, belonged to the so-called occipital artery, as they do in anurans. However, in some specimens of *Eusthenopteron* (and in *Rhizodopsis*, Säve-Söderbergh, 1936) the otical shelf (infrapharyngohyal) is pierced close to the palatine foramen by a vertical canal (Figs 92, 93, Volume 1) and anterodorsal to that another short canal (Figs 86, 88, Volume 1) runs from the orbit to the anterolateral end of the groove for the "occipital" artery. Judging from the position and course of these two canals it seems likely that they transmitted an ascending vessel (Fig. 78, Volume 1) which connected the carotis interna with the "occipital" artery. This presumed ascending vessel was situated much as, and is probably homologous with, the proximal part of the orbital artery in *Amia* which also pierces the otical shelf. If this is true a typical orbital artery with its three branches was present in *Eusthenopteron* (Figs 105B, 107B) although it has fused with the "occipital" artery. In order to come to the conditions in anurans we have only to assume that the proximal part of the orbital artery has been reduced.

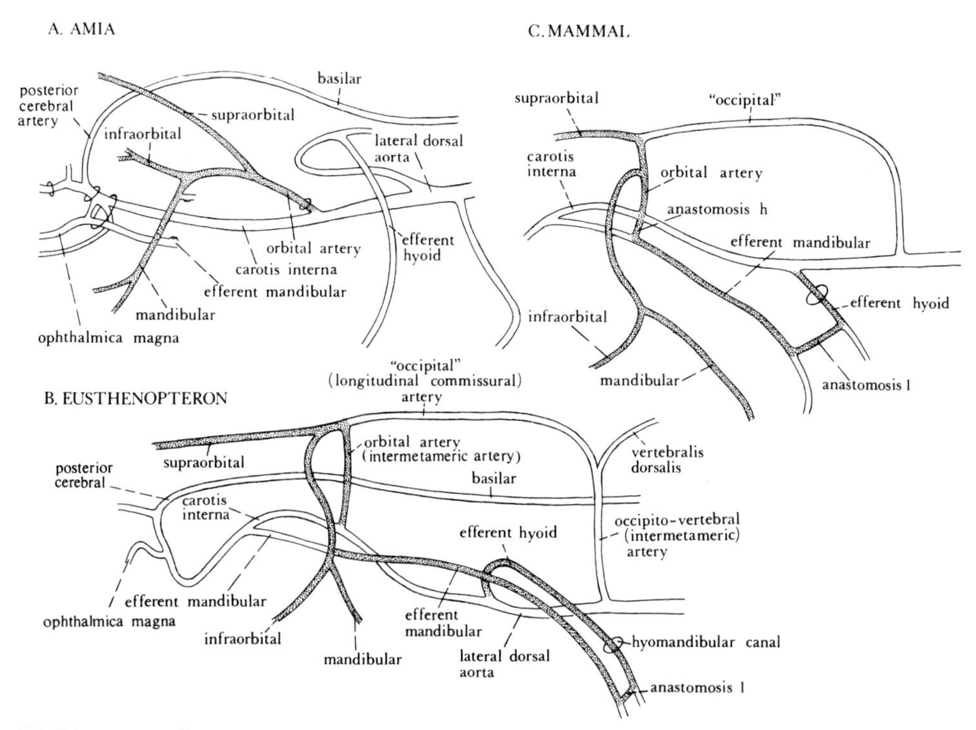

Fig. 105. Diagrams to illustrate composition of stapedial artery (dotted) in mammals. A, *Amia calva*. Orbital artery with its three principal branches (supraorbital, infraorbital, and mandibular), efferent mandibular and hyoid arteries, and certain other vessels (cf. Fig. 61, Volume 1). B, *Eusthenopteron foordi*. C, mammal.

As suggested above* the proximal part of the orbital artery is probably formed by an intermetameric artery. Since the "occipital" artery connects this intermetameric artery with another intermetameric vessel, namely the so-called occipito-vertebral artery, it is readily seen that the supraotic "occipital" artery in anurans and *Eusthenopteron* (Fig. 105B) most likely is a longitudinal anastomosis formed in the same way as the dorsal vertebral artery which is a posterior continuation of the "occipital" artery. Because the use of the term "occipital artery" most conveniently is to be restricted to the ascending intermetameric vessels in the neck region (e.g. the occipito-vertebral artery in anurans) the application of this term to the longitudinal, supraotic commissural vessel in anurans and *Eusthenopteron* is inappropriate.

Let us now turn to the stapedial artery in mammals which has been regarded as an orbital artery (e.g. by Goodrich, 1930). As evidenced by studies of embryos of *Mus* (Fig. 106) this artery in mammals, when complete, is a complex structure composed of four portions (Tandler, 1902; see also Lindahl and Lundberg, 1946; Padget, 1948 Bugge, 1970): (1) a proximal portion which pierces the stapes and is formed by the proximal part of the efferent hyoid artery; (2) an anastomosis between the efferent hyoid and mandibular arteries; (3) the proximal part of the efferent mandibular artery; and (4) a distal portion which carries supraorbital, infraorbital and mandibular

* p. 179.

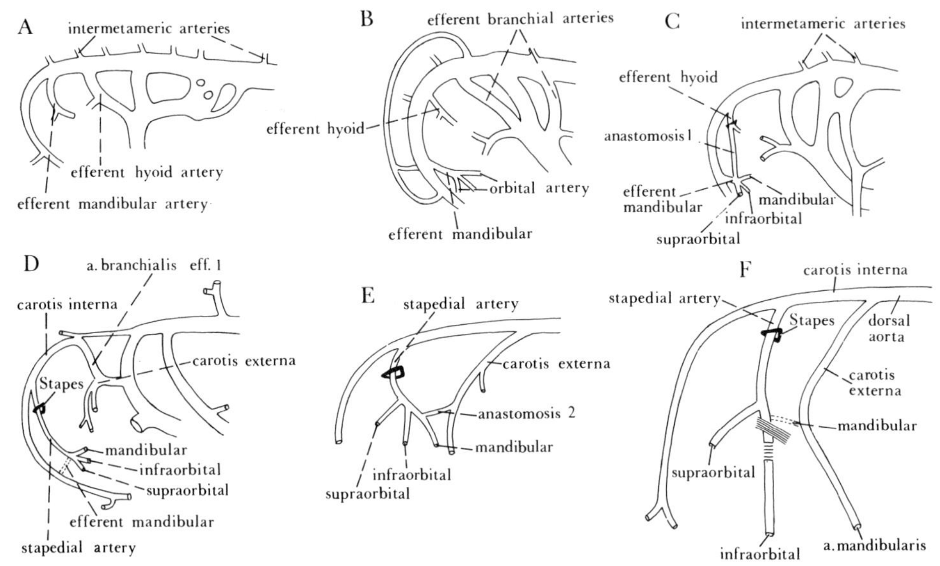

Fig. 106. Rattus norvegicus. A–E, five stages in embryonic development of stapedial artery. F, adult. From Tandler (1902).

branches and therefore may be referred to as the orbital artery. Later in ontogeny, in connection with the formation of secondary anastomoses and atrophy of parts of the original complex stapedial artery, considerable modifications occur (Tandler, 1899; Hafferl, 1933). Due to such changes all the three branches of the orbital artery may be annexed to the system of the ventral aorta and become branches of the external carotid. This has happened in man (Tandler, 1902; Goodrich, 1930; Padget, 1948) in which the proximal portion of the stapedial artery is a temporary embryonic vessel.

Looking for an explanation of the remarkable composition of the stapedial artery in mammals we may again turn to *Eusthenopteron*.

The efferent hyoid artery in *Eusthenopteron* probably traversed the hyomandibular canal and continued in the groove (Fig. 107A, Volume 1) running from the distal end of this canal to the ventral margin of the hyomandibula which is situated in the spiraculo-hyomandibular recess of the palatoquadrate. In this area the efferent hyoid artery evidently ran close to the efferent mandibular artery (Fig. 134, Volume 1) which emerging from the internal carotid a little behind the basal process coursed downwards and backwards in the groove on the inner side of the palatoquadrate to the spiraculo-hyomandibular recess (Fig. 107, Volume 1). An anastomosis could then easily be formed between the two efferent arteries in the recess and if so the three proximal portions of the mammalian stapedial artery were present in *Eusthenopteron* and formed a continuous vessel (Fig. 105). However, in *Eusthenopteron* (Fig. 78, Volume 1) the efferent mandibular artery passed rather close to the place where the presumed orbital artery was given off. With regard to the easiness by which branches of the orbital artery may be captured by other vessels it is not surprising that the orbital artery in mammals has (after suppression of the anterior part of the spiracular

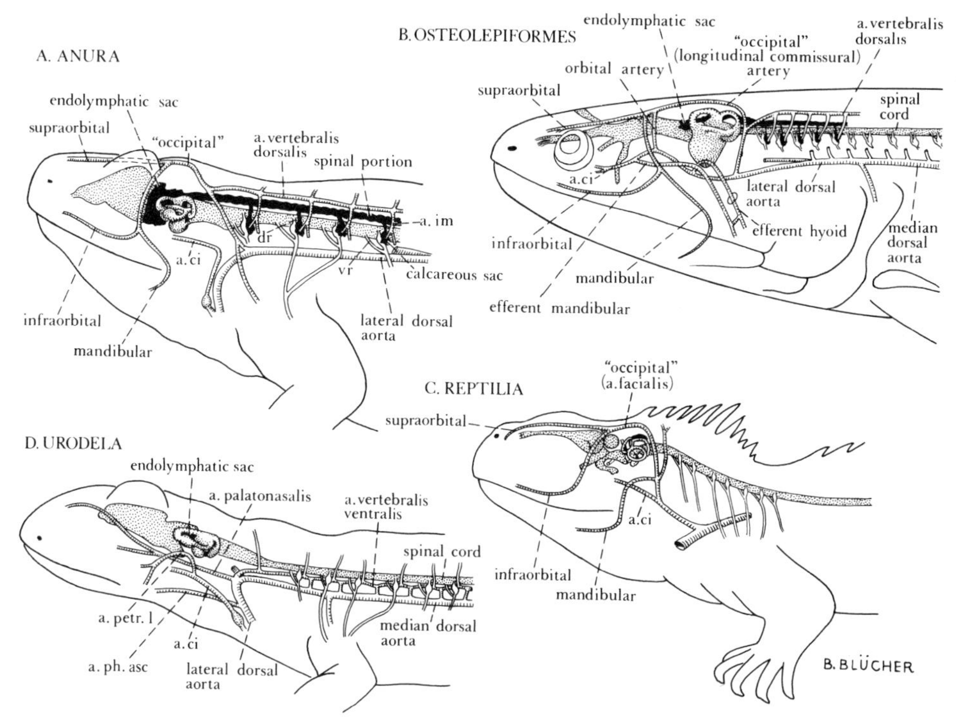

Fig. 107. Diagrams to illustrate fundamental similarities in course of "occipital" artery and in certain other regards between, A, anurans (*Rana*), B, osteolepiforms (*Eusthenopteron*) and, C, reptiles (*Sphenodon*) and, F, different conditions in urodeles. From Jarvik (1975, modified).

a. ci, internal carotid artery; a. im, intermetameric artery; a. petr. l, a petrosa lateralis; a. ph. sc, a. pharyngea ascendens; dr, vr, dorsal and ventral roots of spinal nerves.

gill tube; cf. Fig. 134, Volume 1) become annexed to the efferent mandibular artery (anastomosis h, Fig. 105C) to form the distal part of the stapedial artery. That this has happened and that the stapedial artery in mammals is derivable from the conditions in *Eusthenopteron* is supported by the fact that an "occipital" artery corresponding to that in *Eusthenopteron* and anurans seems to be present in certain mammals. This is the case in *Dasypus* (Tandler, 1899) in which the "occipital" artery reaches the orbit where it has captured the supraorbital branch of the original orbital artery. Accordingly the presence of a supraotic commissural ("occipital") artery must be a primitive mammalian feature. Since a corresponding vessel (usually called "facial" or "temporal" artery) is present also in reptiles (Fig. 107C; Jarvik, 1975) it is evident that the development of a strong supraotic commissure which in the orbit has captured one or more branches of the orbital artery is a characteristic feature of all eutetrapods and the osteolepiform–eutetrapod stock as a whole.

Posterior Cephalic Veins in the Urodelomorpha

In this respect, again, the porolepiforms and urodeles differ fundamentally from osteolepiforms and eutetrapods. As established by the sectioned specimen of *Glyptolepis groenlandica* (Fig. 108) there are no grooves or canals for the "occipital" artery in porolepiforms and such a commissural vessel was certainly lacking, as it is in urodeles (Fig. 107D). In urodeles and conceivably also in porolepiforms those parts of the orbit and snout which in the osteolepiform–eutetrapod stock are supplied by the supra- and infraorbital branches of the orbital artery receive their blood in a quite different fashion, namely by the a. palatonasalis which emerges from the lateral dorsal aorta close to, or jointly with, the vertebral artery and runs forwards ventral to the otic capsule (Francis, 1934; Jarvik, 1975). No branch is sent to the lower jaw which in urodeles is supplied by branches of other vessels (the a. pharyngea ascendens and a. petrosa lateralis). Also in other respects the vessels of urodeles differ widely from those of anurans. This is true of the vessels associated with the saccus endolymphaticus. In urodeles there is an a. vertebralis ventralis which is a ventral vessel running below the spinal roots and ganglia. As pointed out by Schöne (1902) this artery cannot be homologized with the a. vertebralis dorsalis in anurans (and *Eusthenopteron*) which runs dorsal to the spinal roots and supplies the spinal portion of the endolymphatic sac and the calcareous sacs (Fig. 107). In urodeles such a spinal portion is lacking and consequently the v. vertebralis interna dorsalis and its paired anterior continuation, the v. cranialis occipitalis (Gaupp; Jarvik, 1975, fig. 5) which in anurans (and *Eusthenopteron*) drain the posterior parts of the saccus endolymphaticus have no equivalents. In urodeles (*Ambystoma*; Roofe, 1935; Herrick, 1935), in contrast, the saccus endolymphaticus, which consists mainly of the pars anterior, is drained by the large jugular sinus, which passing the jugular foramen empties into the jugular vein (Fig. 108C). In the sectioned specimen of *Glyptolepis* there is no supraotic cavity

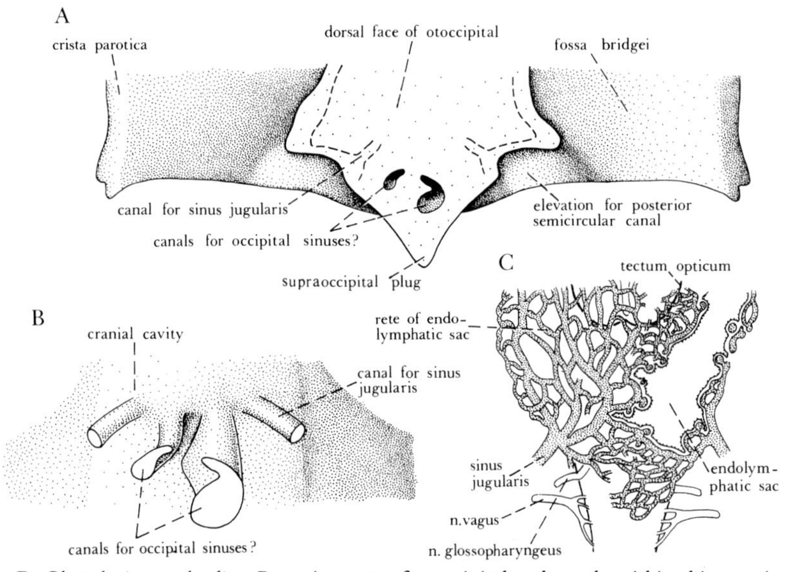

Fig. 108. A, B, *Glyptolepis groenlandica.* Posterior part of otoccipital and canals within this part in dorsal aspects. Graphical restorations on basis of grinding series. C, *Ambystoma tigrinum.* Diagram of dorsal endocranial veins in posterior part of head. After Herrick (1935) and Roofe (1935).

corresponding to that in *Eusthenopteron*. However, two wide paired canals (Fig. 108A, B) without equivalents in *Eusthenopteron*, emerge from the dorsal part of the medullary portion of the cranial cavity and open outwards. Each canal in the anterior pair runs in the posterolateral direction and passing posterodorsal to the space for the posterior semicircular duct it opens into the posteromedial part of the fossa bridgei, where it is continued backwards by a groovelike depression. It seems likely that this canal carried a vessel corresponding to the jugular sinus in urodeles and that this vein also in porolepiforms drained the endolymphatic sac. The posterior pair of canals is more difficult to interpret. These two canals arise close together and close posteromedial to the origins of the canals in the anterior pair. They run in the posterodorsal direction and open on the dorsal side of the supraoccipital plug underneath the median extrascapular. A remarkable fact is that the right canal is much wider than the left one, and moreover, the right canal is widened in its middle part and is continued backwards by a short groove on the dorsal side of the supraoccipital plug. These asymmetrical canals were obviously occupied by some saclike structures and they may have contained posterior diverticles of the endolymphatic sac. A more likely alternative is perhaps that they contained posterior supracephalic parts of the venous system which were continuous with subcutaneous veins corresponding to those in urodeles (Schmalhausen, 1968) and were of importance for the hearing in the way suggested by Schmalhausen.*

* p. 158.

11 Intermandibular Division and the Origin of Tetrapod Tongues

Composition of Intermandibular Division

The material of *Glyptolepis groenlandica* collected in 1956* (Jarvik, 1962; 1963; 1963a; 1972) gave important information as to the structure of the part of the skeleton situated in the area between the lower jaws in porolepiforms. These new data and the evidence provided by the grinding series and other material of *Eusthenopteron foordi* induced me to make a thorough analysis of the intermandibular division as a whole in gnathostomes (Jarvik, 1963, with references). The analysis of this previously neglected part of the head gave many new results of general interest, and several which in an unexpected way support the view (Jarvik, 1942) that the tetrapods are diphyletic. However, the intermandibular division turned out to be complex in structure and this is probably one of the reasons why these results have usually been overlooked or ignored by the monophyletists.

The following three main constituents may be distinguished in the intermandibular division (Fig. 109).

(1) The gular portion of the operculogular membrane, composed of the mandibular gill cover and the anterior parts of the hyoid gill cover, with skin, exoskeletal supporting elements (submandibulars, branchiostegal rays), visceral muscles innervated by the visceral trunks of the n. trigeminus and the n. facialis, vessels, etc.

(2) The subbranchial unit, composed of the hypobranchial muscles with their endoskeletal supporting elements (the subbranchial series), nerves, vessels, connective tissue, etc. The hypobranchial musculature is of myotomic derivation and thus mesomesenchymatic and is innervated by branches of the n. hypoglossus. The subbranchial series is also mesomesenchymatic and because it has arisen to support myotomic muscles it has been referred to the myotomic skeleton.

(3) The hyobranchial apparatus, including the parts of the hyobranchial skeleton situated between the jaws with pertinent visceral muscles, nerves, vessels, etc. The hyobranchial skeleton is, like other parts of the visceral endoskeleton, derived from the neural crest and thus ectomesenchymatic, whereas the visceral muscles originate from the metameric mesodermal visceral tubes.

* Volume 1, p. 244.

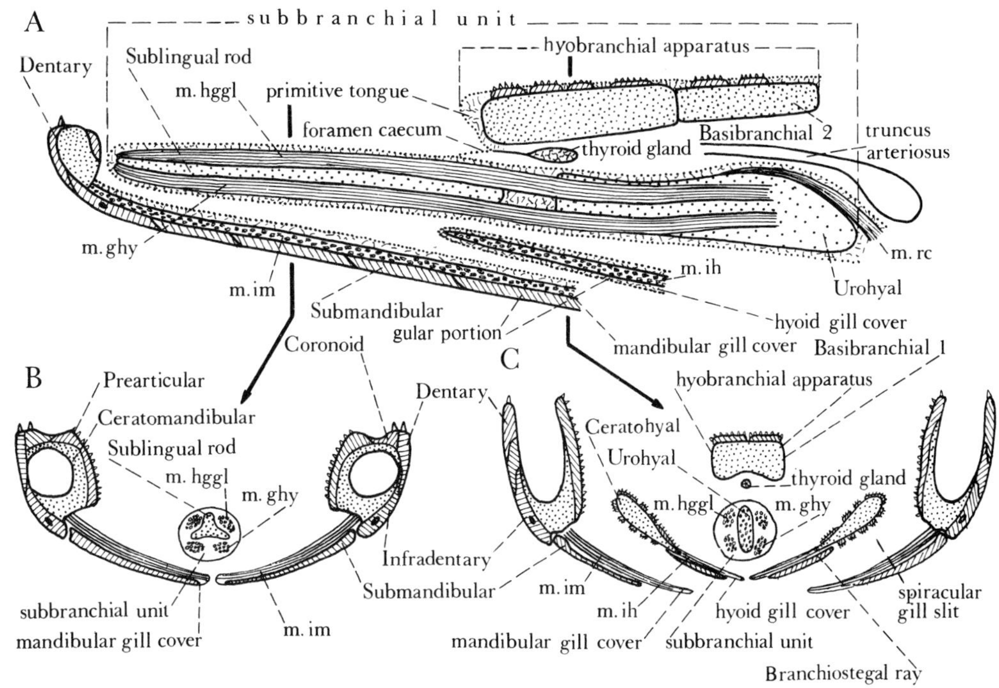

Fig. 109. Diagrammatic longitudinal (A) and transverse (B, C) sections to illustrate composition of intermandibular division of head. Skeleton of *Eusthenopteron* used as basis. From Jarvik (1963).

Explanations of abbreviations of names of muscles or areas of attachment in Figs 109–118;

m.bhe, branchiohyoideus externus; m.ggl, genioglossus; m.ghy, geniohyoideus; m.hggl, hyogenioglossus; m.hgl, hyoglossus; m.ih, m.iha, m.ihp, interhyoideus, interhyoideus anterior, and posterior; m.im, intermandibularis (mylohyoideus in mammals); m.ima, m.iml, m.imp, intermandibularis anterior, lateralis and posterior; m.rc, rectus cervicis; m.sar.1, m.sar.2–4, subarcualis rectus 1, and 2–4; m.trv.1, 2, m.trv.4, transversalis ventralis, 1, 2, and 4.

Another constituent of the intermandibular division is the thyroid gland* which arises as a median invagination of the anterior part of the pharyngeal floor in much the same way as do the lungs farther to the rear (Jarvik, 1968, fig. 6). The point of invagination may be marked by a foramen (foramen caecum in man) but usually the gland looses its connection with the pharyngeal floor and migrates backwards between the hypobranchial apparatus above and the subbranchial unit below, often following the course of the truncus arteriosus.

Gular Portion and Mandibular Gill Cover

The gill cover in teleostome fishes often continues forwards almost to the symphysis of the lower jaws (Jarvik, 1963). A remarkable fact is, however, that the gill cover in its anterior part has always fused with its antimere in the median line, and between the anterior parts of the jaws the most ventral part of the intermandibular division is therefore made up by the gill covers exclusively. Since this fusion usually has occurred

* see p. 202 for salivary glands.

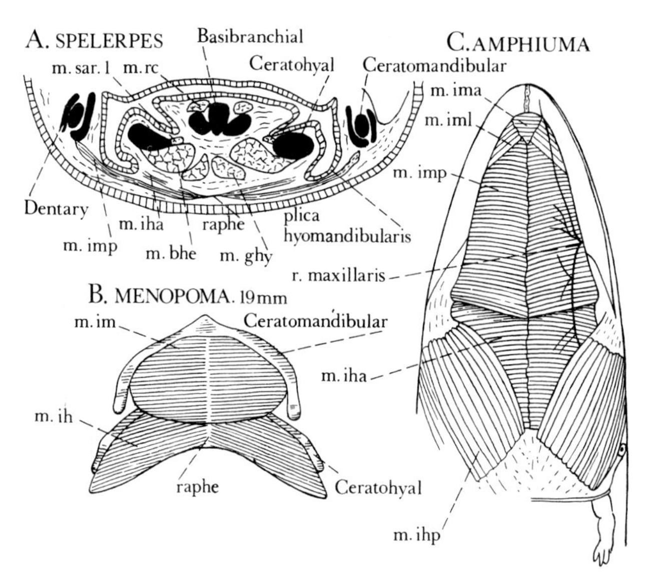

Fig. 110. Intermandibular musculature in urodeles. A, *Spelerpes bislineatus*, metamorphic stage. Transverse section of ventral part of head (after L. Smith). B, *Menopoma*, larval stage. Ventral parts of mandibular and hyoid arches with muscles (after Edgeworth). C, *Amphiuma means*. Head in ventral view. Skin removed (after Luther). From Jarvik (1963). For explanation of lettering see Fig. 109.

only in the deeper (inner) parts, which consist mainly of muscles, the gill covers, as seen from the outside (Fig. 180B, Volume 1), appear to be separate right up to the symphysis and often the one gill cover slightly overlaps that of the other side (Jarvik, 1950, p. 85). The part of the gill cover situated along the lower jaw, and thus in the intermandibular division, is usually termed the branchiostegal membrane but in order to avoid confusion the whole structure with its exoskeletal supporting elements, muscles, nerves, etc. was called the operculogular membrane. Posterodorsal or opercular, and anteroventral or gular, portions were distinguished.

In tetrapods the opercular portion has become partly reduced in connection with the transformations of the gill apparatus and the formation of tympanic membrane, whereas the gular portion has been retained with surprisingly small changes.

Up to 1963 it was generally assumed that the operculogular membrane as a whole is a hyoid gill cover. However, as is well established by their ontogenetic development, innervation and relations to the visceral arches, the muscles of the operculogular membrane belong to two different cephalic metameres, the mandibular and the hyoid (Fig. 110B; Fig. 70, Volume 1). The muscles of the anteroventral (gular) part, usually two in number (m. intermandibularis anterior and posterior) are derived from the mandibular muscle plate and are innervated by branches of the n. trigeminus, whereas those of the posterodorsal (opercular) part (m. interhyoideus anterior and posterior) are derivatives of the hyoid muscle plate and are innervated by the r. hyoideus of the n. facialis. Moreover the m. intermandibularis posterior, which is the most important of the two intermandibularis muscles, arises from the Meckelian element, that is the ceratomandibular (or from dermal bones of the lower jaw) and meets its fellow in a median raphe in the same way as the m. interhyoideus anterior arises from the ceratohyal and joins its antimere in the median line. Since, in addition, the mandibular and hyoid muscle sheets are separated by a deep fold, the plica hyomandibularis, enclosing a pocket which, as proved by the conditions in *Eusthenopteron*, represents the persisting ventral part of the spiracular gill slit, it is evident that the intermandibularis muscles (anterior and posterior) must be the musculature of the original mandibular gill cover. This must mean that the mandibular gill cover has been retained and is represented in the main by the gular portion of the operculogular membrane. Accordingly this membrane is a complex structure formed by the fusion of the mandibular and hyoid gill covers.

The intermandibular musculature which thus is the persisting musculature of the mandibular gill cover is generally well developed. In fishes (Fig. 70, Volume 1), amphibians (Fig. 110), and reptiles, as well as in birds and mammals (Fig. 118) its main constituent, the m. intermandibularis posterior—in mammals generally called the m. mylohyoideus—is a thin muscular sheet which arises from the inner side of the lower jaw and meets its fellow in a median raphe (Luther, 1909; 1914; 1938; Edgeworth, 1935). Thence it follows that a mandibular gill cover is present in all gnathostomes and if we disregard the loss of the supporting dermal bones (the submandibulars) in tetrapods and the occurrence of the secondary superficial muscles in mammals (e.g. the platysma) there are but few differences between the various groups. The tough band of connective tissue (the median raphe) which joins the intermandibularis muscles of both sides is dorsally continuous with the connective tissue, which sheathes the overlying hypobranchial musculature, and with the median raphe of the interhyoid musculature. Because of these fusions the ventral division of the spiracular gill slit situated in the plica hyomandibularis has in all extant gnathostomes become completely closed. However, in early larvae of urodeles (Drüner, 1901) the epithelium of the posterior part of the plica hyomandibularis may be in contact with the epithelium of the skin. This suggests a not too remote ancestral stage in which the ventral division of the spiracular gill slit opened outwards in this area, and there is some evidence that such an opening was really present in porolepiforms (Jarvik, 1963, p. 27; 1972, p. 122).

A remarkable fact in porolepiforms is that the submandibulars, which possibly together with the infradentaries (Fig. 58) are the supporting exoskeletal elements of the mandibular gill cover, in their proximal (lateral) parts present a thickened articular ridge (Jarvik, 1972, p. 121). This ridge fits well into a corresponding articular groove made up by the infradentaries and to some extent by the ceratomandibular (Meckelian bone); and obviously the part of the mandibular gill cover formed by a skin flap and supported in its lateral part by the submandibulars was movable (Fig. 180, Volume 1). In the osteolepiforms the conditions are similar and it is possible that the peculiar longitudinal portions of the posterior intermandibularis muscle in urodeles (Fig. 110C) and anurans which are situated somewhat as the submandibular series in their porolepiform and osteolepiform ancestors have something to do with the original movability of the mandibular gill cover. A detail worth mentioning in this connection is also that the right operculogular membrane in urodele larvae overlaps that of the left side, as in porolepiform (Fig. 180, Volume 1; Jarvik, 1963, fig. 11). This is remarkable since in fishes in general the condition usually is the reverse, the left membrane overlapping the right one (Jarvik, 1950, p. 85).

The intricate evolutionary processes collectively referred to as the rostral prolongation* have strongly influenced the intermandibular division. As is well shown for instance in *Amia* (Fig. 70, Volume 1) the ventral parts of the mandibular and hyoid arches in early embryonic stages form transverse bars. However, these bars soon assume an oblique position at the same time as the ceratomandibulars grow forwards. In connection with these changes the distance between the mandibular and hyoid arches increases considerably which results in the formation of a large prehyoid part of the intermandibular division. Moreover, the anterior interhyoideus muscles separate from the remaining part of the interhyoid musculature and grow forwards dorsal to the posterior intermandibularis muscles of the mandibular gill cover. Probably due to these changes the anterior (gular) branchiostegal rays of the hyoid gill cover loose their connection with the hyoid arch and become situated partly in front of that arch, below the posterior intermandibularis muscle. In consequence of the rostral prolongation and the elongation of the ceratomandibulars and the lower jaws as a whole also the plica hyomandibularis and the ventral division of the spiracular gill slit exhibit prehyoid portions (Fig. 113). The subbranchial unit, too, has been strongly influenced and extends forwards into the prehyoid part of the intermandibular division (Figs 113–118). This is of particular interest since the main part of the tongue in tetrapods is formed in the prehyoid portion of this unit.

HYOBRANCHIAL APPARATUS, SUBBRANCHIAL UNIT AND ORIGIN OF TETRAPOD TONGUES

Porolepiform–Urodele Stock (Urodelomorpha)

In the structure of the gular portion the porolepiforms and the osteolepiforms on the whole are similar, and they agree well with other teleostomes and with tetrapods. As regards the structure of the subbranchial unit and the hyobranchial apparatus, in contrast, there are considerable differences and since the porolepiforms (*Glyptolepis*) agree with urodeles, whereas the osteolepiforms (*Eusthenopteron*) in important regards are as eutetrapods, we have to treat the porolepiform–urodele and the osteolepiform–eutetrapod stocks separately.

In the hyobranchial apparatus the porolepiforms agree with urodeles but differ from osteolepiforms mainly in the following respects (Jarvik, 1962; 1963; 1963a; 1972):

(a) The basibranchial series consists of one basibranchial only (Figs 111, 112). The wide median canal of basibranchial 1 in osteolepiforms has no equivalent in porolepiforms.

(b) Due to reduction in their anteroventral parts the posterior branchial arches do not reach the basibranchial and secondary interarcual articulations have developed. As illustrated in Fig. 111 the phylogeny of urodeles is characterized by a gradual reduc-

* p. 20.

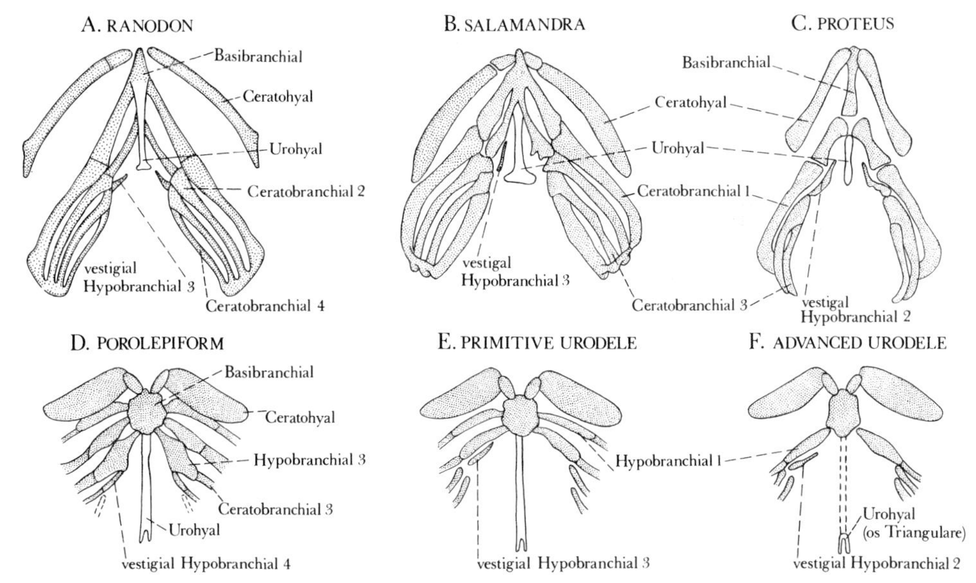

Fig. 111. A–C, hyobranchial skeleton in three urodeles. A, *Ranodon sibiricus*, larva. Ventral aspect. B, *Salamandra maculosa*, larva, c. 35 mm. Dorsal aspect. C, *Proteus anguineus*, adult. Ventral aspect. D–F, diagrams illustrating changes in hyobranchial skeleton from porolepiform to primitive and advanced urodele. From Jarvik (1972).

tion of the branchial arches from behind, a process which has started already in their porolepiform ancestors. The differences in the development of the hyobranchial skeleton between a primitive urodele (e.g. *Ranodon*) and an advanced urodele (e.g. *Proteus*) are in fact of the same kind and not greater than between a primitive urodele and a porolepiform.

(c) In sharp contrast to osteolepiforms the external side of the ceratohyal in porolepiforms lacks teeth and shows impressions for the attachment of muscles corresponding to those in urodele larvae (Fig. 112). One such impression is situated exactly as the area of attachment of the m. subarcualis rectus 1 in urodeles and there are strong indications that there was also a large muscle corresponding to the m. branchiohyoideus externus in urodeles. This is of great interest since this muscle is a larval structure which among recent vertebrates has been encountered only in urodele larvae. Morevoer, there was certainly a m. interhyoideus anterior which from the posterior end of the ceratohyal passed ventral to the m. branchiohyoideus to join its fellow in a median raphe dorsal to the m. intermandibularis posterior.

Also in the structure of the subbranchial unit the porolepiforms agree in a remarkable way with urodeles. The subbranchial series thus consists of a single element, the urohyal (Figs 111, 112). This element is in both groups a long slender rod which extends backwards from the ventral side of the basibranchial and already in *Glyptolepis* it shows a tendency to fuse with the visceral structure. The posterior end of the urohyal (os terminale or triangulare in adult urodeles) is in certain urodeles bifurcate as in *Glyptolepis*.

The urohyal in urodeles is still sometimes called basibranchial 2 (e.g. Balinsky,

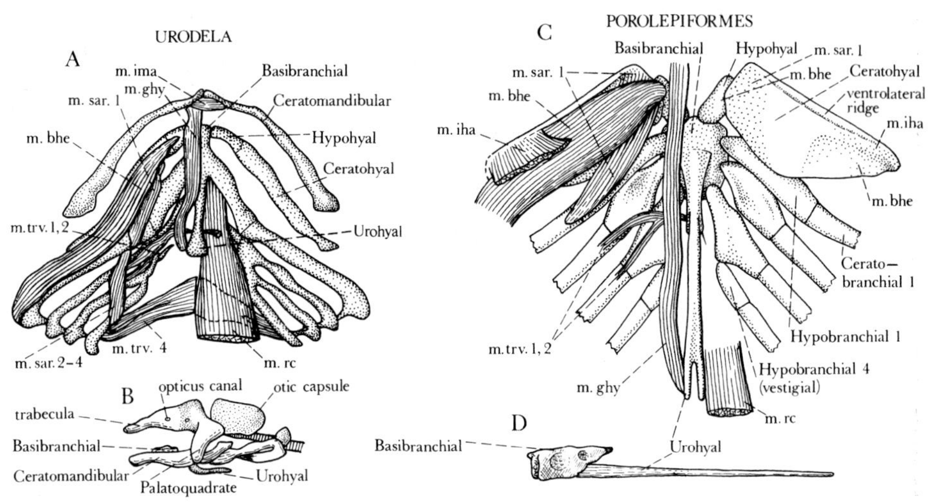

Fig. 112. Diagrams to illustrate similarities in hyobranchial apparatus, subbranchial series, and hypobranchial musculature between urodeles and porolepiforms. A, *Hynobius nebulosus*, larva 22 mm. Meckelian cartilages, hyobranchial skeleton, and urohyal with certain muscles in ventral aspect. After Fox (1959). B, *Ambystoma mexicanum*, chondrocranium in lateral aspect. Cartilage mesomesenchymatic in origin dotted. From Hörstadius and Sellman (1946). C, D, *Glyptolepis groenlandica*. C, hyobranchial skeleton and urohyal with certain muscles in ventral aspect. D, basibranchial and urohyal in lateral aspect. From Jarvik (1963; 1972). For explanation of lettering see Fig. 109.

1970; Chibon, 1974) although it was established long ago by Stone and others* (Figs 35C, 112B) that it does not belong to the hyobranchial apparatus at all. In contrast to the ectomesenchymatic hyobranchial skeleton derived from the neural crest it is mesomesenchymatic ("endomesodermal") in origin. After a perusal of the pertinent literature it could be established (Jarvik, 1963) that the urohyal and the subbranchial series as a whole are intimately related to the hyobranchial musculature. I, therefore, suggested that the subbranchial elements had arisen in the mesomesenchymatic tissues associated with these muscles and that their primary function was to be a support for this musculature which is derived from the metameric myotomes. The subbranchial series, which has been referred to the myotomic skeleton,† and the hypobranchial muscles form a unit (the subbranchial unit; Fig. 109; Fig. 11, Volume 1) different in origin and primarily separate from the overlying and underlying visceral structures.

The intimate relations between the subbranchial series (the urohyal) and the hypobranchial muscles (m. rectus cervicis, m. geniohyoideus) are distinct in urodele larvae (Jarvik, 1963). According to Rempel (1943, p. 109) the ventral processes of the myomeres which form the m. rectus cervicis in early stages of *Triturus* "grow forward under the branchial area and over the lateral surface of the pericardium finally to attach upon the urobranchial cartilage". The muscle continues to grow forwards, and in more advanced stages (Edgeworth, 1935; Rempel, 1943; Fox, 1959) some of its fibres are attached to the posterior part of the urohyal (the os triangulare in the adult), whereas the main portion reaches the anterior end of that element and is inserted into

* p. 57; † p. 58; Volume 1, p. 43.

the angle between the urohyal and hypobranchial 1 (Fig. 112A). Later in ontogeny the rectus cervicis grows still further forwards and, what is remarkable for a hypobranchial muscle, in the adult (see also Özeti and Wake, 1969) its anterior portion passes dorsal to both hypobranchial 1 and the hypohyal and is inserted into the dorsal side of the apex of the basibranchial. Judging from the descriptions by Kallius (1901) and others it is fibres of this portion which upon the formation of the definitive tongue enter this organ and form the so called m. hyoglossus (Fig. 114C4). Because the m. hyoglossus in anurans and amniotes lies ventral to the hyobranchial skeleton and according to current views (like the m. genioglossus) is a derivative of the m. geniohyoideus (and not of the m. rectus cervicis) I, in 1963, was doubtful if the so-called m. hyoglossus in urodeles was homologous with the so-named muscle in eutetrapods. When dealing with the hypobranchial muscles in *Eusthenopteron* (Jarvik, 1963, p. 44) I suggested that the hyo- and genioglossus muscles in eutetrapods are derivatives of a muscle, the hyogenioglossus, which in *Eusthenopteron* traversed the wide median canal of basibranchial 1. However, as established later (Jarvik, 1972) such a canal is lacking in porolepiforms. If we now assume that this hypothetical muscle was present also in porolepiforms it obviously had to take another course than in osteolepiforms when growing forwards. It is therefore possible that the posterior part of the m. hyogenioglossus in urodeles is represented by that portion of the m. rectus cervicis which grows forwards dorsal to hypobranchial 1 and the hypohyal and that both the hyoglossus and the genioglossus in urodeles are derivatives of m. hyogenioglossus. Be this as it may a common characteristic feature of all extant tetrapods is the presence of a paired m. genioglossus which from the area of the jaw symphysis grows upwards and backwards into the tongue (Figs 114C1–C4, 116) and probably, this muscle and the m. hyoglossus, which both are lacking in recent fishes, were initiated in both the osteolepiforms and the porolepiforms.

The most ventral of the two or, if the m. hyogenioglossus is accepted, three principal hypobranchial muscles is the m. geniohyoideus, which like the m. rectus cervicis is present in all gnathostomes (Figs 109, 112–114, 116–118). This muscle, which also is paired, is always attached anteriorly to the lower jaw close to the symphysis but its posterior extent varies and it may be a m. geniobranchialis (branchiomandibularis) as in *Polypterus* and ganoids or a m. geniocoracoideus (coracomandibularis) as in selachians, dipnoans and *Latimeria*. In urodeles (Edgeworth, 1935; Özeti and Wake, 1969) it is inserted into the posterior end of the urohyal or inscriptions in the corresponding position.

Because of the similarities in the construction of both the hyobranchial skeleton and the urohyal it is to be concluded that the hypobranchial muscles in porolepiforms were developed much as in urodeles (Figs 112–114). However, to these similarities indicating a close relationship we have to add most remarkable resemblances in the structure of the floor of the prehyoid part of the mouth cavity.

A quite unique character of larval urodeles (Figs 113, 114; Kallius, 1901; Jarvik, 1972; fig. 103) is the presence of a crescent-shaped glandular field in the floor of the mouth cavity. This field lies close behind the jaw symphysis, in front of and partly below the primitive or larval tongue (Fischzunge, Kallius) formed by the thickened mucous membrane of the basibranchial. As established by the grinding series of

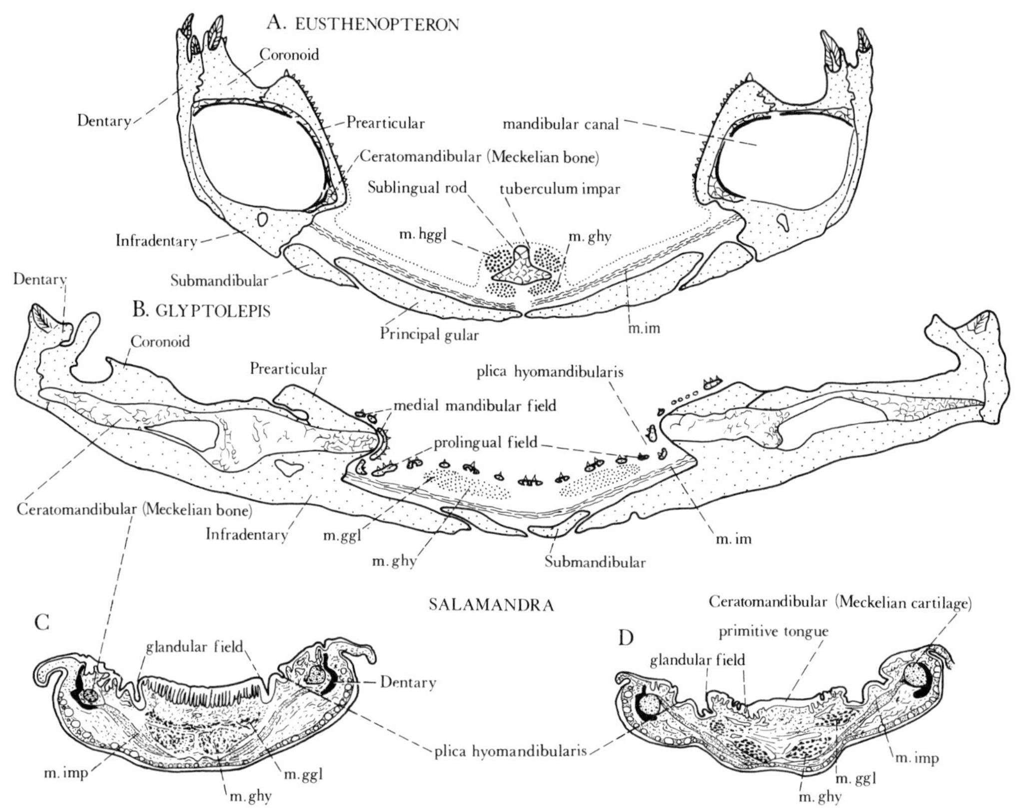

Fig. 113. Anterior part of intermandibular division in transverse sections of, A, osteolepiform, B, porolepiform and, C, D, urodele (larva 50–52 and 55 mm. After Kallius). From Jarvik (1963). For explanation of lettering see Fig. 109.

Glyptolepis (Figs 113B, 114) the porolepiforms in the corresponding area present a large number of small ossicles. These ossicles may carry delicate teeth and have therefore been referred to as the prolingual dental plates (Jarvik, 1962; 1963; 1972). However, more important is that they, in contrast to the typical dental plates of the mouth cavity, show cavities opening both outwards and inwards. Most likely these cavities and probably also the spaces between the ossicles were occupied by glands. Since these ossicles are arranged in a crescent-shaped field it is evident that this field, the prolingual field, is homologous with the glandular field in urodele larvae.

Another unique larval feature in urodeles (Fig. 114A) is the presence of a thin elongated dental plate referred to as the "operculare", "goniale" or "coronoid" (Hertwig, 1874; Gaupp, 1905; Stadtmüller, 1936). This plate which arises by fusion of the basal plates of small teeth, lies on the inner side of the Meckelian cartilage, posterolaterally to the glandular field, and is separated from that by the prehyoid portions of the plica hyomandibularis and the spiracular gill slit. Also in porolepiforms the plica hyomandibularis and the spiracular gill slit continued forwards into the prehyoid part of the intermandibular division and lateral to these structures the medial sides of the Meckelian bone and the prearticular show a vertical

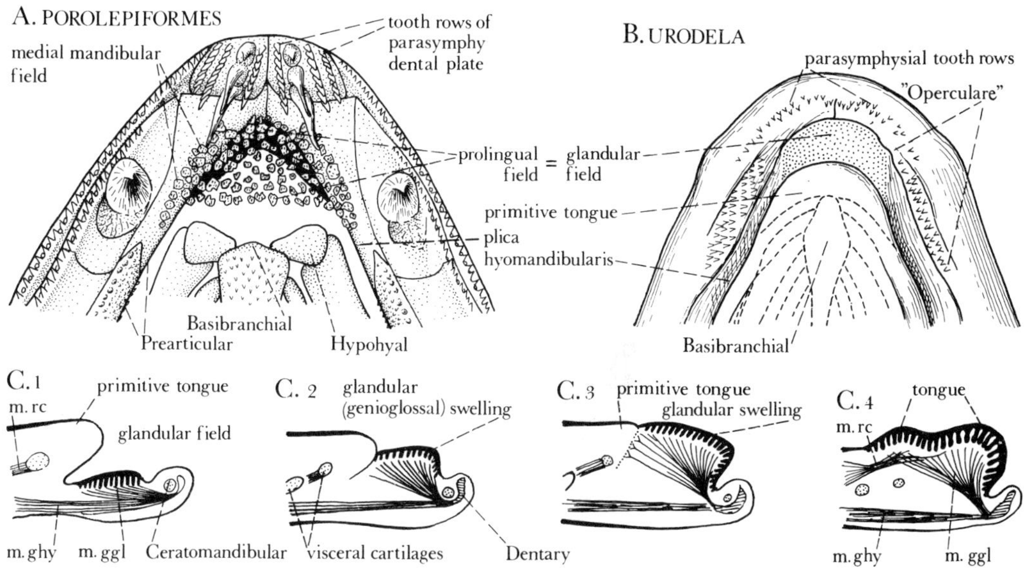

Fig. 114. Evolution of tongue in Urodelomorpha. A, porolepiform (*Glyptolepis*, *Holoptychius*). Anterior parts of lower jaws and skeleton of intermandibular division in dorsal aspect. B, *Triturus punctatus*, 19 mm. Anterior parts of lower jaws and floor of mouth cavity in dorsal aspect. From Jarvik (1963). C.1–C.4, diagrammatic longitudinal sections of floor of mouth cavity of urodele larvae to illustrate four stages in development of tongue. From Kallius (1901). For explanation of lettering see Fig. 109.

elongated area of dental plates, the medial mandibular field (Figs 113B, 114A). In contrast to the dentigerous ossicles of the prolingual field the dental plates of this field are compact in structure and show a distinct lamination; and no doubt they are homologous with the teeth which form the "operculare" in larval urodeles.

In front of the medial mandibular field in porolepiforms there is a large parasymphysial dental plate carried by a modified anteromedial lamina of the dentary. In larval urodeles (*Hynobius*; Jarvik, 1972, fig. 106) the dentary anteriorly is provided with a parasymphysial process carrying teeth. Moreover, in the corresponding area in larvae of *Triturus* (Fig. 114B) there are several teeth arranged in longitudinal rows suggestive of the parallel rows of small teeth of the parasymphysial dental plate in porolepiforms (Fig. 114A). These conditions suggest that vestiges of the both the parasymphysial dental plate and the lamina of the dentary that carries this plate of their porolepiform ancestors are present in urodele larvae (as to other similarities in jaw structure see Jarvik, 1972, p. 269).

As described by Kallius (1901), the glandular field plays an important part in the formation of the definitive tongue in the ontogeny of urodeles (Fig. 114C1–C4). In connection with the growth of the m. genioglossus the floor of the mouth cavity in the area of the crescent-shaped glandular field thickens, rises and ultimately fuses with the primitive tongue which latter is only a thickening of the mucous membrane of the anterior part of the basibranchial. The definitive tongue of urodeles (Fig. 119F2) thus includes the primitive tongue of the larva to which is added anterolaterally and anteriorly a crescent-shaped portion of the floor of the mouth cavity corresponding in size to, and including the glandular field.

It must now be emphasized that this peculiar mode of tongue formation among the tetrapods occurs only in urodeles which are the only tetrapods in which a glandular field is found. Moreover the tooth-bearing "operculare" is known only in this group of tetrapods, in which it is a larval structure. On the other hand, the medial mandibular and prolingual fields of the porolepiforms shown to be homologous with the "operculare" and the glandular field, respectively, in larvae of urodeles, are structures which have no equivalents in osteolepiforms or any other group of fishes. Also in these regards we are thus concerned with most striking and undisputable resemblances between porolepiforms and urodeles indicating a close relationship. No doubt the main anterior part of the definitive tongue of the urodeles has arisen from the prolingual field and the underlying soft tissues in the floor of the mouth cavity of the porolepiforms.

Osteolepiform–Eutetrapod Stock (Osteolepipoda)

In the osteolepiform–eutetrapod stock (Jarvik, 1963, with references) the specializations of the hyobranchial skeleton characteristic of porolepiforms and urodeles have not occurred and there is no "operculare" and no glandular field. The tongue is formed in a different way in relation to an anterior element of the subbranchial series, the sublingual rod (or its derivatives, Fig. 115), not present in porolepiforms and urodeles.

In osteolepiforms (*Eusthenopteron*, Figs 115, 117; Fig. 112, Volume 1) the subbranchial series in addition to the sublingual rod includes a large urohyal, which in contrast to that in porolepiforms (*Glyptolepis*) and urodeles, in its main posterior part is developed as a high median plate. In anurans and amniotes the subbranchial series has become modified and has partly fused with the overlying hyobranchial skeleton.

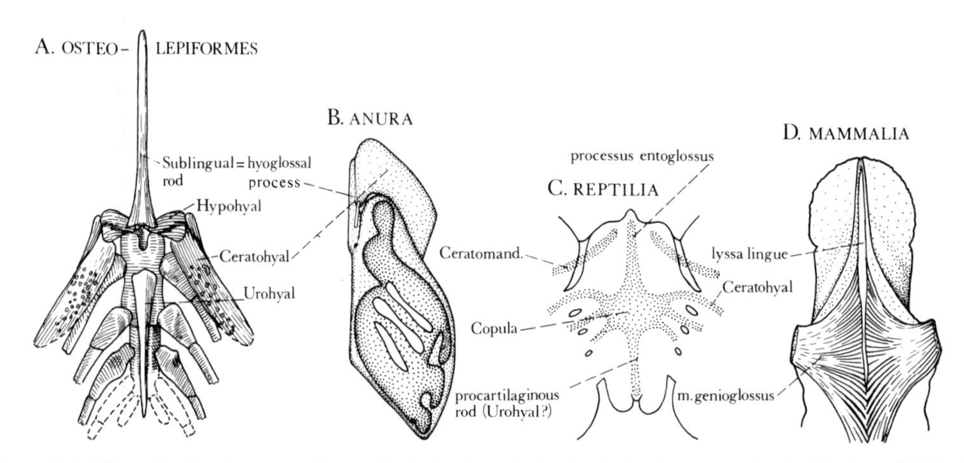

Fig. 115. Sublingual rod and presumed homologies in Osteolepipoda. A, *Eusthenopteron foordi*, Hyobranchial skeleton and subbranchial series in ventral aspect. B, *Xenopus laevis*. Right half of hyobranchial skeleton with hyoglossal process in dorsal aspect. From Sedra and Michael (1957). C, *Lacerta muralis*. Diagram of hyobranchial skeleton of larva in dorsal aspect. After Kallius (1901). D, *Melursus ursinus*. Tongue in ventral aspect dissected to show lyssa lingue (after Weber, 1927). From Jarvik (1963).

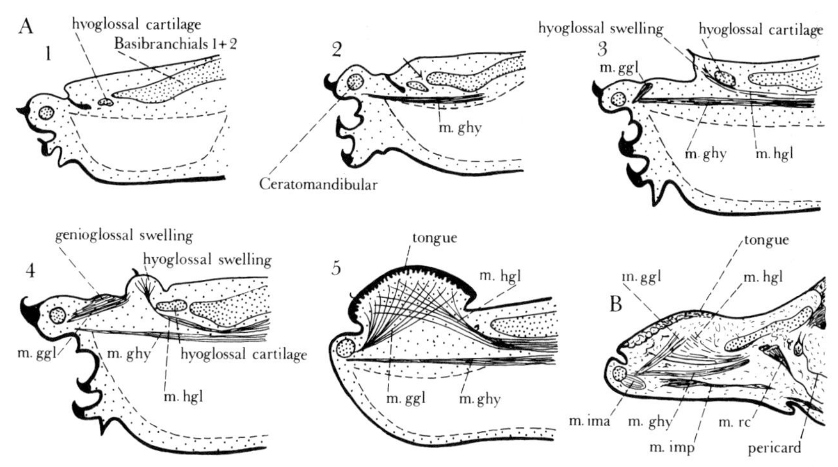

Fig. 116. A, *Alytes obstreticans*. Five stages in evolution of tongue (after Kallius). B, *Leiopelma archeyi*. Intermandibular division in medial longitudinal section (after Stephenson). From Jarvik (1963). For explanation of lettering see Fig. 109.

In anurans, in which two basibranchials may be distinguished, only the posterior part of the sublingual rod has been retained, either as a small hyoglossal cartilage (Fig. 116) or as a hyoglossal process (Fig. 115B). The hyoglossal cartilage has been called "basihyal" or "copula 1" (e.g. Gaupp, Kallius, Edgeworth, Stadtmüller) but as shown by Stone (1929) it is mesomesenchymatic ("endomesodermal"). This condition together with its relations to the hypobranchial muscles and the fact that it in larval stages of *Alytes* lies anteroventral to the basibranchial series (Fig. 116A1, 2) as does the corresponding part of the sublingual rod in *Eusthenopteron* (Fig. 117B), prove that it belongs to the subbranchial series. The urohyal in anurans is represented by two ventral processes of the basibranchial series. The foremost of these processes, the urohyal process (Fig. 117A), has been given many names (Copulastiel, urobranchial prong, basibranchial 1, basibranchial 2) but as established by several writers (Stone, Ichikawa, Stokes James) it is mesomesenchymatic and no doubt it is derived from the anterior part of the osteolepiform urohyal. The posterior median process, which as shown by Swanepoel (1970; called "basibranchial 2") also is mesomesenchymatic, is known only in microhylid frogs (Trewavas, Pentz, Roux). This process may form a large descending crest and is certainly a vestige of the posterior high part of the osteolepiform urohyal as are probably also the parahyoid of Fuchs and Blume's processus laryngeus in *Hyla* (Jarvik, 1963).

In amniotes the hyobranchial skeleton enters into the formation of the so-called hyoid (Zungenbein, Fürbringer, 1922; hyoid bone in man, Fig. 118). However, this element is a complex structure and most likely it includes parts which in contrast to the ectomesenchymatic visceral skeleton are mesomesenchymatic and belong to the subbranchial series. In order to explain this it may be convenient to start with the conditions in *Eusthenopteron*.

In *Eusthenopteron* (Fig. 117B–D) there was certainly a paired m. geniohyoideus, which from the urohyal ran forwards ventral to basibranchial 1 and the sublingual rod

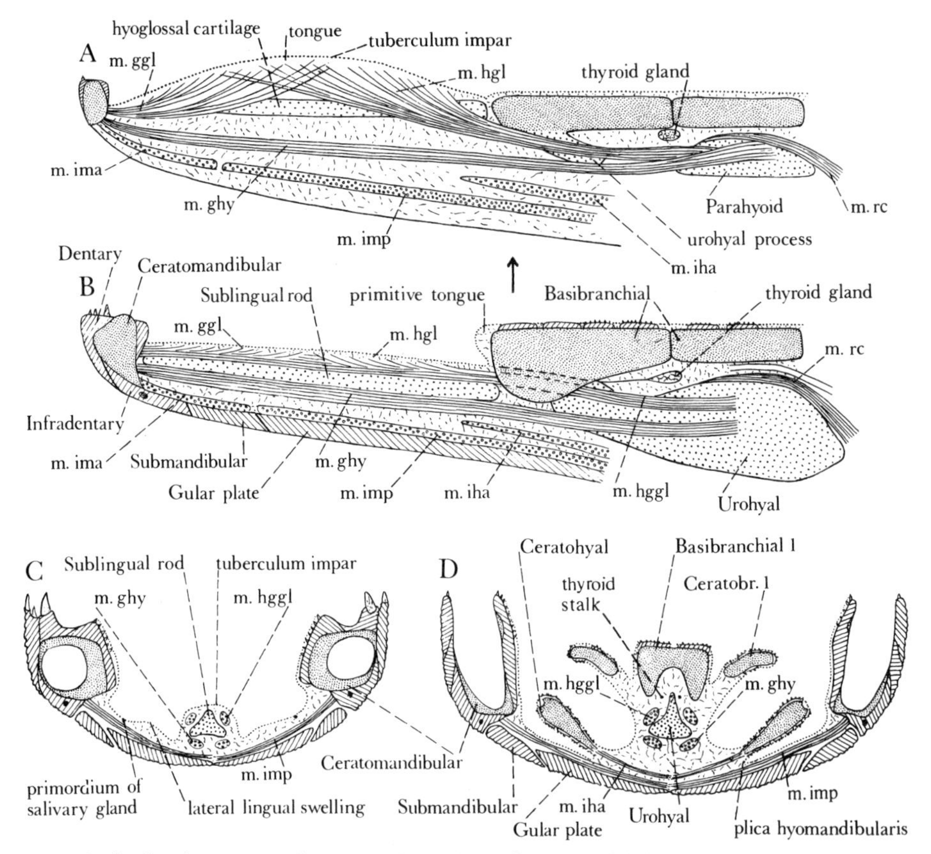

Fig. 117. Longitudinal and transverse diagrammatic sections of intermandibular division to illustrate evolution of tongue in Osteolepipoda. Skeleton of *Eusthenopteron* used as basis (cf. Fig. 109). From Jarvik (1963). For explanation of lettering see Fig. 109.

to the symphysis of the lower jaws. In the prehyoid part of the floor of the mouth cavity the muscles were situated close together dorsal to the medial parts of the fused mandibular gill covers, and in this regard *Eusthenopteron* agrees in a striking way with man (Fig. 118A, C). The m. rectus cervicis obviously passed upwards in front of the pericardium much as in frogs (Fig. 116B) to be inserted into the dorsal part of the urohyal. However, in addition to these two paired muscles, present in all gnathostomes, there was probably a third paired hypobranchial muscle, the hyogenioglossus (Jarvik, 1963, p. 44). This muscle also passed forwards from the urohyal and together with its antimere it most likely traversed the median canal of basibranchial 1, probably accompanied by an anterior continuation of the truncus arteriosus (Figs 130, 366B, Volume 1) and, in larval stages, by the stalk of the thyroid gland. Farther forwards, in front of the basibranchial series and the site of the thyroid invagination, the hyogenioglossus muscles were separated by the median crest of the sublingual rod and most likely fibres representing the genioglossus and hyoglossus muscles in initial stages of development were present (Fig. 117B).

The anterior, prehyoid, parts of the hypobranchial muscles in *Eusthenopteron* constitute, together with the sublingual rod as well as the lining mucous membrane and other pertinent soft parts, the anterior or prehyoid portion of the subbranchial unit (Fig. 117C). This portion, which formed a pronounced median elevation in the part of the floor of the mouth cavity arisen in connection with the rostral prolongation, was developed as, and is certainly homologous with, the tuberculum impar (Figs 117–119) found in ontogenetic stages of reptiles, birds and mammals, and present also in anurans. Accordingly this structure, which in contrast to statements in current textbooks plays an important part at the formation of the tongue, is characteristic of the osteolepiform–eutetrapod stock.

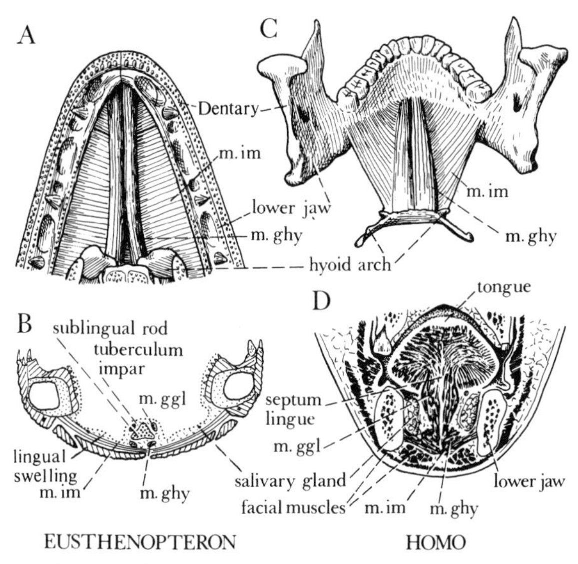

Fig. 118. Representations to illustrate similarities between man and his Devonian forerunners in intermandibular division. A, C, muscles of floor of mouth cavity in dorsal aspect of Devonian osteolepiform and man. B, D, transverse sections. In man (C, D) musculature of mandibular gill cover and geniohyoid muscles have retained their original position, tuberculum impar has grown to form main part of tongue, and has a new mucous membrane derived from lateral lingual swelling; sublingual rod reduced but may persist as a row of small cartilages in septum lingue; salivary glands have increased in size and fill spiracular gill slit. From Jarvik (1963a). For explanation of lettering see Fig. 109.

Exactly as did the median elevation in *Eusthenopteron*, the tuberculum impar in extant eutetrapods extends from the site of the thyroid invagination (foramen caecum) forwards to the jaw symphysis; it is also in eutetrapods situated dorsal to the medial parts of the fused mandibular gill covers, it contains the prehyoid parts of the hypobranchial muscles, and as in *Eusthenopteron* these muscles are, in reptiles and birds supported by a long median endoskeletal rod (Fig. 115C; Jarvik, 1963, fig. 23). This rod may be independent (see Toerien, 1971) but usually it forms a process, the processus lingualis (Fürbringer, Edgeworth, Toerien) or entoglossus (Kallius, Gaupp), of the hyobranchial skeleton. Most likely this rod is homologous with the sublingual rod in *Eusthenopteron*. More precisely the processus lingualis seems to be derived from the median crest of the rod, whereas the horizontal plate of the rod may be retained as the hypoglossum in chelonians. In mammals (Kallius, 1910; Fürbringer, 1922) a short processus lingualis may be present, or is, in some orders, represented by the lytta or lyssa lingue (Fig. 115D; see also Tokarski, 1904) which is a peculiar median tube of tough connective tissue containing cartilaginous structures. Also the row of small cartilages frequently found in or below the septum lingue in man (Gentscheff, 1934) are probably vestiges of the sublingual rod in the osteolepiform ancestors.

The urohyal is imperfectly known in amniotes but is conceivably represented by the ventral median crest and the posterior process of the "copula" in *Lacerta* described by Kallius (1901). Also "copula 2" (Fürbringer, Toerien) or the urohyal (Marinelli, 1936) in birds may belong to this category and similar structures seem to be present in mammals as well (Fürbringer).

The tongue of the anurans and the amniotes (Kallius, 1901; 1905; 1910) may be easily derived from an intermandibular division built as that in the osteolepiforms (Jarvik, 1963). In addition to the tuberculum impar and the sublingual rod or its derivatives a distinctive character of the osteolepiform–tetrapod stock is the presence of a well developed m. hyoglossus in addition to the m. genioglossus. In connection with the growth of these two muscles the part of the floor of the mouth cavity formed by the tuberculum impar became thickened and the dorsal surface rose to the level of the dorsal side of the basibranchial series (Figs 116, 117). Moreover, the primitive tongue (in contrast to urodeles) was reduced and the original fold between the mucous membranes of the tuberculum impar and the basibranchial series became smoothed out and disappeared, as did generally the marking (foramen caecum) of the thyroid invagination in the bottom of that fold. The hyoglossus and genioglossus muscles, which in the adult cross each other in a characteristic way, were both attached to the sublingual rod or its derivatives and a consequence of the growth of these muscles was that the sublingual rod moved upwards (as does the hyoglossal cartilage in the ontogeny of anurans, Fig. 116) until it was on a level with and could fuse with the basibranchial series following posterior to it. However, the fate of the sublingual rod was somewhat different in the anurans and the amniotes and there are also other differences.

In anurans the sublingual rod became much reduced and its anterior part supporting the m. genioglossus has always disappeared. The posterior part, on the other hand, which is associated with the m. hyoglossus, is often retained as the hyoglossal cartilage (Fig. 116) or (in *Xenopus*) the hyoglossal process (Fig. 115B). The loss of the anterior part of the sublingual rod allowed the m. genioglossus to grow upwards independently and the muscle caused a paired bulge, the genioglossal swelling (Figs 116A4, 119B), in the floor of the mouth cavity. This bulge is separated from the hyoglossal swelling by a transverse groove which, however, disappears before the adult stage is reached. As in urodeles the tongue is formed almost entirely of the anterior portion of the subbranchial unit, which, however, in contrast to urodeles, includes both a well developed m. hyoglossus and a sublingual rod and is a tuberculum impar similar to that in amniotes. Only certain ventrolateral parts and sometimes a small posterior area of the dorsum lingue are visceral in origin and are supplied by the n. glossopharyngeus.

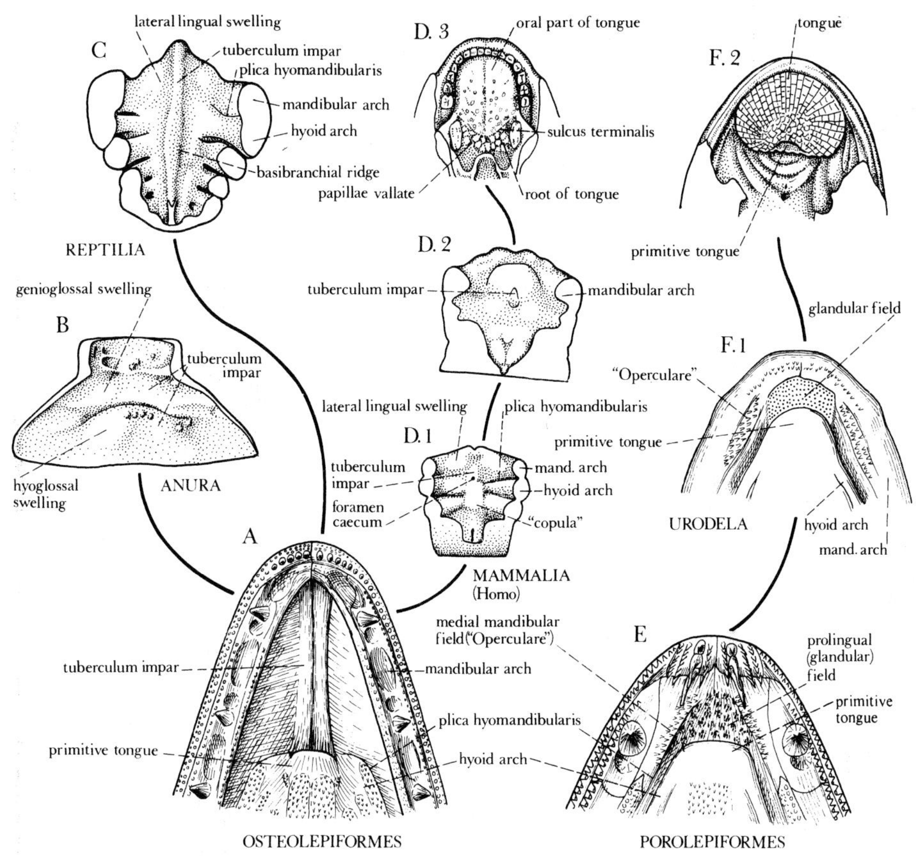

Fig. 119. Diagrams of evolution of tongue in, A–D, Osteolepiformes–Eutetrapoda and, E, F, Porolepiformes–Urodela. From Jarvik (1963).

In amniotes the tongue is more complex in structure than in anurans. Of considerable interest is that a large posterior portion forming the root or pharyngeal part of the tongue has been added (Fig. 119D3). However, as shown by Kallius (1910) there are distinct differences in this portion of the tongue between reptiles and mammals; and his statement that the tongue of the mammals cannot possibly be derived from that in reptiles is in accordance with and supports my view that the mammals have evolved independently from the osteolepiforms* (Fig. 141).

The pharyngeal part (root) of the tongue, which is derived from the posterior portion of the subbranchial unit and the overlying part of the hyobranchial apparatus (cf. Fig. 109A), arises behind the foramen caecum. It is well shown in man (Gray, 1973) in which it is separated from the anterior part of the tongue by a V-shaped groove (sulcus terminalis). Behind that groove and the foramen caecum there is in the human embryo a large median swelling confusingly called the "copula" (Fig. 119D1). This swelling arises between the ventromedial parts of the hyoid and the first and

* p. 170.

second branchial arches and (it may be suspected) dorsal to the intervening part of the basibranchial series, and seems to be a thickening of the mucous membrane of the dorsal side of the hyobranchial apparatus. In the adult this membrane, which is innervated by the n. glossopharyngeus and the n. vagus, covers the pharyngeal part of the tongue.

The anterior or oral part of the amniote tongue is, as in anurans, mainly a derivative of the tuberculum impar, that is the anterior portion of the subbranchial unit. The hypobranchial muscles of that unit are innervated by the n. hypoglossus and conceivably the mucous membrane of the tuberculum impar is supplied by sensory fibres in that nerve, as is probably the main anterior part of the dorsum lingue in both urodeles and anurans (Jarvik, 1963, pp. 48–49, 55, 61–62). However, in amniotes the mucous membrane of the inner sides of the mandibular gill covers becomes thickened and forms a characteristic paired lateral lingual swelling (Fig. 119C, D2) which gradually overgrows the tuberculum impar in the posteromedial direction. In consequence of this overgrowth the mucous membrane of the tuberculum impar (probably innervated by sensory fibres in the n. hypoglossus) becomes gradually reduced and replaced by a mucous membrane derived from the mandibular gill covers (the lateral lingual swellings) and consequently innervated by the n. trigeminus and—because of the fusion between the mandibular and hyoid gill covers—by the n. facialis. Possibly the reduction in ontogeny of the dorsal roots and ganglia (Froriep's ganglia) of the n. hypoglossus and the first cervical nerves may be partly due to this gradual reduction of the original mucous membrane of the tuberculum impar (Jarvik, 1963, p. 55).

As we have now seen the tuberculum impar in amniotes forms all the muscles and their supporting elements (derivatives of the sublingual rod) of the oral part of the tongue. Only the mucous membrane of this part of the tongue, which is derived from the lateral lingual swellings and innervated by the trigeminus and the facialis, is visceral in origin. Posteriorly this membrane, on either side of the foramen caecum, is continued by the mucous membrane of the pharyngeal part of the tongue which is innervated by the glossopharyngeus and vagus.

Finally it may be mentioned that the salivary glands in mammals arise lateral to the lateral lingual swelling as invaginations of that part of the mucous membrane of the inner side of the mandibular gill cover, which is situated along and close to the lower jaw. These glands are thus to be regarded as derivatives of the lateral wall of the persisting ventral division of the spiracular gill slit (Figs 117C, 118).

12 The Orbitotemporal Region

Common Features

Common features in the posterior part of the orbitotemporal region of osteolepiforms and porolepiforms are a well developed basipterygoid process (posterior end of infrapharyngomandibular) articulating with the basal process at the proximal end of the pars pterygoquadrata (epimandibular), and a suprapterygoid process articulating with the processus ascendens palatoquadrati (suprapharyngomandibular; Jarvik, 1954; 1972). In stegocephalians the basipterygoid and ascending processes are generally well developed. The presence of an articular pit (fovea ovalis; Bystrow and Efremov, 1940) on the inner side of the palatoquadrate close behind the articular fossa for the basipterygoid process in *Benthosuchus* indicates that a paratemporal articulation* was retained in this Triassic form. In extant anurans (cf. Shishkin, 1967) and urodeles the basipterygoid process, following a general trend met with in fishes, has been lost and the palatoquadrate articulates with the otical shelf (Figs 49, 136; paratemporal articulation or commissure, Jarvik, 1972). In urodeles there is a suprapterygoid process articulating with the ascending process; in anurans this articulation is a larval feature or is absent (Swanepoel, 1970). In extant reptiles the basipterygoid process is generally well developed (de Beer, 1937). The ascending process forms a slender vertical rod (epipterygoid or columella cranii) which ends freely dorsally and the suprapterygoid process is lacking. In mammals (de Beer, 1937; Starck, 1967) the basipterygoid process is represented by the alar process. The ala temporalis (Fig. 60; alisphenoid) which arises separately but usually fuses with the tip of the alar process is interpreted as the ascending process. However, it sometimes (e.g. in *Myotis*; Frick, 1954) possesses a notch for the r. mandibularis trigemini and seems to be a more complex structure than the ascending process in the osteolepiform ancestors (cf. Figs 109, 129, Volume 1).

Also in the presence of a large pit for the anterior end of the notochord and of a processus connectens osteolepiforms agree with porolepiforms. In *Ichthyostega* the canal for the notochord decreases rapidly in width forwards but how far the notochord reached could not be established. Nor has it been possible to make out the details in the construction of the vestigial intracranial joint (fissura preoticalis; Figs 171B, 172, Volume 1). In all other tetrapods this joint has been lost, partly at least, in connection with the transformation of the subcranial muscle into cartilage (the polar

* Volume 1, p. 148.

cartilage) or bone. In urodeles the polar cartilage is probably represented by the basitrabecular commissure of Regel (1968) which is situated much as the subcranial muscle in porolepiforms, and other striking similarities with porolepiforms in this area have been demonstrated (Fig. 136; Jarvik, 1972). In eutetrapods the polar cartilage has been found independent in some birds (Bjerring, 1967; Toerien, 1971) but usually it has fused with the trabecula, the parachordal or other structures. The crowding together of several elements of different origin (hypophysial cartilages, exchordals, infrapharyngomandibular, infrapharyngohyal, polar cartilage, acrochordal) in the area at and behind the hypophysis renders a safe identification difficult and further investigations are needed. In mammals (*Talpa*; de Beer, 1937) the polar cartilage is possibly represented by the sphenocochlear commissure, whereas the alicochlear commissure which de Beer thought to be the polar cartilage (see also Starck, 1967) more likely is part of the infrapharyngohyal.

In most other respects osteolepiforms differ considerably from porolepiforms in the structure of the orbitotemporal region. The following three areas may be considered.

Hypophysial Area

The high and narrow fossa hypophyseos in osteolepiforms (*Eusthenopteron*, Figs 87, 196B, Volume 1) certainly lodged the neurohypophysis and, ventral to that, the adenohypophysis. No distinct compartment for the pars tuberalis is to be seen, but not unlikely it formed a paired processus lateralis as in amniotes (Wingstrand, 1966). A long buccohypophysial duct pierced the parasphenoid. In porolepiforms (Fig. 120A, B; Fig. 196A, Volume 1) the conditions are very different and remarkable similarities with urodeles in the hypophysial area as a whole have been demonstrated (Jarvik, 1972). The buccohypophysial duct was short and moreover it was paired (cf. Fig. 17).

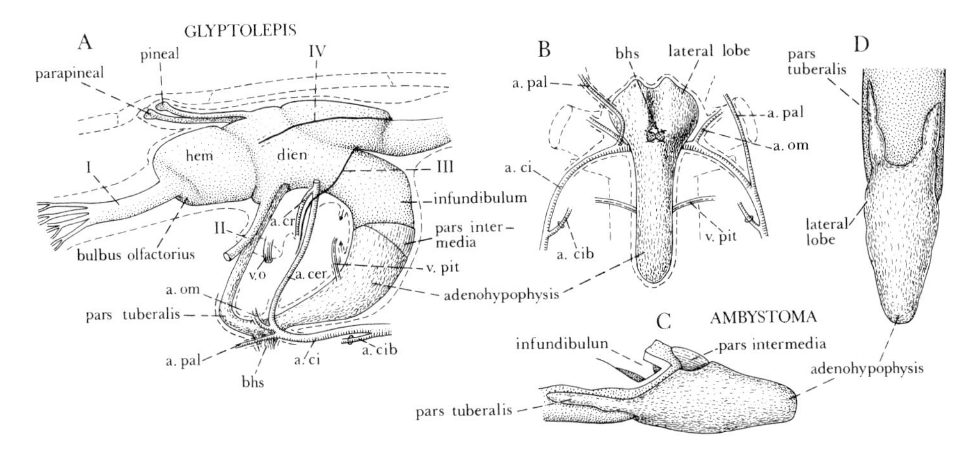

Fig. 120. Porolepiformes–Urodela. A, B, *Glyptolepis groenlandica*. Restoration of anterior part of brain, hypophysis and vessels in lateral and ventral aspects. C, D, *Ambystoma punctatum*, larva 40 mm. Hypophysis with adjoining part of brain in lateral and ventral aspects (after Atwell, 1921). From Jarvik (1972).

a.cer, cerebral artery; a.ci, a. carotis interna; a.cib, medial branch of a. carotis interna; a.cr, a. centralis retinae; a.om, a. ophthalmica magna; a.pal, a. palatina; bhs, left buccohypophysial duct; dien, diencephalon; hem, hemisphere; v.o, optic vein; v.pit, pituitary vein.

The pars tuberalis was a prominent tube-shaped structure emerging from a lateral lobe of the adenohypophysis exactly as in urodeles (Fig. 120C, D).

Neuroepiphysial Complex

In porolepiforms (*Porolepis*, *Glyptolepis*) the pineal and parapineal organs display a remarkable asymmetry (Figs 120A, 121A, 122A). The parapineal organ was situated on the left side, anterolateral to the pineal organ as it is in certain larval stages of primitive urodeles (Figs 121B, C, 122B; Jarvik, 1972; Bjerring, 1975). In osteolepiforms (*Eusthenopteron*), in contrast, both organs were situated in the median line (Fig. 122C) as in both larval and adult anurans (Fig. 122D).

Bulbus and Tractus Olfactorius

In urodeles the bulbus olfactorius is incorporated into the anterolateral part of the hemisphere of the forebrain as it must have been also in porolepiforms (Figs 120–122;

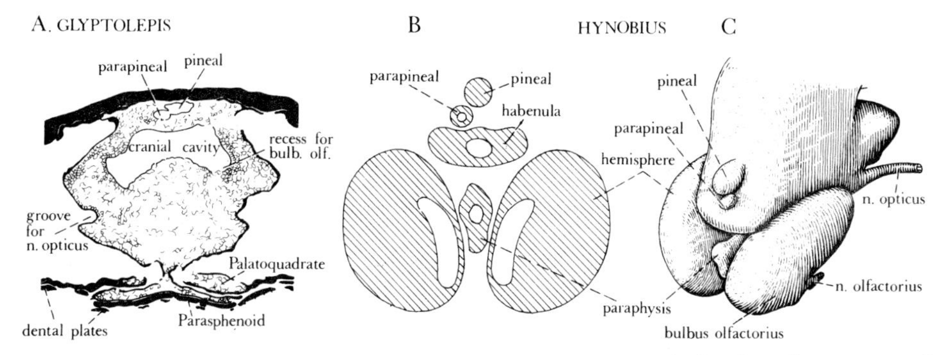

Fig. 121. Porolepiformes–Urodela, neuroepiphysial complex. A, *Glyptolepis groenlandica*. Part of transverse section of orbitotemporal region. From grinding series. B, C, *Hynobius retardatus*, embryo, 15·2 mm. B, diagrammatic transverse section. C, drawing of model of brain in anterolateral aspect. From Bjerring (1975).

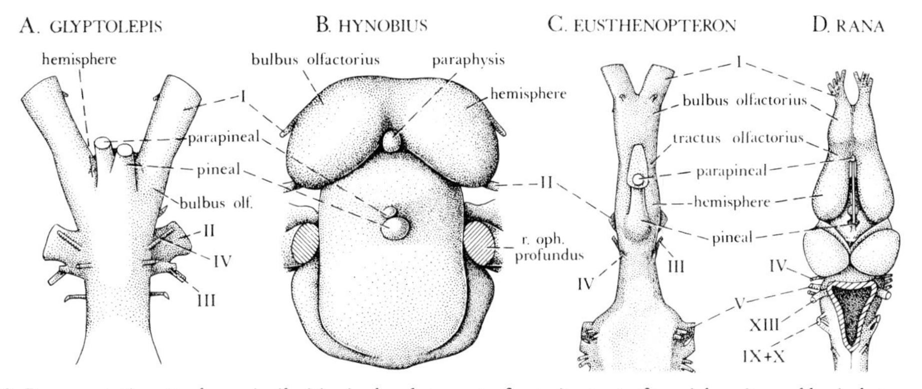

Fig. 122. Representations to show similarities in development of anterior part of cranial cavity and brain between, A, B, porolepiforms and urodeles and, C, D, between osteolepiforms and anurans. A, *Glyptolepis groenlandica*. Anterior part of cranial cavity with canals in dorsal view. B, *Hynobius retardatus*, embryo 15·2 mm. Drawing of same model as in Fig. 121 C. Dorsal view. C, *Eusthenopteron foordi*. Anterior part of cranial cavity with adjoining canals in dorsal view. D, *Rana*. Anterior part of brain in dorsal view. From Jarvik (1972).

Jarvik, 1942; 1972). In this regard we are concerned with a unique specialization indicating a close relationship. In all other gnathostomes the bulbus olfactorius lies in front of the forebrain and is separated from that by a more or less long tractus olfactorius. This was certainly the case also in *Eusthenopteron*, which in this regard, too, agrees well with anurans (Fig. 122C, D). In temnospondylous stegocephalians the orbitotemporal region usually is much elongated, but in a different way in different forms, a condition which has caused remarkable differences in the relative length of the tractus (Fig. 123; Jarvik, 1967a). In *Lyrocephalus*, for example, in which the elongation has occurred mainly in the postorbital zone of growth, the tractus has become elongated, whereas in *Benthosuchus* which has a long preorbital zone the elongation has affected the n. olfactorius.

A remarkable elongation of the tractus olfactorius is characteristic of amniotes. In connection with this elongation the brain has been displaced backwards into the posterior part of the cranial cavity and in sharp contrast to the straight brain stem in fish and amphibians the brain stem is flexed (Fig. 124; Jarvik, 1967a). Disregard of this striking difference and of the considerable elongation of the orbitotemporal region that is characteristic of many early tetrapods (Figs 123, 127) have contributed to the misinterpretations of the dermal bones of the cranial roof in osteolepiforms by Westoll, Romer, Parrington and others. In the forms shown in Fig. 126, representing osteolepiforms, anurans as well as reptiles and mammals, the orbitotemporal region is of moderate length. It is covered by a principal bone which in mammals and obviously also in the other three groups is the frontal, a bone in mammals developing in relation to the frontal diploic vein (Fig. 125). The parietal in these forms (also *Amia*, Fig. 36, Volume 1) is a structure, dorsal to the medial parts of the otic region and as explained above* (Fig. 61) it is related to three cranial tecta. Also in those early tetrapods, e.g. *Lyrocephalus* (Fig. 127), in which the otic region because of the elongation of the orbitotemporal region appears to be short, the parietal (including extrascapular components) is related to that region (Jarvik, 1967a; 1972, pp. 147–150).

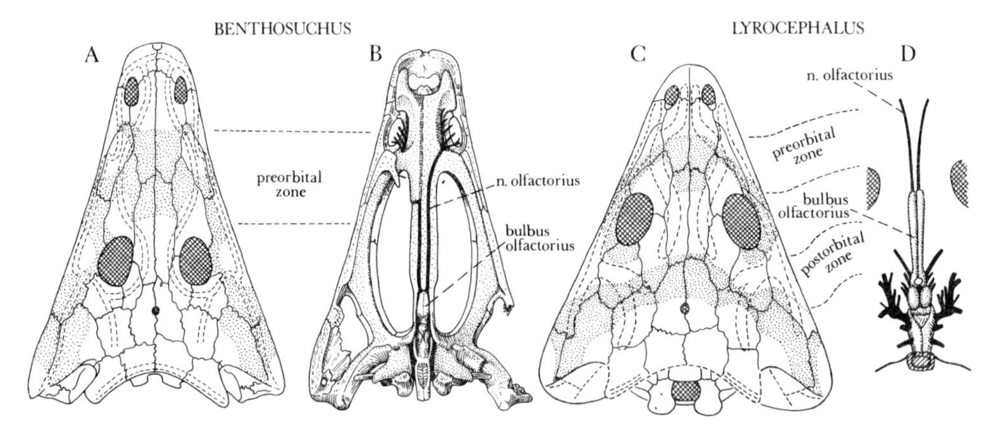

Fig. 123. Representations to illustrate correspondance between elongations of nervus and bulbus olfactorius and zones of intensive growth in temnospondyls. A, B, *Benthosuchus sushkini* (after Säve-Söderbergh, 1937 and Bystrow and Efremov, 1940). C, D, *Lyrocephalus euri* (after Säve-Söderbergh 1937). From Jarvik (1967a).

* pp. 103–104.

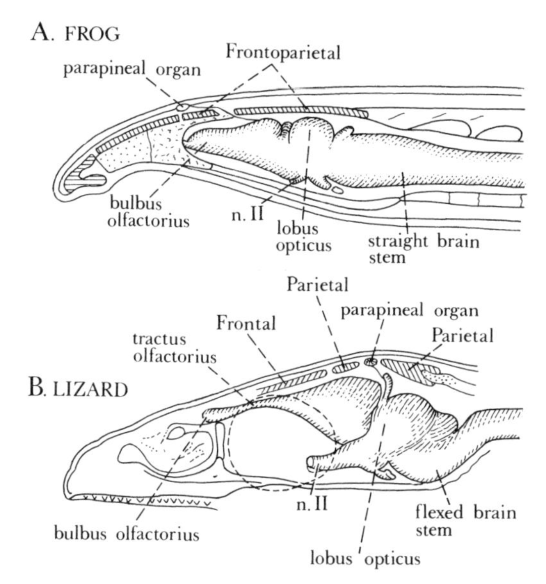

Fig.124. Diagrams to illustrate differences in position of neuroepiphysial complex between, A, batrachomorphs (anuran, mainly *Rana*) with straight brain stem and, B, reptilomorphs (lacertilian, mainly *Lacerta*) with flexed brain stem. From Jarvik (1967a).

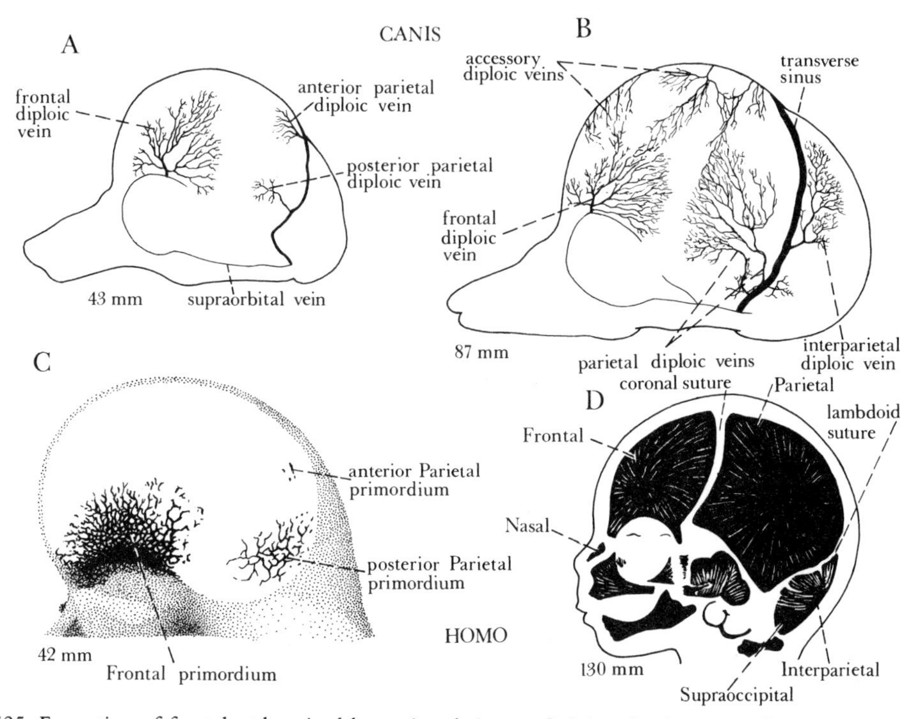

Fig. 125. Formation of frontal and parietal bones in relation to diploic veins in mammals. A, B, two stages in development of diploic veins in *Canis* (after Dziallas, 1952). C, D, two stages in development of frontal, parietal and intertemporal bones in human embryo (after Dziallas, 1954; Starck, 1955). From Jarvik (1967a).

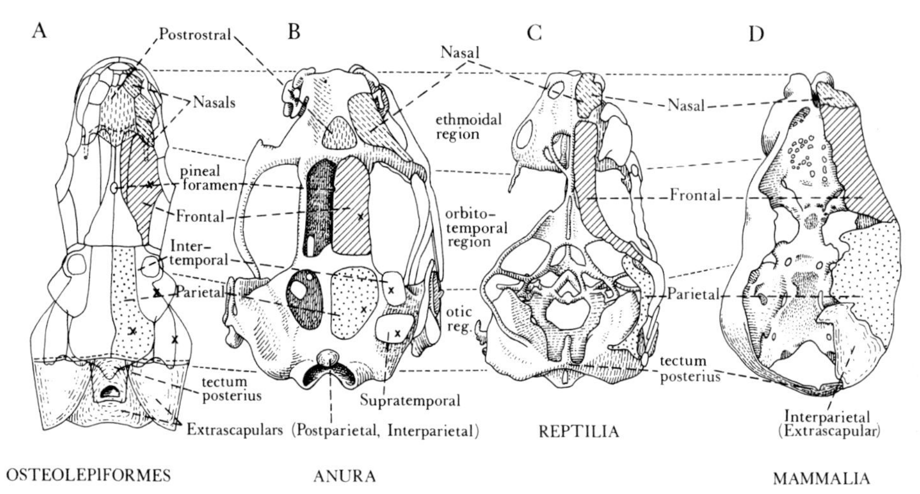

Fig. 126. Neurocranium with dermal bones of skull roof in dorsal aspect to demonstrate similarities in proportions and in position of frontal, parietal, and interparietal (extrascapular) bones in Osteolepipoda. A, osteolepiforms (*Eusthenopteron*), B, anurans (*Rana*, from Jarvik, 1968a), C, reptiles (*Lacerta*, embryo 47 mm, after Gaupp, 1905) and mammals (*Didelphys*, embryo 45·5 mm, after Töplitz, 1920). x marks position of centre of radiation (A) or area of origin (B) of dermal bones.

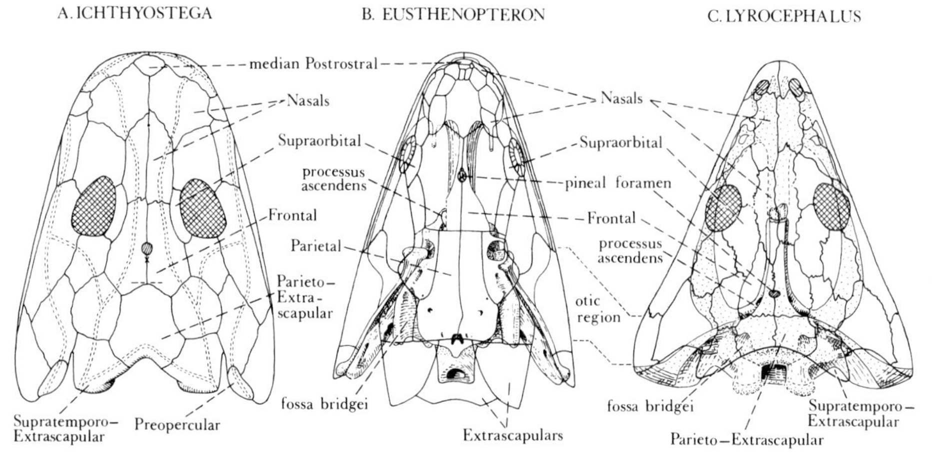

Fig. 127. Interpretation of dermal bones of skull roof in Osteolepipoda. A, *Ichthyostega* sp. Skull in dorsal aspect. Position of ventral opening of buccohypophysial canal indicated by x. Approximate position of fissura preoticalis marked with interrupted line. B, C, position of dermal bones in relation to neurocranium, palatoquadrate and hyomandibula (stapes) in, B, Devonian osteolepiform (*Eusthenopteron*) and, C, Triassic temnospondyl stegocephalian (*Lyrocephalus*). From Jarvik (1967a).

13 The Snout

The Porolepiform–Urodele and the Osteolepiform–Anuran Types

On the basis of excellent and unique fossil material, serial sections and wax models, the snout has been described in detail in representatives of both osteolepiforms (*Eusthenopteron*) and porolepiforms (*Porolepis, Glyptolepis, Holoptychius*; (Jarvik, 1942; 1972). With due consideration of the conditions in other porolepiforms (*Powichthys*, Jessen, 1975) and osteolepiforms (*Osteolepis, Thursius, Gyroptychius, Latvius, Glyptopomus, Megalichthys, Panderichthys*, etc; Jarvik, 1948; 1950; 1966; Vorobyeva, 1977) it has been shown that these two groups differ fundamentally from each other in the structure of this part of the head, as they do also in the anatomy of the postethmoidal parts and in the structure of the paired fins. After a thorough analysis of the structure of the snout in the various groups of lower gnathostomes (1942) it could be established that osteolepiforms agree closely with anurans, and only with that group, whereas porolepiforms in essential regards are as urodeles. On this basis osteolepiform–anuran and porolepiform–urodele types of snout could be distinguished. The main differences between these two types may be summarized as follows (Figs 128–137):

(1) In osteolepiforms and anurans the internasal wall forms a solid nasal septum which may be narrow as in *Eusthenopteron* and *Rana* (Fig. 133) or broad as in *Megalichthys* and *Ascaphus*.

In porolepiforms as well as in urodeles the internasal wall presents a deep internasal cavity (Figs 128, 131). This cavity, which lodged the internasal gland, is paired in porolepiforms as it is also in those urodeles in which a vestige of the internasal ridge, the cartilago infranasalis, is retained. The so-called septum nasi in urodeles is a transverse wall, the lamina precerebralis, which arises late in ontogeny and separates the internasal and cranial cavities.

(2) The anterior part of the nasal cavity in osteolepiforms is subdivided into three compartments corresponding to the recessus superior, recessus medium and recessus inferior of the anurans (Fig. 133). This subdivision is effected by formations homologous with the crista intermedia, lamina superior, lamina inferior and septomaxillary of the anurans. In porolepiforms and urodeles (Fig. 131) these formations are lacking and there is no subdivision of the anterior part of the nasal cavity into compartments.

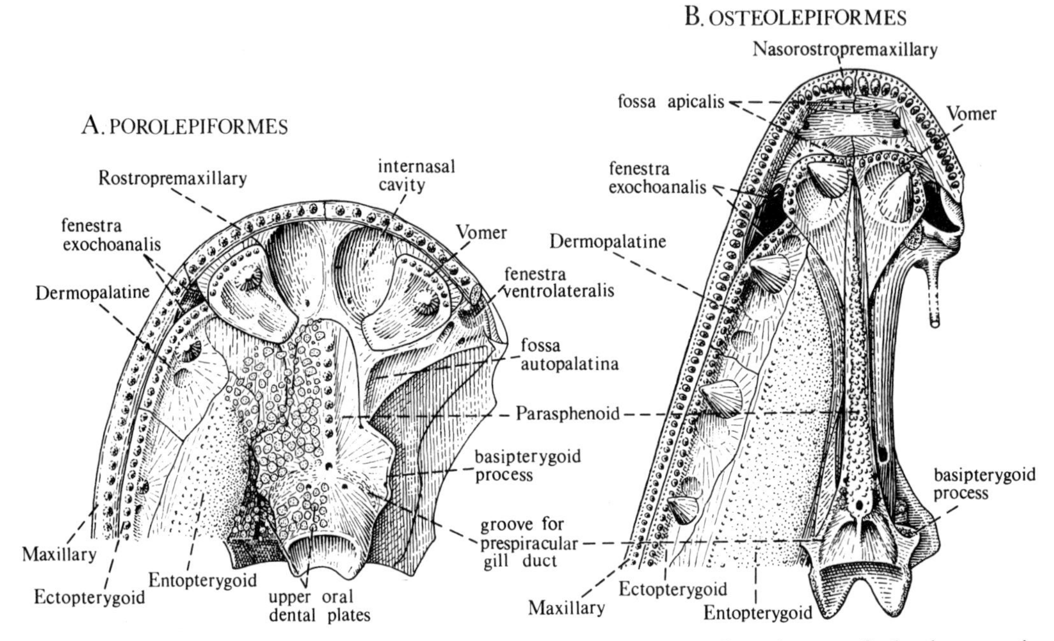

Fig. 128. Representations to illustrate similarities and differences in structure of anterior part of palate between, A, porolepiforms (mainly *Porolepis*) and, B, osteolepiforms (*Eusthenopteron*). From Jarvik (1972). For further explanation see Figs 189, 195, Volume 1 and Figs 82, 124, Volume 1.

(3) The septomaxillary in anurans is homologous with the processus dermintermedius of the lateral rostral in osteolepiforms (Figs 133–135, 137D).

In porolepiforms the lateral rostral has fused with the premaxillary and there is no processus dermintermedius. The so-called septomaxillary in urodeles is a nariodal homologous with the nariodal in porolepiforms (Figs 131, 137E; Jarvik, 1972, pp. 152–154, 190–192).

(4) Between the nariodal and the lachrymal in porolepiforms and certain urodeles there is an exoskeletal opening, the fenestra exonarina posterior (Fig. 132). In osteolepiforms and anurans there is no such opening (Fig. 134).

(5) In porolepiforms and urodeles the inner side of the posterior part of the lateral nasal wall presents a crest, the crista rostrocaudalis (Fig. 131). This crest has no equivalent in osteolepiforms and anurans.

(6)The crista rostrocaudalis in porolepiforms and urodeles forms the dorsal boundary of the lateral recess of the nasal cavity. This recess is bounded laterally by a part of the lateral nasal wall and ventrally by a part of the solum nasi (lamina ectochoanalis) which posteriorly forms a process, Seydel's palatal process, extending backwards into the fenestra ventrolateralis (Fig. 136).

In osteolepiforms and anurans (Fig. 134) the lateral recess is bounded by an incurved part of the lateral nasal wall only; there is no palatal process and no fenestra ventrolateralis.

(7) The fenestra ventrolateralis in porolepiforms, and when present in urodeles (Fig. 136D), includes the fenestra endochoanalis and the fenestra endonarina posterior.

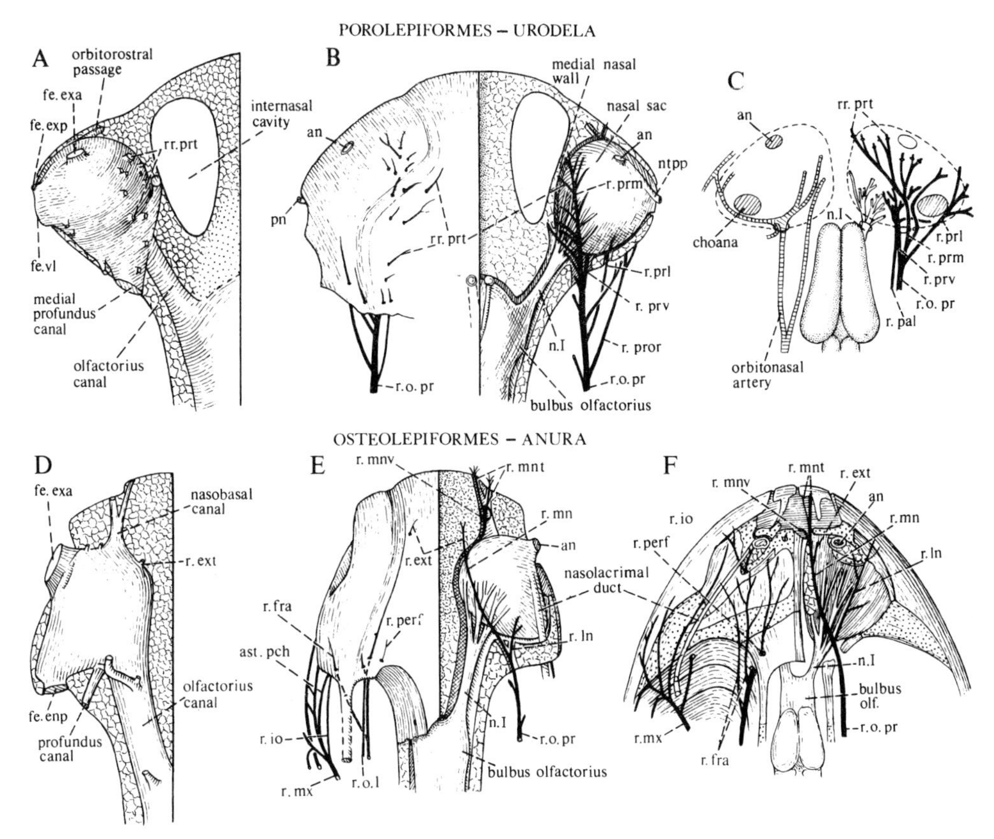

Fig. 129. Representations to illustrate differences in snout anatomy between Porolepiformes–Urodela and Osteolepiformes–Anura. Dorsal views. A, D, cast of left nasal cavity and canals of, A, porolepiform (*Porolepis*) and, D. osteolepiform (*Eusthenopteron*). From Jarvik (1972). B, E, restorations of nerves, nasal sac and certain other soft parts of, B, porolepiform (*Porolepis*) and, E, osteolepiform (*Eusthenopteron*). From Jarvik (1942; 1966; 1972). C, F, nerves and other soft parts of snout in, C, urodele (*Salamandra*) and, F, anuran (*Rana*). From Jarvik (1942, after Francis and Gaupp). For explanation of lettering see Fig. 130.

In osteolepiforms the fenestra endonarina posterior is an opening for the ductus nasolacrimalis in the postnasal wall, separated from the fenestra endochoanalis by an endoskeletal bridge. In anurans the lateral wall of the fenestra endonarina posterior is reduced and the fenestra is represented by a notch in the lateral side of the postnasal wall (Fig. 133).

(8) The posterior nasal tube in osteolepiforms which emerged from the nasal capsule through the fenestra endonarina posterior was developed as a nasolacrimal duct (Figs 129, 135). Just as in anurans, in which the duct passes the notch in the postnasal wall, it ran straight backwards from the nasal sac to the eye ball. The anterior part of the duct was situated close dorsal to the processus dermintermedius. During the transformation of that process into the horseshoe-shaped septomaxillary in the phylogeny of the anurans the duct migrated through the bone from dorsal to ventral. This migration is recapitulated in the ontogeny of anurans.

In porolepiforms and urodeles (Fig. 132) the conditions are different. The fenestra

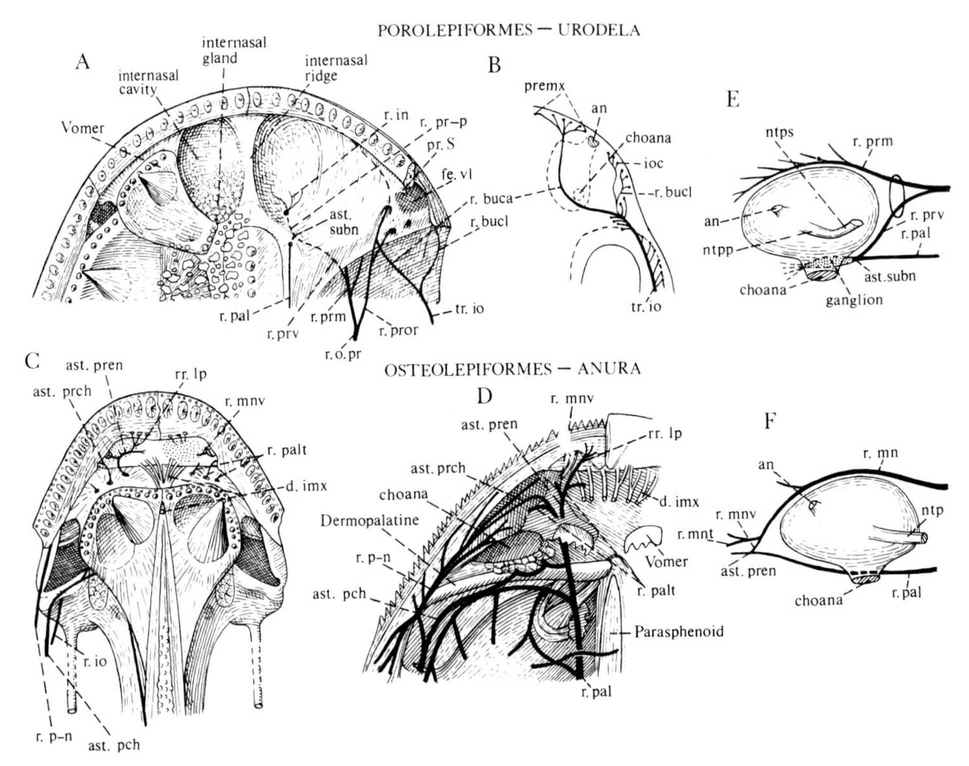

Fig. 130. Representations to illustrate differences in snout anatomy between, A, B, E, Porolepiformes–Urodela and, C, D, F, Osteolepiformes–Anura. A–D, ventral aspects. A, porolepiform (*Porolepis*); B, hynobiid urodele; C, osteolepiform (*Eusthenopteron*); D, anuran (*Rana*). From Jarvik (1942; 1972). E, F, diagrams of nasal sac and certain nerves in lateral aspect to show differences in development and position of profundus–palatine anastomoses; E, porolepiform–urodele condition; F, osteolepiform–anuran condition. From Jarvik (1964).

an, anterior nostril; ast.pch, ast.prch, ast.pren, postchoanal, prechoanal and prenasal anastomoses; ast.subn, subnasal anastomosis; d.imx, efferent ducts of intermaxillary gland; fe.enp, fenestra endonarina posterior; fe.exa, fe.exp, fenestra exonarina anterior and posterior; fe.vl, fenestra ventrolateralis; ioc, neuromasts of infraorbital sensory line; ntp, posterior nasal tube (nasolacrimal duct); ntpp, ntps, primary and secondary portions of posterior nasal tube; n.I, nervus olfactorius; pn, posterior nostril; premx, neuromasts in premaxillary area belonging to terminal arch line; pr.S, Seydel's palatal process; r.buca, anterior branch of r. buccalis lateralis to anterior neuromasts of infraorbital sensory line; r.bucl, lateral branch of r. buccalis (probably including fibres of r. maxillaris); r.ext, r. externus narium of n. profundus; r.fra, rami frontales anteriores of n. profundus; r.in, twigs of combined profundus–palatine nerve to internasal gland; r.io, r. infraorbitalis of r. maxillaris; r.ln, r.mn, r. lateralis and medialis narium; r.mnt, lateral and medial terminal branches of r. medialis narium; r.mnv, ventral terminal branch of r. medialis narium anastomosing with r. palatinus; r.mx, r. maxillaris profundi et terminali; r.o.l, r. ophthalmicus lateralis; r.o.pr, r. ophthalmicus profundi; r.pal, r. palatinus; r. palt, lateral, anterolateral and medial terminal branches of r. palatinus; r.perf, r. frontalis perforans profundi; r.p-n, r. palatonasalis; r.prl, r.prm, lateral and medial terminal profundus branches; r.pror, terminal profundus branch passing orbitorostral passage; r.pr-p, combined profundus-palatine nerve; r.prv, ventral terminal profundus branch anastomosing with r. palatinus; rr.lp, branches of prenasal anastomosis to palatal lamina of ethmoidal shield; rr.prt, twigs of medial terminal profundus branch; tr.io, truncus infraorbitalis.

exonarina posterior has migrated downwards to be situated posteroventral to the anterior external nostril and rather close to the margin of the upper jaw. Due to this migration the infraorbital sensory canal has been pushed downwards and makes a sharp bend downwards in the anterior part of the lachrymal in porolepiforms and in the corresponding place in urodeles. The nasolacrimal duct in larvae of hynobiid

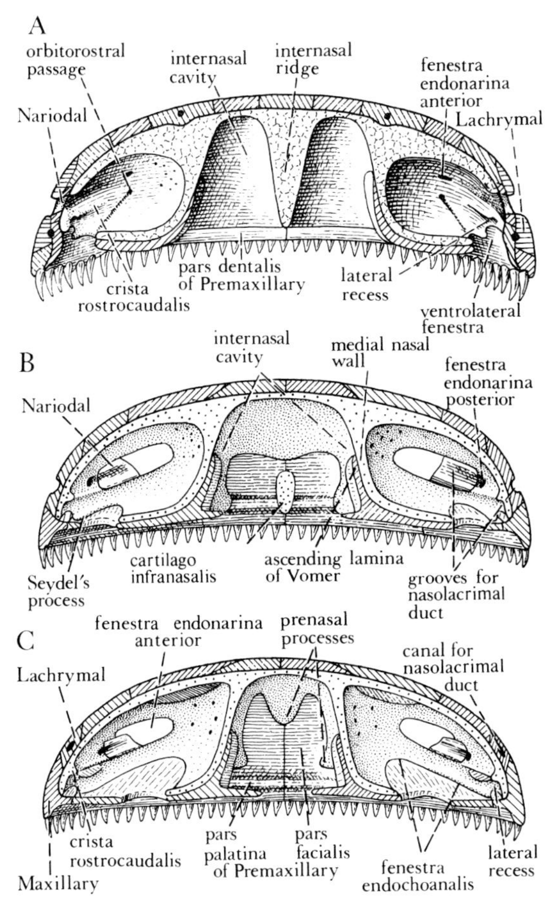

Fig. 131. Three diagrammatic drawings showing probable evolution of urodele snout from early Devonian porolepiform. All figures show snout from behind with hindmost part, including postnasal wall, removed. Cross-sections through endocranial bone tissue reticulated, through dermal bones ruled; cartilage dotted. A, *Porolepis spitsbergensis* with certain details from *Glyptolepis* and *Holoptychius*. B, primitive and, C, more advanced urodele stage. Structures illustrated in B, and C are such which have been observed in larval or primitive adult recent urodeles. From Jarvik (1942; 1972) (for further explanations see Jarvik, 1972, Fig. 77).

urodeles has a remarkable course. Its proximal and first formed (primary) portion runs in the posteroventral direction towards the opening between the nariodal and lachrymal which corresponds to the fenestra exonarina posterior in porolepiforms; in this part of its course it may be grasped from the outside by the nariodal. However, at this opening the duct makes a characteristic sharp bend and is continued by its distal (secondary) portion which runs obliquely upwards to the eye, in early larval stages in a groove on the outside of the lachrymal. In porolepiforms (Fig. 132) the conditions were similar. In *Porolepis* the posterior nasal tube ran very much as the proximal and first formed portion of the nasolacrimal duct in larval urodeles and is certainly homologous with that portion. However, the tube was still short and its external orifice, the posterior external nostril, was situated in the fenestra exonarina posterior. In *Glyptolepis*, in contrast, the tube had acquired an equivalent also to the distal (secondary) portion of the nasolacrimal duct and exactly as in larval urodeles the tube (duct) made a sharp bend in the area of the fenestra exonarina posterior and continued obliquely upwards towards the eye in a groove on the outside of the lachrymal.

As in larval urodeles the nasolacrimal duct in *Glyptolepis* ran close to the underlying infraorbital sensory canal. However, (as in *Eusthenopteron*) the duct and the sensory canal are quite separate and since the distal part of the duct in *Glyptolepis* no doubt has been formed by a posterior prolongation of a short posterior external nasal tube as present in *Porolepis*, the view (Schmalhausen, 1968; Medvedeva, 1975, fig. 78) that the nasolacrimal duct in urodeles is formed by a part of the infraorbital sensory canal cannot possibly be true. The fact that the anterior portion of the infraorbital sensory line in urodeles arises from a separate placode is only what is to be expected, this portion being the terminal arch line.*

The facts that the nasolacrimal duct in *Glyptolepis* made a sharp bend in the area of the fenestra exonarina posterior, exactly as it does in larval urodeles and that, as was predicted in 1942, its secondary portion ran obliquely backwards and upwards towards the eye in a groove on the outside of the lachrymal form a positive proof of the view (supported by much other conclusive evidence) that urodeles are not only related to, but descendants of, primitive porolepiforms (Figs 140, 141).

(9) In osteolepiforms and anurans (Figs 129, 133) the profundus canal in the postnasal wall is of moderate size, there is only one profundus canal in the tectum nasi, and there are passages for the r. maxillaris in the ventrolateral part of the snout. Moreover there is a nasobasal canal or foramen and the vomer is pierced by a palatine canal which in the osteolepiforms divides within the bone into three canals indicating that the r. palatinus broke up into three terminal branches as in anurans (Fig. 130).

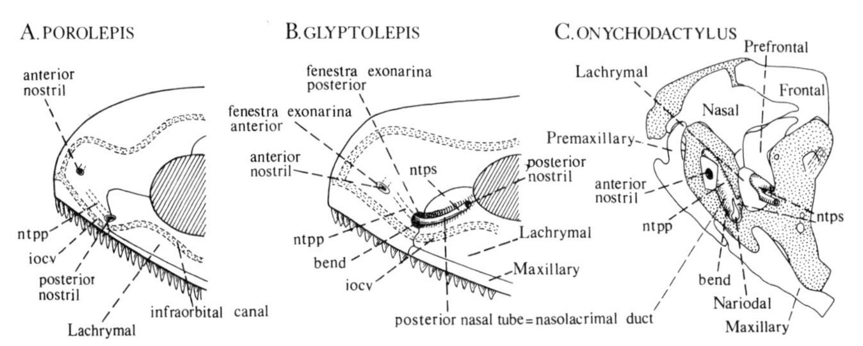

Fig. 132. Transformation of posterior nasal tube into nasolacrimal duct in urodele phylogeny. Even in the earliest known porolepiforms (*Porolepis*, A; Fig. 181A, Volume 1) the fenestra exonarina posterior had moved ventrally and the infraorbital canal runs in a characteristic curve (*iocv*) as it still does in larval urodeles. The posterior nasal tube was in *Porolepis* still short and opened outwards by a posterior nostril in the fenestra exonarina posterior. This short portion running in a groove on the inner side of the nariodal constitutes the primary portion (ntpp) of the nasolacrimal duct. In *Glyptolepis* (B; but not in *Holoptychius*) a secondary portion (ntps) running towards the eye in a groove on the outside of the lachrymal has been added. In larval urodeles (C), too, the distal (secondary) portion of the posterior nasal tube, or as it may now be termed, the nasolacrimal duct, runs in a groove on the outside of the lachrymal and in urodeles the duct (tube) made a characteristic bend when passing the fenestra exonarina posterior, as it did in *Glyptolepis*. In this case we are concerned with a remarkable recapitulation in ontogeny of the phyletic development of the urodeles from the porolepiform ancestors.

* p. 53.

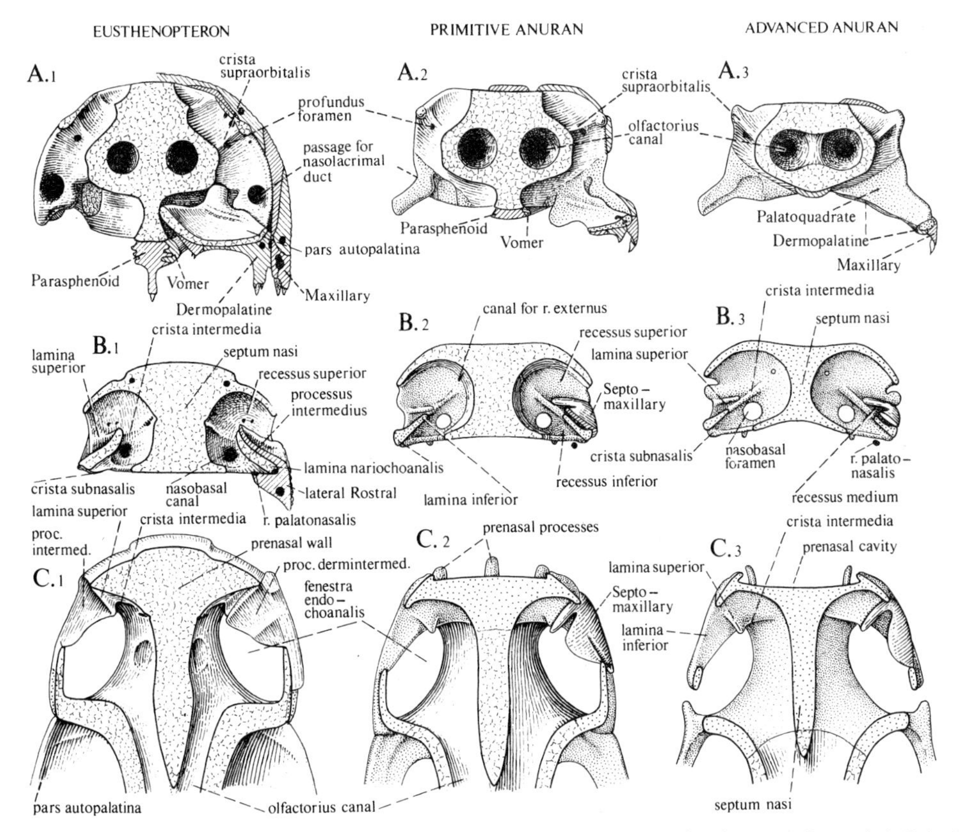

Fig. 133. Drawings to illustrate probable transformations in snout anatomy in development from, A.1, B.1, C.1, osteolepiforms (*Eusthenopteron*) via A.2, B.2, C.2, primitive to A.3, B.3, C.3 advanced anurans. Structures illustrated in A.2–C.2 have been observed in larval or primitive anurans. From Jarvik (1942).

In porolepiforms and urodeles (Fig. 129) the postnasal wall is pierced by a wide medial and one or more lateral profundus canals, the tectum nasi shows several profundus foramina and nasobasal canal is lacking. No palatine canal in the vomer corresponding to that in osteolepiforms and anurans is present. In porolepiforms there is an orbitorostral passage (Fig. 131) but no passages for maxillaris branches in the ventrolateral part of the snout, conditions which agree with the facts that nerves in urodeles course forwards in a position corresponding to the orbitorostral passage and that the r. maxillaris is weakly developed.

(10) If we put nerves in all the canals and other passages found in the snout of osteolepiforms (*Eusthenopteron*), taking into consideration the size, direction, and morphological relations of these structures, it will be seen that the nerves ran and branched almost exactly as in an extant anuran like *Rana* (Figs 129, 130).

If we do the same with *Porolepis* or *Glyptolepis* we will find a corresponding close agreement with *Salamandra* and hynobiid urodeles (Figs 129, 130).

(11) In osteolepiforms, as in anurans, there is an anterior palatal fenestra. In osteolepiforms this fenestra is found in the roof of a characteristic fossa, the fossa apicalis (Figs 128, 134). This fossa, bounded mainly by dermal bones, is situated below

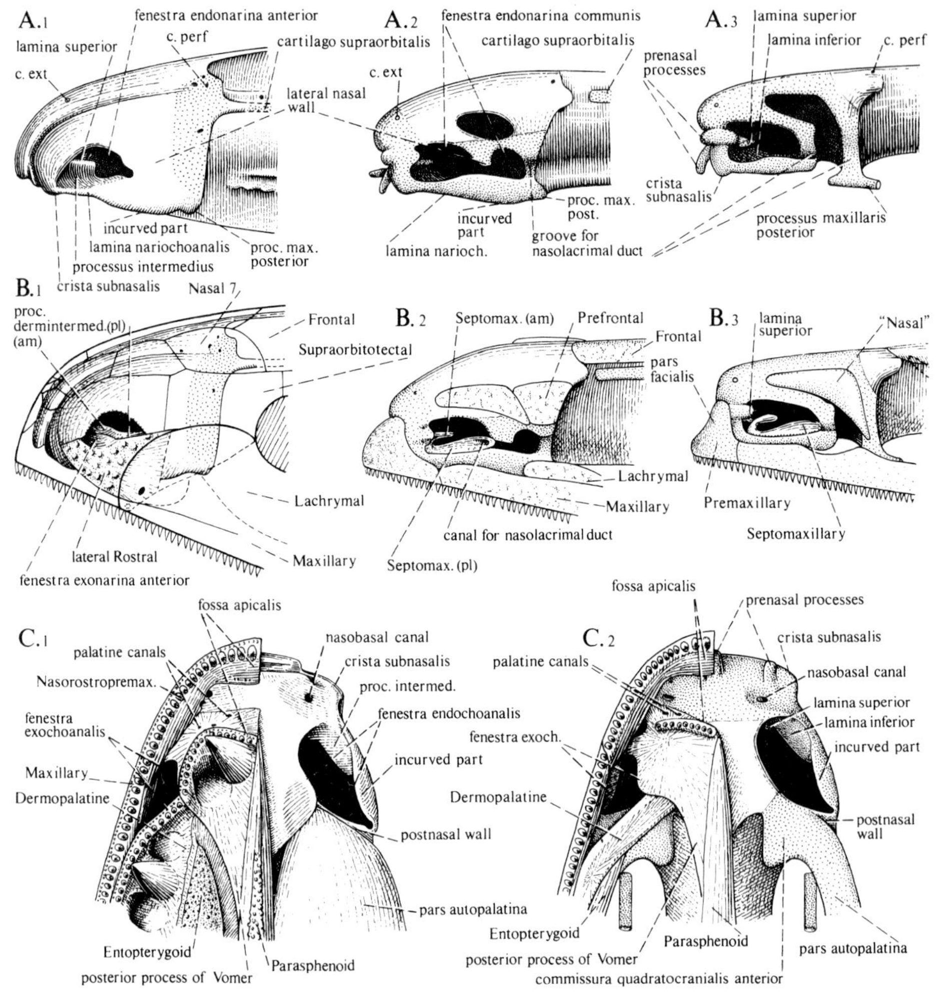

Fig. 134. Further drawings to illustrate transformations in snout anatomy in development from osteolepiforms (*Eusthenopteron*) to advanced anurans. From Jarvik (1942).

the divisio prenasalis communis of the neurocranium and is bounded posteriorly by the more or less transverse portions of the descending tooth-bearing laminae of the vomers. When complete this wall, as in *Eusthenopteron* (Fig. 82, Volume 1), is pierced by a median intervomerine canal with a funnel-shaped anterior opening into the fossa apicalis. The presence of the intervomerine canal and the position and configuration of the fossa apicalis suggest that this fossa lodged a gland homologous with the intermaxillary gland in anurans (Fig. 130; Fig. 124, Volume 1). Conceivably this gland originally arose in the prenasal pit and most likely it also in the ostelepiforms was innervated by twigs of the prenasal nerve-anatomosis formed by a dorsal profundus branch passing forwards dorsal to the nasal sac and joining the r. palatinus in front of that sac (Fig. 130F). Also nasal muscles corresponding to those in anurans were probably present in osteolepiforms, too (Jarvik, 1972, p. 192).

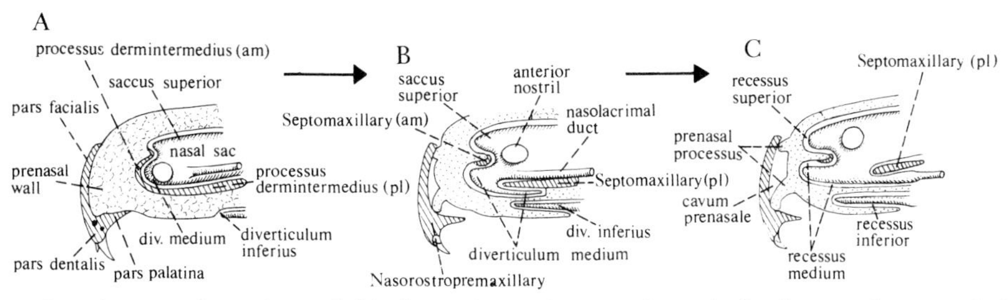

Fig. 135. Three drawings illustrating probable changes in anterior part of snout in development from osteolepiforms (*Eusthenopteron*) to anurans. These changes are partly due to retrogressive development of skeleton. Note that also in this figure structures illustrated in B are such which have been observed in larval or primitive anurans. From Jarvik (1942).

am, pl, anteromedial and posterolateral shanks of anuran septomaxillary and corresponding parts of lateral rostral in osteolepiforms.

In porolepiforms and urodeles (Fig. 136) there is no fossa apicalis in front of the descending tooth-bearing laminae of the vomers and no intervomerine canal. The anterior palatal fenestra is situated mainly below the internasal wall and is the opening of the cavum internasale, which also in urodeles may be paired and protected laterally by an ascending lamina of the vomer (Fig. 131). In urodeles the internasal cavity is occupied by a gland which I (1966; 1972), following the early anatomists (Wiedersheim, Gaupp), prefer to call the internasal gland in contrast to the intermaxillary gland in anurans. In porolepiforms the anterior and deepest part of the internasal cavity is, when the mouth is closed, occupied by the tips of the anterior tusks of the parasymphysial dental plate. Posteriorly there is sufficient place for an internasal gland and very likely a small gland was present in the early porolepiforms (Fig. 130). If so this gland was innervated in the same way as in urodeles, namely by twigs of a subnasal nerve anastomosis formed by a profundus branch which (in contrast to anurans) passed ventral to the nasal sac and joined the palatine nerve medial to the choana (Fig. 130E). There is also some evidence that nasal muscles similar to those in urodeles were present already in porolepiforms.

The facts now presented and several other not recorded in this brief review (see Jarvik, 1942; 1966; 1972) demonstrate a most close agreement in the anatomy of the snout as a whole between osteolepiforms and anurans, on the one hand, and between porolepiforms and urodeles, on the other. The only possible conclusions to be drawn from this strong evidence are those I came to in 1942 and further emphasized in 1966 and 1972, namely that the anurans are closely related to and descendants of osteolepiforms, whereas the urodeles are allied to the porolepiforms and have evolved from primitive porolepiforms, and that the Amphibia accordingly are diphyletic in origin (Figs 140, 141).

General Concluding Remarks

The discovery, in 1956, of well preserved material of a porolepiform, *Glyptolepis groenlandica*, made it possible to extend the comparative analyses also to the postethmoidal parts of the head. Because the results gained in 1942 were based on thorough analyses of a large and complex region of the head in various groups and not on one or

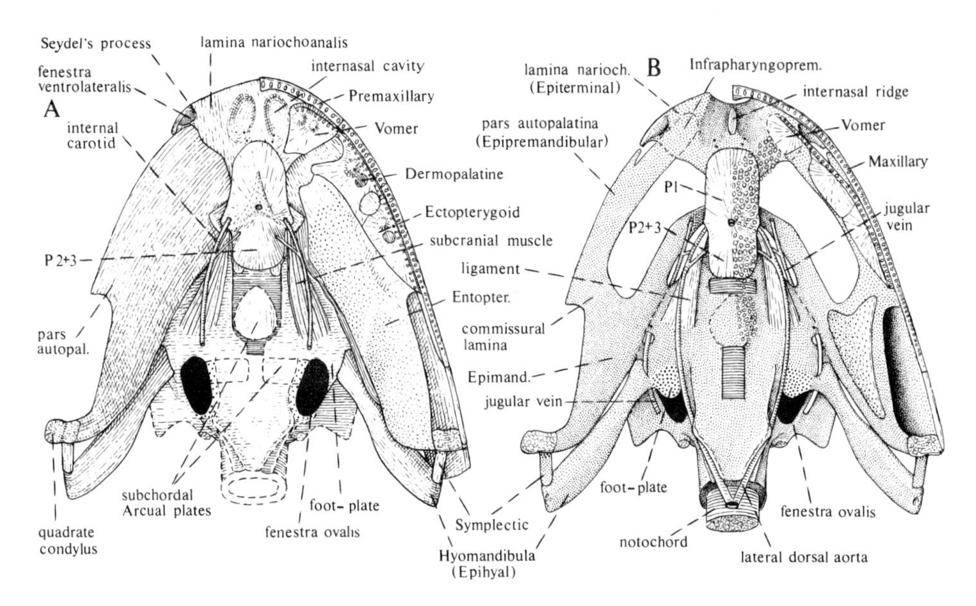

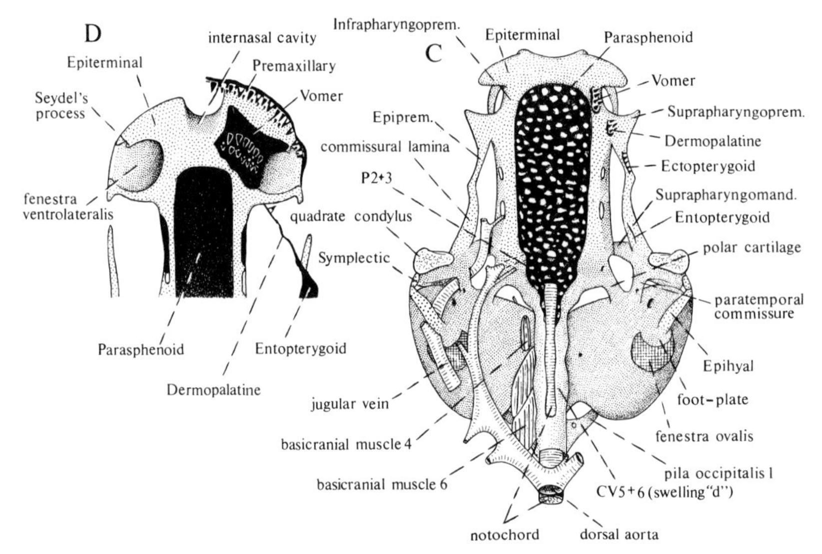

Fig. 136. A–C, diagrammatic representations to demonstrate similarities between, A, porolepiforms (*Glyptolepis*) and, C, primitive urodeles (*Ranodon*) in structure of palatal aspect of skull and, B, probable changes in phylogeny. D, *Ranodon sibiricus*, larva 76 mm. Anterior part of palate in ventral aspect. From Jarvik (1972) (urodeles after Regel and Lebedkina).

a few isolated characters and one-sided comparisons, it was hardly surprising that the new evidence fitted in with and thus strongly supported my earlier results.

Thus it was shown (1962; 1963; 1963a) that the porolepiforms in the intermandibular division, another complex part of the head, in a great number of characters agree with urodeles and only with that group, whereas the osteolepiforms are quite different and agree with anurans. Because it also was established that the tongue in urodeles and anurans has arisen independently from the conditions in the porolepiform and osteolepiform ancestors, respectively, the fact that "the mechanism of the tongue of frogs . . . is fundamentally distinct from that of salamanders" (Regal and Gans, 1976, p. 718) is only what is to be expected. Of considerable interest is, moreover, the fact that also the amniotes in the structure of the intermandibular division and the origin of the tongue are similar to osteolepiforms and anurans. These conditions are in agreement with the view (indicated already in 1942) that the amniotes are descendants of osteolepiforms and that it is justifiable to distinguish an osteolepiform-eutetrapod stock.

Also as to the position of the bulbus olfactorius the porolepiforms and urodeles are unique and differ from osteolepiforms and other eutetrapods.* Moreover there are significant resemblances between porolepiforms and urodeles in the anatomy of the hypophysial area and in the structure of the neuroepiphysial complex.† Most spectacular and important differences in the vascular supply of the orbit and lower jaw and the drainage of the endolymphatic sac between urodeles and anurans have been demonstrated‡; and also in these regards the anurans and other eutetrapods agree with the osteolepiforms whereas the porolepiforms, as far as they are known, are as urodeles. Finally there are important differences in the sound conducting apparatus between porolepiforms and urodeles, on the one hand, and osteolepiforms and eutetrapods on the other, and as we have seen the three ear ossicles in mammals may be derived directly from the conditions in *Eusthenopteron*.

If it is added that the fore and hind limbs in eutetrapods can be easily derived from the pectoral and pelvic fins in osteolepiforms, that porolepiforms in the structure of these fins and the endoskeletal shoulder girdle differ fundamentally from osteolepiforms, and that the limbs in urodeles have arisen independently of those in eutetrapods, it is evident that the view that the amphibians and the tetrapods as a whole are diphyletic now rests on a solid basis.

These new views have been accepted by an increasing number of students, but they have also been criticized or ignored by students who still think that the "Amphibia" and the Tetrapoda as a whole are descendants of some hypothetical proto-tetrapod and accordingly are monophyletic in origin (Watson, 1940; Westoll, 1943; 1961; Kesteven, 1950; Szarski, 1962; 1977; Parson and Williams, 1963; Romer, 1968; Schaeffer, 1969; Torrey, 1971; Panchen, 1972; Jurgens, 1973; Estes and Reig, 1973; and others). Most of the objections have been thoroughly discussed and refuted in previous papers (Jarvik, 1962; 1964; 1965; 1966). As emphasized in these papers it is of course impossible to base any conclusions, as has often been done, on wrong homologizations, misleading redrawings of figures, or descriptions of porolepiforms and osteolepiforms that have been proved to be erroneous in all relevant regards. It is meaningless to discuss these and other obvious mistakes any more and only a few remarks will be given.

(1) Statements in the literature to the effect that there are no important differences in snout structure between osteolepiforms and porolepiforms, and that the differences, which I claimed to exist disappear if a greater number of forms than was available to

* pp. 205–206; † pp. 205; ‡ pp. 183–185.

me in 1942 are studied, are also after 1966 used as important arguments for monophyletism. There is certainly some variation both in porolepiforms and osteolepiforms but hitherto not a single form which in any way bridges the gap between the two groups has been described. Porolepiforms and osteolepiforms are quite distinct and they differ fundamentally from each other not only in snout structure but also in the cranial anatomy as a whole as well as in fin and scale structure. They represent two quite distinct taxa which have been separate since pre-Devonian times (Figs 139, 140).

(2) Urodeles are by tradition considered to be more primitive than anurans, and on the basis of this prejudice many attempts have been made to eliminate the differences between these two distinct groups of extant Amphibia. Because it is generally agreed that tetrapods have evolved from "rhipidistids" it would of course be natural to start with them when discussing the origin and classification of extant amphibians, but this is generally not done. Neither Estes in his discussion of salamander origins (1965) or other modern students of urodele phylogeny (e.g. Wake, 1963; 1966), nor the many students who in recent years have studied the phylogeny of the anurans (see articles by L. Trueb, J. D. Lynch and others, in Vial, 1973) have tried to utilize the new possibilities provided by the exploration of the porolepiforms and osteolepiforms and the safe knowledge we now have of essential parts of the cranial antomy in both these groups of "rhipidistids". These writers therefore have to base their statements as to primitiveness and specialization on other criteria, and as pointed out by Sokol (1977) tradition certainly plays an important part. The remarkable course of the nasolacrimal duct in porolepiforms and larval hynobiids (Fig. 132) clearly supports the traditional view (Noble, 1931) that the hynobiids are among the most primitive urodeles; but also the innervation of the snout, the drainage of the endolymphatic sac and other particulars show that such forms as *Salamandra* and *Ambystoma*, too, have retained primitive conditions. Also the evolution of the hyobranchial apparatus from the porolepiform stage (Fig. 111) is in agreement with these conclusions at the same time as it is clear that the perennibranchiats (*Proteus*, *Necturus*) in this regard are advanced (cf. Vandel, 1966).

However, is it really true that the ascaphids (*Ascaphus*, *Leiopelma*) and the tongueless (aglossan) pipoids (e.g. *Xenopus* and *Pipa*), also referred to as archaic frogs, are primitive anurans which in certain regards approach the urodeles? According to Ritland (1955) *Ascaphus*, characterized by strange larval adaptions for life in mountain torrents, is not primitive and the tongue-less frogs (Aglossa) which in the innervation of the snout (Paterson, 1939; Jarvik, 1942, p. 332) and in certain other regards are said to be urodelelike "are not primitive frogs by any criteria", but "specialized offshoots of the ranoidean . . . lineage" (Sokol, 1977, p. 507). The fact that *Rana* even as to details in the anatomy of the snout and in other respects agrees closely with *Eusthenopteron* can only be taken to mean that the ranids in these regards are primitive. If then, as claimed by Sokol, the pipoids are derivatives of ranoid frogs it is evident that the aberrant innervation of the snout and other alleged urodele similarities (e.g. in the nasal sac; Thrams, 1963; Jarvik, 1942, p. 325) in these highly specialized tongue-less frogs have arisen within the Anura, independently of the conditions in urodeles.

Jurgens (1973), who accepts the traditional view that *Ascaphus* and *Leiopelma* are the most primitive living frogs, has to admit that "the cranial structure does not deviate significantly from the general anuran pattern". As a matter of fact *Ascaphus*, in the snout, in the cranial anatomy as a whole, as well as in the postcranial anatomy, is an unmistakable anuran and it differs fundamentally from the urodeles in numerous respects. However, in the snout of *Ascaphus* (and certain other frogs) Jurgens has found a lateral profundus branch which he homologizes with the lateral terminal profundus branch in urodeles (present also in porolepiforms) and terms the r. lateralis nasi proper. In plain language this homologization implies not only that this profundus branch has been inherited from a presumed common ancestor of urodeles and anurans, but also that this rather inconsiderable anatomical detail was in existence and did not change during the vast period of time when the whole snout was remodelled in an anuran fashion and the jumping legs and other characteristics of the anurans arose. This is of course quite unreasonable and the problem is rather to explain why *Ascaphus* in this particular respect differs from *Rana* and other typical anurans. Unacceptable, too, is Jurgens' final conclusion that the apparent differences in snout structure between urodeles and anurans "can be ascribed to the actual degree of terrestralisation". As is well known many prerequisites for a life on land (choana, tetrapod limb, etc.) were already present in the fish stage and are not due to "terrestralisation". Similarly the crista intermedia, lamina superior, lamina inferior and numerous other characteristics of the anurans were developed already in their osteolepiform ancestors, whereas the crista rostrocaudalis and other characteristics of urodeles had arisen in their porolepiform forerunners and were present as early as in the Lower Devonian.

(3) A widespread conception among palaeontologists is that the ancestors of the three groups of living amphibians (Urodela, Anura, Apoda) are to be found among the Palaeozoic tetrapods. The search for these ancestral forms has resulted in many suggestions and much controversy, but hitherto no indisputable ancestor to any of these groups has been presented. The main reason for this is no doubt the incompleteness of the fossil record, but another important fact is that the endoskeleton of the head in the Palaeozoic tetrapods—partly at least due to retrogressive development—is often imperfectly preserved or lacking. Detailed analyses of the cranial anatomy as rendered possible by the excellent fossil material of osteolepiforms (*Eusthenopteron*) and porolepiforms (*Glyptolepis*) have therefore not been practicable and the students of the early tetrapods have to base their conclusions in the first place on the exoskeleton (which also may be more or less reduced), and what may be preserved of the postcranial skeleton.

The introduction of the "lissamphibian" hypothesis (Parsons and Williams, 1963; cf. Shishkin, 1970; 1973) induced fresh life into the discussion since it now—in the opinion of the partisans of this hypothesis—became necessary to postulate a common ancestor to all the three groups of modern amphibians (urodeles, anurans, apodans) among the Palaeozoic tetrapods. Parsons and Williams presented a list of 19 characters to be expected in the common "lissamphibian" ancestor and since no less than 13 of these characters were found in the temnospondyls Estes (1965) argued that the temnospondyls, in particular the dissorophids, are the most plausible "lissamphibian" relatives. However, most of the characters included in Parsons and Williams' list, as well as in the tabulation given by Estes, such as "body of moderate length", "appendicular skeleton well developed", are of no value for assessing relationship. More to the point is certainly the unexpected discovery (Bolt, 1969; 1977; see also Boy, 1978) of bicuspid, pedicellate teeth just in the dissorophids. This is of interest since the presence of such teeth is considered to be an important common character of all modern amphibians (Parsons and Williams, 1962), and according to Estes and Reig (1973, p. 45) together with an "operculum-plectrum complex", is the only osteological character that links "the three modern orders most intimately". However, the urodeles lack tympanic membranes and differ also in other important regards from the anurans in the structure of the middle ear* (Jarvik 1965; 1972); and the apodans lack also the opercular plate and muscle. Accordingly only one osteological character remains: the presence of bicuspid pedicellate teeth. It is of course not possible to base any safe conclusions as to relationship on a single character and considering the justified objections raised by Lehman (1968) and the facts that such hinged teeth occur also in other groups of vertebrates, whereas many extant and obviously also several fossil "lissamphibians" (urodeles, Estes, 1965; 1969; Estes and Hoffstetter, 1976; anurans, Estes and Reig, 1973) have non-pedicellate teeth it is evident that hinged teeth cannot be a reliable "lissamphibian" character either (as to bicuspid teeth in various fishes see Bystrow, 1939, fig. 8). The dissorophids only provide another example of the fact that such teeth can easily arise independently, as they certainly have also in the three groups of modern amphibians, none of which, as we shall see, are descendants of dissorophids (cf. Romer, 1968, p. 94).

Frogs of modern type appear as early as the Lower Jurassic but the discovery (Piveteau, 1937) of a single specimen of a froglike animal, *Triadobatrachus* (*Protobatrachus*), in the Eotriassic of Madagascar brings the ancestry of the anurans back to the beginning of the Mesozoic. Estes and Reig (1973, p. 49) accept *Triadobatrachus* as "a representative of an ancestral proanuran stock" but they also state "Unfortunately, there is no special resemblance of *Triadobatrachus* to any Paleozoic amphibian group". As is quite plain this statement implies that the anurans cannot be derived from dissorophids or any other Palaeozoic tetrapods, and as I have repeatedly emphasized (1955; 1960; 1964; 1968a) the conclusion must be that they have evolved from the osteolepiforms independently of the known temnospondyls. In agreement with this conclusion Carroll and Currie (1975, p. 230) state that the skull of dissorophids "shows no specific similarities to any of the living groups". They also state that "the skull roof, palate and braincase are totally unlike those of living apodans". This shows clearly that not the apodans either can be descendants of dissorophids or other temnospondyls and although apodans in certain regards resemble microsaurs their origin is unknown (Jarvik, 1968a). Finally, as to the urodeles, they cannot possibly be descendants of any temnospondyls either. According to a widespread opinion they are related to the lepospondyls but this view rests only on similarities in vertebral structure and is untenable.†

These facts show plainly that the "lissamphibian" hypothesis lacks support from palaeontological evidence. Moreover, the common characters in the soft anatomy reported by Parsons and Williams, Cox and other, such as the green rods in the retina, the fat bodies (cf. M. H. Wake, 1968), the amphibian papilla in the labyrinth (cf. Baird, 1974; 1974a) and the cutaneuos respiration‡ cf. Gans 1970) may, like the pedicellate

* pp. 158–161; † cf. p. 158; ‡ see Volume 1, p. 226 for *Ichthyostega*.

teeth have arisen independently in the various groups or are for other reasons untenable (Jarvik, 1965). The concept of the "Lissamphibia" evidently rests on a weak basis* and is definitely refuted by the comparative analyses of the porolepiforms and osteolepiforms and the demonstration of numerous fundamental anatomical differences between urodeles and anurans. However, there are certainly also many important differences between urodeles and anurans which cannot be discussed on a palaeozoological basis. Unfortunately many of the early anatomists treated the amphibians as a whole and the result was that even apparent differences were overlooked or ignored. This fallacy has recently been commented upon by Baird (1974), who reports certain differences in the labyrinth. Other more distinct differences have been recorded, for example, in the musculature by Cords (1921), in the heart by Foxon (1955), Simons (1959) and S. C. Turner (1967), in the brain by Clairambault (1963), Kuhlenbeck *et al.* (1966) and Senn and Farmer (1977), in the intracerebral vessels by Sterzi (1904) and Craige (1938), in the function of the tongue by Regal and Gans (1976; see also Regal, 1966), in the foetal lung by Pattle (1969), in the patterns of lymphoid tissue by Goldstine *et al.* (1975); and according to Cohan (1975) the urodeles and anurans are quite different immunologically. Finally, Nieuwkoop and Sutasurya (1976) arrive at the following conclusions:

> The markedly different mode of mesoderm formation in anurans and urodelan amphibians which is related to the double-layered nature of the anuran blastula, in contrast to the single-layered nature in the urodeles, but particularly the fundamentally different place and mode of origin of the primordial germ cells in the two groups of amphibians strongly pleads in favour of a very ancient bifurcation in the phylogenetic history of the two groups, even suggesting a polyphyletic origin from different ancestral fishes.

If these conclusions are true, which I am unable to judge, we certainly are concerned with differences of a profound nature which is in agreement with the fact that the porolepiforms and osteolepiforms stand wider apart than generally has been assumed, and separated from each other before the Devonian.

When studying the snout (1942, pp. 632–637) I was surprised to find that many of the Triassic temnospondyls in the anterior part of the palate agree more closely with *Eusthenopteron* than with any of the Palaeozoic temnospondyls. In these Triassic forms (Fig. 137), exemplified by *Mastodonsaurus* (*Herpetosaurus*), the vomer, as in *Eusthenopteron*, is a triangular bone. It is provided with a posterior process along the narrow parasphenoid, and with large tusks and tooth-bearing laminae including transverse and longitudinal portions. Moreover, there is a well developed fossa apicalis in front of the transverse tooth-bearing portion and an anterior palatal fenestra, paired in *Mastodonsaurus* but unpaired and very similar to that in *Eusthenopteron* in other forms. Also an intervomerine pit and an intervomerine canal are present. In view of this amazingly close agreement and other resemblances it is evident that these Triassic forms not only must be near relatives of *Eusthenopteron*, but also that they have evolved from osteolepiforms independently of the Palaeozoic temnospondyls, in which the construction of the anterior part of the palate is different. These important facts were ignored by Romer (1947) and the family tree presented by him (p. 306) seems mainly to be a grouping after the geologic age. Also the ichthyostegids and the

* see p. 157 for vertebral column.

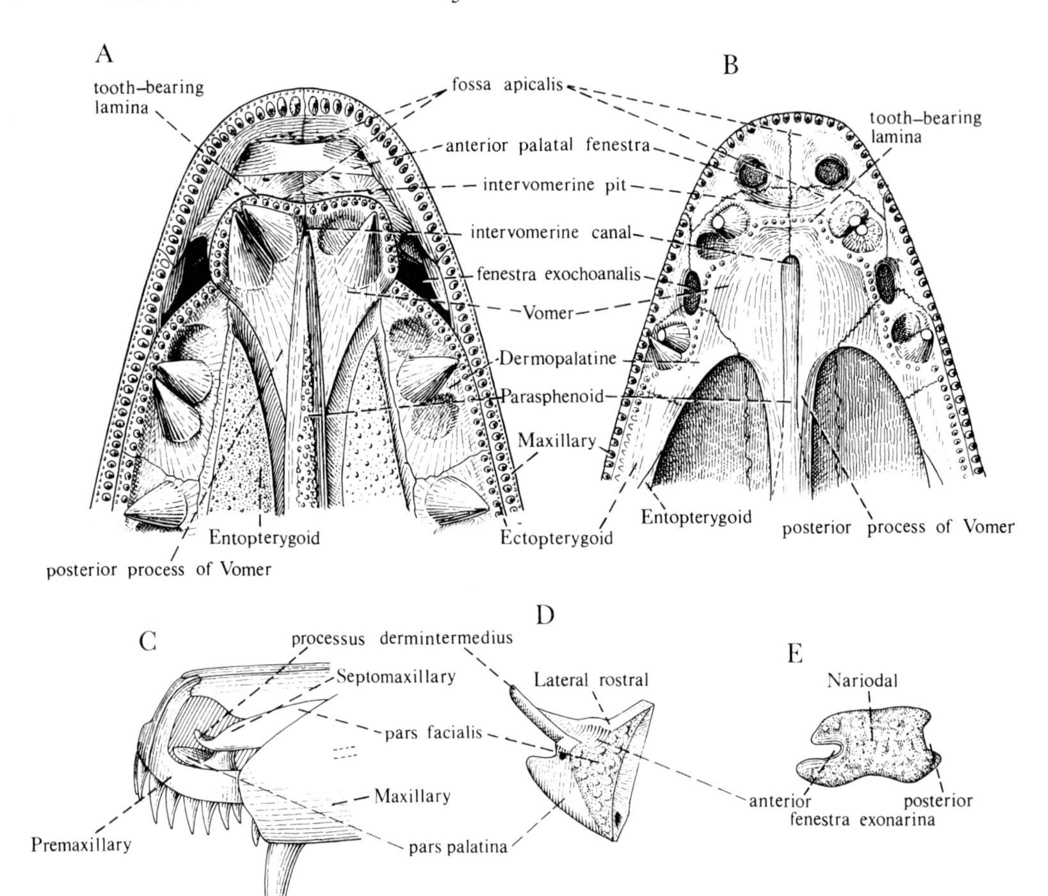

Fig. 137. A, B, anterior parts of palate in ventral aspect to show remarkable similarities between, A, Devonian osteolepiform (*Eusthenopteron*) and, B, Triassic temnospondyl (*Heptasaurus*). C, D, representations to illustrate similarities in structure between, C, septomaxillary in Permian therocephalian reptile (*Scylacosaurus*) and, D, lateral rostral in Devonian osteolepiform (*Eusthenopteron*; cf. Tatarinov, 1976, fig. 8) and, E, different development of nariodal of porolepiforms (*Holoptychius*) and urodeles. From Jarvik (1942; 1972).

acanthostegids have branched off from the osteolepiforms independently of the later appearing temnospondyls and possibly independently of each other (Fig. 140). Of interest in this connection are also the studies of the cranial anatomy of certain temnospondyls by Säve-Söderbergh (1936) and Shishkin (1968; 1970; 1973). Säve-Söderbergh demonstrated remarkable similarities to frogs, e.g. in the arterial system, and according to Shishkin (1968, p.2) "the cranial arterial system in the primitive batrachomorphs resembled that of the anurans to an even greater extent than has been assumed by Säve-Söderbergh". These results are in full agreement with the view that the anurans as well as the temnospondyls are descendants of osteolepiforms, but they do not imply that the anurans have evolved from any of the known temnospondyls.

As we have seen the extant amniotes are certainly related to the osteolepiforms and this may be true of most if not all of the many fossil forms (anthracosaurs, captorhinomorphs, mammallike reptiles, etc.; see Romer, 1966; Carroll, 1969; 1970; Olson, 1971) which most conveniently are classified as Reptilomorpha (Säve-Söderbergh, 1934). In certain regards, however, the reptilomorphs differ from *Eusthenopteron* and other well known osteolepiforms and judging from the conditions in the anthracosaurs (embolomeres, seymouriamorphs) they early in phylogeny had undergone certain modifications which have not yet been traceable back to the osteolepiform stage (Jarvik, 1967a; 1968a). The presence of embolomerous vertebrae in early reptilomorphs may be of little importance* but in the palate, notably the fenestra exochoanalis, vomer and parasphenoid (Fig. 54), they have departed from the conditions known so far in the osteolepiforms. Moreover, the brain has become displaced backwards into the posterior part of the braincase, which has been modified. Because of this modifications the brain stem—in sharp contrast to conditions in osteolepiforms, temnospondyls and anurans—has become flexed in a characteristic way in the area of the medulla oblongata, and the pineal foramen has moved backwards to a point between the parietals (Fig. 124). None of these characters is foreshadowed in the well known osteolepiforms, and the reptilomorphs cannot be derived directly from them. Under these circumstances and because they no doubt are related to osteolepiforms it may be reasonable to assume that the reptilomorphs are descendants of some still unknown osteolepiformlike ancestors. However, as indicated above† it is possible that the mammals and the mammallike reptile (theropsids) have evolved from these ancestors independently of other amniotes (Figs 140, 141; as regards the amniote foetal membranes see Szarski, 1968).

The suggestion that several groups of eutetrapods have arisen independently from osteolepiforms or osteolepiformlike ancestors implies that the tetrapods are polyphyletic. That this is so is supported by the following considerations.

In Devonian times the porolepiforms and osteolepiforms had a world-wide distribution represented by numerous genera and species. However, common to the members of the various populations of these early fishes was that they already possessed many characters essential for a life on land. At the beginning of the Devonian both the osteolepiforms and the porolepiforms had in fact reached a stage in organization where they could turn to a terrestrial life without much further change. In their internal structure they were already tetrapods although they lived in the water and looked like fishes. The transformation from fish to tetrapod therefore implies no radical transformations and most likely took place gradually. In view of this remarkable preadaption for terrestrial life found in the piscine forerunners of the tetrapods it is reasonable to assume that both the porolepiforms and the osteolepiforms utilized their intrinsic possibilities for a life on land several times. However, once in the new environments the descendants of the porolepiforms apparently were not very successful, for, as far as we know, porolepiforms only gave rise to the urodeles (the origin of the Apoda is unknown). The osteolepiforms, in contrast, became much diversified and the new living space was rapidly filled by a crowd of various stegocephalians, anurans, reptiles, birds, mammals and, finally, human beings.

* p. 152; † pp. 170, 201.

IV Recapitulation and Comment

14 General Principles and Methods

Principles for assessing relationship

Two fishes are described in the first section, the extant *Amia calva* and the extinct *Eusthenopteron foordi*, which dates from the Devonian Period, 350 million years ago. The purpose of these descriptions is not only to illuminate the structure of an early fossil vertebrate, but to demonstrate certain facts and principles of fundamental importance for palaeozoological research.

We have already stated that *Eusthenopteron* is a fish. This statement obviously rests on comparison with extant vertebrates and in the same way it is possible to identify fins, scales, visceral arches, nerve foramina and other anatomical features. As a matter of fact, all reliable interpretations of fossil structures must be founded ultimately on comparisons with recent forms. If we cannot relate structural features displayed by the fossils to conditions in living forms we are restricted to speculations and hypotheses.

The comparisons with *Amia* have established that *Eusthenopteron*, in spite of its great geologic age, was already much advanced in cranial anatomy and other particulars, and that in most regards it agrees surprisingly well with the modern ganoid fish. All organs and organ systems of the extant fish were developed in the fossil fish; given the opportunity to study transverse sections through the head of a fresh specimen of *Eusthenopteron*, we would certainly find practically all the numerous structures seen in corresponding sections of *Amia* (Figs 9–12, Volume 1).

Among the numerous characters common to *Eusthenopteron* and *Amia*, some (e.g. the outer dental arcade) are to be regarded as basic teleostome characters, whereas others (e.g. upper and lower jaws and outwardly directed gills) are basic gnathostome characters. However, gnathostomes also have many characters in common with cyclostomes (brain, cranial nerves, six eye muscles, etc.); these are thus basic vertebrate characters. Furthermore, there is reason to assume that the early ontogenetic development in *Eusthenopteron* proceeded in the same way as in recent vertebrates and, therefore, basic characters inherited from the hypothetical common ancestors of Craniota and Acrania (*Amphioxus*) are involved.

It is certainly true that *Amia* and *Eusthenopteron* are essentially similar, but they also differ in many details. It is differences of this kind, rather than basic structures, we must deal with in discussions of relationships and phylogeny. Accordingly, we must have a detailed—and safe—knowledge of the fossils and a corresponding profound knowledge of the various groups of extant vertebrates we have to compare them with. Moreover, it is necessary to consider larval stages of the modern forms, and it is to be required that phylogeny be in accord with ontogeny. Haekel's biogenetic law, that ontogeny recapitulates phylogeny, has been shown to be valid, but only, of course, when modern forms are compared with their true ancestors.

If we want to argue for a close relationship between a fossil and a modern group of vertebrates we must also carefully consider other groups. Most important, we have to rely only on such specializations that are common to the two groups thought to be related and which are not shared with any other groups. Only when all available data, morphological as well as embryological, fit into the picture, are we justified in claiming a relationship (Jarvik, 1968a; 1972, p. 178). When establishing relationship we cannot rely on one or two characters, as done by the cladists.* Nor can we, as in numerical taxonomy, base any safe conclusions on lists of similarities chosen more or less at random from different parts of the animals. Strictly speaking, we should consider the anatomy of the whole body with all its organs and organ systems, but this is hardly possible even when dealing only with extant groups. The nature and incompleteness of the fossil material considerably further restricts our possibilities in these regards, but sometimes we can concentrate on regions of sufficient complexity which are well shown in the fossils; and then we must demonstrate an agreement in the anatomy as a whole in the region we have chosen for analysis. It was these principles for establishing relationship I used in the analysis of the snout in 1942.

Thus, for example, it was shown: (1) that osteolepiforms in a great number of characters show exactly the same specializations in the snout as anurans; (2) that there is agreement also in seemingly inconsiderable structural features, and that thus no weighting of characters was necessary; (3) that osteolepiforms and anurans in these respects differ fundamentally from other groups of lower gnathostomes; (4) that osteolepiforms agree with anurans also in characters shared with other groups; (5) that osteolepiforms in important ways agree better with larval than with adult anurans, and that the ontogenetic development in certain instances implies a recapitulation of phylogeny; (6) that the differences that exist, and which to a considerable extent are due to retrogressive development of the skeleton, could be satisfactorily explained. It was on this basis, and considering the anatomy of the snout as a whole, I concluded that anurans are closely related to osteolepiforms and descendants of osteolepiforms or osteolepiformlike ancestors. Since it was shown that urodeles in the same way agree with porolepiforms the inevitable conclusion was that the Amphibia are diphyletic in origin. Analyses of other complex regions carried out later (1963; 1972; 1975) have supported these conclusions.

Critical Remarks

The terms monophyly, diphyly and polyphyly are often debated by the evolutionary biologists and more or less sophisticated definitions have been proposed. For our

* cf. p. 232.

purposes this is hardly necessary. When Stensiö in 1927 claimed that extant cyclostomes are diphyletic, he simply expressed the view that lampreys and hags have evolved independently of each other from different groups of early fossil cyclostomes; and similarly the statement that the Amphibia are diphyletic only implies that they are descendants of two different groups of fish. The latter statement is, however, valid only with regard to the urodeles and anurans. Because the third group of extant amphibians, the apodans, cannot be derived either from porolepiforms or from osteolepiforms, it may very well originate from a third still unknown group of fish, and if so we must say that the amphibians are polyphyletic. It is also possible that some eutetrapod groups have arisen independently from different osteolepiforms, and this is what I mean when suggesting that the eutetrapods are possibly polyphyletic. With this usage of the terms di- and polyphyly we of course do not deny that all cyclostomes or all teleostomes may have common origins and ultimately are monophyletic (Fig. 139). In this connection, however, it may be appropriate to use the term monophyletist in a restricted sense for those who believe that the extant cyclostomes have a common ancestor among the early Palaeozoic cyclostomes, or that tetrapods have evolved from a hypothetical prototetrapod.

We may now ask "which are the arguments for monophyletism in this restricted sense?"

As to extant cyclostomes the main argument for monophyletic origin is, according to Romer and others (see Jarvik, 1965), the rasping tongue. However, these students overlook that fundamental differences exist between lampreys and hagfishes in practically all organ systems, differences which show that these groups must have evolved separately for a long time. Furthermore, as is fatal for their contention, hags have no rasping tongue; they have a grabbing-biting jaw apparatus very different from the jaw apparatus in lampreys (Fig. 341, Volume 1). Moreover, lampreys share many specializations with cephalaspids and because hags lack these specializations they could not also have originated from cephalaspidlike ancestors. This implies that the extant cyclostomes are diphyletic (whether hags are descendents of heterostracans, as claimed by Stensiö, or from a still unknown cyclostome group is irrelevant).

The main argument for the view that the tetrapods are monophyletic is the pentadactyl limb, and it is categorically declared (see Jarvik, 1965) that this structure can have arisen only once. However, as we have seen, the hind and fore limbs have arisen independently from the different patterns of the pelvic and pectoral fins in the piscine ancestors, and accordingly, in two different ways in tetrapods. Moreover such competent embryologists as Sewertzoff, Steiner and Holmgren have shown that the limbs in urodeles and eutetrapods arise in different ways, a fact which is in agreement with the view that the eutetrapods are derived from fishes independently of urodeles.

What then are the other arguments for tetrapod monophyletism? Are there really any such arguments? Romer's statement, accepted by several writers (see Jarvik, 1964), that the differences in snout structure between osteolepiforms and porolepiforms is due simply to differences in breadth is incomprehensible. The general belief among the monophyletists (e.g. Romer, 1968; Szarski, 1977) that there is no clear distinction between porolepiforms and osteolepiforms is groundless because the descriptions on which this belief is based have been shown to be erroneous in all relevant regards (Jarvik, 1966). This shows plainly how weak the foundation of the monophyletic hypothesis really is.

When the monophyletists and other evolutionary biologists run short of real morphological arguments, they often turn to other subjects such as function or mode of living, or they take refuge in paedomorphosis, the concept of the "Lissamphibia" or various evolutionary theories.

Evolutionary theories have succeeded each other, but in no case I have found them useful for my purposes, and the terminology employed by the various schools has been avoided. This is true also of Hennig's cladistian theory which rapidly has been accepted by many paleontologists and zoologists who obviously think that it is an instrument by which they can easily and safely solve intricate problems of relationship and phylogeny, often without the burden of much knowledge. This usually has resulted in pure armchair speculations, but even the far more serious attempts have not been successful. Thus, for example, the systematic position and relationship of the acanthodians and dipnoans have been carefully "tested" with the aid of cladistic principles. However, such "tests" give a deceptive semblance of reliability, and in both cases the conclusions have turned out to be untenable. Acanthodians are certainly not teleostomes; they are related to sharks. The analysis of the cranial anatomy of dipnoans has shown that these fishes are closer to plagiostomes than to "crossopterygians". Moreover, in constructing cladograms the cladists are forced to make statements of relationship for which there often is little or no evidence. For example, because it is impossible to say what the "sister group" of coelacanthiforms is, it is also impossible to include that group in a cladogram.

As to the lissamphibian hypothesis,* the monophyletists (e.g. Reig in Vial 1973, p. 205; Carroll and Currie, 1975) now tend to exclude the apodans from the lissamphibian community. Reig states that "the lissamphibian hypothesis is far from being a corroborated hypothesis", but nevertheless he and others (Estes and Reig, 1973; Carroll and Currie, 1975; Bolt, 1977) think it plausible that urodeles and anurans, at least, have evolved from Permian dissorophids. These writers disregard the numerous important differences between urodeles and anurans which suggest the two groups became separated long before the first appearance of the proanurans in the early Triassic. Following Romer, these students also ignore the many specializations characteristic of urodeles already developed in the early Devonian porolepiforms, as well as the strong ties between anurans and osteolepiforms. As stressed above† the concept of the "Lissamphibia" lacks support and is to be rejected.

The terms primitive and advanced (or their synonyms) are often used by evolutionary biologists and a deep-rooted conviction is that early extinct forms are more primitive than those living today. It may then be justified to ask "in which respects is *Eusthenopteron* more primitive than *Amia*?". No satisfactory answer can be given to that question. In the regular position of the visceral arches and the presence of dental plates on the outside of the ceratohyal *Eusthenopteron* appears to be primitive, and primitive, too, are probably also the intracranial joint and the vestiges of other intervertebral joints. However, as to the orbital artery *Amia* appears to be more primitive than *Eusthenopteron*; in the latter the three original main branches of the

* see also p. 221; † pp. 222–223.

orbital artery have been captured by a secondary vessel (the "occipital" artery*). In the subdivisions of the anterior part of the nasal cavity and the presence of the processus dermintermedius *Eusthenopteron* is specialized, but this does not mean that it is more advanced than *Amia* in nasal anatomy. We are here concerned with different kinds of specializations and with regard to the much debated concept of homology it is of little interest to state that, for example, the nasal sacs in *Eusthenopteron* and *Amia* are homologous. Of greater interest to us are the different modifications these organs have undergone. It is the different specializations of homologous organs or structures that are of importance when we compare different groups and try to make out relationships.

The well ossified endoskeleton of *Eusthenopteron* is probably a primitive feature. The neurocranium consists almost entirely of bone, whereas in *Amia*, following a general trend, it has undergone a retrogressive development, and is to a considerable extent made up of cartilage or membrane. As in the vertebral column† the membranous parts have often been overlooked and no attempts to homologize these parts in various forms have been made. Much attention has, in contrast, been paid to the ossified parts, which although also products of disintegration are given special names indicating homology with similarly situated endoskeletal bones in other groups, and accordingly are attributed a great phylogenetic importance (Patterson, 1975, p. 566). However, for reasons given below‡ I prefer to regard these ossifications as vestiges of an originally more completely ossified endocranium which have been retained where special strength is needed or for other mechanical reasons. In *Amia*, for example, three of the recti eye muscles are attached to a small process, which is cartilaginous in the larva but ossified in the adult, forming the so called basisphenoid. Is this prehypophysial structure really homologous with the posthypophysial basisphenoid in man and other mammals (de Beer, 1937; Starck, 1967); and which is the phylogenetic importance of the bones of Bridge (Fig. 40, Volume 1) in the lower jaw of *Amia*?

The importance of functional aspects is often stressed in the literature. A proper understanding of functioning in fossil vertebrates would certainly be of great interest. However, for studies of function in fossil vertebrates two conditions must be fulfilled. We must have a safe and detailed knowledge of the fossils and we must know how the mechanism we are interested in operates in living forms. As regards *Eusthenopteron*, we probably have a sufficient knowledge of the skeletal elements involved in, for example, the biting mechanism, but even if we disregard the immovable intracranial joint, it is certainly not easy to make out safely the actions of the many muscles associated with the hyobranchial skeleton and the subbranchial series. If we were to attempt such a functional interpretation we would easily get into unfruitful speculations; and because other early fishes are less well known than *Eusthenopteron*, I have avoided such speculations.

Finally, with respect to environment and mode of life it is certainly possible to extract some information about climate and other conditions from the rocks at Escuminac Bay at the time when *Eusthenopteron* was embedded, but we can never be sure that this fish lived in the environments indicated by the embedding rocks, and this in particular since *Eusthenopteron* and other osteolepiforms have recently been

* p. 180; † p. 156; ‡ p. 259

recorded from marine deposits (Gregory *et al.*, 1977; Lehman, 1977). Still worse, we cannot say anything at all about the environments in the remote past when the structures characteristic of *Eusthenopteron* developed. We cannot explain why the tetrapod limb, the nasolacrimal duct, the choana, and other prerequisites for a terrestrial life arose in the fish ancestors of the tetrapods. Theories about these things found in the literature are pure guesswork.

15 Vertebrate Phylogeny

General Considerations

As we have seen, the Devonian fish *Eusthenopteron* was highly specialized, and although living about 350 million years earlier it was in most respects just as advanced as the extant ganoid *Amia*. This early specialization is, however, not characteristic only of *Eusthenopteron*. The thorough studies of the early Palaeozoic vertebrates during the last five decades have established the following fundamental facts which may be of importance for students of vertebrate evolution and evolution in general.

(1) The early vertebrates are no primitive prototypes. They are all highly specialized.

(2) At their first appearance in the fossil record, whether in the Ordovician, Silurian or Devonian, the vertebrates were divided into several distinct groups, each characterized by a number of specialized features. The number of these groups may be disputed but if we place such diverse groups as the placoderms on par with more uniform groups such as the porolepiforms and disregard some forms which have been considered incertae sedis the number will amount to 11. To this number two groups (apodans, polypterids) not represented in the known early vertebrate fauna have been added in Figs 139, 140.

It is not an isolated phenomenon that the various early vertebrate groups appear fully developed; the sudden appearance of new major groups of reptiles (ichthyosaurs, turtles, etc.) in the Triassic (Fig. 142, Colbert, 1965) or of practically all the modern orders of mammals in the early Cainozoic (Lillegraven, 1972) are other examples (cf. also the presence of most invertebrate groups in the Cambrian).

(3) Some of these early groups (struniiforms, acanthodians, and probably the placoderms) are extinct, but most of them have survived and constitute the modern vertebrate fauna.

(4) This latter fact, and the good knowledge we now have of many early vertebrates, has made it possible to make direct comparisons between early (mainly Devonian) and extant representatives of the same group. In this way we can make out what has happened in a group during a vast period of time, and thereby acquired a new and much widened perspective of vertebrate evolution.

(5) By this comparative method a remarkable conservatism has been demonstrated. Disregarding the retrogressive development of the skeleton and other general trends,

such groups as coelacanthiforms and dipnoans have changed very little since the Devonian, and even small and seemingly unimportant structures have been retained with a surprising tenacity. Many other groups (palaeoniscids, ichthyosaurians, turtles and others) also display a remarkable reluctance to change and this is true also of those groups in which the external appearance has been much modified. The nerves supplying the gills in lampreys thus run and branch in exactly the same way as they did in the early Devonian cephalaspids; striking similarities in the cranial anatomy between extant frogs and their Devonian forerunners and between salamanders and the early Devonian porolepiforms also have been demonstrated.

(6) This implies that the main evolutionary lines, if followed backwards in time to the Devonian or to the Silurian, do not approach each other to the extent that might be expected from some current views on vertebrate evolution, but run more or less in parallel (Fig. 140). A Devonian coelacanthiform was in all important respects a typical coeleacanthiform, and anatomically it differs from contemporaneous dipnoans in about the same way as *Latimeria* differs from *Neoceratodus*.

These facts, gained by the studies of the early fossil vertebrates, show that the vertebrates had passed the most important stages in their evolution before they begin to appear as fossils. What has happened during the last 400 or 500 million years is inconsiderable compared with what must have happened previously, when not only the qualities common to all vertebrates, but also the distinctive characters of the various groups, arose. Unfortunately, the fossils give us no information about these early and comparatively most important stages in the history of the vertebrates.

In view of these facts it may be convenient to distinguish two periods in the evolution of each of the main vertebrate groups: one period documented by the fossil record and one unrecorded period (Fig. 139). The latter may be divided into an early phase, during which the characters common to all vertebrates arose, and a late phase distinguished by branching evolution. The length of the recorded period may, in favourable cases, amount to almost 500 millions years, and if we accept the view that organic life on earth arose about 3000 million years ago, the length of the unrecorded period will then amount to about 2500 million years. The following considerations (Jarvik, 1964; 1968a) may give some further hints as to the length of the unrecorded period.

We know by fossils that the Heterostraci were present in the late Cambrian and represented by several types (arandaspids, astraspids, eriptychiids) in the Middle Ordovician (Fig. 138). The common ancestor of these types must have been living earlier, and still older were the common ancestors of all heterostracans (E) and of all pteraspidomorphs (D). The Pteraspidomorpha and Cephalaspidomorpha are more closely related to each other than to any other group, and share a common ancestor. They are, to use a modern term, "sister groups", and their common ancestor (C) must of course have preceded the common ancestors of both cephalaspidomorphs (F) and pteraspidomorphs (D). Moreover, it is clear that the common ancestor (B) of the Gnathostomata and the Cyclostomata must have been in existence still earlier and that in its turn it was preceded by the common ancestor (A) of the Acrania (*Amphioxus*) and the Craniota (Vertebrata). By these considerations we are pressed farther and farther backwards in time, and we are led to the conclusion that the vertebrates must be a

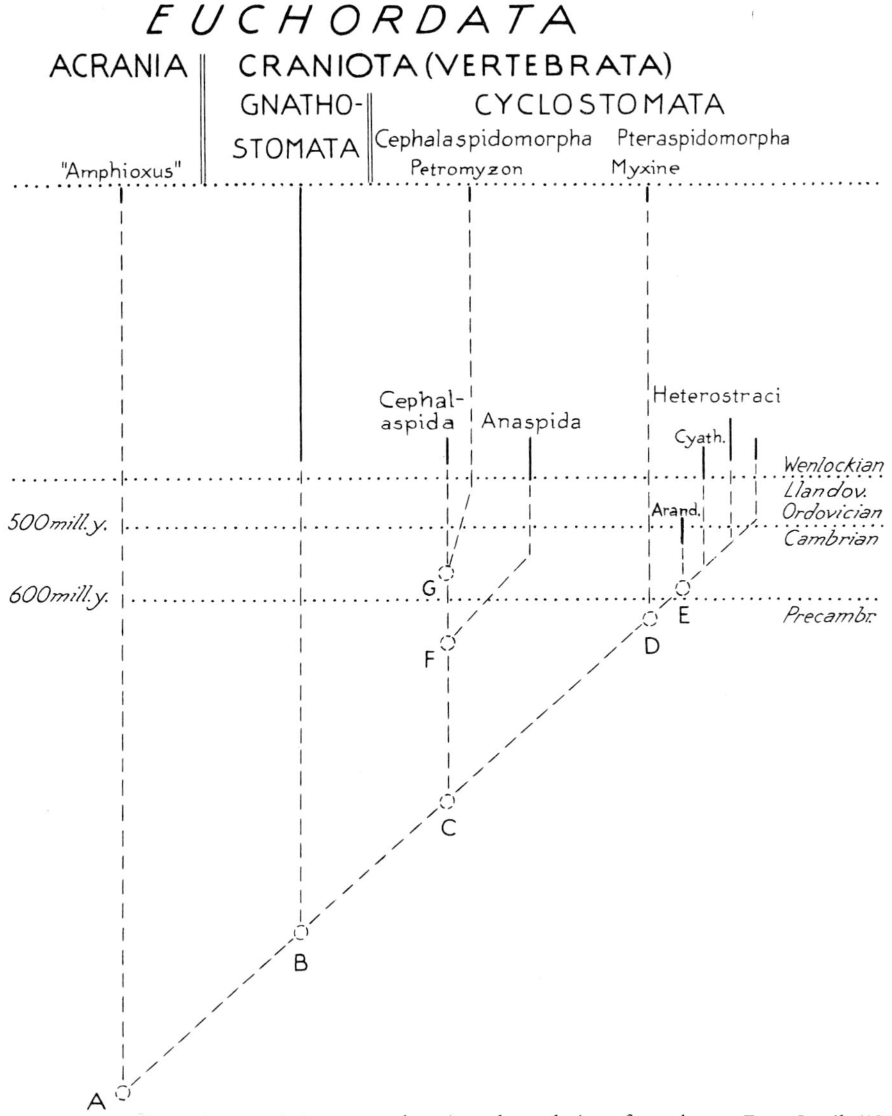

Fig. 138. Diagram illustrating certain important dates in early evolution of vertebrates. From Jarvik (1968a).

more ancient group than is generally believed. However, we do not know anything with certainty about their origin or when they arose. Neither do we know if the vertebrates are related to any other of the phyla which constitute the animal kingdom, but as maintained elsewhere (Jarvik, 1964, p. 87) it seems likely that the specific vertebrate characters began to develop very early in the history of life, and that the decisive division of the animal kingdom into phyla took place earlier and more contemporaneously than is generally assumed. These problems will be further elucidated when turning next to the early phase in the unrecorded period of vertebrate evolution.

UNRECORDED PERIOD

Early Phase

The question of vertebrate or chordate origins has for more than a century been penetrated by a great number of students and is still attracting considerable interest (Berrill, 1955; Raw, 1960; Tarlo, 1960; Gutman, 1969; Denison, 1971; Bone, 1972; Romer, 1972; Jollie, 1973; Jefferies, 1975; Willmer, 1975; Jefferies and Lewis, 1978 and others). Starting, obviously, from the unproved hypothesis that animal life has arisen only once on our planet, these writers have tried to relate the vertebrates to any of the invertebrate phyla, and relationships have, by one or the other, been claimed to all the principal invertebrate groups. No convincing and generally accepted solution has, however, emerged from these efforts. It may therefore be justified to ask whether it really is true that all animals are monophyletic in origin. As indicated above, and claimed by Nursall (1962) and others, another possible alternative would be that the various animal phyla have developed more or less independently from organic matter formed by biopoiesis in different parts of the earth. This is hardly more unreasonable than to postulate that life has arisen independently on other planets. Nursall thinks that the chordates may have evolved independently of the invertebrate phyla and Bjerring (unpublished) goes a step further excluding also the tunicates from the ancestry of the vertebrates.

The Chordata are usually divided into three groups: Tunicata or Urochordata (including Appendicularia, Ascidiacea and Thaliacea), Cephalochordata or Acrania (represented by *Amphioxus*) and Craniota or Vertebrata (including cyclostomata and Gnathostomata). The term Vertebrata is sometimes used in a broader sense (Goodrich, 1930) including also the Cephalochordata; and the Hemi-, Uro- and Cephalochordata are often collectively referred to as protochordates. However, in accordance with current usage, the term Vertebrata has been retained as a synonym of Craniota. The Acrania and the Craniota constitute the Euchordata (Jarvik, 1968a).

Adult tunicates or sea squirts such as the ascidians certainly differ much from vertebrates and in this connection it is their free swimming larva that is of interest. This larva or "tadpole" is a small simple creature with a long tail (in appendicularians retained in the adult). The tail is supported by a liquid skeleton, the notochord, which consists of tubular or sheetlike cells surrounding an extracellular vacuole and differs fundamentally from the notochord in both *Amphioxus* and vertebrates (Flood, 1975, p. 100). Another important difference is that notochordal sheaths are lacking (Olsson, 1965). Dorsal to the notochord there is a hollow neural tube traversed by Reissner's fibre and composed of a non-nervous ependymal tube and one or more nerve cords (Olsson, 1969). Tiny motor nerves are given off to the muscle cells which form the longitudinal muscle bands. Because both the neural tube and the notochord arise early in ontogeny (Brien, 1948; Drach, 1948) much in the same way as in *Amphioxus* and vertebrates, it is generally taken for granted that we are concerned with homologous structures; if so the tunicates must be related to the euchordates. According to several students (Berrill, Raw, Romer and others) the ascidian tadpoles have gained sexual maturity and have developed into vertebrates. The application of this hypothesis of neoteny (or paedogenesis), criticized and rejected by Jägersten (1972), is, in my opinion, only a way to escape from the many difficulties we are facing if we try to derive the vertebrates from tunicate tadpoles. The tunicate notochord is a unique structure and an important fact is that it occurs only in the tail. In this respect the tunicates differ fundamentally from the euchordates in which the formation of the notochord starts in the future head region.* Moreover, the non-nervous cells interpreted as ependyme cells surround the nerve cords (Olsson, 1969), whereas the ependyme cells in *Amphioxus* and vertebrates are internal structures which line the central canal of the spinal cord. Furthermore, there are no motor or sensory columns and the motor nerves to the muscles emerge from the anterior part of the tube close behind the brain vesicle (Jägersten, 1972), and, most important, there are no myotomes or sclerotomes and no traces of metamerism, either in the muscular or the nervous systems. In view of these and other important differences (see Drach, 1948) we cannot—in spite of the similarities in ontogeny and the presence of Reissner's fibre—quite exclude the possibility that we are concerned with parallelism. Be this as it may, tunicates are certainly of little help in unravelling the evolutionary events in the early history of the euchordates. Such attempts must be based on data provided by the ontogenetic development of extant euchordates.

Regardless of whether the euchordates are related to the tunicates, it seems likely that the earliest ontogenetic stages (egg cleavage, blastula and gastrula formation) to some extent reflect the early phylogenetic evolution of the euchordates (Fig. 139). Further changes also recorded in ontogeny resulted in a creature, the proto-euchordate, in essential regards resembling the *Amphioxus* larva shown in Fig. 3A (Jarvik, 1972, p. 294). This animal (Hatschek, 1881; Conklin, 1932) possessed a neural tube, notochord, and gut; all, in contrast to the ascidian tadpole, extending forwards to the anterior end of the animal. There was certainly also a mouth opening, and after the anus had been formed food particles could be transported backwards by the action of cilia. Ectodermal cilia enabled the animal to move. Moreover, parachordal mesoderm was present and was completely subdivided into paired metameric compartments. In the middle parts of these compartments, organs for excretion and reproduction developed, and dorsal to this organ-forming area, in the medial wall of the future somites, myotomes and sclerotomes could be distinguished. The sclerotomes formed a skeletogenous layer surrounding the neural tube and the notochord, and metameric muscles derived from the myotomes arose. Rhythmic contractions of these muscles enabled the animal to move forwards by undulating movements of the body (Fig. 69), and the external cilia became unnecessary and were lost. However, the body of this hypothetical proto-euchordate was still short including in the main only those metameric mesodermal compartments which enter into the formation of the vertebrate head. The trunk and the tail are to be considered as secondary acquisitions formed in connection with prolongation backwards of the neural tube, notochord and gut, as well as successive addition of metameric compartments (somites).

* p. 242.

Late Phase

Some Early Stages in Vertebrate Evolution

The late phase of the unrecorded period (Fig. 139) is characterized by branching evolution. Two lines emerged from the populations of the hypothetical proto-euchordates. In the one line leading to extant cephalochordates (*Amphioxus*) a number of peculiar specializations of little interest to us arose. The other line branched into

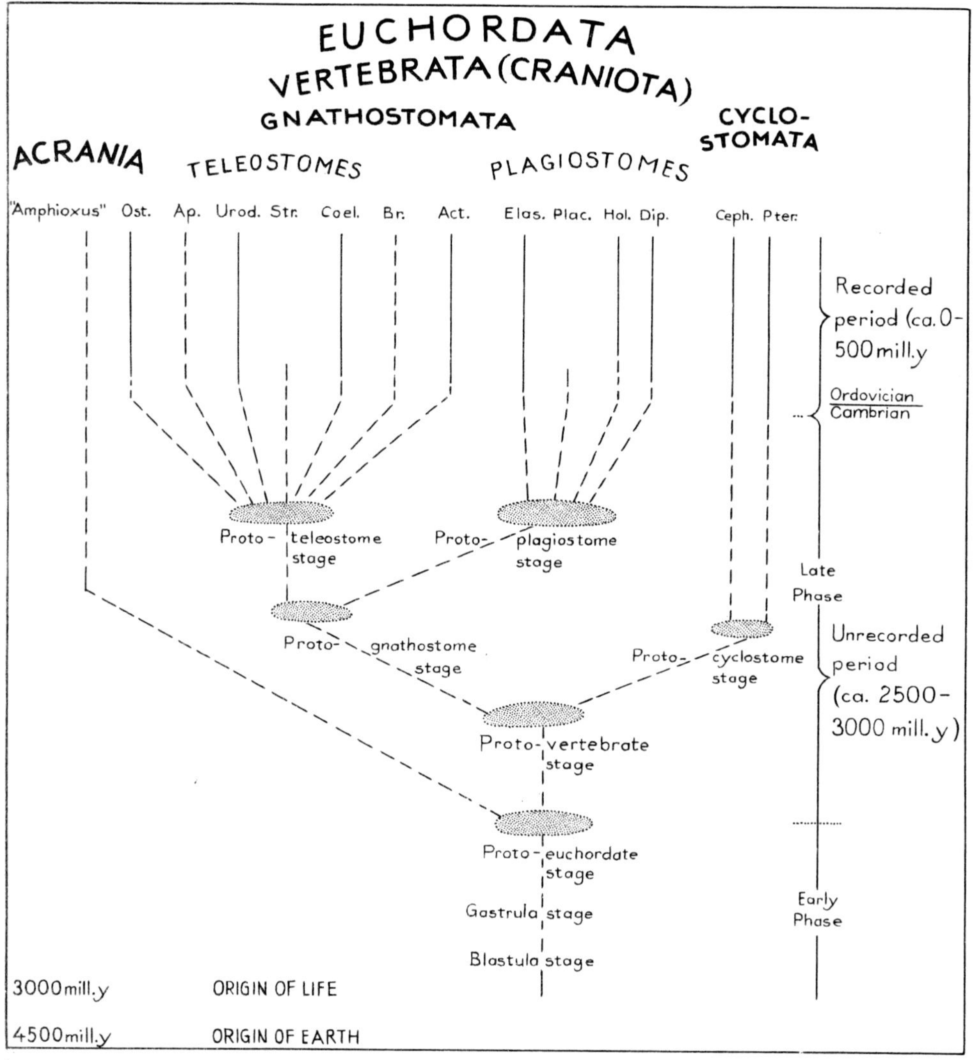

Fig. 139. Diagrammatic representation to illustrate probable evolution of vertebrates from origin of life. Evolution during unrecorded period based mainly on ontogenetic data.

several lines ultimately giving rise to the extant vertebrate fauna. For the sake of simplicity, five hypothetical ancestral groups (proto-vertebrates, proto-cyclostomes, proto-gnathostomes, proto-plagiostomes and proto-teleostomes) may be distinguished.

During the evolution from the proto-euchordate to the proto-vertebrate, all the characters common to cyclostomes and gnathostomes, but not shared by the cephalochordates (Acrania), arose. The anterior part of the neural tube evolved gradually, in the way recorded in ontogeny, into a typical vertebrate brain with five divisions. The brain and the body as a whole were prolonged forwards and both phases in the rostral prolongation (Fig. 11) had occurred when the proto-vertebrate stage was reached. There were paired olfactory organs, olfactory nerves and bulbi, the latter with mitral cells. The pituitary gland, composed of neuro- and adenohypophysial components, was present, as were the neuroepiphysial complex and habenular ganglia. The optic and trochlear decussations had occurred and a typical vertebrate eye moved by six extrinsic eye muscles had developed. Moreover, there was a membranous labyrinth with an endolymphatic duct arising in ontogeny in the same way as it still does in the lamprey as well as in man (Fig. 27). There were also spinal nerves with dorsal and ventral roots, dorsal and ventral cranial nerves, taste buds and a lateralis system. A new acquisition was the neural crest, and migrating crest cells contributed to the spinal and cranial ganglia and formed the visceral endoskeleton. There was certainly also an exoskeleton in the skin and a neurocranium. However, in view of the presence of two different types of crania in extant cyclostomes, both difficult to interpret, we can say very little about the cranium in the proto-vertebrate. However, it was certainly not only in the brain, nerves and sensory organs that the proto-vertebrate was much advanced. This was no doubt true also of other organs and organ systems and we can safely conclude that most of the essential characteristics of the vertebrate body even on biochemical and molecular levels already were developed in this ancient form. The epoch between the proto-euchordate and the emergence of the fully developed proto-vertebrate is therefore one of the most important epochs in the history of the vertebrates, presumably spanning a considerable space of time.

The proto-vertebrates certainly possessed intermetameric gill slits, but nothing with certainty can be said about their gills. When proceeding to the two lines emerging from the proto-vertebrates we have to assume that the proto-gnathostomes had outwardly directed gills, whereas the gills in the proto-cyclostomes were directed inwards towards the pharynx. This is a fundamental difference between gnathostomes and cyclostomes (Fig. 363, Volume 1), and, as explained above*(Fig. 366, Volume 1), the presence of inwardly directed gills in cyclostomes prevented the formation of biting jaws of the gnathostome type. The cyclostomes therefore had to solve these problems in other ways, and because the jaw apparatus in lampreys is fundamentally different from that in hags (Fig. 341, Volume 1; there are also other important differences), we have to assume that the cyclostomes divided into cephalaspidomorph and pteraspidomorph lines soon after the proto-cyclostome stage was reached; each line was then subject to branching evolution.

* Volume 1, p. 480.

The gnathostomes are divided into two principal groups, teleostomes and plagiostomes, and accordingly we have to assume that two evolutionary lines emerged also from the proto-gnathostomes. Further branching late in the unrecorded period resulted in the formation of the various groups which appear in the fossil record in the early Palaeozoic.

As to the differences between the two main gnathostome groups a tentative explanation of the fact that plagiostomes, in contrast to teleostomes, lack outer dental arcade has been given,* whereas the differences embodied in the concepts terminal and subterminal mouth remain unexplained. The problem of the origin of the mouth was discussed by Dohrn and other early embryologists and has recently been touched upon by Regel (1966) and Bjerring (1973; 1977). According to Bjerring there are significant differences in this regard between plagiostomes (including dipnoans) and teleostomes. However, further investigations are needed before we can say what the differences really are, and because we cannot, for example, explain the peculiar fact that nerves and vessels pass from the hyoid to the mandibular arch in the area of the jaw joint, I have found it better to avoid the problem of the origin of the mouth in the discussion of the origin and composition of the vertebrate head (see, however, p. 192 and Fig. 135, Volume 1).

Origin and composition of the gnathostome head, with special reference to *Eusthenopteron* and *Amia*

Ontogenetic Preliminaries

As shown by ontogenetic data, the main posterior part of the head region is the most ancient part of the vertebrate body; it constituted the whole animal in the proto-euchordate ancestor. By processes collectively referred to as rostral prolongation† (Fig. 11) the head grew forwards in front of the optic chiasma (this marks the position of the original anterior end of the head region and therefore is an important landmark), at the same time as the trunk and finally the tail were successively added in the posterior direction.

Early in phylogeny the neural tube, the notochord, and the archenteron all reached the anterior end of the animal and were flanked by the parachordal mesoderm which was completely divided into metameric portions. The dorsal parts of these portions, which in gnathostomes form thick-walled vesicles called somites, differentiated into three organ-forming areas: sclerotome, myotome and dermatome. In connection probably with the formation of the cephalic flexure, which is characteristic of the vertebrate embryo, the notochord-forming capacity of the most anterior, rostral, portion of the chordamesoderm was suppressed,‡ and as a consequence the notochord in vertebrates ends behind the hypophysis. The rostral portion of the chordamesoderm forms together with the surrounding parachordal mesoderm the "prechordal plate". However, as recently maintained by Bjerring (1973), in agreement with several early embryologists and further confirmed above, the foremost part, Platt's vesicle, of the parachordal mesoderm in the "prechordal plate" is a true somite. This terminal somite

* p. 94; † p. 20; ‡ p. 7.

is the foremost in a continuous series developed on each side of the notochord (including its anterior suppressed portion) and the neural tube, and we may distinguish four prootic somites: terminal, premandibular, mandibular and hyoid. Behind the hyoid somite follows a series of branchial somites; in *Amia*, *Eusthenopteron*, and most other gnathostomes five in number. These nine somites have—because of their relations to the branchial units—been regarded as cephalic somites, in contrast to the trunk and (behind the anus) tail somites.

The parasomitic mesoderm in gnathostomes (the lateral plate) and the corresponding part of the parachordal mesoderm in cyclostomes is pierced by intermetameric gill slits; these, like the somites, arise in sequence from front to back. The areas between the gill slits are occupied in the embryo by the metameric primordia of the branchial units. Each such primordium includes (Fig. 7): (1) an outer tube, laterally ectodermal and medially endodermal in origin; (2) an inner mesodermal visceral tube; and in the space between the two tubes; (3) migratory ectomesenchyme derived from the neural crest. The visceral tubes lying immediately in front of the foremost gill slit, between the gill slits and behind the last gill slit are each connected with a somite by way of a constricted somitic stalk (Fig. 4). Because of this, it is evident that these tubes are metameric structures. Thence it follows that the primordia of the branchial units and the derivatives of these primordia which constitute the branchial units in the adult (skeletal visceral arches, visceral muscles, nerves, vessels, gills) are also metameric, as are of course, the derivatives of the somites and the somitic stalks.

Since the time of Balfour it has been customary to use the same name for the somite and the visceral arch in each metamere. Accordingly there are four prootic visceral arches, terminal, premandibular, mandibular and hyoid, and usually five branchial arches. In order to avoid confusion the gill slits are most conveniently named starting from the spiracular gill slit, that is, the gill slit between the mandibular and hyoid arches. Behind that gill slit follow the posthyoid or branchial gill slits (usually four in number). In front of the spiracular gill slit there is a prespiracular gill slit between the mandibular and premandibular visceral arches, and a pre-prespiracular gill slit between the premandibular and terminal arches (Fig. 135, Volume 1). The latter is occupied by the olfactory organ (Figs 22, 24).

Sclerotomic Derivatives

Just as the myotomes form metameric muscle blocks or myomeres, the sclerotomes in the trunk, as well as in the head, produce metameric scleromeres. The scleromeres often become confluent and form a skeletogenous sheath surrounding the neural tube and the notochord (Fig. 13, Volume 1). When skeletal arcual elements appear in this sheath the vertebrae thus formed necessarily must alternate with the metameric muscles; in consequence the vertebrae are usually dimetameric structures derived from half-scleromeres belonging to two adjoining metameres (Fig. 39). We are here concerned with a primary subdivision or segmentation of the metameric material (not a "resegmentation" = "remetamerism") and it is important to distinguish between the metamerism of the vertebrate body (usually called segmentation) and the segmentation of the vertebral column.

With respect to the vertebral column there are no important differences between head and trunk. Dimetameric cranial vertebrae separated by more or less distinct intervertebral (intrametameric) vertebral joints have been identified in the metameric part of the basi cranii, forwards to the optic chiasma, and although fusions and other modifications occur arcual elements can be recognized in each of the cranial vertebrae. As in the trunk, the basi- and interventrals have usually fused into ventral vertebral arches which may form the independent ventral (hypochordal) arcual plates. The dorsal arcual plates are separate parts of the wall between the neural and notochordal canals. The neural arches are represented in the head by the cranial pilae. Because of the strong development of the brain, the cranial pilae (except most posteriorly) are in each vertebra far apart; most likely each pila of the most anterior cranial vertebra (CV 1) has been displaced outwards by the developing eye cup (a part of the brain wall) to form part of the sclera (Fig. 44; cf. Bjerring, 1977, fig. 13). Behind the first vertebra the tips of the pilae are, in each cranial vertebra, connected by a transverse bridge or tectum. Each tectum includes a median element serially homologous with the supraneurals in the vertebral column and paired lateral elements probably representing cranial supradorsals (Fig. 61). By the fusions of the lateral and median tectal elements longitudinal lateral and median taeniae are formed. These taeniae are most distinct in embryonic stages, whereas in the adult the tectal elements have fused forming a more or less complete cranial roof. However, sometimes the spaces between consecutive tecta, that is, between the dorsal parts of the cranial vertebrae, may be retained, and there may be a corresponding suture in the overlying part of the exoskeletal cranial roof. This occurs, for example, between the fronto-ethmoidal and parietal shields, or between the latter and the extrascapular series in *Eusthenopteron* (as regards man see Fig. 125).* With the aid of these sutures, marking the position of the cranial intervertebral joints, an attempt has been made to give a definition of the dermal bones of the cranial roof according to their position in relation to the underlying cranial tecta (Fig. 61). Subdivisions corresponding to the vertebral segmentation are clearly discernible, in embryos (*Amia*) as well as in the adult (*Eusthenopteron*), also in the median part of parasphenoid covering the ventral side of the cranial portion of the vertebral column (Fig. 56).

Myotomic Derivatives

Recent important discoveries (Bertmar, Regel, Bjerring) have considerably widened our knowledge of the cranial myomeres which have become much modified and besides muscles may form skeletal elements. In the posterior part of the head region the hypaxonic portions of some (usually three) of the cephalic myomeres have grown downwards and forwards forming the hypobranchial muscles. These muscles are separated from their antimeres by a median subbranchial series of elements which most likely is derived from myotomic mesomesenchyme and therefore has been referred to the myotomic skeleton.

Farther forwards in the head, where only epaxial portions are developed, these

* see also p. 268.

portions are divided into a ventral basicranial and a dorsal tectocranial series. An important fact is that paired longitudinal basicranial muscles have been identified in the cephalic metameres forwards to the second (Figs 23, 41, 42). However, if it is true that the m. obliquus inferior represents the basicranial muscle in the terminal metamere, basicranial muscles are present in all cranial metameres. The basicranial muscles may be retained in the adult (m. obliquus inferior, basicranial muscle 2 in *Scomber*, the excessively developed basicranial muscle 3 or subcranial muscle in "crossopterygians", and probably some of the muscles in the otic region in osteolepiforms and porolepiforms). However, the basicranial muscles may also be wholly or partly replaced by, or possibly transformed into, myotomic skeletal elements. Such elements are the three basiotic laminae, the polar cartilage and probably the eye stalk in sharks. Because these skeletal elements, like the basicranial muscles (and the myomeres in the trunk), alternate with the dimetameric cranial vertebrae (Fig. 44) and fuse with them this leads to an obliteration of the original intervertebral joints and a consolidation of the basis cranii. In *Eusthenopteron* and other "rhipidistians" the intervertebral joint within the third metamere persists in the adult as the intracranial joint, but when this joint closed in tetrapods a vestige was retained as a transverse fissure in the basis cranii; this is visible in *Ichthyostega* (fissura preoticalis; Figs 171, 172, Volume 1). This fissure in *Ichthyostega* is suggestive of, and obviously serially homologous with, the transverse grooves found farther back, and with the distinct transverse fissures within the fourth and fifth metameres in the basis cranii of *Eusthenopteron*. This condition supports the view that these fissures represent vestigial intervertebral joints probably spanned by vestigial metameric basicranial muscles.

Tectocranial muscles 6, 7 and 8 are generally well developed, and a remarkable fact is that these muscles, in early embryonic stages of *Amia*, are inserted into a depression in the occiput, the fossa tectosynotica; they later migrate into the fossa bridgei which they occupy in the adult. A corresponding migration has probably occurred also in *Eusthenopteron* and *Latimeria*, whereas in early coelacanths (*Nesides*) and in *Polypterus* the muscles have retained their original insertion. Tectocranial muscle 5 has hitherto been found only in lampreys, whereas tectocranial muscles 2, 3 and 4 in vertebrates in general form all the extrinsic eye muscles, except the m. obliquus inferior (Bjerring, 1975; 1977).

Dermatomic Derivatives and Exchordals

The dermatomes may also produce skeletal elements in the head, namely the otic cartilages in the fourth and fifth metameres, and possibly parts of the sclera. Other skeletal elements in the axial parts of the neurocranium are the exchordals (Figs 43, 44) which probably are derivatives of the rostral portion of the chordamesoderm.

The axial portion of the neurocranium (Fig. 44) is accordingly composed of: 1) the cephalic portion of the vertebral column derived from the metameric sclerotomes; 2) myotomic derivatives; 3) dermatomic derivatives and; 4) the exchordals. To this axial portion dorsal elements of the visceral arches have been added (Fig. 47), although in a different way in the various groups.

Visceral Endoskeleton

The visceral arches in gnathostomes originally had a strict metameric disposition and occupied a more or less transverse position as they do in embryos. However, during their phylogenetic development they became directed obliquely backwards and this change and the rostral prolongation have caused considerable modifications (Fig. 56). Mainly due to the rostral prolongation the dorsal parts of the terminal and premandibular visceral arches have been displaced forwards and the infrapharyngomandibular in all gnathostomes, except dipnoans, forms an almost longitudinal bar in the basis cranii (the ectomesenchymatic part of the trabecula). In *Eusthenopteron* and *Amia* the dorsal elements of the terminal arch—as in other visceral arches probably three in number (Bjerring, 1977)—enter together with the supra- and infrapharyngopremandibulars into the formation of the nasal capsule. In *Amia*, in contrast to *Eusthenopteron*, the suprapharyngomandibular (pila lateralis) is incorporated into the braincase. Another remarkable difference is that the supra- and infrapharyngohyals in *Amia* have been plastered to the axial portion of the neurocranium farther forwards than in *Eusthenopteron*. In consequence the processus ascendens posterior of the parasphenoid with the groove for the spiracular gill tube in *Amia* covers the external side of the lateral commissure, whereas in *Eusthenopteron* the corresponding structures (and the spiracular canal) are found farther forwards, in front of the lateral commissure. In *Eusthenopteron* the infrapharyngohyal (otical shelf) has turned forwards from its original transverse position, whereas the suprapharyngeal element in the fourth or hyoid arch has retained its original metameric position. This element has fused with the anterior part of the otic capsule formed by the preotic process, which probably is a lateral process of interdorsal 4 in cranial vertebra 3+4 (Fig. 44). Due to the change in direction of the visceral arches similar anterior displacements of the infrapharyngeal elements have occurred also in the mandibular and first branchial arches. The suprapharyngeal elements in these arches are still strictly metameric in position and articulate with lateral processes of the interdorsals (the suprapterygoid process in the third and the postotic process in the fifth metamere) in the metameres to which they belong.

An important new acquisition in gnathostomes is the palatoquadrate; it is composed of the epal elements of the mandibular and premandibular arches and a commissural lamina probably derived from premandibular branchial rays. In *Eusthenopteron* this structure includes also the suprapharyngomandibular and mandibular branchial rays. In both *Amia* and *Eusthenopteron* the epihyal has fused with a laterohyal component derived from hyoidean branchial rays to form the hyomandibula. In both fishes there is a stylohyal, whereas the sympletic is present only in *Amia*.

The ventral parts of the visceral arches are well developed in the mandibular, hyoid and branchial arches and usually include ceratal and hypal elements. In the terminal arch no ventral elements are known and have most likely been suppressed in connection with the stomodeal invagination and the formation of the mouth. This may be generally true also of the premandibular arch, but as discussed above,* the ceratopre-

* p. 86; Volume 1, pp. 409, 431.

mandibular is possibly present in dipnoans. Probably in connection with the rostral prolongation the ceratomandibular (Meckelian element) has been prolonged forwards. There is usually also a rather large space between the mandibular and hyoid arches in the floor of the mouth cavity. This space is in the porolepiform–urodele stock occupied by the glandular field, and in the osteolepiform–eutetrapod stock by the tuberculum impar. However, the ventral parts of the hyoid and branchial arches also have been prolonged forwards, and this explains why the ceratal elements in each arch are opposed to the epal elements in the next anterior arch.

Dermal Bones

Most likely the exoskeleton arose early in the phylogeny of the vertebrates and presumably the skin of the proto-gnathostome contained some kind of primary scale components, whereas small denticles were evenly distributed in the mucous membranes of the oralobranchial cavity and the gill slits (Figs 56, 69). According to the partly hypothetical interpretation of the dermal bones of the head presented above* and disregarding the special problem of the lateral sensory components (see Ørvig, 1972), we may, according to their position, distinguish between two categories of dermal bones: (1) bones developed in relation to the cranial portion of the vertebral column and (2) bones belonging to the metameric branchial units.

The first category includes the dermal bones of the cranial roof and the median part of the parasphenoid. In the cranial roof (Fig. 61) we have distinguished seven series of transverse elements, each developed in relation to one of the tecta of the dimetameric cranial vertebrae. The boundaries between these series mark the position of the intervertebral joints in the underlying cranial portion of the vertebral column. The median part of the parasphenoid which covers the ventral side of that portion is also composed of skeletal units corresponding to the dimetameric vertebral segments (Fig. 56). When the parasphenoid is short, as for example in *Eusthenopteron*, the posterior of these units are still separate (e.g. the subotical plates underneath CV3+4).

The lateral parts of the parasphenoid are formed by addtion of metameric elements belonging to the second category (Jarvik, 1954). In *Eusthenopteron* horizontal and ascending infrapharyngeal dental plates of the mandibular arch have been added to the parasphenoid stem and form the anterior ascending process of the bone with the groove for the vestigial prespiracular gill tube. This process supports the basipterygoid process which is the posterior end of the infrapharyngomandibular. In *Amia*, in which this gill tube is lost, both the anterior ascending process and the basipterygoid process are retained, although they are much reduced. However, the stem has been prolonged backwards by addition of further segmental units. Moreover, equivalents to the paraotic and spiracular dental plates in *Eusthenopteron*, which are horizontal and ascending infrapharyngohyal dental plates, have been incorporated forming the processus ascendens posterior. On the outside of this process the groove for the spiracular gill tube is found. This groove in *Amia* and palaeoniscids must not be confused with the groove for prespiracular gill tube in porolepiforms and osteolepiforms.†

* pp. 90–105; † p. 260.

The vomer is in both *Eusthenopteron* and *Amia* formed by horizontal (ventral) infrapharyngeal dental plates of the premandibular arch and is continued backwards by the dermopalatine and ectopterygoid which are modified horizontal epal dental plates of that arch (Fig. 56). The entopterygoid and (in *Amia*) dermometapterygoid have arisen by fusion of dental plates on the ventral side of the commissural lamina of the palatoquadrate. The coronoids in the lower jaw are formed by dorsal and the prearticular by medial ceratomandibular dental plates. The outer dental arcade* (Fig. 53) which is characteristic of the teleostomes is derived from lateral dental plates of three different visceral arches, the premaxillary being formed by epiterminal, the maxillary by epipremandibular and the dentary by lateral ceratomandibular dental plates.

The three tooth-bearing dermal bones (premaxillary, maxillary, dentary) of the outer dental arcade may also be regarded as external dermal bones. These bones and all other external bones of the head with the exception of most of the dermal bones of the cranial roof and the anteroventral parts of the exoskeletal shoulder girdle have been tentatively interpreted as supporting elements of four gill covers (Fig. 58): (1) the terminal gill cover following distally to the premaxillary a bone derived from lateral epiterminal dental plates; (2) the premandibular gill cover carried by the epipremandibular (pars autopalatina); (3) the mandibular gill cover developed in relation to the cerato- and epimandibulars (Meckelian element and pars pterygoquadrata); and (4) the hyoidean gill cover including the operculogular series formed by branchiostegal rays. In relation to the each of the three first named gill covers there is a sensory line, respectively named the terminal, premandibular and mandibular arch line (Fig. 33). A fact which at first sight may seem peculiar is that not only the ventral part of the mandibular arch line (the oral line), but also the ventral part of the hyoid arch line (the mandibular canal) are related to the mandibular gill cover. However, this is probably a consequence of the remarkable but still unexplained condition of the passage of the r. mandibularis VII and the r. mandibularis lateralis from the hyoid to the mandibular arch in the area of the jaw joint. It is also to be noticed that elements of the hyoid and mandibular gill covers in *Eusthenopteron* have fused in this area forming the complex submandibulo-branchiostegal plate, and that in *Amia* both the interopercular and the dorsalmost branchiostegal ray are joined by ligaments to the posterior margin of the angular bone of the lower jaw.

Cranial Nerves and Brain

The brain may be divided into premedullary and medullary portions.† The premedullary portion is chiefly a result of the rostral prolongation (Fig. 11) which includes two phases, one related to the sense of sight and one to the sense of smell. The organs of sight (lateral and median eyes) are photoreceptors formed by paired lateral and median evaginations of the wall of the brain. The lateral eyes arise at the anterior end of the original neural tube and each eye is connected with the main part of the brain by the optic nerve (n. II) which is a complex structure including a tube formed by the brain

* p. 92; † p. 19.

wall and a brain tract, the fibres of which cross those of the other side at the optic chiasma. The median evagination often divides into an anterior parapineal and a posterior pineal organ which secondarily (e.g. in *Amia*) may assume an asymmetrical position. The organ of smell, which is placodal in origin, belongs, according to the views advanced above* (Figs 21, 22) to the metameric system of the head.

The medullary portion of the brain posteriorly merges into the spinal cord. It includes: (1) the medulla oblongata which, by definition,† extends forwards to the tuberculum posterius, and (2) the ventral part of the diencephalon (hypothalamus) delimited anteriorly by the optic chiasma. Accordingly the medullary portion extends forwards to the original anterior end of the neural tube and most likely it includes the most primitive part of the brain.

Early in ontogeny the hypothalamus shows a ventral evagination, the distal part of which differentiates into the neurohypophysis. A most remarkable fact is that this structure was associated, early in phylogeny, with a structure of different origin, the adenohypophysis; together they form the hypophysis or pituitary gland. As to the origin of the adenohypophysis there are two opinions. According to the current view the adenohypophysis is derived from ectoderm at the bottom of Rathke's pouch in the roof of the stomodeum invagination (Fig. 15). It is assumed that the buccohypophysial duct is the persisting invagination canal and is thus ectodermal. However, investigations of *Acipenser, Lepisosteus* and *Amia* have clearly shown that the adenohypophysis in these forms is derived from the area of the neuropore and that the buccohypophysial duct which pierces the parasphenoid (in *Lepisosteus*) is endomesodermal (Fig. 16). In view of these and other facts it is the derivation of the adenohypophysis from Rathke's pouch that seems to be in need of confirmation.‡

Disregarding the neurohypophysis and certain other specializations the medullary brain portion is essentially similar to the spinal cord. Also in the medullary portion the four columns of His-Herrick are thus present, and of great interest is that two of these columns, the visceral sensory (VS) and the visceral motor (VM), extend forwards to the area of the optic chiasma. Moreover, metameric dorsal and ventral nerves are given off, and on each of the dorsal nerve roots there is a root ganglion derived from neural crest ectomesenchyme and serially homologous with the spinal ganglia. Autonomic ganglia are also present in the head and an important fact is that a typical spinal nerve has been identified as far forwards as in the second (premandibular) metamere (Fig. 19). However, three special columns have been added in the head (Figs 10, 20): (1) a special visceral motor (sVM) column related to the metameric visceral arch musculature; (2) a special visceral sensory (sVS) column which via trunk ganglia derived from epibranchial placodes receives impulses from taste buds; and (3) a special somatic sensory (sSS) column (the acoustico-lateralis area) which is connected with lateralis ganglia derived from dorsolateral placodes and receives impulses from the neuromasts of the acoustico-lateralis system. These ganglia and the dorsal cranial nerves belonging to them are metameric structures, but a well known fact is that the ganglia and nerve roots in the vagus area have been crowded together from behind (Figs 45, 46, Volume 1). A corresponding displacement to the rear has, however, probably

* pp. 34–35; † p. 30; ‡ p. 27.

occurred also in the prootic metameres and the motor nuclei in the brain may furthermore be subject to considerable migrations in various directions. Due to these conditions the metameric order has been somewhat disturbed, but in spite of these difficulties it has been possible to identify not only the dorsal and ventral cranial nerves in each metamere, but also the root, trunk and lateralis ganglia of each of the dorsal nerves (Figs 21–24).

The metamerism is also clearly expressed in the innervation, composition, and course of the sensory lines of the head. The dorsolateral placodes may be difficult to identify, but the lateralis ganglia which develop from these placodes display a distinct metameric arrangement. However, as indicated, these ganglia also have been crowded together from in front in the preauditory and from behind in the postauditory region. As a consequence the metameric lateralis nerves in the preauditory region appear to be branches of the nervus facialis. The sensory lines with their neuromasts develop from special lateralis or neuromast placodes, which together with the innervation give valuable clues to the composition of the lateral line system. Chiefly on this basis four important visceral arch lines, the terminal, premandibular, mandibular and hyoid, have been distinguished (Fig. 33). Due to the backwards inclination of the gill arches and the formation of the gill covers these lines have been somewhat displaced and in connection with the formation of the mouth the vental part of the hyoid line has, as in *Amia* and *Eusthenopteron*, been associated with the mandibular gill cover. The remaining sensory lines of the head, i.e. the sensory lines of the cranial roof, include longitudinal lines composed of metameric sections and metameric transverse lines. In connection with fusion of dermal bones considerable displacements of sensory lines (canals as well as pit-lines) may occur.

Two lateralis placodes deserve special attention. The most prominent of these placodes, the otic placode proper, which belongs to the fourth metamere, invaginates and forms the otic vesicle. The invagination canal is soon closed but near to that canal the otic vesicle very early in ontogeny forms a dorsal diverticle. This diverticle, which is characteristic of all vertebrates, develops into the endolymphatic duct (Fig. 27). This duct is in *Amia* a simple tube, whereas in *Eusthenopteron* its dorsal part has expanded into a large endolymphatic sac extending backwards into the neural canal. The external opening of the endolymphatic duct in sharks is a secondary formation and not, as generally assumed, the opening of the original invagination canal.* Although it is a basic vertebrate structure, the importance of the endolymphatic duct is still obscure and this is true also of the spiracular sense organ present in *Amia, Eusthenopteron* and several other fishes.* This strange organ probably arises by invagination of a lateralis placode in the third metamere and may possibly be serially homologous with the otic invagination in the fourth metamere.

Visceral Musculature, Aortic Arches and Branchiothyria

The visceral arch muscles, too, are strictly metameric. These muscles are derived from the mesodermal visceral tubes present in all cranial metameres and originally continuous with the somites (Fig. 9). From the premandibular arch backwards these tubes are

* p. 44.

transformed into so-called muscle plates which differentiate into the various visceral muscles; they are supplied by special visceral motor (sVM) fibres in the visceral trunks of the dorsal cranial nerves. In all these regards there are no essential differences between *Amia* and *Eusthenopteron*. As to the fate of the mesodermal visceral tube in the terminal metamere, nothing is known with certainty, but it has been suggested that certain muscles in holocephalans are derivatives of the terminal visceral tube.

Metameric aortic arches are characteristic of the head region, but there are also ascending intermetameric arteries serially homologous with those in the trunk. Longitudinal anastomoses are also formed. Another characteristic feature of the head region is the presence of intermetameric gill slits. A gill slit together with its walls constitutes a branchiothyrium (Fig. 7). The pre-prespiracular branchiothyrium is the foremost of these paired branchiothyria. It is situated between the terminal and premandibular arches and is probably formed by the non-sensory part of the nasal sac and the choanal tube. It contains the olfactory epithelium which has been interpreted as the terminal trunk ganglion (Figs 21, 22). The prespiracular branchiothyrium between the premandibular and mandibular visceral arches is represented by dorsal and ventral divisions. The dorsal division, the prespiracular tube, was lodged in the groove on the outside of the processus ascendens anterior of the parasphenoid in *Eusthenopteron* and porolepiforms (Fig. 134, Volume 1). This tube is present in larval anurans and urodeles (Fig. 135, Volume 1) but seems to be wholly reduced in *Amia*. The ventral division is the plica mandibularis (Fig. 135, Volume 1). The spiracular branchiothyrium, between the mandibular and hyoid arches, is interrupted behind the jaw joint and consists also of dorsal and ventral divisions. The dorsal division or spiracular tube was, in *Eusthenopteron*, wide, whereas in *Amia* it is a blind narrow tube. In both fishes it ascends outside the otic region of the neurocranium and is associated with the spiracular sense organ in the spiracular canal. The ventral division is represented by the plica hyomandibularis.

Finally, the thymus glands are also metameric; they continue the series of nephrotomes forwards in a remarkable way. To sum up: the head is phylogenetically the oldest part of the vertebrate body. It includes brain tube, notochord, and gut, all originally extending to the anterior pole of the animal and certain other median elements (thyroid gland, truncus arteriosus). Disregarding these median elements and the ectodermal parts of the skin, practically the whole head (neurocranium, visceral endoskeleton, dermal bones, cranial nerves, ganglia, lateralis system including membranous labyrinth and spiracular sense organs, taste bud system with olfactory organ, somatic and visceral musculature, vessels, gill covers, gills) is composed of metameric elements, which in *Eusthenopteron*, *Amia*, and most other gnathostomes are derived from nine cephalic metameres. The derivatives of the sclerotomes in the head also form dimetameric vertebrae and the vertebral segmentation has influenced the disposition of the dermal bones on both the dorsal and the ventral side of the cranial portion of the vertebral column.

RECORDED PERIOD

General Remarks

When the vertebrates begin to appear as fossils, in the Cambrian, Ordovician, Silurian or Devonian, they are as we have seen, already much advanced and divided into several distinct groups. The recorded period is characterized by further branching evolution and numerous fossil fishes and tetrapods have been described and classified. We may easily be dazzled by the great variety of fossil forms, yet the fossil record is far too incomplete and most fossil vertebrates still too imperfectly known to allow us to follow, step by step, the evolution of the various groups through the geologic ages. We must also remember that not only the qualities common to all vertebrates but also the distinctive features of each principal group arose in the unrecorded period. When studying vertebrate evolution with the aid of fossils we are therefore concerned only with modifications of the ancestral basic plan of each group.

Students in many fields of research in Biology and Medicine often need information about vertebrate phylogeny. The easiest way to get such information is to consult modern textbooks of Comparative Anatomy or Vertebrate Palaeontology, and such sources are frequently utilized. Diagrammatic family trees in well known and influential textbooks state that the "Agnatha" are the most primitive vertebrates and that evolution has proceeded via Placodermi, Osteichthyes, Amphibia, and Reptilia to Mammalia (Figs 1, 2, Volume 1). It is easy to see that these and other more or less sophisticated family trees found in even the most modern literature are erroneous in fundamental respects: (1) to use the term Agnatha ("jawless") for all cyclostomes is misleading.* In recent vertebrates we may distinguish three different types of jaw apparatus and, as may be evident from Fig. 341, Volume 1, the biting type found in gnathostomes cannot possibly be derived from either of the fundamentally different types encountered in petromyzontids or myxinoids. Moreover, the presence of inwardly directed gills definitely debar cyclostomes from ancestry to placoderms or other gnathostomes. Expressions like "primitive jawless vertebrates" with reference to cyclostomes are nonsensical and to start discussions of phylogenetic problems from the preconceived opinion that cyclostomes are the most primitive vertebrates is certainly not to be recommended.

The main reason why cyclostomes are thought to be primitive is geological. The first vertebrates to appear in the geologic sequence (in the late Cambrian and the Ordovician) and the dominant ones in the Silurian are cyclostomes, and it is because of this that many palaeontologists, even today, seem to believe that cyclostomes must be more primitive than gnathostomes and ancestral to them. However, if gnathostomes and cyclostomes are "sister groups" they therefore are of the same age. If so, the occurrence of unquestionable cyclostomes in the Cambrian can only mean that gnathostomes must have been in existence at the same time, although they have yet to be found as fossils (Fig. 138).

* Volume 1, p. 440.

(2) The view that placoderms are the most primitive gnathostomes lacks foundation (Jarvik, 1964; 1968a).

Because cyclostomes are not primitive progenitors of gnathostomes, and because teleostomes can be derived neither from placoderms nor from selachians—which latter according to a deep-rooted conviction are the most primitive gnathostomes—it is evident that the views about vertebrate phylogeny embodied in the family trees and diagrams in the modern literature break down. In view of this it has appeared more and more desirable to make a clean sweep of all the unproved statements and preconceived ideas that still form the basis of current conceptions of vertebrate phylogeny and to try to construct a grouping more consistent with available facts (Fig. 140).

We may picture the evolution of the vertebrates as a tree or a bush, but if we want to discuss the evolutionary processes on the basis of fossils, we have to remember that it is only the topmost twigs that reach the recorded period (Fig. 139) and that only fragments of these twigs are available for study. With the aid of the methods for establishing relationships outlined above* we have succeeded in combining important such fragments into evolutionary lines. Moreover chiefly utilizing embryological data it has been possible to reconstruct in broad features the evolution of the vertebrates during the long unrecorded period, and to make out the origin and composition of the vertebrate head and the paired limbs. However, as may be gathered from the subsequent brief review we still grope in the dark as to the origin and relationships of several of the principal groups which appear in the early Palaeozoic, notwithstanding that some of these groups have survived to present times and accordingly can be studied more thoroughly than is possible on fossil remains.

Cyclostomata

When turning first to cyclostomes† this of course must not be taken to mean that cyclostomes are the most primitive vertebrates. The reason is that it was in this major vertebrate division that the methods for establishing relationship mentioned above were successfully used for the first time.

Cephalaspidomorphi

In his paper on the early Devonian cephalaspids from Spitsbergen Stensiö (1927) improved the techniques for the study of fossils, and on the basis of serial sections, wax models and mechanical preparations he gave surprisingly detailed and safe descriptions of the previously unknown internal structure of the cephalic shield in cephalaspids. After making out the construction of these ancient creatures Stensiö proceeded to thorough comparisons with extant vertebrates, and by this method he was able to interpret the numerous anatomical structures displayed by the fossils and to reconstruct brain, membranous labyrinth, nerves, vessels, gills and other anatomical features. It then turned out that the cephalaspids anatomically agree surprisingly

* pp. 229–230; † see also Volume 1.

well with recent petromyzontids and only with that group; and this led Stensiö to the inevitable conclusion that the cephalaspids and petromyzontids are closely related and are to be classified together.

Stensiö's well founded views have been generally accepted and disregarding the "electric fields" his results have been confirmed and extended by later investigations by himself and by others.* This is true also of his much debated interpretations of the visceral arches and the cranial nerves in cephalaspids. Vestiges of a premandibular arch, situated as that in cephalaspids, have been identified in early ammocoete stages of the lamprey.† Also, it has been shown that the nerves supplying the inwardly directed gills in cephalaspids ran and branched exactly as they still do in *Ammocoetes*; and this is another example of the remarkable reluctance to change that characterizes vertebrate evolution in the recorded period. Moreover, remains of Stensiö's canals for "electric nerves" have been identified in the membranous labyrinth of the lamprey, the importance of the venous nodules has been stressed‡ and it has been shown that the peculiar mode of eye accommodation in the lamprey was initiated in the cephalaspids.

Pteraspidomorphi

It is well established that extant petromyzontids (lampreys) and myxinoids (hags) differ fundamentally in the structure of the jaw apparatus, nasohypophysial complex and numerous other anatomical regards. This must mean that these two groups of extant cyclostomes have evolved independently for a very long time, and inasmuch as myxinoids cannot possibly be derived from osteostracans they must be descendants of another group of early cyclostomes, which implies that the extant cyclostomes are diphyletic in origin. This is what Stensiö claimed in 1927, but when he suggested that myxinoids are related to heterostracans and are to be classified together with them he met opposition.

However, since the "rasping tongue" has been disposed of as argument for monophyletism only one argument against Stensiö's views remains. It is said (Romer, 1968; Miles, 1971) that heterostracans like gnathostomes are diplorhinous in contrast to other cyclostomes which are thought to be monorhinous. However, as shown by Stensiö (1968) and further elucidated above§ all vertebrates have paired olfactory organs (and olfactory nerves) and are diplorhinous although the nasal sacs in cyclostomes lie close together. This is true also of the better known heterostracans and as demonstrated in Volume 1 both the nasohypophysial complex and the jaw apparatus in some of them, at least, is of the myxinoid type. Because, in addition, several other striking similarities between myxinoids and heterostracans have been demonstrated the only possibility, on present evidence, is to accept Stensiö's view and classify the Myxinoidea together with the Heterostraci as Pteraspidomorphi.

Gnathostomata: Plagiostomi

The plagiostomes, characterized by subterminal mouth and the absence of the outer

* Volume 1, pp. 450–476; † Volume 1, pp. 476–478; ‡ Volume 1, p. 474; § Volume 1, p. 497.

dental arcade, reached the peak of their evolution in Palaeozoic times and are in the recent vertebrate fauna—besides a few holocephalans and dipnoans—represented only by sharks and rays.

Placodermi

The Devonian fish fauna was dominated by the placoderms, which is a most diversified group of fish. It includes about ten subgroups, all, as far as known with certainty, restricted to the Devonian period and thus relatively short-lived. Stensiö (1959) made thorough studies of the pectoral fin and shoulder girdle and on the basis also of embryological evidence he came to the conclusion that the most primitive fin so far known in gnathostomes, the holosomactidial fin (Fig. 68), is found in pachyosteomorph arthrodires, whereas one of the most specialized pectoral appendages, the monomesorhachic fin, distinguishes the antiarchs. The holosomactidial fin is common in late Devonian forms, which have been considered to be most advanced euarthrodires ("pachyosteomorph level", Miles, 1971). Because of this Stensiö's views have been strongly opposed, in particular by Romer (1968, p. 32) who claims that an acceptance of Stensiö's thesis "makes it necessary to assume that the geological record presents the arthrodires upside down". Since all vertebrate groups known from the Devonian to present times are distinguished by a remarkable conservatism, and since several distinct placoderm subgroups, such as radotinids, petalichthyids, rhenanids and antiarchs, have obviously changed very little during their short geological history, it is difficult to believe that the pachyosteomorphs make an exception. The views (Heintz, Romer and others) that the euarthrodires, with narrow-based (merosomactidial, Fig. 68) fins and a long trunk armour, have rapidly changed into forms with long-based (holosomactidial) fins and a short trunk armour (brachythoracids) is therefore unlikely; and most seriously, it is directly contrary to embryological evidence. This view is therefore unacceptable as are in general phylogenetic conclusions based solely on geological data. Also the attempts by Denison, Miles, Westoll and others to classify the placoderms according to a few exoskeletal features such as tesserae, extent of trunk armour or number of paranuchal plates, are to be rejected. Such conclusions must rest on the structure as a whole, but as emphasized by Stensiö, to whom we above all owe our present knowledge of the placoderms, this knowledge is still insufficient. In spite of the great progress made during the last few decades we have to admit that we know almost nothing about the origin of the placoderms and very little about their interrelationships. It is, however, possible that the holocephalans share a common origin with the ptyctodontids, and it cannot be excluded that the batoids (rays) may be in some way related to the rhenanids.

Holmgren's views (1940), based on detailed anatomical and embryological studies, that rays have evolved independently of sharks since Devonian times and that they possibly are related to placoderms have gained some support by the results of recent studies of the acanthodian fishes (Jarvik, 1977).

Acanthodii

According to most writers acanthodians are related to elasmobranchs. Nelson (1968) demonstrated that the infrapharyngobranchials in *Acanthodes* are directed backwards, as in sharks, but the main reason why the early palaeozoologists claimed relationships with elasmobranchs is the striking similarities in the structure of the palatoquadrate and its relations to the neurocranium. The double jaw joint characteristic of *Acanthodes* as well as fossil and primitive extant sharks is an important shared unique specialization, as is also the postorbital articulation. Another significant feature common to *Acanthodes* and sharks is the orbital articulation (Jarvik, 1977). The basipterygoid process is lacking and the orbital articular facet lies in *Acanthodes*, exactly as in notidanid and squaloid sharks, close in front of the palatobasal process. A most important fact is, moreover, that the orbital process of the palatoquadrate in *Acanthodes* is identical in structure with that in, for instance, *Heptanchus* or *Squalus*. The palatine process, too, is very similar to that in the said sharks and as in them it bends medially in its anterior part, a condition indicating that a palatoquadrate commissure homologous with that in sharks was present.*

That a palatoquadrate commissure existed is proved by the specimen of another acanthodian shown in Fig. 274, Volume 1. As also evidenced by this specimen the mouth is subterminal, as in sharks, the nasal openings have a similar position, and there is an upper dental arcade, as in sharks composed of tooth whorls. If it is added that the outer dental arcade is lacking and that acanthodians differ fundamentally from teleostomes in the histological structure and growth of the teeth and that they agree more with elasmobranchs than with teleostomes in the structure of spines and scales, the only possible conclusion must be that acanthodians are plagiostomes and closely related to sharks.

However, this is not all. A detailed analysis revealed (Jarvik, 1977) that *Acanthodes* also in the structure of the neurocranium agrees closely with sharks and differs from teleostomes. The fossa hypophyseos is thus very much as in sharks, the efferent pseudobranchial artery entered that fossa dorsal to the trabecula, which is a distinctive elasmobranch feature, the internal carotid and the jugular vein ran as in sharks and there is an endolymphatic fossa. Moreover, the hyomandibula resembles that in sharks and as in them it articulates with the neurocranium ventral to the jugular vein. The spiracular gill tube was also in *Acanthodes* situated between the hyomandibula and the palatoquadrate (and not in the hypophysial region, as has been suggested), and epibranchial spinal muscles were present. Furthermore, the so-called gill rakers in *Acanthodes* are most likely endoskeletal structures homologous with the endoskeletal cores of the gill rakers which are characteristic of extant plagiostomes (sharks, holocephalans, dipnoans).

Because, in addition, the dermal fin rays are ceratotrichialike, the fins are aplesodic and except probably for a marginal fringe completely covered by fin scales, the endoskeleton of the pectoral fin is tribasal as in advanced sharks, the tail is similar to that in certain sharks, and finally, because the ontogenetic development of sharks as

* Volume 1, pp. 333, 337–339.

shown by Holmgren (1942) in certain regards implies a recapitulation of the phylogenetic development all requirements for establishing relationships are fulfilled.

The acanthodians are plagiostomes, and as evidenced by an overwhelming number of unanimous data (including several unique specializations) they are closely related to the selachians. They therefore are to be referred to the Elasmobranchii.*

Remarks. A notable fact is that acanthodians and sharks share several important characters which are lacking in rays. Such characters are, for example, orbital process and orbital articulation, palatobasal process, epibranchial muscles and aplesodic fins. Considering also Holmgren's statements as to the phylogeny of rays it may then be asked if sharks are not more closely related to acanthodians than to rays. In view of the fact that the palatobasal process and the orbital articulation in acanthodians are located in the posterior part of the orbit, it is even tempting to suggest that acanthodians stand closer to notidanid and squaloid sharks in which the conditions are exactly the same, than to palaeoselachians, galeomorphs and *Chlamydoselachus* in which the palatobasal process and the orbital articulation are found in the anterior part of the orbit. However, these suggestions would require a considerable rearrangement of the elasmobranch taxa which I find premature in view of the prevailing uncertainty as to elasmobranch interrelationships and the differences existing between acanthodians and modern elasmobranchs.

Dipnoi

Disregarding the effects of the retrogressive development of the skeleton, the dipnoans have changed very little since the Devonian when they reached the zenith of their evolution and were represented by a great number of genera and species. Current views say that the dipnoans are teleostomes and in the opinion of many they are closely related to the "crossopterygians" and are often classified together with them as "Sarcopterygii". However, as far back as 1942 I found that the dipnoans differ fundamentally in snout structure from each of the "crossopterygian" groups known at that time (porolepiforms, osteolepiforms and coelacanthiforms); the increased knowledge of *Neoceratodus*, *Latimeria*, and of the Devonian "crossopterygians" and dipnoans gained since the beginning of the 1940s has made me more and more convinced (Jarvik, 1952; 1964; 1967a; 1968; 1968a) that the dipnoans cannot be so closely related to the "crossopterygians" as believed by Westoll, Romer, Miles and others. Looking for other alternatives it was not difficult to see that dipnoans in many important regards agree better with extant elasmobranchs and holocephalans than with teleostomes. As early as in 1881 Retzius claimed relationship with elasmobranchs because of similarities in the membranous labyrinth (also Werner, 1930; 1960), and a great number of other elasmobranchlike anatomical features were reported by the early anatomists. The reason why I now, after some hesitation in 1968, have decided to include the dipnoans in the Plagiostomi is not only that they, in contrast to teleostomes, have a subterminal mouth and lack the outer dental arcade. Several other important elasmobranchlike characters have been discovered and moreover the

* see Volume 1, p. 324 for classification.

attempts to prove relationship with teleostomes have failed.* However, in spite of the many plagiostome features it is, on the basis of present evidence, impossible to relate dipnoans to holocephalans or to any other of the better known plagiostome groups. The Plagiostomi is, however, a most heterogenous assemblage, and as a final remark we may say that dipnoans do not differ more from holocephalans or placoderms than the holocephalans or placoderms differ from sharks.

Gnathostomata: Teleostomi

The Teleostomi, including actinopterygians, brachiopterygians (polypterids) and the four "crossopterygian" groups, together with all the tetrapods, were, in contrast to cyclostomes and plagiostomes, relatively rare in the Devonian. However, two of the groups, namely the osteolepiforms and the actinopterygians, possessed intrinsic qualities for further evolution, and contemporaneously with the radiation of the eutetrapods on land in post-Devonian times the actinopterygians underwent an equally fascinating expansion in the water; today these two groups are the most diverse vertebrates and have been so for vast periods of time.

Actinopterygii

Recent members of this group agree in two probably important respects. They are the only fishes that have an air bladder† (Jarvik, 1968) and the forebrain is specialized in much the same way (Nieuwenhyus, 1963; 1966). These and certain other common features in the soft anatomy indicate that the actinopterygians are a natural group of common descent. Skeletal features which can be studied in the fossil members of the group favour this opinion. There are thus general similarities in the structure of the girdles and paired fins‡ (Jessen, 1972) and the presence of the supraangular (Nelson, 1973), the presupracleithrum (Nybelin, 1976) and of the epimyelencephalic hemopoietic organ§ may be primitive actinopteygian characters.

After the pioneering works by L. Agassiz, R. H. Traquair, A. S. Woodward and others, E. Stensiö in his studies of Triassic fishes from Spitsbergen (1921; 1925) and East Greenland (1932) introduced a new era in the exploration of the fossil actinopterygians. Stensiö's anatomical investigations were continued by E. Nielsen. His outstanding descriptions (1942; 1949) of Triassic fishes from East Greenland made on excellent fossil material and with the aid of Sollas's grinding method are still the corner-stones in our knowledge of fossil actinopterygians. Many other students have made significant contributions and recently several neo-zoologists have joined in the animated debates on the problems of the classifications and interrelationships of this large group of fish (for references see Patterson, 1975).|| It is certainly not easy to master the crowd of fossil and extant actinopterygians known to science and in this connection only a few remarks will be given.

Because palaeoniscids are the oldest actinopterygians and the only ones known from the Devonian it has usually been assumed that they are primitive and ancestral to

* Volume 1; pp. 426–436; † Volume 1, p. 301; ‡ Volume 1, p. 314; § Volume 1, p. 322; || see also Volume 1, p. 310 for palaeoniscids.

all other actinopterygians. These conclusions are premature. The palaeoniscids are a distinct early specialized group and they have changed very little during their long geologic history. In a preceding chapter* palaeoniscids were compared with *Amia* and it was concluded that the amiids can be traced back with confidence only to the Jurassic, and that it is hardly possible to derive them from any well known palaeoniscids. Comparisons with sturgeons, *Lepisosteus*, or teleosts would probably give similar results and we have to conclude that the origin of the extant actinopterygians is obscure. In view of these conditions it may be inappropriate to rely too much on palaeoniscids in order to unravel actinopterygian interrelationships.

Another reason for the difficulty in recognizing evolutionary lineages within the actinopterygians is the existence of parallel evolution. Following a general trend in fishes the scales in several actinopterygians groups have changed from rhombic to cycloid (Gross, 1966; Schultze, 1977) and as in coelacanths the dermal fin rays have decreased in number.

Also as in other vertebrates, the skeleton has undergone a retrogressive development and the neurocranium, which in early forms is well ossified, may disintegrate into a variable number of endoskeletal ossifications separated by more or less extensive cartilaginous or membranous parts. These ossifications are often attributed a great phyletic importance and it has been claimed that the actinopterygian braincase was originally composed of elements homologous with the endoskeletal ossifications (prootic, opisthotic, etc.) found in advanced forms. This view has, however, no support in ontogeny, and it is to be doubted that considerations of these ossifications will aid materially in elucidating the interrelationships and classification of the actinopterygians. As is well known the embryonic components in the neurocranium (cranial pilae, swellings in urodeles and anurans, polar cartilage, basiotic laminae, etc.) are either embryonic arcual elements of cranial vertebrae or dermatomic and myotomic derivatives; and there is no correspondance between these components and the endoskeletal ossifications.

As evidenced by ontogenetic data and supported by the conditions in *Eusthenopteron*, *Ichthyostega* and other adult gnathostomes, the cranial vertebrae are separated by intervertebral joints. Such a joint, the intracranial joint, is retained in adult osteolepiforms but when it gradually disappears, at the transition from fish to tetrapod, it is for some time discernible as a transverse basicranial fissure, the fissura preoticalis, as found in *Ichthyostega*. However, in *Eusthenopteron* and *Ichthyostega* there are similar fissures or traces of such fissures farther back in the basis cranii. This fact, that a series of basicranial fissurae representing vestiges of intracranial joints is clearly discernible as well as other conditions (Bjerring, 1978), refutes the opinion, that the intracranial joint—and accordingly also the fissura preoticalis in the corresponding position in *Ichthyostega*—has migrated backwards to occupy the position of the fissura oticalis ventralis anterior in palaeoniscids. This fissure appears to have the same position in the Gogo palaeoniscid *Moythomasia* as in the post-Devonian forms (as to *Mimia* see Bjerring, 1978) and accordingly the paired dental plate behind that fissure in *Moythomasia* (Gardiner and Bartram, 1977) corresponds to the suboccipital plates in *Eusthenopteron* and not to the subotic plates.

* Volume 1, pp. 309–323.

Other general trends concern the reductions of the endoskeletal basipterygoid process, the processus ascendens anterior of the parasphenoid, and of the prespiracular gill slit. In *Eusthenopteron* (Fig. 134, Volume 1) the spiracular gill tube was lodged in a well defined space outside the otic region, between the mandibular and hyoid arches. Its space was wide, but exactly as in palaeoniscids and *Amia*, in which the tube due to reduction is narrow, it was connected dorsally with the spiracular canal housing the spiracular sense organ. The osteolepiforms and the porolepiforms are the only gnathostomes in which the prespiracular gill tube between the mandibular and premandibular arches was retained in the adult (Fig. 134, Volume 1). As described in 1954 the prespiracular tube in them was lodged in a groove on the outside of the processus ascendens anterior of the parasphenoid which supports the basipterygoid process. This groove is situated outside the hypophysial area, in the posterior part of the orbitotemporal region. It lies far in front of the wide space for the spiracular tube in the otic region and is separated from that space not only by the intracranial joint but also (in osteolepiforms) by the paratemporal articulation. That a prespiracular gill tube really existed in porolepiforms and osteolepiforms is supported by the fact that a distinct but transient such tube occurs in larval stages of their tetrapod descendants (Fig. 135, Volume 1), the urodeles (Regel, 1966) and the anurans (Regel, 1973). However, in spite of all these facts Gardiner, Miles and Patterson, obviously misled by the conditions in palaeoniscids, claim that the prespiracular groove in porolepiforms and osteolepiforms was developed for the spiracular gill tube. The reason for this mistake is apparently that the processus ascendens anterior of the parasphenoid in palaeoniscids lies close in front of and sometimes is continuous with the processus ascendens posterior at the base (see discussions in Nielsen, 1942; Poplin, 1974). In palaeoniscids in which both the processus ascendens anterior and the basipterygoid process are less well developed than in osteolepiforms and porolepiforms, no traces of a groove for the prespiracular gill tube have been found. Nor has such a tube been described in *Amia*, in which the processus ascendens anterior and the basipterygoid process are present although more vestigial than in the palaeoniscids. However, that a prespiracular gill slit originally existed also in actinopterygians is shown by the existence of transient prespiracular gill tubes in embryonic stages of *Lepisosteus* (van Schrick, 1927) and *Acipenser* (Neumayer, 1932).

Brachiopterygii, Coelacanthiformes

A tenacious misconception is that the brachiopterygians or polypterids are actinopterygians. As a matter of fact they do not share a single of the actinopterygian characteristics mentioned above. The polypterids thus have lungs,* the telencephalon is everted but of a special type, and the cerebellum is formed by invagination. It is doubtful if the so-called valvula cerebelli is homologous with that in actinopterygians† Also the pituitary gland differs markedly from that in other fishes (T. Kerr, 1968). Moreover the supraangular is lacking for which reason Nelson (1973) referred polypterids to "sarcopterygians", there is no presupracleithrum but a spiracu-

* Volume 1, p. 299; † Volume 1, pp. 294–295.

lar series of bones without equivalents in actinopterygians, a condition which according to Nybelin (1976) favours the opinion that the polypterids is an isolated group. An important fact is also that there is no hemopoietic organ dorsal to the medulla oblongata; a quite different structure is there instead, the cisterna spinobulbaris.* Furthermore polypterids as to the endoskeleton and soft anatomy of the paired fins differ fundamentally from actinopterygians† (Jessen, 1972; 1973). In addition polypterids, in contrast to well known palaeoniscids, lack a posterior myodome, there is no spiracular canal, the spiracular gill tube is wide, and the ascending process of the parasphenoid is a complex structure. Furthermore the nasal sac differs widely from that in actinopterygians, and if it is added that polypterids differ from actinopterygians in the structure of the dorsal and caudal fins‡ it is evident that polypterids cannot be classified as actinopterygians.

This raises the question which are the closest relatives of the polypterids or, in other words, which among extant vertebrates is their "sister group". No affirmative answer can be given to this question. It may be true that the polypterids share at least one unique specialization with the coelacanthiformes, namely the structure of the nasal sac.§ There are certainly also other resemblances, e.g. in the development of the fossa bridgei, the operculogular series and the lower jaw, but are these resemblances really sufficient for assessing relationship?

According to Nelson (1970) "polypterids are among the few Recent fishlike vertebrates that can be compared in some detail with *Latimeria*", and he claims that the subcranial muscle in that fish "probably was derived from one or more anterior body myomeres such as occur in Recent polypterids". This latter statement and Nelson's suggestion that this musculature "may have shifted forward, segmenting off as a distinct muscle, in relation to the appearance of the new and relatively advanced intracranial joint of primitive sarcopterygians" have been disputed by Bjerring (1973) but are still accepted by Patterson (1975) and Miles (1977).

In support of these views Nelson states that the insertion of the subcranial muscle in fossil "rhipidistians" and coelacanths "is restricted to an area behind the ascending process of the parasphenoid" and that "it is essentially the same insertion as that of the most anterior myomere of *Polypterus*" in which that myomere inserts "on the parasphenoid as far forwards as the ascending process". These statements include a fallacy. The ascending process in "rhipidistans" to which Nelson refers is the processus ascendens anterior which is situated outside the posterior part of the orbitotemporal region. The ascending process in *Polypterus*, in contrast, is the processus ascendens posterior which lies farther to the rear, outside the otic region. Moreover this process in *Polypterus* has been prolonged backwards by incorporation of elements belonging to the posthyoidean branchial arches, and, in fact, the body myomeres in *Polypterus* do not extend farther forwards than, for instance, in *Amia*. Also, why assume a shift forward when the trunk myomeres of *Polypterus*, in Nelson's opinion, extend as far forwards as did the subcranial muscle in "rhipidistians"? How is the sudden appearance of such an intricate structure as the intracranial joint to be explained? As we have seen, this joint most likely is a persisting intervertebral joint and the subcranial muscle is a member of the series of basicranial muscles.

* Volume 1, p. 297; † Volume 1, p. 301; ‡ Volume 1, p. 308; § Volume 1, p. 299.

When discussing the relationships of coelacanths we have firstly to consider the remarkable reluctance to change that distinguishes this group. In this connection it may be enough to remind of the small affacial process of the otoccipital carrying a small rounded dental plate present in the Devonian *Nesides* as well as in the extant *Latimeria*. This inconsiderable structure, which I (1964) have compared with an old-fashioned front collar-stud, has been on the side of the coelacanthid braincase for about 350 million years, a condition which is well worth keeping in mind when discussing the evolution of coelacanths and of vertebrates in general.

Secondly it is to be emphasized that the coelacanths show a great number of peculiar and aberrant features; e.g. the kidneys are ventral in position and situated in the tail (Millot and Anthony, 1973a); the heart is simple in structure (Anthony *et al.*, 1965; Millot *et al.*, 1978); the eggs are large (Millot and Anthony, 1974); and the innervation of the sensory lines of the cheek has a characteristic structure (Millot and Anthony, 1965). Some of these features are known only in *Latimeria*, whereas others, as evidenced by the fossils, were present already in the Devonian coelacanths. This applies to the peculiar rostral organ with its three paired outlets, the fossa ventrolateralis in the postnasal wall, the intermetameric position of the intracranial joint, the antotic process, the affacial process with its dental plate, the prefacial eminence, the occipital commissure connecting the membranous labyrinths and the course of the n. abducens in the otic capsule (see also Bjerring, 1973).

We cannot say where, for instance, the peculiar rostral organ comes from or when it arose, but there is certainly nothing in it to prove that the vertebrates, as postulated by Schaeffer (1952) for coelacanths and by Westoll (1949) for dipnoans, underwent a rapid development in the early Devonian. In other words, there is no reason to believe that the rate of evolution at the transition from the unrecorded to the recorded period changed suddenly to the slow rate characteristic of the latter period. It probably took a considerable time in phylogeny to produce such specialized structures as the rostral organ and other unique features present in the earliest known coelacanths. If we imagine that the coelacanths and polypterids are "sister groups" we have to assume that their common ancestor was living before the appearance of the coelacanths in the fossil record that is in the Silurian or earlier. However, if we try to reconstruct such an ancestral form, we will probably—with regard to the many specializations in both coelacanths and polypterids—find it to be a generalized teleostome type. It must conceivably not be very different from the ancestral types which may be reconstructed on the presumption that, for example, polypterids and palaeoniscids, or porolepiforms and osteolepiforms, are "sister groups". This is our present dilemma, and it is because of these conditions that it is so difficult to make out the interrelationships of the principal teleostome groups. In fact we cannot say anything with certainty about the rate of evolution in the remote past when the numerous specialisations which characterize each of the teleostome groups arose, and we have to admit that we can say very little if anything at all about the origin and interrelationships of the coelacanths and other early teleostome groups. In view of these facts I have found it most reasonable to keep these groups apart, as I have done in diagrams since 1959 (Jarvik, 1959; 1960, fig. 28).

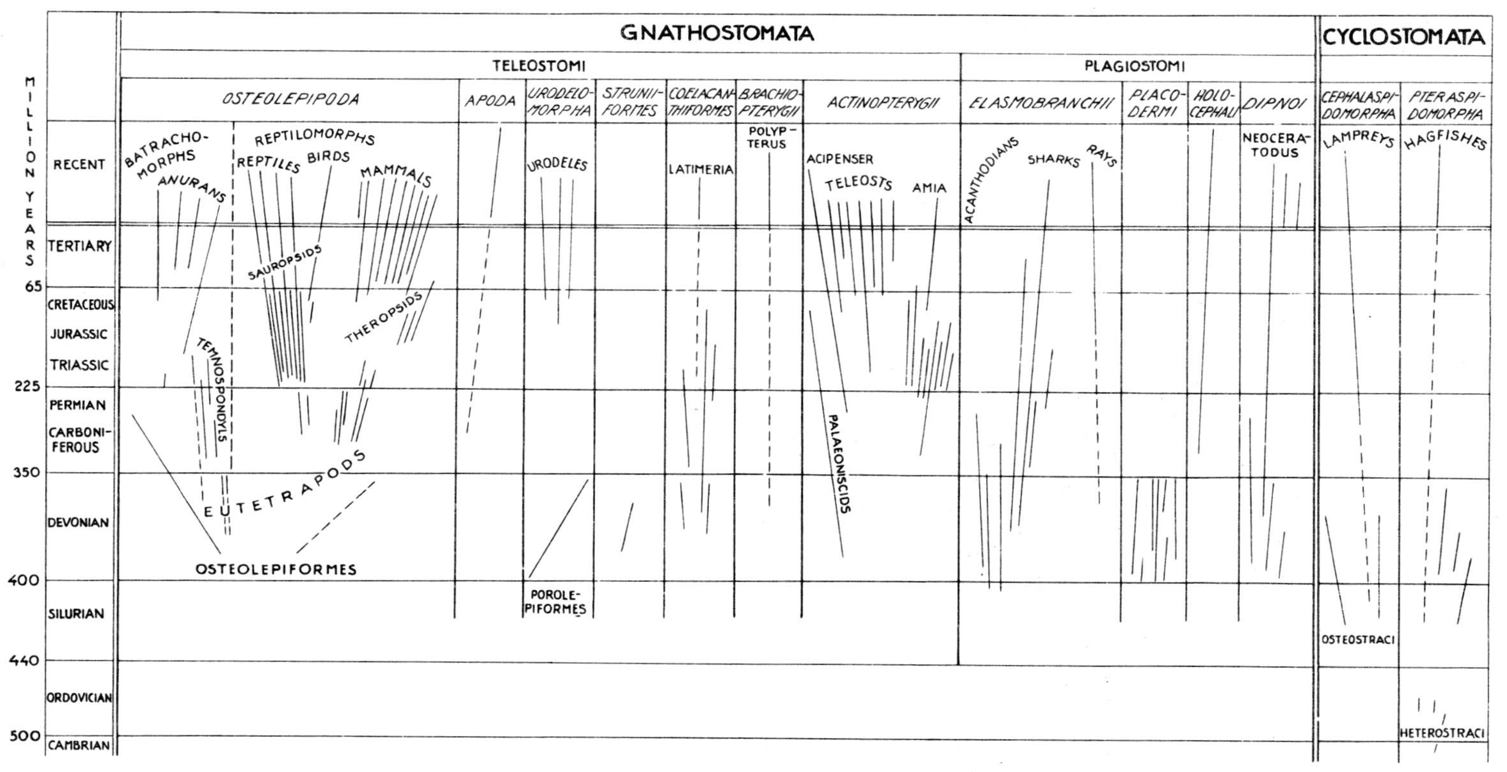

Fig. 140. Diagrammatic representation of evolution of vertebrates. Modified from Jarvik (1959; 1964; 1968a).

Osteolepiforms, Porolepiforms and Tetrapods

The porolepiforms and osteolepiforms are often brought together as "Rhipidistia" or "Choanata" (Miles, 1977). This procedure may have been justified when these fishes were imperfectly known but is, on present evidence, quite unacceptable, and has precluded the understanding of the origin of the tetrapods. Porolepiforms and osteolepiforms certainly agree in several respects. However, if we disregard primitive features such as the intracranial joint and associated structures there are hardly more than the presence of the choana and similarities in the submandibular and operculogular series, and in the lower jaw that suggest relationship; can these resemblances be attributed more importance than the similarities in the structure of the nasal sac, the operculogular series, and the lower jaw between coelacanths and polypterids? When claiming relationship we cannot rely on a few characters. We have to consider the structure of the animals as a whole, and with respect to the porolepiforms and the osteolepiforms we have to explain the fundamental differences between these two groups in the structure of the snout as a whole, in the hypophysial and neuroepiphysial areas, in the occiput, in the intermandibular division, in the paired fins and in other regards. Because the many features characteristic of the porolepiforms were present already in the Lower Devonian porolepids it is evident that these specializations must have arisen in the Silurian or earlier.

However, once established the specializations characteristic of the porolepiforms have been retained with a remarkable tenacity in one group of vertebrates, the

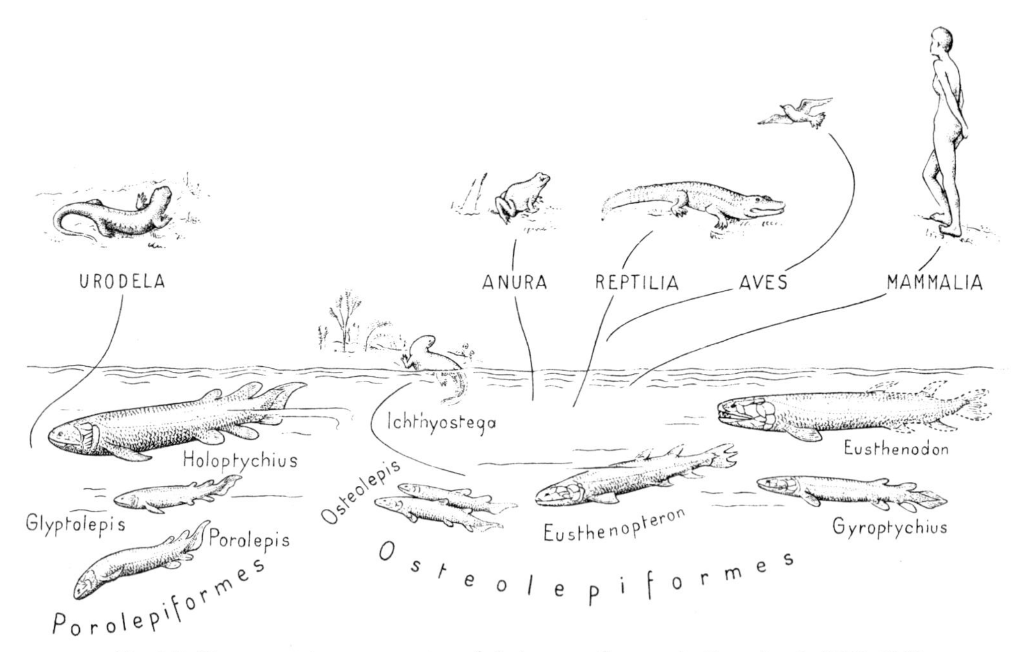

Fig. 141. Diagrammatic representation of phylogeny of tetrapods. From Jarvik (1959; 1960).

urodeles (Jarvik, 1942; 1972). Because of the profound similarities in cranial anatomy that have been demonstrated the only possible conclusion must be that the urodeles are closely related to and descendants of porolepiforms. The urodeles are therefore to be classified together with porolepiforms and for this systematic unit the term Urodelomorpha has been proposed (Fig. 140; Jarvik, 1968a). The problem touched upon in 1942 (p. 639) if any of the late Palaeozoic lepospondyls belong to the Urodelomorpha is still unsolved and the origin of the apodans (cf. Kuhlenbeck, 1970; Carroll and Currie, 1975) is unknown (Jarvik, 1968a).

The osteolepiforms, in particular *Eusthenopteron* have been studied by me for more than forty years. These studies (Jarvik, 1937–1975) and the results gained by Stensiö (1963a), Jessen (1966), Bjerring (1967–1977) and others, have unequivocally demonstrated that most specific anatomical features in osteolepiforms occur also in the Anura and in important regards in the Amniota, but not in the porolepiforms or other fishes, and not in the urodela either. Moreover, it has been shown (Säve-Söderbergh, 1932, 1935; 1936; Jarvik, 1942; 1952; 1967a) that the ichthyostegalians, and other temnospondyl stegocephalians also agree well with the osteolepiforms and anurans. The anthracosaurs, too, in important regards are osteolepiformlike. This evidence can only be taken to mean that all these groups of tetrapods, the Eutetrapoda (Säve-Söderbergh, 1934), are closely related to osteolepiforms and descendants of them, or at any rate, of osteolepiformlike ancestors. However, when dealing with the phylogeny of the Eutetrapoda we have to consider their two main stems (Fig. 140), the Batrachomorpha and the Reptilomorpha (Säve-Söderbergh, 1934; 1935; Jarvik, 1967a; 1968a; Shishkin, 1968).

The Batrachomorpha comprise two main tetrapod groups, the Temnospondyli and the Anura (Jarvik, 1968a). The temnospondyls are in the late Upper Devonian represented by two distinct types, ichthyostegids and acanthostegids. These two types of early tetrapods are in certain regards much modified and none of them can have given rise to the later appearing temnospondyls or other tetrapods. Moreover, they differ widely from each other in the structure of the posterior part of the skull, and obviously they have become modified in different ways. These facts show that the differentiation of the temnospondyls was well on the way in late Devonian times and, moreover, they indicate that ichthyostegids and acanthostegids branched off from ancestral osteolepiforms at an early date and possibly independently of each other. Another remarkable fact is that several of the latest (Triassic) temnospondyls in important regards* agree more closely with well known osteolepiforms like *Eusthenopteron* than with the Devonian and other Palaeozoic temnospondyls, and very likely they have evolved independently from some *Eusthenopteron*-like forebears.

The Anura cannot be derived from dissorophids or other temnospondyls† (Jarvik, 1967a) and form another separate line of evolution (Figs 140, 141). A very close agreement has been found to exist between this extant group and their osteolepiform forerunners. The similarities even in minor structural details between the common frog (*Rana*) and *Eusthenopteron*, e.g. in the structure of the snout as a whole, and in the posterior part of the head, are really remarkable. Because of the close affinities

* p. 223; † p. 222.

between osteolepiforms and anurans and because Palaeozoic intermediate forms are unknown it is obvious that discussions of the phylogeny of the Anura on a palaeontological basis must rest chiefly on direct comparisons between extant anurans and their osteolepiform ancestors. The lissamphibian hypothesis has been rejected.

The Reptilomorpha differ in certain ways from the well known osteolepiforms* and cannot be directly derived from them. Under these circumstances and because they no doubt are allied to osteolepiforms it may be reasonable to assume that reptilomorphs are descendants of some still unknown osteolepiformlike ancestors as tentatively indicated by a broken line in Fig. 140.

According to traditional views, expressed in diagrammatic family trees in textbooks (Romer, 1970; Torrey, 1971), the mammals have evolved from primitive reptiles and the reptiles (including birds) from primitive amphibians; and the urodeles are the most primitive living amphibians. The discovery that urodeles and anurans have developed from different fish groups has weakened the foundation of these views and the term Amphibia has no meaning in phyletic taxonomy. Porolepiforms and urodeles have nothing to do with the direct ancestry of the reptilomorphs. Searching for the ancestors of extant reptiles, birds, mammals and the numerous fossil forms regarded as reptilomorphs, attention has focused on the late Palaeozoic anthracosaurs (embolomeres, seymouriamorphs and certain other groups = Batrachosauria, Efremov, 1946; Panchen, 1977) and it has been suggested that the anthracosaurs are descendants of ichthyostegidlike forerunners. The resemblances between the pelvic girdle in *Ichthyostega* and the embolomere *Archeria* are certainly striking (Figs 160–162, Volume 1) but because the differences in the structure of the

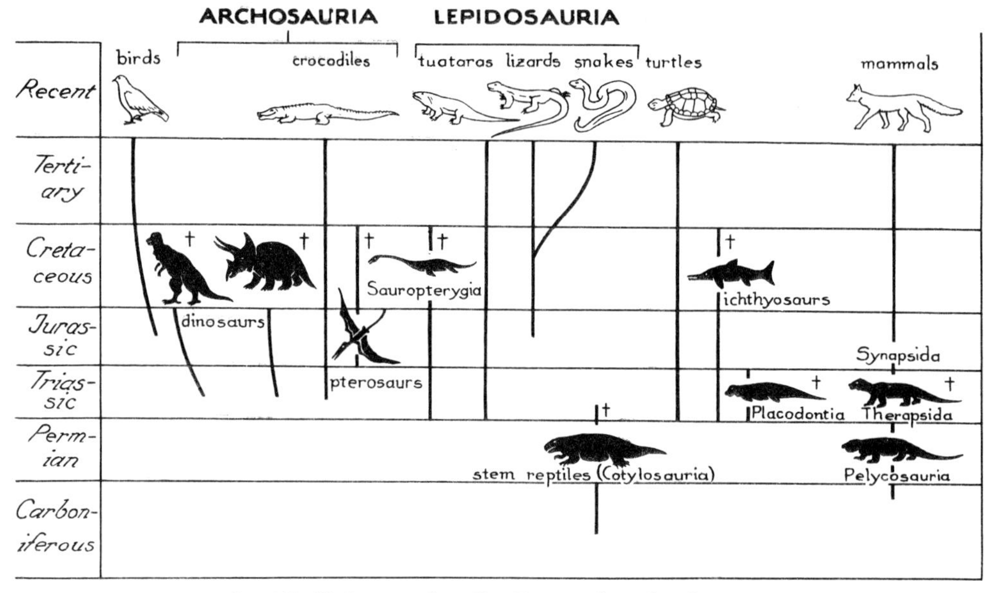

Fig. 142. Phylogeny of reptiles. From Kuhn-Schnyder (1977).

* p. 225.

vertebrae and the skull are great (cf. Romer, 1947) it is impossible to say what this resemblance means. As to the cranial anatomy the anthracosaurs (see Panchen, 1977a, with references) are less well known than *Eusthenopteron* and our knowledge is still too incomplete to form a safe basis for discussions of the controversial problems as to the origin, classification and interrelationships of the reptilomorphs (Ginsburg, 1970; Olson, 1971). In these regards we are in fact in the same dilemma as when we try to unravel the interrelationships of the early fishes. Like them the various reptile groups appear suddenly in the geologic record and several of them (mesosaurs, ichthyosaurs, chelonians) show the same reluctance to change as many of the fish groups. Accordingly if we try to construct a family tree of reptiles (Fig. 142), avoiding preconceived opinions and arbitrary arrangements according to the geologic age, and keep strictly to well established anatomical facts we will find that the evolutionary lines run in parallel as in the general diagram of vertebrate evolution (Fig. 140). It may now be asked if all these groups really are descendants of anthracosaurs. Cannot one or more of them have evolved directly from the osteolepiforms?

In the beginning of the 1950s the grinding series of the head of *Eusthenopteron* was finished. Illustrative wax models were then made of it and the horizon of our understanding of the structure and modifications of the eutetrapods was widened greatly. By studying the ontogeny we may sometimes form an opinion of how these modifications have arisen. When inspecting the models of *Eusthenopteron* I soon began to feel dubious about the reliability of the famous Reichert-Gaupp theory of the ear ossicles in mammals, and I became more and more convinced that this chain of bones can be derived directly from elements of the hyoid arch in *Eusthenopteron*.* This is the main reason why I as early as in 1959 indicated and expressed in diagrammatic representations, that the mammals and the theropsid stem as a whole have evolved from the osteolepiforms independently of the reptiles (Figs 140, 141).

Eusthenopteron and Man

In comparisons with mammals it is often profitable to use our own species, *Homo sapiens*, which is better known anatomically than any other mammal; besides in some ways it is surprisingly primitive.

Eusthenopteron certainly differs greatly from man and other mammals and there is nothing in the external appearance which betrays that we are concerned with one of our closest relatives among fishes. However, if we turn to internal structures we will find many fundamental resemblances. The widened knowledge of the osteolepiforms has made it easier to understand certain structural peculiarities in our own bodies.

If you feel with the hand in your neck you will touch a median elevation, the external occipital protuberance, which is formed by the occipital bone. This bone is a complex structure including an endoskeletal component formed by ossification of the tectum posterius, and a dermal bone, the interparietal. The endoskeletal component in man as well as its equivalent in *Eusthenopteron* (the supraoccipital plug; Fig. 97, Volume 1) serves as the attachment area for a ligament: the ligamentum supraspinale in man (Kopsch, 1947); the supraneural ligament in *Eusthenopteron* (Fig. 131, Volume

* pp. 161–175.

1). The interparietal in man usually arises from two pairs of centres. However, sometimes (de Beer, 1937) there may be a median and a paired lateral interparietal. This is also of great interest because in the corresponding place in *Eusthenopteron*, dorsal to the tectum posterius (supraoccipital plug), lies the extrascapular series also composed of median and paired lateral elements. In many mammallike reptiles such a series of three bones also is present.

The parietal, following in front of (in man dorsal to) this transverse series of bones is in *Eusthenopteron* as well as in man composed of anterior (in man dorsal) and posterior (ventral) parietals, and the wide gap between the parietals and the frontals in human embryos and the coronal suture in the adult (Fig. 125) appear to be vestiges of the gap between the frontoethmoidal and parietal shields in *Eusthenopteron*. This gap or suture marks the position of an ancient intervertebral joint (the intracranial joint), as does probably the suture between the parietal shield and the extrascapular series in *Eusthenopteron*, and thence possibly also the lambdoid suture between the parietal and the interparietal in man. The origin and distribution of the branches of the diploic veins in mammals (Fig. 125), in relation to which the frontal, parietal and interparietal are formed seem also to reflect the original subdivision of the cranial roof in *Eusthenopteron* into frontoethmoidal and parietal shields.

Eusthenopteron certainly possessed an olfactory epithelium and the sensations of smell were conducted to the brain via mitral cells in the olfactory bulb as they still are in man. Choana and nasolacrimal duct were present as well, and the eye was moved by six extrinsic eye muscles innervated by three ventral cranial nerves. Prerequisites for the mammalian tympanic membrane and ear ossicles were also present and equivalents to the manubrium and the lateral process of the malleus in man were developed (Fig. 96).

Also of considerable interest is that prerequisites for the formation of our tongue, and thence in important regards for our faculty of speech, were present in *Eusthenopteron*. Also remains of the sublingual rod which in our Devonian forerunners supported the future tongue musculature are occasionally retained in man as a series of cartilaginous nodules in the septum lingue. Moreover, the musculature of the original mandibular gill cover persists in man and connects the rami of the lower jaw in much the same way as in *Eusthenopteron* (Fig. 118). An interesting fact is also that our salivary glands in ontogeny arise in the mucous membrane of the inner side of that gill cover and we can perhaps say that, when our appetite is whetted, the saliva pours forth out of the depths of the ancient spiracular gill slit.

The most prominent feature of man is no doubt his large and elaborate brain. However, this big brain would certainly never have arisen—and what purpose would it have served—if our arm and hand had become specialized as strongly as has, for instance, the foreleg of a horse or the wing of a bird. It is the remarkable fact that it is the primitive condition, inherited from our osteolepiform ancestors and retained with relatively small changes in our arm and hand, that has paved the way for the emergence of man. We can say, with some justification, that it was when the basic pattern of our five-fingered hand for some unaccountable reason was laid down in the ancestors of the osteolepiforms that the prerequisites for the origin of man and the human culture arose.

Bibliography

Adam, H. and Strahan, R. (1963). Systematics and geographical distribution of myxinoids. *In* "The biology of *Myxine*" (Eds A. Brodal and R. Fänge), 1–8. Universitetsforlaget, Oslo.

Addens, J. L. (1933). The motor nuclei and roots of the cranial and first spinal nerves of vertebrates. *Z. Anat. EntwGesch.* **101**, 307–410.

Adelmann, H. B. (1932). The development of the prechordal plate and mesoderm of *Amblystoma punctatum*. *J. Morph.* **54**, 1–54.

Agassiz, L. (1833–1844). "Recherches sur les Poissons Fossiles", Vols I–V. Neuchâtel.

Agassiz, L. (1844–1845). "Monographie des Poissons Fossiles du vieux grès rouge, etc." Neuchâtel.

Aldinger, H. (1930). Über das Kopfskelett von *Undina acutidens* Reis und den kinetischen Schädel der Coelacanthiden. *Zentbl. Miner, Geol. Paläont.* **1**, 22–48.

Aldinger, H. (1931). Über einige Besonderheiten im Schädeldach von *Eusthenopteron foordi* Wh. (Pisces, Crossopterygii). *Zentbl. Miner. Geol. Paläont.* **6**, 300–305.

Aldinger, H. (1932). Über einen Eugnathiden aus der unteren Wolgastufe von Ostgrönland. *Meddr Grønland* **86**, 1–51.

Aldinger, H. (1937). Permische Ganoidfische aus Ostgrönland. *Meddr Grønland* **102**, 5–392.

Allen, W. F. (1905). The blood-vascular system of the Loricati, the mailcheeked fishes. *Proc. Wash. Acad. Sci.* **7**, 27–157.

Allin, E. F. (1975). Evolution of the mammalian middle ear. *J. Morph.* **147**, 403–438.

Allis, E. (1889). The anatomy and development of the lateral line system in *Amia calva*. *J. Morph.* **2**, 463–540.

Allis, E. P. (1897). The cranial muscles and cranial and first spinal nerves in *Amia calva*. *J. Morph.* **12**, 487–809.

Allis, E. P. (1897a). The morphology of the petrosal bone and of the sphenoidal region of the skull of *Amia calva*. *Zool. Bull.* **1**, 1–26.

Allis, E. P. (1898). On the morphology of certain of the bones of the cheek and snout of *Amia calva*. *J. Morph.* **14**, 425–466.

Allis, E. P. (1898a). The homologies of the occipital and first spinal nerves of *Amia* and teleosts. *Zool. Bull.* **2**, 83–97.

Allis, E. P. (1899). On certain homologues of the squamosal, intercalar, exoccipital and extrascapular bones of *Amia calva*. *Anat. Anz.* **16**, 49–72.

Allis, E. P. (1900). The pseudobranchial circulation in *Amia calva*. *Zool. Jb. (Anat)* **14**, 107–134.

Allis, E. P. (1903). The skull, and the cranial and first spinal muscles and nerves in *Scomber scomber*. *J. Morph.* **18**, 45–328.

Allis, E. P. (1908). The pseudobranchial and carotid arteries in the gnathostome fishes. *Zool. Jb. (Anat.)* **27**, 103–134.

Allis, E. P. (1909). The cranial anatomy of the mail-cheeked fishes. *Zoologica, Stuttg.* **22**, 1–219.

Allis, E. P. (1912). The branchial, pseudobranchial and carotid arteries in *Heptanchus (Notidanus) cinereus*. *Anat. Anz.* **41**, 478–92.

Allis, E. P. (1912a). The pseudobranchial and carotid arteries in *Esox, Salmo* and *Gadus*, together with a description of the arteries in the adult *Amia*. *Anat. Anz.* **41**, 113–142.

Allis, E. P. (1919). The lips and the nasal apertures in the gnathostome fishes. *J. Morph.* **32**, 145–205.

Allis, E. P. (1922). The cranial anatomy of *Polypterus*, with special reference to *Polypterus bichir*. *J. Anat.* **56**, 189–291.

Allis, E. P. (1923). The cranial anatomy of *Chlamydoselachus anguineus*. *Acta zool., Stockh.* **4**, 123–221.

Allis, E. P. (1931). Concerning the homologies of the hypophysial pit and the polar and trabecular cartilages. *J. Anat.* **65**, 247–265.

Allis, E. P. (1934). Concerning the course of the latero-sensory canals in Recent fishes, prefishes and *Necturus*. *J. Anat.* **68**, 361–415.

Allis, E. P. (1938). Concerning the development of the prechordal portion of the vertebrate head. *J. Anat.* **72**, 584–607.

Andrews, S. M. (1973). Interrelationships of crossopterygians. *In* "Interrelationships of fishes". (Eds P. H. Greenwood *et al.*), 137–177. Academic Press, London.

Andrews, S. M. (1977). The axial skeleton of the coelacanth, *Latimeria*. *In* "Problems in vertebrate evolution" (Eds S. M. Andrews *et al.*), Linn. Soc. Symp. Vol. 4, 271–288.

Andrews, S. M. and Westoll, T. S. (1970). The postcranial skeleton of *Eusthenopteron foordi* Whiteaves. *Trans. R. Soc. Edinb.* **68(9)**, 207–329.

Andrews, S. M. and Westoll, T. S. (1970a). The postcranial skeleton of rhipidistian fishes excluding *Eusthenopteron*. *Trans. R. Soc. Edinb.* **68(12)**, 391–489.

Anthony, J., Millot, J. and Robineau, D. (1965). Le coeur et l'aorte ventrale de *Latimeria chalumnae* (poisson coelacanthidé). *C. r. hebd. Séanc. Acad. Sci., Paris* **261**, 223–226.

Anthony, J. and Robineau, D. (1976). Sur quelques caractères juvéniles de *Latimeria chalumnae* Smith (Pisces, Crossopterygii Coelacanthidae). *C. r. Acad. Sci., Paris* **283**, 1739–1742.

Artedi, P. (1738). "Ichthyologia". Lugduni Batavorum. (Reprint 1962. Wheldon and Wesley, New York.)

Atwell, W. J. (1921). The morphogenesis of the hypophysis in the tailed Amphibia. *Anat. Rec.* **22**, 373–390.

Atz, J. W. (1952). Internal nares in the teleost, *Astroscopus*. *Anat. Rec.* **113**, 105–115.

Atz, J. W. (1952a). Narial breathing in fishes and the evolution of internal nares. *Q. Rev. Biol.* **27**, 365–377.

Baird, I. L. (1974). Some aspects of the comparative anatomy and evolution of the inner ear in submammalian vertebrates. *Brain, Behav. Evol.* **10**, 11–36.

Baird, I. L. (1974a). Anatomical features of the inner ear in submammalian vertebrates. *In* "Handbook of sensory physiology" (Eds W. D. Keidel and W. D. Neff), Vol. 5(1), 159–212.

Balfour, F. M. (1876). The development of elasmobranch fishes. *J. Anat. Physiol., Lond.* **11**, 128–172.

Balfour, F. M. (1878). "A monograph on the development of elasmobranch fishes." Macmillan, London.

Balfour, F. M. (1881). "A treatise on comparative embryology." Vol. 2. Macmillan, London.

Balfour, F. M. (1881a). On the development of the skeleton of the paired fins of Elasmobranchii, considered in relation to its bearings on the nature of the limbs of the Vertebrata. *Proc. zool. Soc., Lond.* **1881**, 656–671.

Balinsky, B. I. (1933). Das Extremitätenseitenfeld; seine Ausdehnung und Beschaffenheit.

Arch. EntwMech. Org. **130**, 704–746.

Balinsky, B. I. (1935). Experimentelle Extremitäteninduktion und die Theorien des phylogenetischen Ursprungs der paarigen Extremitäten der Wirbeltiere. *Anat. Anz.* **80**, 136–142.

Balinsky, B. I. (1937). Über die zeitlichen Verhältnisse bei der Extremitäteninduktion. *Arch. EntwMech Org.* **136**, 250–285.

Balinsky, B. I. 3rd edn (1970). "An introduction to Embryology." W. B. Saunders, London.

Bardack, D. and Zangerl, R. (1968). First fossil lamprey: A record from the Pennsylvanian of Illinois. *Am. Ass. for Adv. Sc.* **162**, 1265–1267.

Bardack, D. and Zangerl, R. (1971). Lampreys in the fossil record. *In* "The biology of lampreys" (Eds M. N. Hardisty and I. C. Potter), Vol. 1, 67–84, Academic Press, London.

Barry, T. H. (1956). The ontogenesis of the sound-conducting apparatus of *Bufo angusticeps* Smith. *Gegenbaurs morph. Jb.* **97**, 477–544.

Barry, T. H. (1963). On the variable occurrence of the tympanum in recent and fossil tetrapods. *S. Afr. J. Sc.* **59**, 160–175.

Bartram, A. W. H. (1977). The macrosemiidae, a Mesozoic family of holostean fishes. *Bull. Br. Mus. nat. Hist. (Geol.)* **29(2)**, 137–234.

de Beaumont, G. (1973). "Guide des vertébrés fossiles." Delachaux et Niestle, Neuchâtel.

de Beer, G. R. (1923). Some observations on the hypophysis of *Petromyzon* and of *Amia*. *Q. Jl microsc. Sci.* **67**, 257–290.

de Beer, G. R. (1924). The prootic somites of *Heterodontus* and of *Amia*. *Q. Jl microsc. Sci.* **68**, 17–38.

de Beer, G. R. (1924a). Studies on the vertebrate head. *Q. Jl microsc. Sci.* **68**, 288–339.

de Beer, G. R. (1937). "The development of the vertebrate skull." Oxford University Press, Oxford.

de Beer, G. R. (1947). The differentiation of neural crest cells into visceral cartilages and odontoblasts in *Amblystoma*, and a re-examination of the germ-layer theory. *Proc. R. Soc. (B)* **134**, 377–398.

de Beer, G. R. and Moy-Thomas, J. A. (1935). On the skull of Holocephali. *Phil. Trans. R. Soc. (B)* **224**, 287–312.

Beltan, L. (1968). La faune ichthyologique de l'Eotrias du N.W. de Madagascar: le neurocrane. Centre national de la recherche sci, Paris.

Bendix-Almgreen, S. E. (1968). The bradyodont elasmobranchs and their affinities; a discussion. *In* "Current problems of lower vertebrate phylogeny" (Ed. T. Ørvig), Nobel Symp. Vol. 4, 153–170. Almqvist and Wiksell, Stockholm.

Bendix-Almgreen, S. E. (1971). The anatomy of *Menaspis armata* and the phyletic affinities of the menaspid bradyodonts. *Lethaia* **4**, 21–49.

Bendix-Almgreen, S. E. (1975). The paired fins and shoulder girdle in *Cladoselache*, their morphology and phyletic significance. *Colloques int. Cent. natn. Rech. scient.* **218**, 111–123.

Bendix-Almgreen, S. E. (1976). Palaeovertebrate faunas of Greenland. *In* "Geology of Greenland" (Eds A. Escher and W. S. Watt), 537–573. Geological Survey of Greenland, Copenhagen.

Berg, L. S. (1958). "System der rezenten und fossilen fischartigen und Fische." VEB, Berlin. Deutscher Verlag d. Wiss. (transl. by W. Gross of L. S. Berg, 1955, *Trudy Zool. Inst. Leningr.* **20**, 1–286; in Russian).

Bergquist, H. (1932). Zur Morphologie des Zwischenhirns bei niederen Wirbeltieren. *Acta zool., Stockh.* **13**, 58–303.

Bergquist, H. (1952). Studies on the cerebral tube in vertebrates. The neuromeres. *Acta zool., Stockh.* **33**, 117–187.

Bergquist, H. (1953). Prechordal plate, cerebral tube and notochord conjoined in 1·7 mm human embryo. *Acta Soc. Med. Upsaliensis* **58**, 342–359.
Bergquist, H. and Källén, B. (1954). Notes on the early histogenesis and morphogenesis of the central nervous system in vertebrates. *J. comp. Neurol.* **100(3)**, 627–660.
Bernacsek, G. M. (1977). A lungfish cranium from the Middle Devonian of the Yukon Territory, Canada. *Palaeontographica* **157**, 175–200.
Bernasconi, A. F. (1951). Über den Ossifikationsmodus bei *Xenopus laevis* Daud. *Denkschr. schweïz. naturf. Ges.* **79(2)**, 193–250.
Bernhauser, A. (1961). Zur Knochen- und Zahnhistologie von *Latimeria chalumnae* Smith und einiger Fossilformen. *Wien. Sber. öst. Akad. mathem.-naturw. Kl., Abt. 1* **170**, 119–137.
Berrill, N. J. (1955). "The origin of vertebrates". Clarendon Press, Oxford.
Bertin, L. (1958). Système nerveux, Vessie gazeuse. *In* "Traité de Zoologie" (Ed. P.-P. Grassé) Vol. 13, 854–922, 1342–1398. Masson, Paris.
Bertmar, G. (1959). On the ontogeny of the chondral skull in Characidae, with a discussion on the chondrocranial base and the visceral chondrocranium in fishes. *Acta zool., Stockh.* **40**, 203–364.
Bertmar, G. (1961). Tungbensbågens skelett hos fiskarna (On the hyoid arch skeleton in fishes). *Zool. Revy*, 45–60. (In Swedish, English summary.)
Bertmar, G. (1962). On the ontogeny and evolution of the arterial vascular system in the head of the African characidean fish *Hepsetus odoe*. *Acta zool., Stockh.* **43**, 255–295.
Bertmar, G. (1963). The trigemino-facialis chamber, the cavum epiptericum and the cavum orbitonasale, three serially homologous extracranial spaces in fishes. *Acta zool., Stockh.* **44**, 329–344.
Bertmar, G. (1963a). Finns det kotor i huvudet? (Are there vertebrae in the vertebrate head?). *Zool. Revy*, 47–54. (In Swedish.)
Bertmar, G. (1965). On the development of the jugular and cerebral veins in fishes. *Proc. zool. Soc. Lond.* **144**, 87–130.
Bertmar, G. (1965a). The olfactory organ and upper lips in Dipnoi, an embryological study. *Acta zool., Stockh.* **46**, 1–40.
Bertmar, G. (1966). On the ontogeny and homology of the choanal tubes and choanae in Urodela. *Acta zool., Stockh.* **47**, 43–59.
Bertmar, G. (1966a). The development of skeleton, blood-vessels and nerves in the dipnoan snout, with a discussion on the homology of the dipnoan posterior nostrils. *Acta zool., Stockh.* **47**, 81–150.
Bertmar, G. (1968). Lungfish phylogeny. *In* "Current problems of lower vertebrate phylogeny" (Ed. T. Ørvig). Nobel Symp. Vol. 4, 259–283. Almqvist and Wiksell, Stockholm.
Bertmar, G. (1968a). Phylogeny and evolution in lungfishes. *Acta zool., Stockh.* **49**, 189–201.
Bertmar, G. (1969). The vertebrate nose, remarks on its structural and functional adaptation and evolution. *Evolution, Lancaster, Pa.* **23**, 131–152.
Bertmar, G. (1972). Secondary folding of olfactory organ in young and adult sea trout. *Acta zool., Stockh.* **53**, 113–120.
Bertmar, G. (1972a). Labyrinth cells, a new cell type in vertebrate olfactory organs. *Z. Zellforsch.* **132**, 245–256.
Bigelow, H. B. and Schroeder, W. C. (1953). Sawfishes, guitarfishes, skates and rays. Fishes of the Western North Atlantic, 2. *Mem. Sears. Fdn mar. Res.* **1**, 1–514.
Bing, R. and Burckhardt, R. (1905). Das Centralnervensystem von *Ceratodus forsteri*. *Med.-naturw. Ges. Denkschr., Jena* **4(1)**, 513–584.
Bjerring, H. C. (1967). Does a homology exist between the basicranial muscle and the polar

cartilage? *Colloques int. Cent. natn. Rech. scient.* **163**, 223–267.
Bjerring, H. C. (1968). The second somite with special reference to the evolution of its myotomic derivatives. *In* "Current problems of lower vertebrate phylogeny" (Ed. T. Ørvig). Nobel Symp. Vol. 4, 341–357. Almqvist and Wiksell, Stockholm.
Bjerring, H. C. (1970). Nervus tenius, a hitherto unknown cranial nerve of the fourth metamere. *Acta zool., Stockh.* **51**, 107–114.
Bjerring, H. C. (1971). The nerve supply to the second metamere basicranial muscle in osteolepiform vertebrates, with some remarks on the basic composition of the endocranium. *Acta zool., Stockh.* **52**, 189–225.
Bjerring, H. C. (1972). The nervus rarus in coelacanthiform phylogeny. *Zool. Scr.* **1**, 57–68.
Bjerring, H. C. (1972a). Morphological observations on the exoskeletal skull roof of an osteolepiform from the Carboniferous of Scotland. *Acta zool., Stockh.* **53**, 73–92.
Bjerring, H. C. (1972b). The rhinal bone and its evolutionary significance. *Zool. Scr.* **1**, 193–201.
Bjerring, H. C. (1973). Relationships of coelacanthiforms. *In* "Interrelationships of fishes" (Eds P. H. Greenwood *et al.*), 179–205. Academic Press, London.
Bjerring, H. C. (1975). Contribution à la connaissance de la neuro-épiphyse chez les Urodèles et leurs ancêtres Porolépiformes, avec quelques remarques sur la signification évolutive des muscles striés parfois présents dans la région neuro-épiphysaire des Mammifères. *Colloques int. Cent. natn. Rech. scient.* **218**, 231–256.
Bjerring, H. C. (1977). A contribution to structural analysis of the head of craniate animals. *Zool. Scr.* **6**, 127–183.
Bjerring, H. C. (1978). The "intracranial joint" *versus* the "ventral otic fissure". *Acta zool., Stockh.* **59**, 203–214.
Blechschmidt, E. (1963). "Der menschliche Embryo. The human Embryo". Friedrich-Karl Schattauer-Verlag, Stuttgart.
Blot, J. (1969). Holocéphales et Elasmobranches: Systématique. *In* "Traité de Paléontologie" (Ed. J. Piveteau) Vol. 4(2), 702–776. Masson, Paris.
Bockelie, T. and Fortey, R. A. (1976). An early Ordovician vertebrate. *Nature, Lond.* **260**, 36–38.
Bolk, L., Göppert, T., Kallius, E. and Lubosch, W. (1931–39) (Eds). "Handbuch der vergleichenden Anatomie der Wirbeltiere." Vols 1–6. Gesamt-Inhalts Übersicht, Gesamt-Sachverzeichnis. Urban and Schwarzenberg, Berlin and Wien.
Bolt, J. R. (1969). Lissamphibian origins: possible protolissamphibian from the Lower Permian of Oklahoma. *Science, N.Y.* **166**, 888–891.
Bolt, J. R. (1977). Dissorophoid relationships and ontogeny, and the origin of the Lissamphibia. *J. Paleont.* **51(2)**, 235–249.
Bone, Q. (1963). The central nervous system. *In* "The biology of *Myxine*" (Eds A. Brodal and R. Fänge), 50–91. Universitetsforlaget, Oslo.
Bone, Q. (1972). The origin of chordates. *In* "Oxford Biology Readers." (Eds. J. J. Head and O. E. Lowenstein), Oxford University Press, London.
Bondy, G. (1907). Beiträge zur vergleichenden Anatomie des Gehörorgans der Säuger. *Arb. anat. Inst., Wiesbaden (Anatomische Hefte)* **35**, 295–404.
Borchwardt, V. G. (1977). Development of the vertebral column in embryogenesis of *Lacerta agilis*. *Zool. Zh.* **56**, 576–587.
Boreske, J. R. (1974). A review of the North American fossil amiid fishes. *Bull. Mus. comp. Zool. Harv.* **146**, 1–87.
Boy, J. A. (1978). Die Tetrapodenfauna (Amphibia, Reptilia) des saarpfälzischen Rotliegenden (Unter-Perm; SW-Deutschland). 1. *Branchiosaurus*. *Mainzer geowiss. Mitt.* **7**, 27–76.

Brachet, A. (1935). "Traité d'Embryologie des Vertébrés." Masson, Paris.
Branson, B. A. and Moore, G. A. (1962). The lateralis components of the acoustico-lateralis system in the sunfish family Centrarchidae. *Copeia* **1**, 1–108.
Braus, H. (1899). Beiträge zur Entwicklung der Muskulature und des peripheren Nervensystems der Selachier. 1, 2. *Morph. Jb.* **27**, 415–496, 501–629.
Braus, H. (1901). Die Muskeln und Nerven der Ceratodusflosse. Ein Beitrag zur vergleichenden Morphologie der freien Gliedmasse bei niederen Fischen und zur Archipterygiumtheorie. *Denkschr. Med.-Naturw. Ges. Jena* **4(1)**, 139–300.
Braus, H. (1904). Die Entwickelung der Form der Extremitäten und des Extremitätenskeletts. *In* "Handbuch der vergleichenden und experimentellen Entwickelungslehre der Wirbeltiere" (Ed. O. Hertwig), Vol. 3(2), 167–336. Fischer, 1906, Jena.
Braus, H. (1919). Brustschulterapparat der Froschlurche. *Sitzb. Heidelberger Akad. Wiss.* (**B**) **15**, 1–50.
Brien, P. (1948). Embranchement des Tuniciers: Morphologie et reproduction. *In* "Traité de Zoologie" (Ed. P.-P. Grassé), Vol. 11, 553–930. Masson, Paris.
Brien, P. (1962). Formation du cloaque urinaire et origine des sacs pulmonaires chez *Protopterus. Annls Mus. R. Afr. cent.* (**8**)**108**, 1–51.
Brien, P. (1962a). Etude de la formation de la structure des écailles des Dipneustes actuels et de leur comparaison avec les autres types d'écailles des Poissons. *Annls Mus. R. Afr. cent.* (**8**)**108**, 53–128.
Brien, P. (1963). L'Éthologie du Protoptère et sa signification dans l'histoire des vertébrés. *Publs. Univ. Elisabethville* **6**, 107–129.
Brien, P. and Bouillon, J. (1959). Ethologie des larves de *Protopterus dolloi* Blgr. et étude de leurs organes respiratoires. *Annls Mus. R. Congo belge.* (**8**)**71**, 23–74.
Bridge, T. W. (1877). The cranial osteology of *Amia calva. J. Anat. Physiol. Lond.* **11**, 605–622.
Broad, D. S. (1973). Amphiaspidiformes (Heterostraci) from the Silurian of the Canadian Arctic Archipelago. *Bull. Geol. Surv. Can.* **222**, 35–46.
Broad, D. S. and Dineley, D. L. (1973). *Torpedaspis*, a new Upper Silurian and Lower Devonian genus of Cyathaspididae (Ostracodermi) from Arctic Canada. *Bull. Geol. Surv. Can.* **222**, 53–82.
Brodal, A. and Fänge, R. (1963) (Eds). "The biology of *Myxine*." Universitetsforlaget, Oslo.
Brohl, E. (1909). Die sogenannten Hornfäden und die Flossenstrahlen der Fische. *Jena. Z. Naturw.* **45**, 345–380.
Broman, I. (1899). Die Entwickelungsgeschichte der Gehörknöchelchen beim Menschen. *Arb. anat. Inst., Wiesbaden* (*Anatomische Hefte*) **11**, 509–661.
Brookover, C. (1910). The olfactory nerve, the nervus terminalis and the preoptic sympathetic system in *Amia calva* L. *J. comp. Neurol.* **20**, 49–118.
Brühl, C. B. (1880). "Zootomie aller Thierklassen für Lernende, nach Autopsien skizziert." Atlas, Lief. 16. Wien.
Bryant, W. L. (1919). On the structure of *Eusthenopteron. Bull. Buff. Soc. Nat. Sc.* **13**, 1–59.
Budgett, J. S. (1907). "The work of John Samuel Budgett" (Ed. J. G. Kerr). Cambridge University Press, Cambridge.
Bugge, J. (1970). The contribution of the stapedial artery to the cephalic arterial supply in muroid rodents. *Acta anat.* **76**, 313–336.
de Burlet, H. M. (1934). Vergleichende Anatomie des statoakustichen Organs. a) Die innere Ohrsphäre. *In* "Handbuch der vergleichenden Anatomie der Wirbeltiere". (Eds L. Bolk, *et al.*), Vol. 2(2), 1293–1380. Urban and Schwarzenberg, Berlin and Wien.
de Burlet, H. M. and Versteegh, C. (1930). Über Bau und Funktion des Petromyzonlabyrinthes. *Acta Oto-Laryngol., suppl.* **13**, 1–58.

Bütschli, O. (1921). "Vorlesungen über vergleichende Anatomie." Vol. 1. Julius Springer, Berlin.

Byczkowska-Smyk, W. (1962). Vascularization and size of the respiratory surface of gills in *Acipenser stellatus* Pall. *Acta biol. cracov.* **5**, 303–315.

Bystrow, A. P. (1939). Zahnstruktur der Crossopterygier. *Acta zool., Stockh.* **20**, 283–338.

Bystrow, A. P. and Efremov, J. A. (1940). *Benthosuchus sushkini* Efr.—A labyrinthodont from the Eotriassic of Sharjenga River. *Trudy paleont. Inst.* **10(1)**, 1–152.

Cafaurek, F. (1883). Das Skelet der jetzt lebenden Knochenganoiden. *Progm k.k. deutsch. Obergym., Kleinseite.* 1–42.

Campbell, K. S. W. (1965). An almost complete skull roof and palate of the dipnoan *Dipnorhynchus sussmilchi* (Etheridge). *Palaeontology* **8**, 634–637.

Campbell, K. S. W. and Bell, M. W. (1977). A primitive amphibian from the late Devonian of New South Wales. *Alcheringa* **1**, 369–381.

Campenhout, E. W. (1935). Experimental researches on the origin of the acoustic ganglion in amphibian embryos. *J. exp. Zool.* **72**, 175–193.

Carroll, R. L. (1969). Problems of the origin of reptiles. *Biol. Rev.* **44**, 393–432.

Carroll, R. L. (1970). The ancestry of reptiles. *Phil. Trans. R. Soc. (B)* **257**, 267–308.

Carroll, R. L. and Currie, P. J. (1975). Microsaurs as possible apodan ancestors. *Zool. J. Linn. Soc.* **57**, 229–247.

Casier, E. (1952). Un Paléoniscoide du Famennien inférieur de la Fagne: *Stereolepis marginis* n. gen., n.sp. *Bull. Inst. roy. Sc. nat. Belg.* **28(47)**, 1–10.

Casier, E. (1954). Note additionnelle relative à "*Stereolepis*" (=*Osorioichthys* nov. nom.) et à l'origine de l'interoperculaire. *Bull. Inst. roy. Sc. nat. Belg.* **30(2)**, 1–12.

Chandler, A. C. (1911). On a lymphoid structure lying over the myelencephalon of *Lepisosteus. Univ. Calif. Publs Zool.* **9**, 85–104.

Chase, J. N. (1963). The labyrinthodont dentition. *Breviora* **187**, 1–13.

Chase, J. N. (1965). *Neldasaurus wrightae*, A new rhachitomous labyrinthodont from the Texas Lower Permian. *Bull. Mus. comp. Zool. Harv.* **133**, 156–225.

Chiarugi, G. (1898). Di un organo epiteliale situato al dinanzi della Ipofisi e di altri punti relativi allo sviluppo della regione ipofisaria in embrioni di *Torpedo ocellata. Monitore zool. ital.* **9**, 37–56.

Chibon, P. (1974). Un système morphogénétique remarquable: La créte neurale des vertébrés. *Ann. Biol.* **13**, 459–476.

Cĭhák, R. (1972). Ontogenesis of the skeleton and intrinsic muscles of the human hand and foot. *Ergebn. Anat. EntwGesch.* **46**, 3–194.

Clairambault, P. (1963). Le tèlencéphale de *Discoglossus pictus* (Oth.). Étude anatomique chez le têtard et chez l'adulte. *J. Gehirnf.* **6**, 87–121.

Cohen, N. (1975). Phylogeny of lymphocyte structure and function. *Am. Zool.* **15**, 119–133.

Colbert, E. H. (1965). The appearance of new adaptations in Triassic tetrapods. *Israel J. Zool.* **14**, 49–62.

Cole, F. J. (1896). On the cranial nerves of *Chimaera monstrosa*, with a discussion of the lateral line system and of the morphology of the chorda tympani. *Trans. R. Soc. Edinb.* **38**, 631–680.

Cole, F. J. (1913). A monograph on the general morphology of the myxinoid fishes, based on a study of *Myxine*. V. *Trans. R. Soc. Edinb.* **49**, 293–344.

Compagno, L. J. V. (1973). Interrelationships of living elasmobranchs. *In* "Interrelationships of fishes" (Eds P. H. Greenwood *et al.*), 15–61. Academic Press, London.

Conel, J. L. (1929). The development of the brain of *Bdellostoma stouti*. I. External growth changes. *J. comp. Neurol.* **47**, 343–379.

Conklin, E. G. (1932). The embryology of amphioxus. *J. Morph.* **54**, 69–118.
Cope, E. D. (1885). The position of *Pterichthys* in the system. *Am. Nat.* **19**, 289–291.
Cordier, R. (1954). Le système nerveux central et les nerfs cérébro-spinaux. *In* "Traité de Zoologie" (Ed. P.-P. Grassé), Vol. 12, 202–332. Masson, Paris.
Cordier, R. and Dalcq, A. (1954). Organe stato-acoustique. *In* "Traité de Zoologie" (Ed. P.-P. Grassé), Vol. 12, 453–521. Masson, Paris.
Cords, E. (1921). Die Hautmuskeln der Amphibien nebst Bemerkungen über Hautmuskeln im allgemeinen. *Zool. Jb.* (*Anat.*) **42**, 283–326.
Corrington, J. D. (1930). Morphology of the anterior arteries of sharks. *Acta zool., Stockh.* **11**, 185–261.
Cox, C. B. (1967). Cutaneous respiration and the origin of the modern Amphibia. *Proc. Linn. Soc. Lond.* **178**, 37–47.
Craigie, E. H. (1938). Vascularity in the brain of the frog (*Rana pipiens*). *J. comp. Neurol.* **69**, 453–470.
Crompton, A. W. (1953). The development of the chondrocranium of *Spheniscus demersus* with special reference to the columella auris of birds. *Acta zool., Stockh.* **34**, 72–146.
Crompton, A. W. (1958). The cranial morphology of a new genus and species of ictidosaurian. *Proc. zool. Soc. Lond.* **130**, 183–216.
Crompton, A. W. (1972). The evolution of the jaw articulation of cynodonts. *In* "Studies in vertebrate evolution" (Eds K. A. Joysey and T. S. Kemp), 231–251. Oliver and Boyd, Edinburgh.
Cuvier, G. 1st edn (1812). "Recherches sur les ossemens fossiles, où l'on rétablit les caractères de plusieurs animaux dont le revolutions du globe ont détruit les espèces", Vols 1–4, Paris.
Dabelow, A. (1928). Über Art and Ursachen der Entstehung des Kiefergelenkes der Säugetiere. *Gegenbaurs morph. Jb.* **62/63**, 493–560.
Daget, J. (1964). Le crane des téléostéens. *Mém. Mus. natn. Hist. nat., Paris* **31**, 163–321.
Daget, J., Bauchot, R. and Arnoult, J. (1964). Développement du chondrocrane et des arcs aortiqués chez *Polypterus senegalus* Cuvier. *Acta zool., Stockh.* **46**, 201–244.
Dalcq, A. and Pasteels, J. (1954). Le développment des vertébrés. *In* "Traité de Zoologie" (Ed. P.-P. Grassé), Vol. 12, 35–201. Masson, Paris.
Damas, H. (1944). Recherches sur le développement de *Lampetra fluviatilis* L. Contribution à l'étude de la céphalogenèse des vertébrés. *Archs Biol., Liège* **55**, 1–284.
Damas, H. (1951). Observations sur le développement des ganglions craniens chez *Lampetra fluviatilis* (L.) *Archs Biol., Paris* **62**, 65–95.
Damas, H. (1954). La branchie prespiraculaire des cephalaspides. *Annls Soc. zool. Belg.* **85**, 89–102.
Daniel, J. F. (1934). "The Elasmobranch Fishes." University of California Press, Berkeley.
Daniel, J. F. and Bennett, L. H. (1931). Veins in the roof of the buccopharyngeal cavity of *Squalus sucklii*. *Univ. Calif. Publs Zool.* **37**, 35–40.
von Davidoff, M. (1880). Beiträge zur vergleichenden Anatomie der hinteren Gliedmasse der Fische. *Morph. Jb.* **6**, 433–468.
Davidson, P. (1918). The musculature of *Heptanchus maculatus*. *Univ. Calif. Publs Zool.* **18**, 151–170.
Dawson, J. A. (1963). The oral cavity, the "jaws" and the horny teeth of *Myxine glutinosa*. *In* "The biology of *Myxine*" (Eds A. Brodal and R. Fänge), 231–255. Universitetsforlaget, Oslo.
Dean, B. (1895). The early development of *Amia*. *Q. Jl microsc. Sci.* **38**, 413–444.
Dean, B. (1985a). "Fishes, living and fossil." Columbia University biological series. Macmillan, New York.

Dean, B. (1896). On the larval development of *Amia calva*. *Zool. Jb. Syst.* **9**, 639–672.
Dean, B. (1906). Chimaeroid fishes and their development. *Publs Carnegie Instn* **32**, 1–194.
Dean, B. (1909). Studies on fossil fishes (sharks, chimaeroids and arthrodires). *Mem. Am. Mus. nat. Hist.* **9**, 211–287.
Dechaseaux, C. (1937). Le Genre *Amia* son histoire paléontologique. *Annls Paléont.* **26**, 1–16.
Dechaseaux, C. (1955). Série des Urodélomorphes: Lepospondyli, Urodèles. *In* "Traité de Paléontologie" (Ed. J. Piveteau), Vol. 5, 275–311. Masson, Paris.
Degener, L. M. (1924). The development of the dentary bone and teeth of *Amia calva*. *J. Morph.* **39**, 113–139.
Dendy, A. (1907). On the parietal sense-organs and associated structures in the New Zealand lamprey (*Geotria australis*) *Q. Jl microsc. Sci.* **51**, 1–31.
Denison, R. H. (1947). The exoskeleton of *Tremataspis*. *Am. J. Sci.* **245**, 337–365.
Denison, R. H. (1951). Evolution and classification of the Osteostraci. The exoskeleton of early Osteostraci. *Fieldiana, Geol.* **11**, 157–218.
Denison, R. H. (1952). Early Devonian fishes from Utah. Part 1. Osteostraci. *Fieldiana, Geol.* **11**, 265–287.
Denison, R. H. (1953). Early Devonian fishes from Utah. Part II. Heterostraci. *Fieldiana, Geol.* **11**, 291–355.
Denison, R. H. (1958). Early Devonian fishes from Utah. Part III. Arthrodira. *Fieldiana, Geol.* **11**, 461–551.
Denison, R. H. (1960). Fishes of the Devonian Holland Quarry Shale of Ohio. *Fieldiana, Geol.* **11**, 555–613.
Denison, R. H. (1963). New Silurian Heterostraci from Southeastern Yukon. *Fieldiana, Geol.* **14**, 105–141.
Denison, R. H. (1964). The Cyathaspididae—A family of Silurian and Devonian jawless vertebrates. *Fieldiana, Geol.* **13**, 311–473.
Denison, R. H. (1966). *Cardipeltis*: An early Devonian agnathan of the order Heterostraci. *Fieldiana, Geol.* **16**, 89–116.
Denison, R. H. (1966a). The origin of the lateral-line sensory system. *Am. Zool.* **6**, 369–370.
Denison, R. H. (1967). Ordovician vertebrates from Western United States. *Fieldiana, Geol.* **16**, 131–192.
Denison, R. H. (1968). The evolutionary significance of the earliest known lungfish, *Uranolophus*. *In* "Current Problems of lower vertebrate phylogeny" (Ed. T. Ørvig). Nobel Symp. Vol. 4, 247–257. Almqvist and Wiksell, Stockholm.
Denison, R. H. (1968a). Early Devonian lungfishes from Wyoming, Utah and Idaho. *Fieldiana, Geol.* **17**, 353–413.
Denison, R. H. (1970). Revised classification of Pteraspididae with description of new forms from Wyoming. *Fieldiana, Geol.* **20**, 1–41.
Denison, R. H. (1971). The origin of the vertebrates: a critical evaluation of current theories. *Proc. N. Am. Palaeont. Conv.* **H**, 1132–1146.
Denison, R. H. (1971a). On the tail of the Heterostraci (Agnatha). *Forma et functio* **4**, 87–99.
Denison, R. H. (1973). Growth and wear of the shield in Pteraspididae (Agnatha). *Palaeontographica (A)* **143**, 1–10.
Denison, R. H. (1974). The structure and evolution of teeth in lungfishes. *Fieldiana, Geol.* **33**, 31–58.
Denison, R. H. (1975). Evolution and classification of placoderm fishes. *Breviora* **432**, 1–24.
Denison, R. H. (1978). Placodermi. *In* "Handbook of Paleoichthyology" (Ed. H.-P. Schultze), Vol. 2, 1–122. Gustav Fischer, Stuttgart and New York.

Desmond, A. J. (1974). On the coccosteid arthrodire *Millerosteus minor*. *Zool. J. Linn. Soc.* **54**, 277–298.

Devillers, C. (1954). Structure et evolution de la colonne vertébrale. *In* "Traité de Zoologie" (Ed. P.-P. Grassé), Vol. 12, 605–672. Masson, Paris.

Devillers, C. (1954a). Origine et evolution des nageoires et des membres. *In* "Traité de Zoologie" (Ed. P.-P. Grassé), Vol. 12, 710–790. Masson, Paris.

Devillers, C. (1958). Le Système latéral. *In* "Traité de Zoologie" (Ed. P.-P. Grassé), Vol. 13, 940–1032. Masson, Paris.

Devillers, C. (1961). Origine de l'oreille moyenne des Mammifères. *In* "Traité de Paléontologie" (Ed. J. Piveteau), Vol. 6(1), 371–407. Masson, Paris.

Dijkgraaf, S. (1963). The functioning and significance of the lateral-line organs. *Biol. Rev.* **38**, 51–105.

Dill, R. E. (1963). The distribution of striated muscle in the epiphysis cerebri of the rat. *Acta anat.* **54**, 310–316.

Dineley, D. L. and Loeffler, E. J. (1976). Ostracoderm faunas of the Delorme and associated Siluro-Devonian formations, North West Territories, Canada. *Palaeontology* **18**, 1–214.

Disler, N. N. (1960). "Lateral line sense organs and their importance in fish behaviour", 1–328. Israel Program for Scientific Translation. Keter Press, Jerusalem (1971).

Dohrn, A. (1884). Studien zur Urgeschichte des Wirbelthierkörpers. IV. Die Entwicklung und Differenzirung der Kiemenbogen der Selachier. *Mitt. zool. Stn Neapel* **5**, 102–151.

Dohrn, A. (1884a). Studien zur Urgeschichte des Wirbelthierkörpers. VI. Die paarigen und unpaaren Flossen der Selachier. *Mitt. zool. Stn Neapel* **5**, 161–195.

Dohrn, A. (1885). Studien zur Urgeschichte des Wirbelthierkörpers. VII. Entstehung und Differenzirung des Zungenbein-Kiefer- Apparates der Selachier. *Mitt. zool. Stn Neapel* **6**, 1–48.

Dohrn, A. (1886). Studien zur Urgeschichte des Wirbelthierkörpers. XI. Spritzlochkieme der Selachier, Kiemendeckelkieme der Ganoiden. Pseudobranchie der Teleostier. *Mitt. zool. Stn Neapel* **7**, 128–176.

Dohrn, A. (1904). Studien zur Urgeschichte des Wirbelthierkörpers. 23. Die Mandibularhöhle der Selachier. 24. Die Prämandibularhöhle. *Mitt. zool. Stn Neapel* **17**, 1–294.

Dollo, L. (1895). Sur la phylogénie des Dipneustes. *Bull. Soc. belge Géol. Paléont. Hydrol.* **9**, 79–128.

Dollo, L. (1907). Les ptyctodontes sont des arthrodères. *Bull. Soc. belge Géol. Paléont. Hydrol.* **21**, 1–12.

Drach, P. (1948). Embranchement des Céphalocordés. *In* "Traité de Zoologie", (Ed. P.-P. Grassé), Vol. 11, 931–1037. Masson, Paris.

Drüner, L. (1901). Studien zur Anatomie der Zungenbein-, Kiemenbogen- und Kehlkopfmuskeln der Urodelen. *Zool. Jb. (Anat.)* **15**, 435–622.

Duke-Elder, W. S. (1958). The eye in evolution. *In* "System of ophthalmology", (Ed. W. S. Duke-Elder), Vol. 1, 1–843. Henry Kimpton, London.

Dunkle, D. H. (1964). Preliminary description of a paleoniscoid fish from the Upper Devonian of Ohio. *Scient. Publs Cleveland Mus. nat. Hist.* **3**(**1**), 5–16.

Dunkle, D. H. and Schaeffer, B. (1973). *Tegeolepis clarki* (Newberry), a Palaeonisciform from the Upper Devonian Ohio Shale. *Palaeontographica* **143**, 151–158.

Dunn, E. R. (1941). The "opercularis" muscle of salamanders. *J. Morph.* **69**, 207–215.

Dutertre, A. P. (1930). Les poissons dévoniens du Boulonnais. *Bull. Soc. géol. Fr.* (**4**)**30**, 571–586.

Dziallas, P. (1952). Die Entwicklung der Venae diploicae beim Haushunde und ihr Einschluss in das knöcherne Schädeldach. *Gegenbaurs morph. Jb.* **92**, 500–576.

Dziallas, P. (1954). Zur Entwicklung des menschlichen Schädeldaches. *Anat. Anz.* **100**, 236–242.

Eakin, R. M. (1973). "The third eye". University of California Press, Berkeley.

Echols, J. (1963). A new genus of Pennsylvanian fish (Crossopterygii, Coelacanthiformes) from Kansas. *Univ. Kans. Publs Mus. nat. Hist.* **12**, 475–501.

Eddy, J. M. P. (1972). The pineal complex. *In* "The biology of lampreys" (Eds M. W. Hardisty and I. C. Potter), Vol. 2, 91–102. Academic Press, London.

Eddy, J. M. P. and Strahan, R. (1968). The role of the pineal complex in the pigmentary effector system of the lampreys, *Mordacia mordax* (Richardson) and *Geotria australis* Gray. *Gen. comp. Endocr.* **11**, 528–534.

Eddy, J. M. P. and Strahan, R. (1970). The structure of the epiphyseal complex of *Mordacia mordax* and *Geotria australis* (Petromyzonidae). *Acta zool., Stockh.* **51**, 67–84.

Edgeworth, F. H. (1926). On the development of the cranial muscles in *Protopterus* and *Lepidosiren. Trans. R. Soc. Edinb.* **54**, 719–734.

Edgeworth, F. H. (1928). The development of some of the cranial muscles of ganoid fishes. *Phil. Trans. R. Soc. (B)* **217**, 39–89.

Edgeworth, F. H. (1935). "The cranial muscles of vertebrates." University Press, Cambridge.

Edinger, T. (1956). Paired pineal organs. *Folia psychiat. neurol. neurochir. neerl. Suppl.* **2**, 121–129.

van Eeden, J. A. (1951). The development of the chondrocranium of *Ascaphus truei* Stejneger with special reference to the relations of the palatoquadrate to the neurocranium. *Acta zool., Stockh.* **32**, 41–176.

Efremov, J. A. (1946). On the subclass Batrachosauria, a group of forms intermediate between amphibians and reptiles. *Izv. Akad. Nauk SSSR (Biol)* **6**, 616–638. (In Russian, English summary.)

Ekman, S. (1941). Ein laterales Flossensaumrudiment bei Haiembryonen. *Nova Acta R. Soc. Scient. upsal.* **IV. 12(7)** 5–44.

Eloff, F. C. (1953). On the occurrence of pineal cartilages in the chondrocranium of a mammal. *J. Linn. Soc.* **42**, 269–272.

El-Toubi, M. R. (1949). The development of the chondrocranium of the spiny dogfish, *Acanthias vulgaris (Squalus Acanthias). J. Morph.* **84**, 227–279.

Emelianov, S. W. (1926). Die Entwicklung der Rippen und ihr Verhältnis zur Wirbelsäule. II. Die Entwicklung der Rippen des *Acipenser, Amia, Lepidosteus* and *Polypterus. Russk. zool. Zh.*, 3–38 (In Russian, German summary.)

Emelianov, S. W. (1935). Die Morphologie der Fischrippen. *Zool. Jb. (Anat.)* **60**, 133–288.

Engler, E. (1929). Untersuchungen zur Anatomie und Entwicklungsgeschichte des Brustschulterapparates der Urodelen. *Acta zool., Stockh.* **10**, 144–229.

Estes, R. (1965). Fossil salamanders and salamander origins. *Am. Zool.* **5**, 319–334.

Estes, R. (1969). The fossil record of amphiumid salamanders. *Breviora* **322**, 1–11.

Estes R. and Berberian, P. (1969). *Amia (= Kindleia) fragosa* (Jordan), a Cretaceous amiid fish, with notes on related European forms. *Breviora* **329**, 1–14.

Estes, R. and Hoffstetter, R. (1976). Les Urodèles du Miocène de La Grive-Saint-Alban (Isère, France). *Bull. Mus. natn. Hist. nat., Paris* **57**, 297–343.

Estes, R. and Reig, O. A. (1973). The early fossil record of frogs. A review of the evidence. *In* "Evolutionary biology of the anurans," (Ed. J. L. Vial), 11–63. University of Missouri Press, Columbia.

Evans, F. G. and Krahl, V. E. (1945). The torsion of the humerus: A phylogenetic survey from fish to man. *Am. J. Anat.* **76**, 303–337.

Eycleshymer, A. C. and Davis, B. M. (1897). The early development of the epiphysis and paraphysis in *Amia*. *J. comp. Neurol.* **7**, 45–70.

Fahrenholz, C. (1925). Über die Entwicklung des Gesichtes und der Nase bei der Geburtshelferkröte (*Alytes obstetricans*). *Gegenbaurs morph. Jb.* **54**, 421–503.

Fahrenholz, C. (1929). Über die "Drüsen" und die Sinnesorgane in der Haut der Lungenfische. *Z. mikrosk.-anat. Forsch.* **16**, 55–74.

Fawcett, E. (1918). The primordial cranium of *Poecilophoca weddelli* (Weddell's seal), at the 27-MM. C.R. length. *J. Anat.* **52**, 412–440.

Fawcett, E. (1923). Some observations on the roof of the primordial human cranium. *J. Anat.* **57**, 245–250.

Fernholm, B. (1969). A third embryo of *Myxine*: Considerations on hypophysial ontogeny and phylogeny. *Acta. zool., Stockh.* **50**, 169–177.

Fernholm, B. and Holmberg, K. (1975). The eyes in three genera of hagfish (*Eptatretus, Paramyxine* and *Myxine*)—a case of degenerative evolution. *Vision Res.* **15**, 253–259.

Fernholm, B. and Olsson, R. (1969). A cytopharmacological study of the *Myxine* adenohypophysis. *Gen. Comp. Endocrinol.* **13**, 336–356.

Fisk, A. (1954). The early development of the ear and acoustico-facialis complex of ganglia in the lamprey *Lampetra planeri* Bloch. *Proc. zool. Soc. Lond.* **124**, 125–151.

Fisk, A. (1957). Experiments on the development of the acoustico-facialis complex and associated structures in the lamprey *Lampetra planeri* (Bloch). *Proc. zool. Soc. Lond.* **128**, 267–278.

Flood, P. R. (1975). Fine structure of the notochord of amphioxus. *Symp. zool. Soc. Lond.* **36**, 81–104.

Fontaine, M., Damas, H., Rochon-Duvigneaud, A. and Pasteels, J. (1958). Classe des Cyclostomes. Formes actuelles: super-ordres des Petromyzonoidea et des Myxinoidea. *In* "Traité de zoologie" (Ed. P.-P. Grassé, Vol. 13(1), 13–172. Masson, Paris.

Forster-Cooper, C. (1937). The Middle Devonian fish fauna of Achanarras. *Trans. R. Soc. Edinb.* **59**, 223–239.

Fox, H. (1954). Development of the skull and associated structures in the Amphibia with special reference to the urodeles. *Trans. zool. Soc. Lond.* **28**, 241–295.

Fox, H. (1959). A study of the development of the head and pharynx of the larval urodele *Hynobius* and its bearing on the evolution of the vertebrate head. *Phil. Trans. R. Soc. (B)* **242**, 151–205.

Fox, H. (1963). Prootic anatomy of the *Neoceratodus* larva. *Acta anat.* **52**, 126–129.

Fox, H. (1963a). The hyoid of *Neoceratodus* and a consideration of its homology in urodele Amphibia. *Proc. zool. Soc. Lond.* **141**, 803–810.

Fox, H. (1963b). Prootic arteries of Dipnoi and Amphibia. *Acta zool., Stockh.* **44**, 345–360.

Fox, H. (1965). Early development of the head and pharynx of *Neoceratodus* with a consideration of its phylogeny. *J. Zool.* **146**, 470–554.

Fox, R. C. (1965). Chorda tympani branch of the facial nerve in the middle ear of tetrapods. *Univ. Kans. Publs Mus. nat. Hist.* **17**, 15–21.

Foxon, G. E. H. (1955). Problems of the double circulation in vertebrates. *Biol. Rev.* **30**, 196–228.

Foxon, G. E. H. (1964). Blood and respiration. *In* "Physiology of the Amphibia" (Ed. J. A. Moore), 151–209. Academic Press, New York and London.

Francis, E. T. (1934). "The anatomy of the salamander" Oxford University Press, Oxford.

François, Y. (1959). La nageoire dorsale; anatomie comparée et evolution. *Ann. Biol.* **35**, 81–113.

François, Y. (1966). Structure et développement de la vertébre de salmo et des téléostéens.

Archs Zool. exp. gén. **107**, 284–324.

Frank, G. H. and Smit, A. L. (1974). The early ontogeny of the columella auris of *Crocodilus niloticus* and its bearing on problems concerning the upper end of the reptilian hyoid arch. *Zoologica Afr.* **9**, 59–88.

Frank, G. H. and Smit, A. L. (1976). The morphogenesis of the avian columella auris with special reference to *Struthio camelus*. *Zoologica Afr.* **11**, 159–182.

Franque, H. (1846). "Afferuntur nonnulla ad Amiam calvam (Lin.) accuratius cognoscendam". G. Schade, Berlin.

Franz, V. (1934). Vergleichende Anatomie des Wirbeltierauges. *In* "Handbuch der vergleichende Anatomie der Wirbeltiere" (Eds L. Boilke *et al.*), Vol. 2, 989–1292. Urban and Schwarzenberg, Berlin and Wien.

Frick, H. (1954). "Die Entwicklung und Morphologie des Chondrokraniums von *Myotis* Kaup." George Thieme Verlag, Stuttgart.

Frick, H. and Starck, D. (1963). Vom Reptil-zum Säugerschädel. *Z. Säugetierk* **28**, 321–341.

Froriep, A. (1891). Zur Entwickelungsgeschichte der Kopfnerven. *Verh. anat. Ges., Jena* **1891**, 55–65.

Fuchs, H. (1905). Herkunft und Entwickelung der Gehörknöchelchen. *Arch. Anat. Physiol.* **Suppl**, 1–176.

Fuchs, H. (1931). Über das Os articulare mandibulae bipartitum einer Echse (*Physignathus lesueurii*), *Gegenbaurs* morph. Jb. **67**, 318–370.

Fürbringer, K. (1903). Beiträge zur Kenntnis des Visceralskelets der Selachier. *Gegenbaurs morph. Jb.* **31**, 360–445.

Fürbringer, K. (1904). Beiträge zur Morphologie des Skeletes der Dipnoer nebst Bemerkungen über Pleuracanthiden, Holocephalen und Squaliden. In Semon, R., Zoologische Forschungsreisen, etc. *Denkschr. med.-naturw. Ges. Jena* **4**, 425–510.

Fürbringer, M. (1897). Ueber die spino-occipitalen Nerven der Selachier und Holocephalen und ihre vergleichende Morphologie. *In* "Festschrift zum siebenzigsten Geburtstage von Carl Gegenbaur", Vol. **3**, 349–788. Wilhelm Engelmann, Leipzig.

Fürbringer, M. (1922). Das Zungenbein der Wirbeltiere insbesondere der Reptilien und Vögel. *Abh. heidelb. Akad. Wiss.* **11**, 1–164.

Gabe, M. (1967). Le tégument et ses annexes. *In* "Traité de Zoologie" (Ed. P.-P. Grassé), Vol. 16, 1–233. Masson, Paris.

Gadow, H. F. (1933). "The evolution of the vertebral column." Cambridge University Press, Cambridge.

Gadow, H. F. and Abbott, E. C. (1895). On the evolution of the vertebral column of fishes. *Phil. Trans. R. Soc.(B)* **186**, 163–221.

Gage, S. P. (1893). The brain of *Diemyctylus viridescens*, from larval to adult life and comparisons with the brain of *Amia* and *Petromyzon*. Wilder Ø. Cent. Book, 259–313.

Gans, C. (1970). Respiration in early tetrapods—the frog is a red herring. *Evolution, Lancaster, Pa.* **24**, 723–734.

Gardiner, B. G. (1963). Certain palaeoniscoid fishes and the evolution of the snout in actinopterygians. *Bull. Br. Mus. nat. Hist. (Geol.)* **8(6)**, 258–325.

Gardiner, B. G. (1967). Further notes on palaeoniscoid fishes with a classification of the Chondrostei. *Bull. Br. Mus. nat. Hist. (Geol.)* **14(5)**, 143–206.

Gardiner, B. G. (1967a). The significance of the preoperculum in actinopterygian evolution. *In* "Fossil Vertebrates" (Eds C. Patterson and P. H. Greenwood). J. Linn. Soc. (Zool.), Vol. 47, 197–209.

Gardiner, B. G. (1973). Interrelationships of teleostomes. *In* "Interrelationships of fishes" (Eds P. H. Greenwood *et al.*), 105–135. Academic Press, London.

Gardiner, B. G. and Bartram, A. W. H. (1977). The homologies of ventral cranial fissures in osteichthyans. *In* "Problems in Vertebrate Evolution" (Eds S. M. Andrews *et al.*), Linn. Soc. Symp., Vol. 4, 227–245.

Gardiner, B. G. and Miles, R. S. (1975). Devonian fishes of the Gogo formation, Western Australia. *Colloque int. Cent. natn. Rech. scient.* **218**, 73–79.

Garman, S. (1904). The chimaeroids (Chismopnea RAF., 1815; Holocephala Müll., 1834), especially *Rhinochimaera* and its allies. *Bull. Mus. comp. Zool. Harv.* **41**, 243–272.

Garman, S. (1913). The Plagiostomia (sharks, skates and rays). *Mem. Mus. comp. Zool. Harv.* **36**, 1–528.

Gasc, J.-P. (1967). Squelette hyobranchial. *In* "Traité de Zoologie" (Ed. P.-P. Grassé), Vol. 16, 550–583. Masson, Paris.

Gaupp, E. (1893). Beiträge zur Morphologie des Schädels. 1. Primordial-Cranium und Kieferbogen von *Rana fusca*. *Morph. Arb.* **2**, 275–481.

Gaupp, E. (1896–1904). A. Ecker's und R. Wiedersheim's Antomie des Frosches, Vol. 1, 1–299 (1896), Vol. 2, 1–548 (1899), Vol. 3, 1–961 (1904). Vieweg u. Sohn, Braunschweig.

Gaupp, E. (1905). Die Entwickelung des Kopfskelettes. *In* "Handbuch der vergleichenden und experimentellen Entwickelungslehre der Wirbeltiere" (Ed. O. Hertwig), Vol. 3(2), 573–874. Fischer, 1906, Jena.

Gaupp, E. (1913). Die Reichertsche Theorie. *Arch. Anat. Physiol.* **1912**, Suppl: 1–416.

Gegenbaur, C. (1865). "Untersuchungen zur vergleichenden Antomie der Wirbelthiere" **2**, 1–176. Leipzig.

Gegenbaur, C. (1870). Über das Skelet der Gliedmaassen der Wirbelthiere im allgemeinen und der Hintergliedmaassen der Selachier insbesondere. *Jen. Z. Naturw.* **5**, 397–447.

Gegenbaur, C. (1872). "Untersuchungen zur vergleichenden Anatomie der Wirbelthiere", 3. Das Kopfskelet der Selachier, ein Beitrag zur Erkenntniss der Genese des Kopfskeletes der Wirbelthiere. Wilhelm Engelmann, Leipzig.

Gegenbaur, C. (1872a). Über das Archipterygium. *Jena Z. Naturw.* **7**, 131–141.

Gegenbaur, C. (1876). Zur Morphologie der Gliedmaassen der Wirbelthiere. *Morph. Jb.* **2**, 396–420.

Gegenbaur, C. (1887). Die Metamerie des Kopfes und die Wirbeltheorie des Kopfskeletes, im Lichte der neueren Untersuchungen betrachtet und geprüft. *Morph. Jb.* **13**, 1–114.

Gentscheff, C. (1934). Über Skelettreste in der menschlichen Zunge. *Virchows Arch. path. Anat. Physiol.* **293**.

George, W. C. (1942). A presomite human embryo with chorda canal and prochordal plate. *Contr. Embryol.* **30**, 1–8.

Gilbert, P. W. (1952). The origin and development of the head cavities in the human embryo. *J. Morph.* **90**, 149–173.

Ginsburg, L. (1970). Les reptiles fossiles. *In* "Traité de Zoologie" (Ed. P.-P. Grassé), Vol. 14(3), 1161–1332. Masson, Paris.

Glaesner, L. (1925). Normentafel zur Entwicklungsgeschichte des gemeinen Wassermolchs (*Molge vulgaris*). *NormTaf. EntwGesch. Wirbeltiere* **14**, 1–49.

Goetsch, W. (1915). Über Hautknochenbildung bei Teleostiern und bei *Amia calva*. *Arch. mikrosk. Anat. EntwMech.* **86(1)**, 435–468.

Goette, A. (1901). Über die Kiemen der Fische. *Z. wiss. Zool.* **69**, 533–577.

Goldstine, S. N., Manickavel, W. and Cohen, N. (1975). Phylogeny of gut-associated lymphoid tissue. *Am. Zool.* **15**, 107–118.

Goodrich, E. S. (1901). On the pelvic girdle and fin of *Eusthenopteron*. *Q. Jl microsc. Sci.* **45**, 311–324.

Goodrich, E. S. (1904). On the dermal fin-rays of fishes, living and extinct. *Q. Jl microsc. Sci.* **47**, 465–522.

Goodrich, E. S. (1906). Notes on the development, structure and origin of the median and paired fins of fish. *Q. Jl microsc. Sci.* **50**, 333–376.

Goodrich, E. S. (1908). On the scales of fish, living and extinct, and their importance in classification. *Proc. zool. Soc. Lond.* **1907**, 751–774.

Goodrich, E. S. (1909). Cyclostomes and fishes. *In* Lankester, E. R. (ed.) "A treatise on zoology" (Ed. E. R. Lankester), Vol. 9. R. and R. Clark, Edinburgh.

Goodrich, E. S. (1911). On the segmentation of the occipital region of the head in the Batrachia Urodela. *Proc. zool. Soc. Lond.* **1910**, 101–120.

Goodrich, E. S. (1918). On the development of the segments of the head in *Scyllium. Q. Jl microsc. Sci.* **63**, 1–30.

Goodrich, E. S. (1919). Restorations of the head of *Osteolepis. J. Linn. Soc.* **34**, 181–188.

Goodrich, E. S. (1925). On the cranial roofing-bones in the Dipnoi. *J. Linn. Soc.* **36**, 79–86.

Goodrich, E. S. (1928). *Polypterus* a palaeoniscid? *Palaeobiologica*, **1**, 87–92.

Goodrich, E. S. (1930) "Studies on the structure and development of vertebrates" Macmillan, London.

Goodrich, E. S. (1931). On the relationship of the ostracoderms to the cyclostomes. *Proc. Linn. Soc. Lond.* **142**, 45–49.

Gorizdro-Kulczycka. Z. (1950). Les dipneustes dévoniens du Massif de S-te Croix. *Acta geol. pol.* **1**, 53–105.

Goronowitsch, N. (1888). Das Gehirn und die Kranialnerven von *Acipenser ruthenus*. Ein Beitrag zur Morphologie des Wirbelthierkopfes. *Morph. Jb.* **13**, 427–574.

Gosline, W. A. (1961). Some osteological features of modern lower teleostean fishes. *Smithson. misc. Collns* **142(3)**, 1–42.

Goujet, D. (1972). Nouvelles observations sur la joue d'Arctolepis (Eastman) et d'autres Dolichothoraci. *Annls Paléont.* **58**, 3–11.

Goujet, D. (1973). *Sigaspis*, un nouvel arthrodire du Dévonien inférieur du Spitsberg. *Palaeontographica (A)* **143**, 73–88.

Goujet, D. (1975). *Dicksonosteus*, un nouvel arthrodire du Dévonien du Spitsberg. Remarques sur le squelette viscéral des Dolichothoraci. *Colloques int. Cent. natn. Rech. scient.* **218**, 81–100.

Graham-Smith, W. (1978). On some variations in the latero-sensory lines of the placoderm fish *Bothriolepis. Phil. Trans. R. Soc. (B)* **282**, 1–39.

Graham-Smith, W. (1978a). On the lateral lines and dermal bones in the parietal region of some crossopterygian and dipnoan fishes. *Phil. Trans. R. Soc. (B)* **282**, 41–105.

Graham-Smith, W. and Westoll, T. S. (1937). On a new long-headed dipnoan fish from the Upper Devonian of Scaumenac Bay P.Q. Canada. *Phil. Trans. R. Soc. (B)* **59**, 241–266.

Grassé, P.-P. (1950–1970) (Ed.). "Traité de Zoologie", Vols 12–17. Masson, Paris.

Gray, H. 29th American edn (1973), "Anatomy of the human body" (Ed. C. M. Goss). Lea and Febiger, Philadelphia.

Greenwood, P. H. (1974). Review of Cenozoic freshwater fish faunas in Africa. *Ann. Geol. Surv. Egypt* **IV**, 211–232.

Gregory, J. T., Morgan, T. G. and Reed, J. (1977). Devonian fishes in Central Nevada. *Univ. Calif., Riverside Campus Mus. Contr.* **4**, 112–120.

Gregory, W. K. (1915). Present status of the problem of the origin of the Tetrapoda, with special reference to the skull and paired limbs. *Ann. N.Y. Acad. Sci.* **26**, 317–383.

Gregory, W. K. (1933). Fish skulls: A study of the evolution of natural mechanisms. *Trans. Am. phil. Soc.* **23**, 75–481.

Gregory, W. K. (1935). Further observations on the pectoral girdle and fin of *Sauripterus taylori* Hall, a crossoperygian fish from the Upper Devonian of Pennsylvania, with special reference to the origin of the pentadactylate extremities of Tetrapoda. *Proc. Am. phil. Soc.* **75**, 673–690.

Gregory, W. K. (1951). "Evolution emerging." Vol. 1 (text); Vol. 2 (illustrations), Macmillan, New York.

Gregory, W. K., Rockwell, H. and Evans, F. G. (1939). Structure of the vertebral column in *Eusthenopteron foordi* Whiteaves. *J. Paleont.* **13**, 126–129.

Gregory, W. K. and Raven, H. C. (1941). Studies on the origin and early evolution of paired fins and limbs. *Ann. N.Y. Acad. Sci.* **42**, 273–360.

Greil, A. (1913). Entwicklungsgeschichte des Kopfes und des Blutgefässystemes von *Ceratodus forsteri*. 2 Theil: Die epigenetischen Erwerbungen während der Stadien, 39–48. *In* "Denkschr. med.-naturw. Ges. Jena," Vol. 4.

Griffiths, I. (1953). On the nature of the fronto-parietal in Amphibia, Salientia. *Proc. zool. Soc. Lond.* **123**, 781–792.

Griffiths, I. (1954). On the "otic element" in Amphibia Salientia. *Proc. zool. Soc. Lond.* **124**, 35–50.

Griffiths, I. (1959). The embryonic origin of the intrinsic limb musculature in Amphibia, Salientia. *Experientia* **15(4)**, 1–5.

Griffiths, I. (1963). The phylogeny of the Salientia. *Biol. Rev.* **38**, 241–292.

Gross, W. (1932). Die Arthrodira Wildungens. *Geol. Pal. Abh.* **19**, 1–61.

Gross, W. (1933). Die Fische des baltischen Devons. *Paleontographica (A)* **79**, 1–74.

Gross, W. (1933a). Die Wirbeltiere des rheinischen Devons. *Abh. preuss. geol. L.-Anst., n.F., H.* **154**, 1–83.

Gross, W. (1936). Beiträge zur Osteologie baltischer und rheinischer Devon-Crossopterygier. *Paläont. Z.* **18**, 129–155.

Gross, W. (1937). Das Kopfskelett von *Cladodus wildungensis,* 1. Endocranium und Palatoquadratum. *Senckenbergiana* **19**, 80–107.

Gross, W. (1938). Das Kopfskelett von *Cladodus wildungensis*, 2. Der Kieferbogen. *Senckenbergiana* **20**, 123–145.

Gross, W. (1941). Über den Unterkiefer einiger devonischer Crossopterygier. *Abh. preuss. Akad. Wiss. Phys.-math. Kl.* **7**, 3–51.

Gross, W. (1947). Die Agnathen und Acanthodier des obersilurischen Beyrichienkalks. *Paleontographica (A)* **46**, 91–161.

Gross, W. (1953). Devonische Palaeonisciden-Reste in Mittel-und Osteuropa. *Paläont. Z.* **27**, 85–112.

Gross, W. (1954). Zur Phylogenie des Schultergürtels. *Paläont. Z.* **28**, 20–40.

Gross, W. (1954a). Zur Conodonten-Frage. *Senckenberg. leth.* **35**, 73–85.

Gross, W. (1956). Über Crossopterygier und Dipnoer aus dem baltischen Oberdevon im Zusammenhang einer vergleichenden Untersuchung des Porenkanalsystems paläozoischer Agnathen und Fische. *K. svenska VetenskAkad. Handl. (***4**)**5**, 1–140.

Gross, W. (1957). Mundzähne und Hautzähne der Acanthodier und Arthrodiren. *Palaeontographica (A)* **109**, 1–40.

Gross, W. (1958). Über die älteste Arthrodiren-Gattung. *Notizbl. hess. Landesamt Bodenforsch.* **86**, 7–30.

Gross, W. (1959). Arthrodiren aus dem Obersilur der Prager Mulde. *Palaeontographica (A)* **113**, 1–35.

Gross, W. (1960). Über die Basis bei den Gattungen *Palmatolepis* und *Polygnathus* (Conodontida). *Paläont. Z.* **34**, 40–58.

Gross, W. (1961). *Lunaspis broilii* und *Lunaspis heroldi* aus dem Hunsrückschiefer (Unterdevon, Rheinland). *Notizbl. hess. Landesamt Bodenforsch.* **89**, 17–43.

Gross, W. (1962). Neuuntersuchung der Stensiöellida (Arthrodira, Unterdevon). *Notizbl. hess. Landesamt Bodenforsch.* **90**, 48–86.

Gross, W. (1962a). Peut-on homologuer les os des arthrodires et des téléostomes? *Colloques int. Cent. natn. Rech. scient.* **104**, 69–74.

Gross, W. (1963). *Gemuendina stuertzi* Traquair. Neuuntersuchung. *Notizbl. hess. Landesamt Bodenforsch.* **91**, 36–73.

Gross, W. (1963a). *Drepanaspis gemuendenensis* Schlüter. Neuuntersuchung. *Palaeontographica (A)* **121**, 133–155.

Gross, W. (1964). Über die Randzähne des Mundes, die Ethmoidalregion des Schädels und die Unterkiefersymphyse von *Dipterus oervigi* n. sp. *Paläont. Z.* **38**, 7–25.

Gross, W. (1965). *Onychodus jaekeli* Gross (Crossopterygii, Oberdevon), Bau des Symphysenknochens und seiner Zähne. *Senckenberg. leth.* **46a**, 123–131.

Gross. W. (1965a). Über den Vorderschädel von *Ganorhynchus splendens* Gross (Dipnoi, Mitteldevon). *Paläont. Z.* **39**, 113–133.

Gross, W. (1966). Kleine Schuppenkunde. *N. Jb. Geol. Paläont. Abh.* **125**, 29–48.

Gross, W. (1967). Über Thelodontier-Schuppen. *Palaeontographica (A)* **127**, 1–67.

Gross, W. (1967a). Über das Gebiss der Acanthodier und Placodermen. *In* "Fossil Vertebrates" (Eds C. Patterson and P. H. Greenwood), J. Linn. Soc. (Zool.), Vol. 47, 121–130.

Gross, W. (1968). Fragliche Actinopterygier-Schuppen aus dem Silur Gotlands. *Lethaia* **1**, 184–218.

Gross, W. (1968a). Porenschuppen und Sinneslinien des Thelodontiers *Phlebolepis elegans* Pander. *Paläont. Z.* **42**, 131–146.

Gross, W. (1969). *Lophosteus superbus* Pander, ein Teleostome aus dem Silur Oesels. *Lethaia* **2**, 15–47.

Gross, W. (1971). Downtonische und dittonische Acanthodier-Reste des Ostseegebietes. *Palaeontographica (A),* **136**, 1–82.

Gross, W. (1971a). *Lophosteus superbus* Pander, Zähne, Zahnknochen und besondere Schuppenformen. *Lethaia* **4**, 131–152.

Gross, W. (1973). Kleinschuppen, Flossenstacheln und Zähne von Fischen aus europäischen und nordamerikanischen Bonebeds des Devons. *Palaeontographica (A)* **142**, 51–155.

Gudger, E. W. (1940). The breeding habits, reproductive organs and external embryonic *development of Chlamydoselachus*, based on notes and drawings by Bashford Dean. "The Bashford Dean Memorial Volume: Archaic fishes", Vol. 7, 523–633. Am. Mus. Nat. Hist., New York.

Gudger, E. W. and Smith, B. G. (1933). The natural history of the frilled shark *Chlamydoselachus anguineus*. "The Bashford Dean Memorial Volume: Archaic fishes", Vol. 5, 245–319. Am. Mus. Nat. Hist., New York.

Guibé, J. (1970). La peau et les productions cutanées. *In* "Traité de Zoologie" (Ed. P.-P. Grassé), Vol. 14(2), 6–32. Masson, Paris.

Günther, A. (1871). Description of *Ceratodus*, a genus of ganoid fishes, recently discovered in rivers of Queensland, Australia. *Phil. Trans. R. Soc.* II, 509–571.

Gustafsson, G. (1935). On the biology of *Myxine glutinosa* L. *Ark. zool.* **28A**, 1–8.

Gutmann, W. F. (1969). Die Entstehung des Vertebraten-Kopfes, ein phylogenetisches Modell. *Senckenberg. biol.* **50**, 433–471.

Hafferl, A. (1933). Das Arteriensystem. *In* "Handbuch der vergleichenden Anatomie der Wirbeltiere" (Eds L. Bolk *et al.*), Vol. 6, 563–684. Urban and Schwarzenberg, Berlin and Wien.

Hagelin, L.-O. (1974). Development of the membranous labyrinth in lampreys. *Acta zool., Stockh.* **55** (Suppl.), 1–218.
Hagelin, L.-O. and Johnels, A. G. (1955). On the structure and function of the accessory olfactory organ in lampreys. *Acta zool., Stockh.* **36**, 113–125.
Haller, v. Hallerstein, V. (1934). Äussere Gliederung des Zentralnervensystems. Kranialnerven. In "Handbuch der vergleichenden Anatomie der Wirbeltiere" (Eds L. Bolk *et al.*) Vol. 2(1), 1–318, 541–684, 817–832, 835–840. Urban and Schwarzenberg, Berlin and Wien.
Halstead, L. B. (1969). "The pattern of vertebrate evolution", Oliver and Boyd, Edinburgh.
Halstead, L. B. (1971). The presence of a spiracle in the Heterostraci (Agnatha). *Zool. J. Linn. Soc.* **50**, 195–197.
Halstead, L. B. (1973). The heterostracan fishes. *Biol. Rev.* **48**, 279–332.
Halstead, L. B. (1973a). Affinities of the Heterostraci (Agnatha). *Biol. J. Linn. Soc.* **5**, 339–349.
Hammarberg, F. (1937). Zur Kenntnis der ontogenetischen Entwicklung des Schädels von *Lepidosteus platystomus*. *Acta zool., Stockh.* **18**, 210–337.
Hansen, G. N. (1971). On the structure and vascularization of the pituitary gland in some primitive actinopterygians. *Biol. Skr.* **18**, 5–60.
Hardisty, M. W. and Potter, I. C. (1971–1972) (Eds). "The biology of lampreys", Vol. 1 (1971); Vol. 2 (1972). Academic Press, London.
Hardisty, M. W. and Potter, I. C. (1971) (Eds). The general biology of adult lampreys. *In* "The biology of lampreys", Vol. 1, 128–206. Academic Press, London.
Hardisty, M. W. and Potter, I. C. (1971a) (Eds). Paired species. *In* "The biology of lampreys", Vol. 1: 249–277. Academic Press, London.
Harman, N. B. (1899). The palpebral and oculomotor apparatus in fishes. *J. Anat. Physiol., Lond.* **34**, 1–40.
Harris, J. E. (1938). 1. The dorsal spine of *Cladoselache*. 2. The neurocranium and jaws of *Cladoselache*. *Scient. Publs Cleveland Mus. nat. Hist.* **8**, 1–12.
Harris, J. E. (1950). *Diademodus hydei*, a new fossil shark from the Cleveland shale. *Proc. zool. Soc. Lond.* **120**, 683–697.
Harrison, R. G. (1903). Experimentelle Untersuchungen über die Entwicklung der Sinnesorgane der Seitenlinie der Amphibien. *Arch mikr. Anat.* **63**, 35–149.
Hatschek, B. (1881). Studien über Entwicklung des Amphioxus. *Arb. zool. Inst. Univ. Wien.* **4**, 1–88.
Hauge, F. S. (1924). The chondrocranium of *Amia calva*. *J. Morph.* **39**. 267–277.
Hay, O. P. (1895). On the structure and development of the vertebral column of *Amia*. *Publs Field Mus. nat. Hist.* **1**, 1–54.
Hazelton, R. D. (1970). A radioautographic analysis of the migration and fate of cells derived from the occipital somites in the chick embryo with specific reference to the development of the hypoglossal musculature. *J. Embryol. exp. Morph.* **24(3)**, 455–466.
Heintz, A. (1932). The structure of *Dinichthys*. *Bashford Dean meml volume* **4**, 111–224.
Heintz, A. (1937). *Lunaspis*-Arten aus dem Devon Spitzbergens. *Skr. Svalbard Ishavet* **72**, 1–23.
Heintz, A. (1939). Cephalaspida from Downtonian of Norway. *Skr. norske Vidensk-Akad., Mat.-naturv. Kl.* **5**, 5–117.
Heintz, A. (1958). The head of the anaspid *Birkenia elegans*, Traq. *In* "Studies on fossil vertebrates" (Ed. T. S. Westoll), 71–85. Athlone Press, London.
Heintz, A. (1962). Les organes olfactifs des Heterostraci. *Colloques int. Cent. natn. Rech. scient.* **104**, 13–29.
Heintz, A. (1963). Phylogenetic aspects of myxinoids. *In* "The biology of *Myxine*" (Eds A. Brodl and R. Fänge), 9–21. Universitetsforlaget, Oslo.
Heintz, A. (1967). A new tremataspidid from Ringerike, South Norway. *In* "Fossil vertebrates" (Eds C. Patterson and P. H. Greenwood). J. Linn. Soc. (Zool.) Vol. 47, 55–68.

Heintz, A. (1967a). Some remarks about the structure of the tail in cephalaspids. *Colloques int. Cent. natn. Rech. scient.* **163**, 21–35.
Heintz, A. (1968). The spinal plate in *Homostius* and *Dunkleosteus*. *In* "Current problems of lower vertebrate phylogeny" (Ed. T. Ørvig). Nobel Symp., Vol. 4, 145–151. Almqvist and Wiksell, Stockholm.
Heintz, A. (1969). New agnaths from Ringerike Sandstone. *Skr. norske Vidensk-Akad. Mat.-naturw. Kl.* **26**, 3–28.
Heintz, A. (1974). Additional remarks about *Hemicyclaspis* from Jeløya, Southern Norway. *Norsk geol. Tidsskr.* **54**, 375–384.
Heintz, N. (1968). The pteraspid *Lyktaspis* n.g. from the Devonian of Vestspitsbergen. *In* "Current problems of lower vertebrate phylogeny" (Ed. T. Ørvig). Nobel Symp., Vol. 4, 73–80. Almquist and Wiksell, Stockholm.
Heintz, N. (1972). The thelodont *Sigurdia lata* n.g., n.sp. *Norsk Polarisnt. Årb.* **1970**, 112–116.
Hemmings, S. K. (1978). The Old Red Sandstone antiarchs of Scotland: *Pterichthyodes* and *Microbrachius*. *Palaeontogr. Soc. (Monogr.)* **131**, 1–64.
Herrick, C. J. (1899). The cranial and first spinal nerves of *Menidia*; a contribution upon the nerve components of the bony fishes. *J. comp. Neurol.* **9**, 153–455.
Herrick, C. J. (1935). The membranous parts of the brain, meninges and their blood vessels in *Amblystoma*. *J. comp. Neurol.* **61**, 297–346.
Hertwig, O. (1874). Ueber das Zahnsystem der Amphibien und seine Bedeutung für die Genese des Skelets der Mundhöhle. *Arch. mikr. Anat.* **11**, (Suppl.).
Hertwig, O. (1901). Die Lehre von Keimblättern. *In* "Handbuch der vergleichenden und experimentellen Entwickelungslehre der Wirbeltiere, (Ed. O. Hertwig), Vol. 1(1), 699–966. Gustav Fischer, Jena.
Hertwig, O. (Ed.) (1901–1906). "Handbuch der vergleichenden und experimentellen Entwickelungslehre der Wirbeltiere", Vols 1–3. Fischer, Jena.
Heyler, D. (1969). Acanthodii. *In* "Traité de Paléontologie" (Ed. J. Piveteau), Vol. 4(3), 21–70. Masson, Paris.
Hill, B. H. (1935). The early development of the thymus glands in *Amia calva*. *J. Morph.* **57**, 61–89.
Hill, B. H. (1935a). The early development of the thyroid gland in *Amia calva*. *J. Morph.* **57**, 533–542.
Hills, E. S. (1941). The cranial roof of *Dipnorhynchus sussmilchi* (Eth. Fil.). *Rec. Aust. Mus.* **21**, 45–55.
Hills, E. S. (1943). The ancestry of the Choanichthyes. *Aust. J. Sci.* **6**, 21–23.
Hinsberg, V. (1901). Die Entwicklung der Nasenhöhle bei Amphibien. *Arch. mikr. Anat.* **58**, 413–482.
Hoffmann, C. K. (1896–97). Beiträge zur Entwicklungsgeschichte der Selachii. *Morph. Jb.* **24**, 209–286; **25**, 250–304.
Holfreter, J. (1938). Veränderungen der Reaktionsweise im alternden isolierten Gastrulaektoderm. *Arch. EntwMech. Org.* **138**, 163–196.
Holly, M. (1933). Cyclostomata. *In* "Das Tierreich" (Eds P. E. Schulze, W. Kükenthal, K. Heider and R. Hesse), Vol. 59. W. de Gruyter, Berlin and Leipzig.
Holmberg, K. (1978). Electron microscopic analysis of the inner synaptic layer in the river lamprey (*Lampetra fluviatilis*). *Acta zool., Stockh.* **59**, 107–117.
Holmberg, K. and Öhman, P. (1976). Fine structure of retinal synaptic organelles in lamprey and hagfish photoreceptors. *Vision Res.* **16**, 237–239.
Holmes, R. and Carroll, R. (1977). A temnospondyl amphibian from the Mississippian of Scotland. *Bull. Mus. comp. Zool. Harv.* **147**, 489–511.

Holmgren, N. (1928). Some observations about the growth of the tail in *Lepidosiren*. *Acta zool., Stockh.* **9**, 321–329.
Holmgren, N. (1931). Notes on the development of the hypohysis in *Acipenser ruthenus*. *Acta zool., Stockh.* **12**, 145–152.
Holmgren, N. (1933). On the origin of the tetrapod limb. *Acta zool., Stockh.* **14**, 186–295.
Holmgren, N. (1939). Contribution to the question of the origin of the tetrapod limb. *Acta zool., Stockh.* **20**, 89–124.
Holmgren, N. (1940). Studies on the head in fishes. Embryological, morphological, and phylogenetical researches. 1. Development of the skull in sharks and rays. *Acta zool., Stockh.* **21**, 51–267.
Holmgren, N. (1941). Studies on the head in fishes. 2. Comparative anatomy of the adult selachian skull, with remarks on the dorsal fins in sharks. *Acta zool., Stockh.* **22**, 1–100.
Holmgren, N. (1942). Studies on the head of fishes. 3. The phylogeny of elasmobranch fishes. *Acta zool., Stockh.* **23**, 129–261.
Holmgren, N. (1942a). General morphology of the lateral sensory line system of the head in fish. *K. svenska VetenskAkad. Handl.* **20(1)**, 3–46.
Holmgren, N. (1943). Studies on the head of fishes. 4. General morphology of the head in fish. *Acta zool., Stockh.* **24**, 1–188.
Holmgren, N. (1946). On two embryos of *Myxine glutinosa*. *Acta zool., Stockh.* **27**, 1–90.
Holmgren, N. (1949). On the tetrapod limb problem—again. *Acta zool., Stockh.* **30**, 485–508.
Holmgren, N. (1949a). Contributions to the question of the origin of tetrapods. *Acta zool., Stockh.* **30**, 459–484.
Holmgren, N. (1952). An embryological analysis of the mammalian carpus and its bearing upon the question of the origin of the tetrapod limb. *Acta zool., Stockh.* **33**, 1–115,
Holmgren, N. (1953). On the "horizontal pit-line" in the teleostomian fishes. *Acta zool., Stockh.* **34**, 147–154.
Holmgren, N. and van der Horst, C. J. (1925). Contribution to the morphology of the brain of *Ceratodus*. *Acta zool., Stockh.* **6**, 59–165.
Holmgren, N. and Pehrson, T. (1949). Some remarks on the ontogenetical development of the sensory lines on the cheek in fishes and amphibians. *Acta zool., Stockh.* **30**, 249–314.
Holmgren, N. and Stensiö, E. (1936). Kranium und Visceralskelett der Akranier, Cyclostomen und Fische. *In* "Handbuch der vergleichenden Anatomie der Wirbeltiere" (Eds L. Bolk *et al.*), Vol. 4, 233–500. Urban and Schwarzenberg, Berlin and Wien.
Holmgren, U. (1965). On the ontogeny of the pineal- and para-pineal organs in teleost fishes. *Prog. Brain Res.* **10**, 172–182.
Hopson, J. A. (1966). The origin of the mammalian middle ear. *Am. Zool.* **6**, 437–450.
Hopson, J. A. (1974). The functional significance of the hypocercal tail and lateral fin fold of anaspid ostracoderms. *Fieldiana, Geol.* **33**, 83–93.
van der Horst, C. J. (1925). The myelencephalic gland of *Polyodon, Acipenser* and *Amia*. *Proc. K. ned. Akad. Wet.* **28**, 432–442.
van der Horst, C. J. (1925a). The cerebellum of fishes. I. General morphology of the cerebellum. *Proc. K. ned. Akad. Wet.* **28**, 735–746.
van der Horst, C. J. (1928). A cutaneous branch of the facial nerve in a teleost. *Proc. K. ned. Akad. Wet.* **31**, 24–30.
Hörstadius, S. (1950). "The neural crest". Oxford University Press, London.
Hörstadius, S. and Sellman, S. (1946). Experimentelle Untersuchungen über die Determination des knorpligen Kopfskelettes bei Urodelen. *Nova Acta Soc. Sci. upsal.* **13(8)**, 1–170.
Hotton, N. III (1952). Jaws and teeth of American xenacanth sharks. *J. Paleont.* **26**, 489–500.
Howell, A. B. (1933). Homology of paired fins in fishes. *J. Morph.* **54**, 451–457.

Howes, G. B. (1887). On the skeleton and affinities of the paired fins of *Ceratodus* with observations upon those of Elasmobranchii. *Proc. zool. Soc. Lond.* **1887**, 3–26.
Hughes, G. M. (1976). On the respiration of *Latimeria chalumnae*. *J. Linn. Soc. (Zool.)* **59**, 195–208.
Hussakof, L. (1912). Notes on Devonic fishes from Scaumenac Bay, Quebec. *Bull. N. Y. St. Mus.* **158**, 111–139.
Huxley, T. H. (1858). On the theory of the vertebrate skull. *Proc. R. Soc. Lond.* **9**, 381–433.
Huxley, T. H. (1861). Preliminary essay upon the systematic arrangement of the fishes of the Devonian epoch. *Mem. geol. Surv. U.K.* **10**, 1–46.
Huxley, T. H. (1876). On *Ceratodus forsteri*, with observations on the classification of fishes. *Proc. zool. Soc. Lond.* **1876**, 24–59.
Iselstöger, H. (1937). Das Neurocranium von *Rhina squatina* und einige Bemerkungen über ihre systematische Stellung. *Zool. Jb. (Anat.)* **62**, 349–394.
Iselstöger, H. (1941). Das Labyrinth von *Pristiophorus japonicus* Günther. *Gegenbaurs morph. Jb.* **86**, 259–286.
Jacobson, C.-O. (1959). The localization of the presumptive cerebral regions in the neural plate of the axolotl larva. *J. Embryol. exp. Morph.* **7**, 1–21.
Jaekel, O. (1899). Über die primäre Zusammensetzung des Kieferbogens und Schultergürtels. *Verh. dt. zool. Ges.* **1899**, 249–258.
Jaekel, O. (1906). Einige Beiträge zur Morphologie der ältesten Wirbeltiere. *Sber. Ges. naturf. Freunde, Berl.*, 180–189.
Jaekel, O. (1925). Das Mundskelett der Wirbeltiere. *Gegenbaurs morph. Jb.* **55**, 401–484.
Jaekel, O. (1927). Der Kopf der Wirbeltiere. *Ergebn. Anat. EntwGesch.* **27**, 811–974.
Jägersten, G. (1972). "Evolution of the metazoan life cycle." Academic Press, London and New York.
Jakubowski, M. (1975). Anatomical structure of the olfactory organs provided with internal nares in the Antarctic fish *Gymnodraco acuticeps* Boul. (Bathydraconidae). *Bull. Acad. pol. Sci. Cl. II. Sér. Sci. biol.* **23**, 115–120.
Janot, C. (1967). A propos des Amiidés actuels et fossiles. *Colloques int. Cent. natn. Rech. scient.* **163**, 139–153.
Janvier, Ph. (1971). La forme et la position du sac nasal chez les Ostéostracés. *C.r. hebd. Séanc. Acad. Sci., Paris* **272**, 2434–2436.
Janvier, Ph. (1974). The sensory line system and its innervation in the Osteostraci (Agnatha, Cephalaspidomorphi). *Zool. Scr.* **3**, 91–99.
Janvier, Ph. (1974a). The structure of the naso-hypophysial complex and the mouth in fossil and extant cyclostomes, with remarks on amphiaspiforms. *Zool. Scr.* **3**, 193–200.
Janvier, Ph. (1974b). Preliminary report on late Devonian fishes from central and eastern Iran. *Rep. Geol. Surv. Iran* **31**, 5–47.
Janvier, Ph. (1975). Les yeux des cyclostomes fossiles et le problème de l'origine des Myxinoides. *Acta zool. Stockh.* **56**, 1–9.
Janvier, Ph. (1975a). Remarques sur l'orifice naso-hypophysaire des céphalaspidomorphes. *Annls Paléont. (Vert.)* **61(1)**, 3–16.
Janvier, Ph. (1975b). Spécialisations précoces et caractères primitifs du système circumlatoire des Ostéostracés. *Colloques. int. Cent. natn. Rech. scient.* **218**, 15–31.
Janvier, Ph. (1975c). Anatomie et position systématique des Galéaspides (Vertebrata, Cyclostomata), Céphalaspidomorphes du Dévonien inférieur du Yunnan (Chine). *Bull. Mus. Hist. nat., Paris,* **278**, 1–16.
Janvier, Ph. (1977). Contribution a la connaissance de la systématique et de l'anatomie du genre

Boreaspis Stensiö (Agnatha, Cephalaspidomorphi, Osteostraci) du Dévonien inférieur du Spitsberg. *Annls Paléont. (Vert.)* **63(1)**, 1–32.

Janvier, Ph. (1978). On the oldest known teleostome fish *Andreolepis hedei* Gross (Ludlow of Gotland), and the systematic position of the lophosteids. *Eesti NSV Tead. Akad. Toim.*, 88–95.

Janvier, Ph. (1978a). Les nageoires paires des Ostéostracés et la position systématique des Céphalaspidomorphes. *Annls Paléont. (Vert.)* **64**, 113–142.

Jaquet, M. (1897–99). Contributions à l'anatomie comparée des systémes squelettaire et musculaire de *Chimaera collei, Callorhynchus antarcticus, Spinax niger, Protopterus annectens, Ceratodus forsteri* et Axolotl. *Archs Sci. méd.* **2**, 174–206; **3**, 300–340; **4**, 241–273.

Jarvik, E. (1937). On the species of *Eusthenopteron* found in Russia and the Baltic states. *Bull. geol. Instn Univ. Upsala* **27**, 63–127.

Jarvik, E. (1942). On the structure of the snout of crossopterygians and lower gnathostomes in general. *Zool. Bidr. Upps.* **21**, 235–675.

Jarvik, E. (1944). On the dermal bones, sensory canals and pit-lines of the skull in *Eusthenopteron foordi* Whiteaves, with some remarks on E. *säve-söderberghi* Jarvik. *K. svenska Vetensk-Akad. Handl.* **(3) 21(3)**, 1–48.

Jarvik, E. (1944a). On the exoskeletal shoulder-girdle of teleostomian fishes, with special reference to *Eusthenopteron foordi* Whiteaves. *K. svenska VetenskAkad. Handl.* **(3) 21(7)**, 1–32.

Jarvik, E. (1947). Notes on the pit-lines and dermal bones of the head in *Polypterus. Zool. Bidr. Upps.* **25**, 70–78.

Jarvik, E. (1948). On the morphology and taxonomy of the Middle Devonian osteolepid fishes of Scotland. *K. svenska VetenskAkad. Handl.* **(3) 25**, 1–301.

Jarvik, E. (1948a). Note on the Upper Devonian vertebrate fauna of East Greenland and on the age of the ichthyostegid stegocephalians. *Ark. Zool.* **41A**, 1–8.

Jarvik, E. (1949). On the Middle Devonian Crossopterygians from the Hornelen Field in Western Norway. *Årbok Univ. Bergen.* **1948** (Naturv.r.), 1–48.

Jarvik, E. (1950). Middle Devonian vertebrates from Canning Land and Wegeners Halvø (East Greenland). 2. Crossopterygii. *Meddr Grønland* **96**, 1–132.

Jarvik, E. (1950a). Note on Middle Devonian crossopterygians from the eastern part of Gauss Halvö, East Greenland. With an appendix: An attempt at a correlation of the Upper Old Red Sandstone of East Greenland with the marine sequence. *Meddr Grønland* **149**, 1–20.

Jarvik, E. (1950b). On some osteolepiform crossopterygians from the Upper Old Red Sandstone of Scotland. *K. Svenska VetenskAkad. Handl.* **(4) 2**, 1–35.

Jarvik, E. (1952). On the fish-like tail in the ichthyostegid stegocephalians with descriptions of a new stegocephalian and a new crossopterygian from the Upper Devonian of East Greenland. *Meddr Grønland* **114**, 1–90.

Jarvik, E. (1954). On the visceral skeleton in *Eusthenopteron* with a discussion of the parasphenoid and palatoquadrate in fishes. *K. svenska VetenskAkad. Handl.* **(4) 5**, 1–104.

Jarvik, E. (1955). The oldest tetrapods and their forerunners. *Scient. Mon., N.Y.* **80**, 141–154.

Jarvik, E. (1955a). Ichthyostegalia. *In* "Traité de Paléontologie" (Ed. J. Piveteau), Vol. V, 53–66. Masson, Paris.

Jarvik, E. (1959). De tidiga fossila ryggradsdjuren. Paleoanatomiska arbetsmetoder och resultat (The early fossil vertebrates. Paleoanatomical methods and results). *Svensk Naturv.* **1959**, 5–80. (In Swedish).

Jarvik, E. (1959a). Dermal fin-rays and Holmgren's principle of delamination. *K. svenska VetenskAkad. Handl.* **(4) 6**, 1–51.

Jarvik, E. (1960). "Théories de l'évolution des vertébrés reconsidérées à la lumière des récentes découvertes sur les vertébrés inférieurs." Masson, Paris.

Jarvik, E. (1961). Devonian vertebrates. *In* "Geology of the Arctic" (Ed. G. O. Raasch), Vol. 1, 197–204. P.P., Toronto.

Jarvik, E. (1962). Les porolépiformes et l'origine des urodèles. *Colloques int. Cent. natn. Rech. scient.* **104**, 87–101.

Jarvik, E. (1963). The composition of the intermandibular division of the head in fish and tetrapods and the diphyletic origin of the tetrapod tongue. *K. svenska VetenskAkad. Handl.* (**4**) **9**, 1–74.

Jarvik, E. (1963a). The fossil vertebrates from East Greenland and their zoological importance. *Experientia* **19**, 1–6.

Jarvik, E. (1964). Specializations in early vertebrates. *Annls Soc. r. zool. Belg.* **94**, 1–95.

Jarvik, E. (1965). Die Raspelzunge der Cyclostomen und die pentadactyle Extremität der Tetrapoden als Beweise für monophyletische Herkunft. *Zool. Anz.* **175**, 101–143.

Jarvik, E. (1965a). On the origin of girdles and paired fins. *Israel J. Zool.* **14**, 141–172.

Jarvik, E. (1966). Remarks on the structure of the snout in *Megalichthys* and certain other rhipidistid crossopterygians. *Ark. Zool.* (**2**) **19**, 41–98.

Jarvik, E. (1967). On the structure of the lower jaw in dipnoans: with a description of an early Devonian dipnoan from Canada, *Melanognathus canadensis* gen. et sp. nov. *In* "Fossil vertebrates" (Eds C. Patterson and P. H. Greenwood). J. Linn. Soc. (Zool.) Vol. 47, 155–183. Academic Press, London.

Jarvik, E. (1967a). The homologies of frontal and parietal bones in fishes and tetrapods. *Colloques int. Cent. natn. Rech. scient.* **163**, 181–213.

Jarvik, E. (1968). The systematic position of the Dipnoi. *In* "Current problems of lower vertebrate phylogeny" (Ed. T. Ørvig) Nobel Symp., Vol. 4, 223–245. Almqvist and Wiksell, Stockholm.

Jarvik, E. (1968a). Apects of vertebrate phylogeny. *In* "Current problems of lower vertebrate phylogeny" (Ed. T. Ørvig). Nobel Symp., Vol. 4, 497–527. Almqvist and Wiksell, Stockholm.

Jarvik, E. (1972). Middle and Upper Devonian Porolepiformes from East Greenland with special references to *Glyptolepis groenlandica* n.sp. *Meddr Grønland* **187**, 1–295.

Jarvik, E. (1975). On the saccus endolymphaticus and adjacent structures in osteolepiforms, anurans and urodeles. *Colloques int. Cent. natn. Rech. scient.* **218**, 191–211.

Jarvik, E. (1977). The systematic position of acanthodian fishes. *In* "Problems in vertebrate evolution" (Eds S. M. Andrews *et al.*), Vol. 4, 199–225. Academic Press, London.

Jefferies, R. P. S. (1975). Fossil evidence concerning the origin of the chordates. Symp. zool. Soc. Lond. Vol. 36, 253–318. Academic Press, London.

Jefferies, R. P. S. and Lewis, D. N. (1978). The English Silurian fossil *Placocystites forbesianus* and the ancestry of the vertebrates. *Phil. Trans. R. Soc.* **282**, 205–323.

Jessen, H. (1966). Struniiformes. *In* "Traité de Paléontologie", (Ed. J. Piveteau), Vol. 4(3), 387–398. Masson, Paris.

Jessen, H. (1966a). Die Crossopterygier des Oberen Plattenkalkes (Devon) der Bergisch-Gladbach-Paffrather Mulde (Rheinisches Schiefergebirge) unter Berücksichtigung von amerikanischem und europäischem *Onychodus*-Material. *Ark. Zool.* (**2**) **18**, 305–389.

Jessen, H. (1967). The position of the Struniiformes (*Strunius* and *Onychodus*) among crossopterygians. *Colloques int. Cent. natn. Rech. scient.* **163**, 173–180.

Jessen, H. (1968). *Moythomasia nitida* Gross und *M.cf. striata* Gross, devonische Palaeonisciden aus dem Oberen Plattenkalk der Bergisch-Gladbach-Paffrather Mulde (Rheinisches Schiefergebirge). *Palaeontographica* **128**, 87–114.

Jessen, H. (1968a). A Devonian osteolepidid fish from British Columbia. *Bull. geol. Surv. Can.* **165**, 65–70.

Jessen, H. (1968b). The gular plates and branchioestegal rays in *Amia, Elops* and *Polypterus*. *In* "Current problems of lower vertebrate phylogeny." (Ed. T. Ørvig). Nobel Symp., Vol. 4, 89–107. Almqvist and Wiksell, Stockholm.

Jessen, H. (1972). Schultergürtel und Pectoralflosse bei Actinopterygiern. *Fossils and Strata* **1**, 1–101.

Jessen, H. (1973). Weitere Fischreste aus dem Oberen Plattenkalk der Bergisch-Gladbach-Paffrather Mulde (Oberdevon, Rheinisches Schiefergebirge). *Palaeontographica* **143**, 159–187.

Jessen, H. (1973a). Interrelationships of actinopterygians and brachiopterygians: evidence from pectoral anatomy. *In* "Interrelationships of fishes" (Eds P. H. Greenwood *et al.*), 227–232. Academic Press, London.

Jessen, H. (1975). A new choanate fish, *Powichthys thorsteinssoni* n.g., n.sp., from the early Lower Devonian of the Canadian Arctic Archipelago. *Colloques int. Cent. natn. Rech. scient.* **218**, 213–222.

Johnels, A. G. (1948). On the development and morphology of the skeleton of the head of *Petromyzon. Acta zool., Stockh.* **29**, 140–278.

Johnston, J. B. (1905). The cranial nerve components of *Pretromyzon. Gegenbaurs morph. Jb.* **34**, 149–203.

Johnston, J. B. (1909). The morphology of the forebrain vesicle in vertebrates. *J. comp. Neurol.* **19**, 458–495.

Johnston, J. B. (1911). The telencephalon of selachians. *J. comp. Neurol.* **21**, 1–113.

Johnston, J. B. (1913). Nervus terminalis in reptiles and mammals. *J. comp. Neurol.* **23**, 97–120.

Johnston, J. B. (1923). Further contributions to the study of the evolution of the forebrain. *J. comp. Neurol.* **35**, 338–475; **36**, 143–192.

Jollie, M. (1962). "Chordate morphology." Reinhold, New York.

Jollie, M. (1968). The head skeleton of a new-born *Manis javanica* with comments on the onotogeny and phylogeny of the mammal head skeleton. *Acta zool., Stockh.* **49**, 228–303.

Jollie, M. (1971). Some developmental aspects of the head skeleton of the 35–37 mm *Squalus acanthias* foetus. *J. Morph.* **133**, 17–40.

Jollie, M. (1973). The origin of the chordates. *Acta zool., Stockh.* **54**, 81–100.

Jollie, M. (1977). Segmentation of the vertebrate head. *Amer. Zool.* **17**, 323–333.

Jollie, M. (1977a). The origin of the vertebrate brain. *Ann. N.Y. Acad. Sci.* **299**, 74–86.

Jordan, D. S. (1925). "Fishes." Appleton, New York.

Jurgens, J. D. (1963). Contributions to the descriptive and comparative anatomy of the cranium of the Cape Fruit-Bat *Rousettus aegyptiacus leachi* Smith. *Annale Univ. Stellenbosch* **38(A)**, 3–37.

Jurgens, J. D. (1973). The morphology of the nasal region of Amphibia and its bearing on the phylogeny of the group. *Annale Univ. Stellenbosch* **46**, 3–136.

Kälin, J. A. (1938). Die paarigen Extremitäten der Fische (Pterygia). *In* "Handbuch der vergleichenden Anatomie der Wirbeltiere" (Eds L. Bolk *et al.*), Vol. 5, 1–70. Urban and Schwarzenberg, Berlin and Wien.

Kallius, E. (1901). Beiträge zur Entwickelung der Zunge. I. Amphibien und Reptilien. *Arb. anat. Inst., Wiesbaden* (*Anatomische Hefte*) **16**, 533–748.

Kallius, E. (1905). Beiträge zur Entwickelung der Zunge. II. Vögel. *Arb. anat. Inst., Wiesbaden* (*Anatomische Hefte*) **28**, 311–579.

Kallius, E. (1910). Beiträge zur Entwickelung der Zunge. III. Säugetiere, *Sus scrofa domestica.*

Arb. anat. Inst., Wiesbaden (*Anatomische Hefte*) **41**, 173–337.

van de Kamer, J. C. (1965). Histological structure and cytology of the pineal complex in fishes, amphibians and reptiles. *In* "Progress in brain research" (Eds J. Kappers and J. P. Schadé), Vol. 10, 30–48. Elsevier, Amsterdam.

van Kampen, P. N. (1905). Die Tympanalgegend des Säugetierschädels. *Gegenbaurs morph. Jb.* **34**, 321–722.

Kappers, C. U. Ariëns (1907). Untersuchungen über das Gehirn der Ganoiden *Amia calva* und *Lepidosteus osseus*. *Abh. senckenb. naturforsch. Ges.* **30**, 447–500.

Kappers, J. Ariëns, (1965). Survey of the innervation of the epiphysis cerebri and the accessory pineal organs of vertebrates. *In* "Progress in brain research" (Eds J. Kappers and J. P. Schadé), Vol. 10, 87–153. Elsevier, Amsterdam.

Kappers, C. U. Ariëns, Huber, G. C. and Crosby, E. C. (1960). "The comparative anatomy of the nervous system of vertebrates, including man." Hafner, New York.

Karatajute-Talimaa, V. (1978). "Silurian and Devonian thelodonts of the USSR and Spitsbergen." Mokslas, Vilnius.

Keibel, F. (1902). Die Entwickelung der äusseren Körperform der Wirbeltierembryonen, insbesondere der menschlichen Embryonen aus den ersten 2 Monaten. *In* "Handbuch der vergleichenden und experimentellen Entwickelungslehre der Wirbeltiere" (Ed. O. Hertwig), Vol. 1 (2), 1–176. Fischer, 1906, Jena.

Kemp, A. (1977). The pattern of tooth plate formation in the Australian lungfish, *Neoceratodus forsteri* Krefft. *Zool. J. Linn. Soc.* **60**, 223–258.

Kermack, K. A. and Mussett, F. (1958). The jaw articulation of the Docodonta and the classification of Mesozoic mammals. *Proc. R. Soc.* (*B*)**149**, 204–15.

Kermack, K. A., Mussett, F. and Rigney, H. W. (1973). The lower jaw of *Morganucodon*. *Zool. J. Linn. Soc.* **53**, 87–175.

Kerr, J. G. (1907). The development of *Polypterus senegalus* Cuv. Budgett memorial volume, 195–284. Cambridge.

Kerr, J. G. (1932). Archaic fishes – *Lepidosiren, Protopterus, Polypterus* – and their bearing upon problems of vertebrate morphology. *Jena Z. Naturw.* **67**, 419–433.

Kerr, T. (1948). The pituitaries of *Amia, Lepidosteus* and *Acipenser*. *Proc. zool. Soc. Lond.* **118**, 973–983.

Kerr, T. (1952). The scales of primitive living actinopterygians. *Proc. zool. Soc. Lond.* **122**, 55–77.

Kerr, T. (1968). The pituitary in polypterines and its relationship to other fish pituitaries. *J. Morph.* **124**, 23–28.

Kesteven, H. L. (1931). The skull of *Neoceratodus forsteri*; A study in phylogeny. *Rec. Aust. Mus.* **18**, 235–265.

Kesteven, H. L. (1942–45). The evolution of the skull and the cephalic muscles. *Mem. Aust. Mus.*, Vol. **8**.

Kesteven, H. L. (1950). The origin of the tetrapods. *Proc. R. Soc. Vict.* **59**, 93–138.

Kiaer, J. (1924). The Downtonian fauna of Norway. 1. Anaspida. *Skr. VidenskSelsk. Christiania* **1(6)**, 1–139.

Kiaer, J. (1928). The structure of the mouth of the oldest known vertebrates, pteraspids and cephalaspids. *Palaeobiologica* **1**, 117–134.

Kiaer, J. (1932). The Downtonian and Devonian vertebrates of Spitsbergen. IV. Suborder Cyathaspida (a preliminary report edicted by A. Heintz). *Skr. Svalb. Ish.* **52**, 1–26.

Kiaer, J. and Heintz, A. (1935). The Downtonian and Devonian vertebrates of Spitsbergen. V. Suborder Cyathaspida, 1. *Skr. Svalb. Ish.* **40**, 1–139.

Killian, G. (1891). Zur Metamerie des Selachierkopfes. *Verh. anat. Ges., Jena* **5**, 85–107.

Kindahl, M. (1938). Zur Entwicklung der Exkretionsorgane von Dipnoern und Amphibien mit Anmerkungen bezüglich Ganoiden und Teleostier. *Acta zool., Stockh.* **19**, 1–190.

Kingsbury, B. F. (1897). The structure and morphology of the oblongata in fishes. *J. comp. Neurol.* **7**, 1–36.

Kingsbury, B. F. (1897a). The encephalic evaginations in ganoids. *J. comp. Neurol.* **7**, 37–44.

Kingsbury, B. F. (1920). The extent of the floor-plate of His and its significance. *J. comp. Neurol.* **32**, 113–133.

Kinsbury, B. F. (1922). The fundamental plan of the vertebrate brain. *J. comp. Neurol.* **34**, 461–484.

Kingsbury, B. F. (1924). The developmental significance of the notochord (chorda dorsalis). *Z. Morph. Anthrop.* **24**, 59–73.

Kingsbury, B. F. (1930). The developmental significance of the floor-plate of the brain and spinal cord. *J. comp. Neurol.* **50**, 177–201.

Kingsbury, B. F. and Adelmann, H. B. (1924). The morphological plan of the head. *Q. Jl microsc. Sci.* **68**, 239–285.

Kingsbury, B. F. and Reed, H. D. (1908). The columella auris in Amphibia. *J. Morph.* **20**, 549–627.

Kingsley, J. S. (1900). The ossicula auditus. *Tufts Coll. Stud.* **1**, 203–274.

Kingsley, J. S. and Thyng, F. W. (1904). The hypophysis in *Amblystoma*. *Tufts Coll. Stud.* **1**, 363–378.

van der Klaauw, C. J. (1922). Über die Entwickelung des Entotympanicums. *Tijdschr. ned. dierk. Vereen.* **18**, 135–176.

van der Klaauw, C. J. (1923). Die Skelettstückchen in der Sehne des Musculus stapedius und nahe dem Ursprung der Chorda tympani. *Z. Anat. EntwGesch.* **69**, 32–83.

van der Klaauw, C. J. (1931). The auditory bulla in some fossil mammals. *Bull. Am. Mus. nat. Hist.* **62**, 1–352.

Kleerekoper, H. (1972). The sense organs. *In* "The biology of lampreys" (Eds M. W. Hardisty and I. C. Potter), Vol. 2, 373–404. Academic Press, London.

Knouff, R. A. (1927). The origin of the cranial ganglia of *Rana*. *J. comp. Neurol.* **44**, 259–361.

Knouff, R. A. (1935). The developmental pattern of ectodermal placodes in *Rana pipiens*. *J. comp. Neurol.* **62**, 17–65.

Kölliker, R. A. (1860). "Ueber das Ende der Wirbelsäule der Ganoiden und einiger Teleostier." Leipzig.

Kopsch, F. (1947). "Rauber-Kopsch: Lehrbuch und Atlas der Anatomie des Menschen." I. Allgemeines – Skeletsystem – Muskelsystem. Thieme, Leipzig.

Krause, R. (1923). "Mikroskopische Anatomie der Wirbertiere in Einzeldarstellungen." Vol. 3, Amphibien; Vol. 4. Teleostier, Plagiostomen, Zyklostomen und Leptokardier, 451–906. de Gruyter, Berlin and Leipzig.

Kravetz, L. (1911). Entwickelung des Knorpelschädels von *Ceratodus*. *Byull. mosk. Obshch. Ispyt. Prir.* **1910**, 332–365.

Krefft, J. L. G. (1870). Description of a gigantic amphibian allied to the genus *Lepidosiren*, from the Vide-Bay district, Queensland (*Ceratodus forsteri*). *Proc. zool. Soc. Lond*, 221–224.

Kryzanovsky, S. G. (1927). Die Entwicklung der paarigen Flossen bei *Acipenser, Amia* und *Lepidosteus*. *Acta zool., Stockh.* **8**, 278–352.

Kryzanovsky, S. G. (1934). Die Atmungsorgane der Fischlarven (Teleostomi). *Zool. Jb. (Anat.)* **58**, 21–60.

Kryzanovsky, S. G. (1934a). Die Pseudobranchie (Morphologie und biologische Bedeutung). *Zool. Jb. (Anat.)* **58**, 171–238.

Kuhlenbeck, H. (1970). A note on the morphology of the hypophysis in the gymnophione

Schistomepum thomense. Okajimas Folia anat. jap. **46**, 307–319.
Kuhlenbeck, H. (1973). "The central nervous system of vertebrates", Vol. 3/2. Karger, Basel.
Kuhlenbeck, H. (1975). "The central nervous system of vertebrates." Vol. 4. Karger, Basel.
Kuhlenbeck, H., Malewitz, T. D. and Beasley, A. B. (1966). Further observations on the morphology of the forebrain in Gymnophiona, with reference to the topologic vertebrate forebrain pattern. *In* "Evolution of the forebrain" (Eds R. Hassler and H. Stephen), 9–19. Thieme, Stuttgart.
Kuhn-Schnyder, E. (1977). Die Geschichte des Lebens auf der Erde. *Mitt. naturf. Ges. Solothurn* **27**, 5–124.
von Kupffer, C. W. (1893). Die Entwicklung des Kopfes von *Acipenser sturio* an Medianschnitten untersucht. Studien zur vergleichenden Entwicklungsgeschichte des Kopfes der Kranioten Vol. 1, 1–95. Lehmann, München and Leipzig.
von Kupffer, C. W. (1894). Die Entwicklung des Kopfes von *Ammocoetes planeri*. Studien zur vergleichenden Entwicklungsgeschichte des Kopfes der Kranioten, Vol. 2, 1–79. Lehmann, München and Leipzig.
von Kupffer, C. W. (1900). Zur Kopfentwicklung von *Bdellostoma*. Studien zur vergleichenden Entwicklungsgeschichte des Kopfes der Kranioten, Vol. 4, 1–86. Lehmann, München and Leipzig.
von Kupffer, C. W. (1905). Die Morphogenie des Centralnervensystems. *In* "Handbuch der vergleichenden und experimentellen Entwickelungslehre der Wirbeltiere" (Ed. O. Hertwig) Vol. 2(3), 1–272. Fischer, 1906, Jena.
Lagios, M. D. (1968). Tetrapod-like organization of the pituitary gland of the polypteriformid fishes: *Calamoichthys calabaricus* and *Polypterus plasma*. *Gen. Comp. Endocrinol.* **11**, 300–315.
Lagios, M. D. (1970). The median eminence of the bowfin, *Amia calva* L. *Gen. Comp. Endocrinol.* **15**, 453–463.
Lagios, M. D. (1972). Evidence for the hypothalami-hypophysial portal vascular system in the coelacanth *Latimeria chalumnae* Smith. *Gen. Comp. Endocrinol.* **18**, 73–82.
Lagios, M. D. (1974). Granular epithelioid (juxtaglomerular) cell and renovascular morphology of the coelacanth *Latimeria chalumnae* Smith (Crossopterygii) compared with that of other fishes. *Gen. Comp. Endocrinol.,* **22**, 296–307.
Lamb, A. B. (1902). The development of the eye muscles in *Acanthias*. *Am. J. Anat.* **1**, 185–202.
Landacre, F. L. (1910). The origin of the cranial ganglia in *Ameiurus*. *J. comp. Neurol.* **20**, 309–411.
Landacre, F. L. (1912). The epibranchial placodes of *Lepidosteus osseus* and their relation to the cerebral ganglia. *J. comp. Neurol.* **22**, 1–69.
Landacre, F. L. (1916). The cerebral ganglia and early nerves of *Squalus acanthias*. *J. comp. Neurol.* **27**, 20–55.
Landacre, F. L. (1921). The fate of the neural crest in the head of the urodeles. *J. comp. Neurol.* **33**, 1–43.
Landacre, F. L. and Conger, A. C. (1913). The origin of the lateral line primordia in *Lepidosteus osseus*. *J. comp. Neurol.* **23**, 576–616.
Lankester, E. R. (1868–70). The Cephalaspidae. *In* "A monograph of the fishes of Old Red Sandstone of Britain" (Eds J. Powrie and E. R. Lankester), Vol. 1. Palaeontogr. Soc. (Monogr.), 1868, 21, 1–33; 1870, 23, 1–62.
Larsell, O. (1950). The nervus terminalis. *Ann. Otol. Rhinol. Lar.* **59**, 414–438.
Larsell, O. (1967). "The comparative anatomy and histology of the cerebellum from myxinoids through birds." The University of Minnesota Press, Minneapolis.
Larsen, L. O. and Rothwell, B. (1972). Adenohypophysis. *In* "The biology of lampreys" (Eds M. W. Hardisty and I. C. Potter), Vol. 2, 1–67. Academic Press, London.

Leach, W. J. (1951). The hypophysis of lampreys in relation to the nasal apparatus. *J. Morph.* **89**, 217–246.

Lebedkina, N. S. (1960). [Development of the parasphenoid of urodele Amphibia.] *Dokl. Akad. Nauk SSSR* **133**, 1476–1479. (In Russian.)

Lebedkina, N. S. (1964). [The development of the dermal bones of the basement of the skull in Urodela (Hynobiidae).] *Trudy zool. Inst. Leningr.* **33** 75–172. (In Russian.)

Le Danois, Y. (1961). Remarques sur les poissons orbiculates du sous-ordre Ostracioniformes. *Mèm. Mus. natn. Hist. nat., Paris. A, Zool.* **11**, 207–338.

Lehman, J.-P. (1947). Description de quelques exemplaires de *Cheirolepis canadensis* (Whiteaves). *K. svenska VetenskAkad. Handl.* **(3)24**, 5–40.

Lehman, J.-P. (1951). Un nouvel Amiidé de l'Eocène du Spitzberg, *Pseudamia heintzi. Tromsø Mus. Årsh.* **70(2)**, 3–11.

Lehman, J.-P. (1952). Etude complémentaire des poissons de l'Eotrias de Madagascar. *K. svenska VetenskAkad. Handl.* (**4**)**2**, 1–201.

Lehman, J.-P. (1955). Amphibiens (Amphibia Linné) Généralités. *In* "Traité de Paléontologie" (Ed. J. Piveteau), Vol. 5, 3–52. Masson, Paris.

Lehman, J.-P. (1956). L'evolution des Dipneustes et l'origine des Urodèles. *Colloques int. Cent. natn. Rech. scient.* **60**, 69–76.

Lehman, J.-P. (1959). Les Dipneustes du Dévonien supérieur du Groenland. *Meddr Grønland* **164**, 1–58.

Lehman, J.-P. (1959a). L'évolution des vertébrés inférieurs. Monographies Dunod, Vol. 22, 1–188. Dunod, Paris.

Lehman, J.-P. (1962). A propos de la double articulation de la cuirasse des Arthrodires. *Colloques int. Cent. natn. Rech. scient.* **104**, 63–68.

Lehman, J.-P. (1966). Actinopterygii, Dipnoi, Crossopterygii, Brachiopterygii. *In* "Traité de Paléontologie" (Ed. J. Piveteau), Vol. 4(3), 1–387, 398–420. Masson, Paris.

Lehman, J.-P. (1967). Quelques remarques concernant *Drepanaspis gemuendenensis* Schlüter. *In* "Fossil vertebrates" (Eds C. Patterson and P. H. Greenwood). J. Linn. Soc. (Zool.) Vol. 47, 39–43. Academic Press, London.

Lehman, J.-P. (1967a). Quelques réflexions a propos de l'evolution. *Annls Biol.* **6**, 537–544.

Lehman, J.-P. (1968). Remarques concernant la phylogénie des Amphibiens. *In* "Current problems of lower vertebrate phylogeny" (Ed. T. Ørvig) Nobel Symp, Vol. 4, 307–315. Almqvist and Wiksell, Stockholm.

Lehman, J.-P. (1969). Incertae sedis *Palaeospondylus gunni* Traquair. *In* "Traité de Paléontologie" (Ed. J. Piveteau), Vol. 4(2), 777–781. Masson, Paris.

Lehman, J.-P. (1973). Les preuves paléontologiques de l'évolution. Collection SUP, le biologiste, Vol. 1, 1–176. Presses universitaires de France, Paris.

Lehman, J.-P. (1973a). Un nouveau coccostéomorphe, *Belgiosteus mortelmansi. Annls Paléont. (Vert.)* **59**, 1–14.

Lehman, J.-P. (1975). Quelques réflexions sur la phylogénie des Vertébrés inférieurs. *Colloques int. Cent. natn. Rech. scient.* **218**, 257–264.

Lehman, J.-P. (1976). Nouveaux Poissons fossiles du Dévonien du Maroc. *Annls Paléont. (Vert.)* **62**, 1–34.

Lehman, J.-P. (1977). Sur la présence d'un Ostéolepiforme dans le Dévonien supérieur du Tafilalet. *C.r. hebd. Séanc. Acad. Sci., Paris* **285**, 151–153.

Lehman, J.-P. (1977a). Nouveaux Arthrodires du Tafilalet et de ses environs. *Annls Paléont. (Vert.)* **63**, 105–132.

Lehmann, W. M. (1956). *Dipnorhynchus lehmanni* Westoll, ein primitiver Lungenfish aus dem rheinischen Unterdevon. *Paläont. Z.* **30**, 21–25.

Lehmann, W. M. and Westoll, T. S. (1952). A primitive dipnoan fish from the Lower Devonian of Germany. *Proc. R. Soc. (B)* **140(900)**, 403–421.
Lehtola, K. A. (1973). Ordovician vertebrates from Ontario. *Contr. Mus. Paleont. Univ. Mich.* **24**, 23–30.
Lekander, B. (1949). The sensory line system and the canal bones in the head of some Ostariophysi. *Acta zool., Stockh.* **30**, 1–131.
Liem, K. F. and Woods, L. P. (1973). A probable homologue of the clavicle in the holostean fish *Amia calva. J. Zool.* **170**, 521–531.
Lillegraven, J. A. (1972). Ordinal and familial diversity of Cenozoic mammals. *Taxon* **21**, 261–274.
Lillegraven, J. A. (1976). Biological considerations of the marsupial-placental dichotomy. *Evolution, Lancaster, Pa.* **29**, 707–722.
Lindahl, P. E. (1944). Zur Kenntnis der Entwicklung von Haftorgan und Hypophyse bei *Lepidosteus. Acta zool., Stockh.* **25**, 97–133.
Lindahl, P. E. (1948). Über Die Entwicklung und Morphologie des Chondrocraniums von *Procavia capensis* Pall. *Acta zool., Stockh.* **29**, 281–376.
Lindahl, P. E. and Lundberg, M. (1946). On the arteries in the head of *Procavia capensis* Pall. and their development. *Acta zool., Stockh.* **27**, 101–153.
Lindström, T. (1949). On the cranial nerves of the cyclostomes with special reference to n. trigeminus. *Acta zool., Stockh.* **30**, 316–458.
Linnaeus, C. (1758). "Systema naturae per regna tria naturae, secundum classes, ordines, genera, species cum characteribus, differentiis, synonymis, locis." Editio decima, reformata, L, 1–824. Laurentii Salvii, Holmiae.
Liu, Y. H. (1965). New Devonian agnathans of Yunnan. *Vertebr. palasiat.* **9**, 130–140.
Liu, Y. H. (1973). [On new forms of Polybranchiaspiformes and Petalichthyida from the Devonian of South-west China.] *Vertebr. palasiat.* **11**, 132–143. (In Chinese.)
Liu, Y. H. (1975). Lower Devonian agnathans of Yunnan and Sichuan. *Vertebr. palasiat.* **13**, 201–216. (In Chinese, English summary.)
Llinás, R. (1969). (Ed.) "Neurobiology of cerebellar evolution and development." Proc. first int. Symposium. American Medical Association, Chicago.
Locy, W. A. (1895). Contribution to the structure and development of the vertebrate head. *J. Morph.* **11**, 495–594.
Lombard, R. E. and Bolt, J. R. (1979). Evolution of the tetrapod ear: an analysis and reinterpretation. *Biol. J. Linnean Soc. Lond.* **11**, 19–76.
Løvtrup, S. (1977). "The phyolgeny of vertebrata." John Wiley, London.
Low, A. (1909). Further observations of the ossification of the human lower jaw. *J. Anat. Physiol., Lond.* **44**, 83–95.
Lowenstein, O. (1970). The electrophysiological study of the responses of the isolated labyrinth of the lamprey (*Lampetra fluviatilis*) to angular acceleration, tilting and mechanical vibration. *Proc. R. Soc. (B)* **174**, 419–434.
Lowenstein, O. and Thornhill, R. A. (1970). The labyrinth of *Myxine*: anatomy, ultrastructure and electrophysiology. *Proc. R. Soc. (B)* **176**, 21–42.
Lubosch, W. (1905). Die Entwicklung und Metamorphose des Geruchsorganes von *Petromyzon* und seine Bedeutung für die vergleichende Anatomie des Geruchsorganes. *Z. Naturw.* **40**, 95–148.
Lund, R. (1977). *Echinochimaera meltoni*, new genus and species (Chimaeriformes) from the Mississippian of Montana. *Ann. Carneg. Mus.* **46**, 195–221.
Luther, A. (1909). Untersuchungen über die vom N. Trigeminus innervierte Muskulatur der Selachier (Haie und Rochen). *Acta Soc. Sci. fenn.* **36**, 1–168.

Luther, A. (1909a). Muskulatur und Skelett des Kopfes des Haies *Stegostoma tigrinum* Gm. und der Holocephalen. *Acta Soc. Sci. fenn.* **37**, 1–60.

Luther, A. (1913). Über die vom N. Trigeminus versorgte Muskulatur der Ganoiden und Dipneusten. *Acta Soc. Sci. fenn.* **41**, 1–72.

Luther, A. (1914). Über die vom N. Trigeminus versorgte Muskulatur der Amphibien. *Acta Soc. Sci. fenn.* **44**, 1–151.

Luther, A. (1938). Die Visceralmuskulatur der Acranier, Cyclostomen und Fische. *In* "Handbuch der vergleichenden Anatomie der Wirbeltiere" (Eds L. Bolk *et al.*), Vol. 5, 468–542. Urban and Schwarzenberg, Berlin and Wien.

Lyarskaya, L. A. (1977). [New data about *Asterolepis ornata* Eichwald from early Frasnien deposits of Balticum.] *In* "Papers on the phylogeny and systematics of fossil jawed and jawless fishes" (Ed. V. V. Menner), 36–45. Akad. Nauk, Moscow. (In Russian.)

McKibben, P. S. (1911). The nervus terminalis in urodele Amphibia. *J. comp. Neurol.* **21**, 261–298.

McMurrich, J. P. (1885). The cranial muscles of *Amia calva* (L), with a consideration of the relations of the post-occipital or hypoglossal nerves in the various vertebrate groups. *Stud. Biol. Lab., Johns Hopkins Univ.* **3**, 121–153.

Marcus, H. (1937). Lungen. *In* "Handbuch der vergleichenden Anatomie der Wirbeltiere" (Eds L. Bolk *et al.*), Vol. 3, 909–988. Urban and Schwarzenberg, Berlin and Wien.

Marinelli, W. (1936). Kranium und Visceralskelett. A. Allgemeine Probleme. *In* "Handbuch der vergleichenden Anatomie der Wirbeltiere" (Eds L. Bolk *et al.*), Vol. 4, 207–232. Urban and Schwarzenberg, Berlin and Wien.

Marinelli, W. and Strenger, A. (1954). "Vergleichende Anatomie und Morphologie der Wirbeltiere," Vol. I (L.) 1–80. F. Deuticke, Wien.

Marinelli, W. and Strenger, A. (1956). "Vergleichende Anatomie und Morphologie der Wirbeltiere", Vol. 2, 81–172. F. Deuticke, Wien.

Marinelli, W. and Strenger, A. (1959). "Vergleichende Anatomie und Morphologie der Wirbeltiere", Vol. 3, 173–308. F. Deuticke, Wien.

Marinelli, W. and Strenger, A. (1973). "Vergleichende Anatomie und Morphologie der Wirbeltiere", Vol. 4, 309–460. F. Deuticke, Wien.

Mark-Kurik, E. (1968). New finds of psammosteids (Heterostraci) in the Devonian of Estonia and Latvia. *Isvest. Akad. Nauk Est. SSR (Chim. Geol.)* **17**, 409–424.

Mark-Kurik, E. (1973). *Actinolepis* (Arthrodira) from the Middle Devonian of Estonia. *Palaeontographica (A)* **143**, 89–108.

Mark-Kurik, E. (1973a). *Kimaspis*, a new palaeacanthaspid from the early Devonian of central Asia. *Isvest. Akad. Nauk Est. SSR (Chim. Geol.)* **22**, 322–330.

Mark-Kurik, E. (1974). Discovery of new Devonian fish localities in the Soviet Arctic. *Isvest. Akad. Nauk Est. SSR (Chim. Geol.)* **23**, 332–334.

Mark-Kurik, E. (1977). [Structure of the shoulder girdle of early ptyctodontids.] *In* "Papers on the phylogeny and systematics of fossil jawed and jawless fishes" (Ed. V. V. Menner), 61–70. Akad. Nauk, Moscow. (In Russian.)

Marshall, A. M. (1879). The morphology of the vertebrate olfactory organ. *Q. Jl micros. Sci.* **19**, 300–340.

Märss, T. (1977). Structure of *Tolypelepis* from the Baltic Upper Silurian. *Isvest. Akad. Nauk Est. SSR (Chim.Geol.)* **26**, 57–68.

Matthes, E. (1934). Geruchsorgan. *In* "Handbuch der vergleichenden Anatomie der Wirbeltiere" (Eds L. Bolk *et al.*) Vol. 2, 879–948. Urban and Schwarzenberg, Berlin and Wien.

Matveiev, B. S. (1925). The structure of the embryonal skull of the lower fishes. *Byull mosk. Obshch. Ispyt. Prir.* **34**, 416–475. (In Russian, English summary.)

Matveive, B. S. (1932). Zur Theorie die Rekapitulation. Über die Evolution der Schuppen, Federn und Haare auf dem Wege embryonaler Veränderungen. *Zool. Jb (Anat.)* **55**, 555–580.

Matveiev, B. S. (1940). Origin of bony scales in fishes. Development of ganoid scales of Polypteri. *Dokl. Akad. Nauk SSSR.* **29**, 651–653.

Maurer, F. (1895). "Die Epidermis und ihre Abkömmlinge." Wilhelm Engelmann, Leipzig.

Maurer, F. (1904). Die Entwickelung des Muskelsystems und der elektrischen Organe. *In* "Handbuch der vergleichenden und experimentellen Entwickelungslehre der Wirbeltiere" (Ed. O. Hertwig), Vol. 3(1), 1–80. Fischer, 1906, Jena.

Maurer, F. (1912). Die ventrale Rumpfmuskulatur der Fische (Selachier, Ganoiden, Teleostier, Crossopterygier, Dipnoer). *Jena Z. Naturw.* **49**, 1–118.

Mayer, P. (1885). Die unpaaren Flossen der Selachier. *Mitt. zool. Stn Neapel* **6**, 217–285.

Meader, R. G. (1939). The forebrain of bony fishes. *Proc. K. ned. Akad. Wet.* **42**, 657–670.

Medvedeva, I. M. (1961). [Some data on the early development of lateral lines in the head of Hynobiidae.] *Dokl. Akad. Nauk SSSR*, **139**, 748–751 (In Russian.)

Medvedeva, I. M. (1964). [The development, origin and homology of the choanae in Amphibia.] *Trudy zool. Inst. Leningr.* **33**, 173–211 (In Russian.)

Medvedeva, I. M. (1975). Olfactory organ in amphibian and its phylogenetic significance. *Trudy zool. Inst., Leningr.* **58**, 1–174 (In Russian, English summary.)

de Meijere, J. C. H. (1931). Haare. *In* "Handbuch der vergleichenden Anatomie der Wirbeltiere" (Eds L. Bolk *et al.*), Vol. 1, 585–632. Urban and Schwarzenberg, Berlin and Wien.

Meurling, P. (1967). The vascularization of the pituitary in elasmobranchs. *Sarsia* **28**, 1–104.

Milaire, J. (1957). Contribution à la connaissance morphologique et cytochimique des bourgeons de membres chez quelques Reptiles. *Archs Biol., Paris* **68**, 429–512.

Milaire, J. (1962). Histochemical aspects of limb morphogenesis in Vertebrates. *In* "Advances in morphogenesis" (Eds M. Abercrombie and J. Brachet), Vol. 2, 183–209. Academic Press, London.

Miles, R. S. (1964). A reinterpretation of the visceral skeleton of *Acanthodes*. *Nature, Lond.* **204**, 457–459.

Miles, R. S. (1965). Some features of the cranial morphology of acanthodians and the relationships of the Acanthodii. *Acta zool., Stockh.* **46**, 233–255.

Miles, R. S. (1965a). Ventral thoracic neuromast lines of placoderm fishes. *Nature, Lond.* **206**, 524–525.

Miles, R. S. (1966). The acanthodian fishes of the Devonian Plattenkalk of the Paffrath Trough in the Rhineland, with an appendix containing a classification of the Acanthodii and a revision of the genus *Homalacanthus*. *Ark. Zool.* **2**(**18**), 147–194.

Miles, R. S. (1966a). *Protitanichthys* and some other coccosteomorph arthrodires from the Devonian of North America. *K. svenska VetenskAkad. Handl.* (**4**)**10**, 1–49.

Miles, R. S. (1967). Observations on the ptyctodont fish, *Rhamphodopsis* Watson. *In* "Fossil vertebrates" (Eds C. Patterson and P. H. Greenwood). J. Linn. Soc. (Zool.), Vol. 47, 99–120. Academic Press, London.

Miles, R. S. (1967a). The cervical joint and some aspects of the origin of the Placodermi. *Colloques int. Cent. natn. Rech. scient.* **163**, 49–71.

Miles, R. S. (1968). Jaw articulation and suspension in *Acanthodes* and their significance. *In* "Current problems of lower vertebrate phylogeny" (Ed. T. Ørvig). Nobel Symp., Vol. 4, 109–27. Almqvist and Wiksell, Stockholm.

Miles, R. S. (1968a). The Old Red Sandstone antiarchs of Scotland: Family Bothriolepididae. *Palaeontogr. Soc. (Monogr.)* **122**, 1–30.
Miles, R. S. (1969). Features of placoderm diversification and the evolution of the arthrodire feeding mechanism. *Trans. R. Soc. Edinb.* **68**, 123–170.
Miles, R. S. (1970). Remarks on the vertebral column and caudal fin of acanthodian fishes. *Lethaia* **3**, 343–362.
Miles, R. S. (1971). *In* J. A. Moy-Thomas and R. S. Miles "Palaeozoic fishes", 2nd edn, Chapman and Hall, London.
Miles, R. S. (1971a). The Holonematidae (placoderm fishes), a review based on new specimens of *Holonema* from the Upper Devonian of Western Australia. *Phil. Trans. R. Soc.* **263**, 101–234.
Miles, R. S. (1973). Relationships of Acanthodians. *In* "Interrelationships of fishes" (Eds P. H. Greenwood *et al.*), 63–103. Academic Press, London.
Miles, R. S. (1973a). Articulated acanthodian fishes from the Old Red Sandstone of England, with a review of the structure and evolution of the acanthodian shoulder-girdle. *Bull. Br. Mus. nat. Hist. (Geol.)* **24**, 111–213.
Miles, R. S. (1973b). An actinolepid arthrodire from the Lower Devonian Peel Sound formation, Prince of Wales Island. *Palaeontographica (A)* **143**, 109–118.
Miles, R. S. (1975). The relationships of the Dipnoi. *Colloques int. Cent. natn. Rech. scient.* **218**, 133–48.
Miles, R. S. (1977). Dipnoan (lungfish) skulls and the relationships of the group: a study based on new species from the Devonian of Australia. *J. Linn. Soc. (Zool.)* **61**, 1–328.
Miles, R. S. and Westoll, T. S. (1968). The placoderm fish *Coccosteus cuspidatus* Miller ex Agassiz from the Middle Old Red Sandstone of Scotland. Part I. Descriptive morphology. *Trans. R. Soc. Edinb.* **67**, 373–476.
Miles, R. S. and Young, G. C. (1977). Placoderm interrelationships reconsidered in the light of new ptyctodontids from Gogo, Western Australia. *In* "Problems in vertebrate evolution" (Eds S. M. Andrews *et al.*). Linn. Soc. Symp. Ser., Vol. 4, 124–196. Academic Press, London.
Miller, H. (1849). "Foot-prints of the Creator; or the *Asterolepis* of Stromness." London and Edinburgh.
Millot, J. (1954). Le troisième Coelacanthe. *Naturaliste malgache, Suppl.* **1**, 1–26.
Millot, J. (1955). Unité spécifique des Coelacanthes actuels. *Nature, Paris* **3238**, 58–59.
Millot, J. and Anthony, J. (1954). Tubes rostraux et tubes nasaux de *Latimeria* (Coelacanthidae). *C.r. hebd. Séanc. Acad. Sci., Paris*, **239**, 1241–1243.
Millot, J. and Anthony, J. (1956). Considerations preliminaires sur la squelette axial et les systeme nerveux central de *Latimeria chalumnae* Smith. *Mém. Inst. scient. Madagascar (A)* **11**, 167–188.
Millot, J. and Anthony, J. (1956a). L'organe rostral de *Latimeria*. *Annls Sci. nat. (Zool.)* **11**, 381–389.
Millot, J. and Anthony, J. (1958). "Anatomie de *Latimeria chalumnae*", Vol. 1. Cent. natn. Rech. scient., Paris.
Millot, J. and Anthony, J. (1960). Un nouvel aspect du Coelacanthe: Le montage complet de son squelette. *Sci. Nat., Paris* **37**, 1–3.
Millot, J. and Anthony, J. (1962). Premières précisions sur l'organisation télencéphale chez *Latimeria chalumnae*. *C. r. hebd. Séanc. Acad. Sci. Paris* **254**, 2067–2068.
Millot, J. and Anthony, J. (1965). "Anatomie de *Latimeria chalumnae*", Vol. 2. Cent. natn. Rech. scient., Paris.
Millot, J. and Anthony, J. (1973). L'appareil excréteur de *Latimeria chalumnae* Smith. *Annls Sci. nat. (Zool.)* **15**, 293–328.

Millot, J. and Anthony, J. (1973a). La position ventrale du rein de *Latimeria chalumnae*. *C. r. hebd. Séanc. Acad. Sci., Paris* **276**, 2171–2173.

Millot, J. and Anthony, J. (1974). Les oeufs du Coelacanthe. *Sci. Nat., Paris* **121**, 3–4.

Millot, J., Anthony, J. and Robineau, D. (1972). Etat commenté des captures de *Latimeria chalumnae* Smith (Poisson, Crossoptérygien, Coelacanthidé) effectuées jusqu áu mois d'octobre 1971. *Bull. Mus. natn. Hist. nat., Paris* **53**, 533–548.

Millot, J., Anthony, J. and Robineau, D. (1978). "Anatomie de *Latimeria chalumnae*", Vol. 3 Cent. natn. Rech. scient., Paris.

Miner, R. W. (1925). The pectoral limb of *Eryops* and other primitive tetrapods. *Bull. Am. Mus. nat. Hist.* **51**, 145–312.

Monath, T. (1965). The opercular apparatus of salamanders. *J. morph.* **116**, 149–170.

Mookerjee, H. K. (1930). On the development of the vertebral column of Urodela. *Phil. Trans. R. Soc(B)* **218**, 415–444.

Morescalchi, A. (1973). Amphibia. *In* "Cytotaxonomy and vertebrate evolution" (Eds A. B. Chiarelli and E. Capanna), 233–348. Academic Press, London.

Moroff, T. (1904). Über die Entwicklung der Kiemen bei Fischen. *Arch. mikrosk. Anat. EntwMech.* **64**, 189–213.

Moss, M. L. (1968). The origin of vertebrate calcified tissues. *In* "Current problems of lower vertebrate phylogeny" (Ed. T. Ørvig). Vol. 4, 359–371. Almqvist and Wiksell, Stockholm.

Moss, M. L. (1972). The vertebrate dermis and the integumental skeleton. *Am. Zool.* **12**, 27–34.

Moulton, J. M. (1974). A description of the vertebral column of *Eryops* based on the notes and drawings of A. S. Romer. *Breviora* **428**, 1–44.

Moy-Thomas, J. A. (1934). The structure and affinities of *Tarrasius problematicus* Traquair. *Proc. zool. Soc. Lond. (B)*, 367–376.

Moy-Thomas, J. A. (1935). Notes on the types of fossil fishes in the Leeds City Museum. II. Acanthodii, Dipnoi, and Crossopterygii. *Proc. Leeds phil. lit. Soc.* (Sci. Sect.), **3**, 111–116.

Moy-Thomas, J. A. (1935a). The coelacanth fishes from Madagascar. *Geol. Mag.* **72**, 213–227.

Moy-Thomas, J. A. (1936). The evolution of the pectoral fins of fishes and the tetrapod fore-limb. *Sch. Sci. Rev.* **36**, 592–599.

Moy-Thomas, J. A. (1936a). The structure and affinities of the fossil elasmobranch fishes from the Lower Carboniferous of Glencartholm, Eskdale. *Proc. zool. Soc. Lond.* (**B**), 762–788.

Moy-Thomas, J. A. (1937). The Carboniferous coelacanth fishes of Great Britain and Ireland. *Proc. zool. Soc. Lond. (B)*, 383–415.

Moy-Thomas, J. A. (1938). The problem of the evolution of the dermal bones in fishes. *In* "Evolution, essays on aspects of evolutionary biology" (Ed. G. R. de Beer), 305–319. Clarendon Press, Oxford.

Moy-Thomas, J. A. (1939). "Palaeozoic fishes." Methuens monographs on biological subjects. Methuen, London.

Moy-Thomas, J. A. (1939a). The early evolution and relationships of the elasmobranchs. *Biol. Rev.* **14**, 1–26.

Moy-Thomas, J. A. (1940). The Devonian fish *Palaeospondylus gunni* Traquair. *Phil. Trans. R. Soc. (B)* **230**, 391–413.

Moy-Thomas, J. A. and Miles, R. S. (1971). "Palaeozoic fishes", 2nd ed. Chapman and Hall, London.

Moy-Thomas, J. A. and Westoll, T. S. (1935). On the Permian coelacanth, *Coelacanthus granulatus*, Ag. *Geol. Mag.*, **72**, 446–457.

Moy-Thomas, J. A. and White, E. I. (1939). On the palatoquadrate and hyomandibula of *Pleuracanthus sessilis* Jordan. *Geol. Mag.* **76**, 459–463.

Müller, E. (1909). Die Brustflosse der Selachier. *Arb. anat. Inst., Wiesbaden* (*Anatomische Heft*) **39**, 469–601.

Müller, E. (1911). Untersuchungen über die Muskeln und Nerven der Brustflosse und der Körperwand bei *Acanthias vulgaris. Arb. anat. Inst., Wiesbaden* (*Anatomische Hefte*) **43**, 1–147.

Müller, E. 2nd edn (1922). "Lärobok i ryggradsdjurens jämförande anatomi", Albert Bonnier, Stockholm. (In Swedish.)

Müller, F. (1968). Zur Phylogenese des sekundären Kiefergelenks. *Revue suisse Zool.* **75**, 373–414.

Müller, F. (1969). Zur Phylogenese des sekundären Kiefergelenks: Zeugniswert diarthognather Fossilien im Lichte neuer ontogenetischer Befunde. *Revue suisse Zool.* **76**, 710–715.

Müller, J. (1834). Vergleichende Anatomie der Myxinoiden, der Cyclostomen mit durchbohrtem Gaumen. *Abh. preuss. Akad. Wiss.* **1834**, 65–340.

Müller, J. (1846). "Über den Bau und die Grenzen der Ganoiden und über das natürliche System der Fische" k.Akad. Wiss, Berlin.

Munk, O. (1964). The eye of *Calamoichthys calabaricus* Smith, 1865 (Polypteridae, Pisces) compared with the eye of other fishes. *Vidensk. Meddr dansk naturh. Foren.* **127**, 113–126.

Munk, O. (1968). The eyes of *Amia* and *Lepisosteus* (Pisces, Holostei) compared with the brachiopterygian and teleostean eyes. *Vidensk. Meddr dansk naturh. Foren.* **131**, 109–127.

Nauck, E. T. (1931). Über umwegige Entwicklung. Untersuchungen über eine ontogenetische Entwicklungsweise und ihre Beziehungen zur Phylogenese. *Gegenbaurs morph. Jb.* **66**, 65–195.

Nauck, E. T. (1938). Extremitätenskelett der Tetrapoden. *In* "Handbuch der vergleichenden Anatomie der Wirbeltiere" (Eds L. Bolk *et al.*), Vol. 5, 71–248. Urban and Schwarzenberg, Berlin and Wien.

Neal, H. V. (1898). The segmentation of the nervous system in *Squalus acanthias. Bull. Mus. comp. Zool. Harv.* **31**, 147–294.

Neal, H. V. (1918). Neuromeres and metameres. *J. Morph.* **31**, 293–315.

Nelsen, O. E. (1953) "Comparative embryology of the vertebrates". Blakiston, New York.

Nelson, G. J. (1968). Gill-arch structure in *Acanthodes. In* "Current problems of lower vetebrate phylogeny" (Ed. T. Ørvig). Nobel Symp., Vol. 4, 129–143. Almqvist and Wiksell, Stockholm.

Nelson, G. J. (1969). Gill arches and the phylogeny of fishes, with notes on the classification of vertebrates. *Bull. Am. Mus. nat. Hist.* **141**, 475–552.

Nelson, G. J. (1970). Subcephalic muscles and intracranial joints of sarcopterygian and other fishes. *Copeia* **1970**, 468–471.

Nelson, G. J. (1972). Cephalic sensory canals, pitlines, and the classification of esocoid fishes, with notes on galaxiids and other teleosts. *Am. Mus. Novit.* **2492**, 1–49.

Nelson, G. J. (1973). Relationships of clupeomorphs, with remarks on the structure of the lower jaw in fishes. *In* "Interrelationships of fishes" (Eds P. H. Greenwood *et al.*), 333–349. Academic Press, London.

Netsky, M. G. and Shuangshoti, O. C. (1970). Studies on the choroid plexus. *In* "Neurosciences research" (Eds S. Ehrenpreis and O. C. Solnitzky), Vol. 3, 131–171. Academic Press, London.

Neumayer, L. (1906). Histogenese und Morphogenese des peripheren Nervensystems, der Spinalganglien und des Nervus Sympathicus. *In* "Handbuch der vergleichenden und experimentellen Entwickelungslehre der Wirbeltiere" (Ed. O. Hertwig), Vol. 2(3), 513–626. Fischer, 1906, Jena.

Neumayer, L. (1932). Studie über die Entwicklung des Kopfes von *Acipenser*. *Acta zool., Stockh.* **13**, 305–404.
Neumayer, L. (1938). Die Entwicklung des Kopfskelettes von *Bdellostoma* St. L. *Archo ital. Anat. Embriol.* **40**, 1–222.
Nevo, A. (1956). Fossil frogs from a Lower Cretaceous bed in Southern Israel (Central Negev). *Nature, Lond.* **178**, 1191–1192.
Newberry, J. S. (1873). Description of fishes of the Devonian system. *Rep. Invest. Div. geol. Surv. Ohio.* **1**(**2**), 290–324.
Newberry, J. S. (1889). The Paleozoic fishes of North America. *Monogr. U.S. geol. Surv.* **16**, 1–340.
Nielsen, E. (1932). Permo-Carboniferous fishes from East Greenland. *Meddr Grønland* **86**, 1–63.
Nielsen, E., (1936). Some few preliminary remarks on Triassic fishes from East Greenland. *Meddr Grønland* **112**, 1–55.
Nielsen, E. (1942). Studies on Triassic fishes from East Greenland. 1. *Glaucolepis* and *Boreosomus*. *Meddr Grønland* **138**, (*Palaeozoologica Groenlandica*, **1**), 1–403.
Nielsen, E. (1949). Studies on Triassic fishes from East Greenland. 2. *Australosomus* and *Birgeria*. *Meddr Grønland* **146** (*Palaeozoologica Groenlandica*, **3**), 1–309.
Nielsen, E. (1955). *Tupilakosaurus*. *In* "Traité de Paléontologie" (Ed. J. Piveteau), Vol. 5, 224–226. Masson, Paris.
Nieuwenhuys, R. (1963). The comparative anatomy of the actinopterygian forebrain. *J. Hirnforsch.* **6**, 171–192.
Nieuwenhuys, R. (1965). The forebrain of the crossopterygian *Latimeria chalumnae* Smith. *J. Morph.* **177**, 1–24.
Nieuwenhuys, R. (1966). The interpretation of the cell masses in the teleostean forebrain. *In* "Evolution of the forebrain" (Eds R. Hassler and H. Stephen), 32–39. Thieme, Stuttgart.
Nieuwenhuys, R. (1967). Comparative anatomy of the cerebellum. *In* "Progress in brain research" (Eds C. A. Fox and R. S. Sneider), Vol. 25, 1–93. Elsevier, Amsterdam.
Nieuwenhuys R., Bauchot, R. and Arnoult, J. (1969). Le développement du télencéphale d'un poisson osseux primitif, *Polypterus senegalus* Cuvier. *Acta zool., Stockh.* **50**, 101–125.
Nieuwenhuys, R. and Bodenheimer, T. S. (1966). The diencephalon of the primitive bony fish *Polypterus* in the light of the problem of homology. *J. Morph.* **188**, 415–450.
Nieuwenhuys, R. and Hickey, M. (1965). A survey of the forebrain of the Australian lungfish *Neoceratodus forsteri*. *J. Hirnforsch* **7**, 433–452.
Nieuwkoop, P. D. and Sutasurya, J. A. (1976). Embryological evidence for a possible polyphyletic origin of the Recent amphibians. *J. Embryol. exp. Morph.* **35**, 159–167.
Nilsson, T. (1943). Über einige postkraniale Skelettreste der triassischen Stegocephalen Spitzbergens. *Bull. geol. Instn Univ. Upsala* **30**, 227–272.
Nilsson, T. (1944). On the morphology of the lower jaw of Stegocephalia. With special reference to Eotriassic stegocephalians from Spitsbergen. II. General part. *K. svenska VetenskAkad. Handl.* (**3**)**21**, 1–70.
Nishi, S. (1938). Muskeln des Rumpfes. Muskeln des Kopfes. Parietale Muskulatur. *In* "Handbuch der vergleichenden Anatomie der Wirbeltiere" (Eds L. Bolk *et al.*), Vol. 5, 351–466. Urban and Schwarzenberg, Berlin and Wien.
Noble, G. K. (1931). "The biology of the Amphibia". McGraw-Hill, New York. (Dover reprint, 1954.)
Norman, J. R. and Greenwood, P. H. 2nd edn (1963). "A history of fishes." Ernest Benn, London.
Norris, H. W. (1925). Observation upon the peripheral distribution of the cranial nerves of

certain ganoid fishes (*Amia, Lepidosteus, Polyodon, Scaphirhynchus* and *Acipenser*). *J. comp. Neurol.* **39**, 345–432.
Norris, H. W. and Hughes, S. P. (1920). The cranial occipital and anterior spinal nerves of the dogfish, *Squalus acanthias*. *J. comp. Neurol.* **31**, 294–392.
Norris, H. W. and Hughes, S. P. (1920a). The spiracular sense-organ in elasmobranches, ganoids and dipnoans. *Anat. Rec.* **18**, 205–209.
Novitskaya, L. (1971). Les Amphiaspides (Heterostraci) du Dévonien de la Sibérie. Cahiers de Paléont., C.N.R.S., Paris, 7–127.
Novitskaya, L. (1973). *Liliaspis*—ein Poraspide aus dem unterdevons des Urals und einige Bemerkungen über die Phylogenie der Poraspiden (Agnatha). *Palaeontographica (A)* **143**, 25–34.
Novitskaya, L. (1975). Sur la structure interne et les liens phylogénétiques des Hétérostraci. *Colloques int. Cent. natn. Rech. scient.* **218**, 31–40.
Nursall, J. R. (1962). On the origins of the major groups of animals. *Evolution, Lancaster Pa* **16**, 118–123.
Nybelin, O. (1956). Les canaux sensoriels du museau chez *Elops saurus* (L.). *Ark. Zool.* (**2**)**10**, 453–458.
Nybelin, O. (1963). Zur Morphologie und Terminologie des Schwanzskelettes der Actinopterygier. *Ark. Zool.* (**2**)**15**, 485–516.
Nybelin, O. (1967). Notes on the reduction of the sensory canal system and of the canal-bearing bones in the snout of higher actinopterygian fishes. *Ark. Zool.* (**2**)**19**, 235–246.
Nybelin, O. (1971). On the caudal skeleton in *Elops* with remarks on other teleostean fishes. *Acta R. Soc. Sci. Litt., Zool., Gothoburg.* **7**, 5–52.
Nybelin, O. (1974). A revision of the leptolepid fishes. *Acta R. Soc. Sci. Litt., Zool., Gothoburg.* **9**, 1–202.
Nybelin, O. (1976). On the so-called postspiracular bones in crossopterygians, brachiopterygians and actinopterygians. *Acta R. Soc. Sci. Litt., Zool., Gothoburg.* **10**, 5–31.
Nybelin, O. (1977). The polyural skeleton of *Lepisosteus* and certain other actinopterygians. *Zool. Scr.* **6**, 233–244.
Nybelin, O. (1979). Contributions to taxonomy and morphology of the genus *Elops* (Pisces, Teleostei). *Acta R. Soc. Sci. Litt., Zool., Gothoburg.* **12**, 1–37.
Obruchev, D. (1943). A new restoration of *Drepanaspis*. *Dokl. Akad. Nauk SSSR* **41**, 268–271.
Obruchev, D. (1944). An attempted restoration of *Psammolepis paradoxa*. *Dokl. Akad. Nauk SSSR* **42**, 143–145.
Obruchev, D. (1964). Branch Agnatha. Branch Gnathostomi: Class Placodermi. *In* "Fundamentals of Paleontology" (Ed. I. A. Orlov), Vol. **11**, 36–259. (English translation, Jerusalem 1967.)
Obruchev, D. and Mark-Kurik, E. (1965). Devonian psammosteids (Agnatha, Psammosteidae) of the USSR. *Eesti NSV Tead. Akad. Toim.*, 1–304. (In Russian, English summary.)
Obruchev, D. and Mark-Kurik, E. (1968). On the evolution of the Psammosteids (Heterostraci). *Eesti NSV Tead. Akad. Toim.* **17**, 279–284.
O'Donoghue, C. H. and Abbott, E. (1928). The blood vascular system of the spiny dogfish, *Squalus acanthias* Linné, and *Squalus sucklii* Gill. *Trans. R. Soc. Edinb.*, **55**, 823–890.
Oksche, A. (1965). Survey of the development and comparative morphology of the pineal organ. *In* "Progress in brain research" (Eds J. Kappers, J. Ariëns and J. P. Schadé), Vol. 10, 3–29. Elsevier, Amsterdam.
Olson, E. C. (1966). The middle ear. Morphological types in amphibians and reptiles. *Am. Zool.* **6**, 399–419.
Olson, E. C. (1971). "Vertebrate paleozoology." Wiley-interscience, New York.

Olsson, R. (1958). Studies on the subcommissural organ. *Acta zool., Stockh.* **39**, 71–102.
Olsson, R. (1965). Comparative morphology and physiology of the *Oikopleura* notochord. *Israel J. Zool.* **14**, 213–220.
Olsson, R. (1969). Phylogeny of the ventricle system. *In* "Cirkumventrikuläre Organe und Liquor" (Ed. G. Sterba), 291–305. Gustav Fischer, Jena.
Olsson, R. (1971). Kordatzoologi. *In* "Biologi" (Eds E. Dahl and B. Noren), Vol. 5, 1–296. Almqvist and Wiksell, Stockholm. (In Swedish.)
Orlov, J. A. (Ed.) (1964). "Osnovy Paleontologii", Vol. 11. Moscow. "Fundamentals of Paleontology." English translation, Jerusalem 1967.
Ørvig, T. (1951). Histologic studies of placoderms and fossil elasmobranchs. 1. The endoskeleton, with remarks on the hard tissues of lower vertebrates in general. *Ark. Zool.* **2(2)**, 321–454.
Ørvig, T. (1957). Remarks on the vertebrate fauna of the Lower Upper Devonian of Escuminac Bay, P.Q. Canada, with special reference to the porolepiform crossopterygians. *Ark. Zool.* **10(6)** 367–426.
Ørvig, T. (1957a). Notes on some Paleozoic lower vertebrates from Spitsbergen and North America. *Norsk geol. Tidsskr.* **37**, 285–353.
Ørvig, T. (1958). *Pycnaspis splendens,* new genus, new species, a new ostracoderm from the Upper Ordovician of North America. *Proc. U.S. natn. Mus.* **108**, 1–23.
Ørvig, T. (1960). New finds of acanthodians, arthrodires, crossopterygians, ganoids and dipnoans in the Upper Middle Devonian calcareous flags (Oberer Plattenkalk) of the Bergisch Gladbach-Paffrath Trough. 1. *Paläont. Z.* **34**, 295–335.
Ørvig, T. (1961). New finds of acanthodians, etc. 2. *Paläont. Z.* **35**, 10–27.
Ørvig, T. (1961a). Notes on some early representatives of the Drepanaspida (Pteraspidomorphi, Heterostraci). *Ark. Zool.* **12(33)**, 515–535.
Ørvig, T. (1962). Y a-t-il une relation directe entre les arthrodires ptyctodontides et les holocéphales? *Colloques int. Cent. natn. Rech. scient.* **104**, 49–61.
Ørvig, (1967). Phylogeny of tooth tissues: evolution of some calcified tissues in early vertebrates. *In* "Structural and chemical organization of teeth," (Ed. A. E. W. Miles), Vol. 1, 45–110. Academic Press, New York and London.
Ørvig, T. (1967a). Some new acanthodian material from the Lower Devonian of Europe. *In* "Fossil vertebrates" (Eds C. Patterson and P. H. Greenwood). J. Linn. Soc. (Zool.), Vol. 47, 131–153. Academic Press, London.
Ørvig, T. (1968). The dermal skeleton; general considerations. *In* "Current problems of lower vertebrate phylogeny" (Ed. T. Ørvig). Nobel Symp., Vol. 4, 373–397. Almqvist and Wiksell, Stockholm.
Ørvig, T. (1969). Thelodont scales from the Grey Hoek Formation of Andrée Land, Spitsbergen. *Norsk geol. Tidsskr.* **49**, 387–401.
Ørvig, T. (1969a). Cosmine and cosmine growth. *Lethaia* **2**, 219–239.
Ørvig, T. (1969b). A new brachythoracid arthrodire from the Devonian of Dickson Land, Vestspitsbergen. *Lethaia* **2(3)**, 261–271.
Ørvig T. (1969c). Vertebrates from the Wood Bay group and the position of the Emsian-Eifelian boundary in the Devonian of Vestspitsbergen. *Lethaia* **2(4)**, 273–319.
Ørvig, T. (1971). Comments on the lateral line system of some brachythoracid and ptyctodontid arthrodires. *Zool. Scr.* **1**, 5–35.
Ørvig, T. (1972). The latero-sensory component of the dermal skeleton and its phyletic significance. *Zool. Scr.* **1**, 139–155.
Ørvig, T. (1973). Acanthodian dentition and its bearing on the relationships of the group. *Palaeontographica (A)* **143**, 119–150.

Ørvig, T. (1975). Description, with special reference to the dermal skeleton, of a new radotinid arthrodire from the Gedinnian of Arctic Canada. *Colloques int. Cent. natn. Rech. scient.* **218**, 41–71.

Ørvig, T. (1976). Palaeohistological notes. 3. The interpretation of pleromin (pleromic hard tissue) in the dermal skeleton of psammosteid heterostracans. *Zool. Scr.* **5**, 35–47.

Ørvig, T. (1976a). Palaeohistological notes. 4. The interpretation of osteodentine, with remarks on the dentition in the Devonian dipnoan *Griphognathus*. *Zool. Scr.* **5**, 79–96.

Ørvig, T. (1977). A survey of odontodes ("dermal teeth") from developmental, structural, functional, and phyletic points of view. *In* "Problems in vertebrate evolution" (Eds S. M. Andrews *et al.*). Linn. Soc. Symp. Ser. Vol.4, 53–75. Academic Press, London.

Ørvig, T. (1978). Microstructure and growth of the dermal skeleton in fossil actinopterygian fishes: *Nephrotus* and *Colobodus*, with remarks on the dentition in other forms. *Zool. Scr.* **7**, 297–326.

Ossian, C. R. and Halseth, M. A. (1976). Discovery of Ordovician vertebrates in the Arbuckle Mountains of Oklahoma. *J. Paleont.* **50**, 773–777.

Özeti, N. and Wake, D. B. (1969). The morphology and evolution of the tongue and associated structures in salamanders and newts (Family Salamandridae). *Copeia* **1969**(**1**), 91–123.

Padget, D. H. (1948). The development of the cranial arteries in the human embryo. *Contr. Embryol.* **32**, 207–260.

Pageau, Y. (1969). Nouvelle faune ichthyologique du Devonien moyen dans les Grès de Gaspé (Québec). II. Morphologie et systématique. *Naturaliste can.* **96**, 399–478, 805–889.

P'an, K., Wang, S. and Liu, Y. (1975). The Lower Devonian Agnatha and Pisce from South China. *Prof. Pap. Stratigr. Palaeont.*, **1**, 135–167. Geological Press, Peking. (In Chinese, English title.)

P'an K. and Wang, S. (1978). [Devonian Agnatha and Pisces of South China.] *In* Symposium on the Devonian System of South China", 1974, 298–333. Geological Press, Peking (In Chinese.)

Pander, C. H. (1858). Über die Ctenodipterinen des devonischen Systems." St Petersburg.

Panchen, A. L. (1972). The interrelationships of the earliest tetrapods. *In* "Studies in vertebrate evolution" (Eds K. A. Joysey and T. S. Kemp), 65–87. Oliver and Boyd, Edinburgh.

Panchen, A. L. (1973). On *Crassygirinus scoticus* Watson, a primitive amphibian from the Lower Carboniferous of Scotland. *Palaeontology* **16**, 179–193.

Panchen, A. L. (1977). The origin and early evolution of tetrapod vertebrae. *In* "Problems in vertebrate evolution" (Eds S. M. Andrews *et al.*). Linn. Soc. Symp. Ser., Vol. 4, 289–318. Academic Press, London.

Panchen, A. L. (1977a). On *Anthracosaurus russelli* Huxley (Amphibia: Labyrinthodontia) and the family Anthracosauridae. *Phil. Trans. R. Soc. (B)* **279**, 447–512.

Parker, T. J. (1886). On the blood-vessels of *Mustelus antarcticus*; a contribution to the morphology of the vascular system in the Vertebrata. *Phil. Trans. R. Soc.*, **177**, 685–732.

Parker, W. N. (1892). On the anatomy and physiology of *Protopterus annectens*. *Trans. R. Ir. Acad.* **30**, 109–230.

Parrington, F. R. (1950). The skull of *Dipterus*. *Ann. Mag. nat. Hist.* (**12**)**3**, 534–547.

Parrington, F. R. (1956). The patterns of dermal bones in primitive vertebrates. *Proc. zool. Soc. Lond.* **127**, 389–411.

Parrington, F. R. (1958). On the nature of the Anaspida. *In* "Studies on fossil vertebrates" (Ed. T. S. Westoll), 108–128. Athlone Press, London.

Parrington, F. R. (1967). The identification of the dermal bones of the head. *In* "Fossil vertebrates" (Eds C. Patterson and P. H. Greenwood). J. Linn. Soc. (Zool.), Vol. 47, 231–239. Academic Press, London.

Parrington, F. R. (1978). A further account of the Triassic mammals. *Phil. Trans. R. Soc. (B)* **282**, 177–204.
Parsons, T. S. and Williams, E. E. (1962). The teeth of Amphibia and their relation to amphibian phylogeny. *J. Morph.* **110**, 375–390.
Parsons, T. S. and Williams, E. E. (1963). The relationships of the modern Amphibia: A re-examination. *Q. Rev. Biol.*, **38**, 26–53.
Paterson, N. F. (1939). The head of *Xenopus laevis*. *Q.Jl microsc. Sci.* **81(2)**, 161–234.
Patten, B. M. (1964). "Foundations of embryology." McGraw-Hill, New York.
Patten, W. (1912). "The evolution of the vertebrates and their kin." Blakiston's, Philadelphia.
Patterson, C. (1964). A review of Mesozoic acanthopterygian fishes, with special reference to those of the English Chalk. *Phil. Trans. R. Soc. (B)* **247**, 213–482.
Patterson, C. (1965). The phylogeny of the chimaeroids. *Phil. Trans. R. Soc.(B)* **249**, 101–219.
Patterson, C. (1968). *Menaspis* and the bradyodonts. *In* "Current problems of lower vertebrate phylogeny." (Ed. T. Ørvig). Nobel Symp., Vol. 4, 171–205. Almqvist and Wiksell, Stockholm.
Patterson, C. (1973). Interrelationships of holosteans. *In* "Interrelationships of fishes" (Eds P. H. Greenwood *et al.*), 233–305. Academic Press, London.
Patterson, C. (1975). The braincase of pholidophorid and leptolepid fishes, with a review of the actinopterygian braincase. *Phil. Trans. R. Soc. (B)* **269**, 275–579.
Patterson, C. (1977). Cartilage bones, dermal bones and membrane bones, or the exoskeleton versus the endoskeleton. *In* "Problems in vertebrate evolution" (Eds S. M. Andrews *et al.*). Linn. Soc. Symp. Ser. Vol. 4, 77–121. Academic Press, London.
Pattle, R. E. (1969). The development of the foetal lung. *In* "Ciba Fdn. Symp. on foetal anatomy " (Eds G. E. W. Wolstenholme and M. O'Connor), 132–142. Churchill, London.
Pearson, A. A. (1936), The acustico-lateral centers and the cerebellum, with fiber connections, of fishes. *J. comp. Neurol.* **65**, 201–294.
Pearson, A. A. (1941). The development of the n. terminalis in man. *J. comp. Neurol.* **75**, 40–66.
Pearson, A. A. (1941a). The development of the olfactory nerve in man. *J. comp. Neurol.* **75**, 199–217.
Pehrson, T. (1922). Some points in the cranial development of teleostomian fishes. *Acta zool., Stockh.* **3**, 1–63.
Pehrson, T. (1940). The development of dermal bones in the skull of *Amia calva*. *Acta zool., Stockh.* **21**, 1–50.
Pehrson, T. (1944). The development of the laterosensory canal bones in the skull of *Esox lucius*. *Acta zool., Stockh.* **25**, 135–157.
Pehrson, T. (1945). Some problems concerning the development of the skull of turtles. *Acta zool., Stockh.* **26**, 157–184.
Pehrson, T. (1947). Some new interpretations of the skull in *Polypterus*. *Acta zool., Stockh.* **28**, 399–455.
Pehrson, T. (1949). The ontogeny of the lateral line system in the head of dipnoans. *Acta zool., Stockh.* **30**, 153–182.
Pehrson, T. (1958). The early ontogeny of the sensory lines and the dermal skull in *Polypterus*. *Acta zool., Stockh.* **39**, 241–258.
Peter, K. (1901). Die Entwickelung des Geruchsorgans und Jakobsonischen Organs in der Reihe der Wirbeltiere. Bildung der äusseren Nase und des Gaumens. *In* "Handbuch der vergleichenden und experimentellen Entwickelungslehre der Wirbeltiere" (Ed. O. Hertwig), Vol. 2(2), 8–21. Fischer, 1906, Jena.
Peyer, B. (1949). Goethes Wirbeltheorie des Schädels. *NeujBl. naturf. Ges. Zürich* **1950**, 1–131.

Pfeiffer, W. (1968). Die Fahrenholzschen Organe der Dipnoi und Brachiopterygii. *Z. Zellforsch* **90**, 127–147.
Pfeiffer, W. (1968a). Das Geruchsorgan der Polypteridae (Pisces, Brachiopterygii). *Z. Morph. Tiere* **63**, 75–110.
Pfeiffer, W. (1969). Das Geruchsorgan der rezenten Actinistia und Dipnoi (Pisces). *Z. Morph. Tiere* **64**, 309–337.
Pietschmann, V. (1935). Cyclostoma. *In* "Kükenthal: Handbuch der Zoologie", (Ed. T. Krumbach), Vol. VI(A1), 127–547. de Gruyter, Berlin.
Piper, H. (1902). Die Entwickelung von Magen, Duodenum, Schwimmblase, Leber, Pankreas und Milz bei *Amia calva*. *Arch. Anat. Physiol. Anat. Abt. Suppl.*, 1–78.
Piveteau, J. (1937). Un amphibien du Trias inférieur. Essai sur l'origine et l'evolution des amphibiens anoures. *Annls Paléont.* **26**, 135–177.
Piveteau, J. (1964–1969) (Ed.). "Traité de Paléontologie." Vol. 4. Masson, Paris.
Plate, L. (1922). "Allgemeine Zoologie und Abstammungslehre", Vol. I. Gustav Fischer, Jena.
Plate, L. (1924). "Allgemeine Zoologie und Abstammungslehre", Vol. II. Gustav Fischer, Jena.
Platt, J. B. (1891). A contribution to the morphology of the vertebrate head, based on a study of *Acanthias wulgaris*. *J. Morph.* **5**, 79–106.
Poll, M. (1965). Anatomie et systématique des Polyptères. *Bull. Acad. r. Belg. Cl. Sci.* **51**, 553–569.
Poplin, C. (1973). Survivance de la "ligne épibranchiale" en particulier du "canal dorso-latéral-antérieur" ("ligne profonde") chez les Vertébrés inférieurs. *Bull. Mus. natn. Hist. nat., Paris* **173**, 117–139.
Poplin, C. (1974). Étude de quelques Paléoniscidés Pennsylvaniens du Kansas. Cah. Paléont., C.N.R.S. Paris, 1–151.
Portmann, A. 5th edn (1976). "Einführung in die Vergleichende Morphologie Der Wirbeltiere", 1–344. Schwabe, Basel.
Potter, I. C. and Robinson, E. S. (1971). The chromosomes. *In* "The biology of lampreys" (Eds M. W. Hardisty and I. C. Potter), Vol. 1, 279–293. Academic Press, London.
Price, G. C. (1896). Zur Ontogenie eines Myxinoiden. *Sber. bayer. Akad. Wiss.* **26**, 69–74.
Ranzi, S. (1926). Ricerche embriologiche e morfologiche sul ductus endolymphaticus (o aquaeductus vestibuli ovvero recessus labyrinti) dei Vertebrati. I. L'aquaeductus vestibuli dei Selaci. *Pubbl. Staz. zool. Napoli* **7**, 169–213.
Rauther, M. (1929–40). Echte Fische. *In* "Bronn's Kl. Ordn. Tierreichs", Vol. 6(1). Akad Verlagsges. Leipzig.
Rauther, M. (1937). Kiemendarmderivate der Cyclostomen und Fische. *In* "Handbuch der vergleichenden Anatomie der Wirbeltiere", (Eds L. Bolk *et al.*), Vol. 2, 211–278. Urban and Schwarzenberg, Berlin and Wien.
Raven, P. (1931). Zur Entwicklung der Ganglienleiste. 1. Die Kinematik der Ganglienleistenentwicklung bei den Urodelen. *Wilhem Roux Arch. Entwmech. Org.* **125**, 210–292.
Raw, F. (1960). Outline of a theory of origin of the vertebrate. *J. Paleont.* **34**, 497–539.
Raynaud, A. and Adrian, M. (1975). Mise en évidence, au moyen de la microscopie électronique, de la pénétration des cellules somitique dans la mésoblaste de l'ébauche des membres des embryons de reptiles (*Anguis fragilis, Lacerta viridis*). *Archs Anat. microsc. Morph. exp.* **64**, 287–316.
Rayner, D. H. (1941). The structure and evolution of the holostean fishes. *Biol. Rev.* **16**, 218–237.
Rayner, D. H. (1948). The structure of certain Jurassic holostean fishes, with special reference to their neurocrania. *Phil. Trans. R. Soc. (B)* **233**, 287–345.

Rayner, D. H. (1951). On the cranial structure of an early palaeoniscid, *Kentuckia*, gen. nov. *Trans. R. Soc. Edinb.*, **62**, 53–83.
Reed, H. D. (1920). The morphology of the sound-transmitting apparatus in caudate Amphibia and its phylogenetic significance. *J. Morph.* **33**, 325–375.
Reese, A. M. (1910). The lateral line system of *Chimaera colliei*. *J. exp. Zool.* **9**(**1**), 349–370.
Regal, P. J. (1966). Feeding specializations and the classification of the terrestrial salamanders. *Evolution, Lancaster, Pa.* **20**, 392–407.
Regal, P. J. and Gans, C. (1976). Functional aspects of the evolution of frog tongues. *Evolution, Lancaster, Pa.*, **30**, 718–734.
Regel, E. D. (1961). [Traces of segmentation in the chordal division of the chondrocranium in *Hynobius kayserlingii*.] *Dokl. Akad. Nauk. SSSR* **140**, 253–255. (In Russian.)
Regel, E. D. (1964). The development of cartilaginous neurocranium and its connection with palatoquadratum in *Hynobius keyserlingii*. *Trudy zool. Inst. Leningr.* **33**, 34–74. (In Russian, English title.)
Regel, E. D. (1966). Prespiracular visceral clefts and their role in the formation of the mouth in Caudata. *Zool. Zh.* **45**, 237–244. (In Russian, English summary.)
Regel, E. D. (1968). The development of the cartilaginous neurocranium and its connection with the upper part of mandibular arch in Siberian salamander *Ranodon sibiricus (Hynobiidae, Amphibia)*. *Trudӯ zool Inst. Leningr.* **46**, 5–85. (in Russian, English title.)
Regel, E. D. (1973). [Prespiracular fissurae of the Anura.] *Dokl. Akad. Nauk SSSR* **208**, 1487–1490. (In Russian.)
Regel, E. D. and Epstein, S. M. (1972). Segmentation of interauricular skull region in Anura. *Zool. Zh.* **51**, 1517–1528. (In Russian, English summary.)
Reichert, C. (1837). Uber die Visceralbogen der Wirbeltiere im allgemeinen und deren Metamorphosen bei den Vögeln and Säugetieren. *Arch. Anat. Physiol.* **1837**, 120–222.
Reighard, J. E. and Mast, S. O. (1908). The development of the hypophysis of *Amia*. *J. Morph.* **19**, 497–509.
Reinbach, W. (1939). Untersuchungen über die Entwicklung des Kopfskeletts von *Calyptocephalus gayi* (mit einem Anhang über das Os supratemporale der anuren Amphibien). *Jena. Z. Naturw.* **72**, 211–362.
Reinbach, W. (1952). Zur Entwicklung des Primordialkraniums von *Dasypus novemcinctus* Linné (*Tatusia novemcincta* Lesson) I, II. *Z. Morph. Anthrop* **44**, 375–444; **45**, 1–72.
Reisinger, E. and Knepper, A. (1972). Phylogenetische Aspecte bei agnathen Wirbeltieren auf Grund der Vornierenentwicklung. *Z. zool. Syst. Evolutionsforsch* **10**(1971), 241–267.
Remane, A. (1936). Wirbelsäule und ihre Abkömmlinge. *In* "Handbuch der vergleichenden Anatomie der Wirbeltiere", (Eds L. Bolk *et al.*), Vol. IV, 1–206. Urban and Schwarzenberg, Berlin and Wien.
Rempel, A. G. (1943). The origin and differentiation of the larval head musculature of *Triturus torosus* (Rathke). *Univ. Calif. Publs Zool.* **51**, 82–127.
Reno, H. W. (1966). The infraorbital canal, its lateral line ossicles and neuromasts, in the minnows *Notropis volucellus* and *N. buchanani*. *Copeia* **1966**, 403–413.
Repetski. J. E. (1978). A fish from the Upper Cambrian of North America. *Science, N.Y.* **200**, 529–531.
Retzius, G. (1881). "Das Gehörorgan der Wirbelthiere", Vol. I. Samson and Wallin, Stockholm.
Retzius, G. (1881a). Das membranöse Gehörorgan von *Polypterus bichir* Geoffr. und *Calamoichthys calabaricus* J. A. Smith. *Biol. Unters.* **4**, 61–66.
Retzius, G. (1884). "Das Gehörorgan der Wirbelthiere", Vol. II. Samson and Wallin, Stockholm.

Richardson, L. R. (1953). *Neomyxine* n.g. (Cyclostomata) based on *Myxine biniplicata* Richardson and Jowett 1951, and further data on the species. *Trans. R. Soc. N.Z.* **81**, 379–383.
Richardson, L. R. and Jowett, J. P. (1951). A new species of *Myxine* (Cyclostomata) from Cook Strait. *Zoology Publs Vict. Univ. Wellington* **1**, 1–5.
Ritchie, A. (1964). New light on the morphology of the Norwegian Anaspida. *Skr. norske VidenskAkad. Mat-naturw. Kl.* **14**, 3–35.
Ritchie, A. (1967). *Ateleaspis tessellata* Traquair, a non-cornuate cephalaspid from the Upper Silurian of Scotland. *In* "Fossil vertebrates" (Eds C. Patterson and P. H. Greenwood). J. Linn. Soc. (Zool.), Vol. 47, 69–81.
Ritchie, A. (1968). New evidence on *Jamoytius kerwoodi* White, an important ostracoderm from the Silurian of Lanarkshire, Scotland. *Palaeontology* **11**(**1**), 21–39.
Ritchie, A. (1968a). *Phlebolepis elegans* Pander, an Upper Silurian thelodont from Oesel, with remarks on the morphology of thelodonts. *In* "Current problems of lower vertebrate phylogeny (Ed. T. Ørvig). Nobel Symp., Vol. 4, 81–88. Almqvist and Wiksell, Stockholm.
Ritchie, A. (1973). *Wuttagoonaspis* gen. nov., an unusual arthrodire from the Devonian of Western New South Wales, Australia. *Palaeontographica (A)* **143**, 58–72.
Ritchie, A. and Gilbert-Tomlinson, J. (1977). First Ordovician vertebrates from the Southern Hemisphere. *Alcheringa*, **1**, 351–368.
Ritland, R. M. (1955). Studies on the post-cranial morphology of *Ascaphus truei*. II. Myology. *J. Morph.* **97**, 215–282.
Rochon-Duvigneaud, A. (1954). L'oeil des Vertébrés. *In* "Traité de Zoologie" (Ed. P.-P. Grassé), Vol. 12, 333–452. Masson, Paris.
Robertson, G. M. (1938). The Tremataspidae. *Am. J. Sci.* (**5**)**35**, 172–206, 273–296.
Robertson, G. M. (1970). The oral region of ostracoderms and placoderms: Possible phylogenetic significance. *Am. J. Sci.* **269**, 39–64.
Robineau, D. and Anthony, J. (1973). Biomécanique du crâne de *Latimeria chalumnae* (Poisson crossoptérygien coelacanthidé). *C.r. hebd. Séanc. Acad. Sci., Paris (D)* **276**, 1305–1308.
Rognes, K. (1973). Head skeleton and jaw mechanism in Labrinae (Teleostei: Labridae) from Norwegian waters. *Årbok Univ. Bergen. Mat-naturv. ser. 1971* (**4**), 1–149.
Romer, A. S. (1922). The locomotor apparatus of certain primitive and mammal-like reptiles. *Bull. Am. Mus. nat. Hist.* **46**, 517–606.
Romer, A. S. (1924). Pectoral limb musculature and shoulder girdle structure in fish and tetrapods. *Anat. Rec.* **27**(**2**), 119–143.
Romer, A. S. (1933). "Vertebrate Paleontology." University Press, Chicago.
Romer, A. S. (1937). The braincase of the Carboniferous crossopterygian *Megalichthys nitidus*. *Bull. Mus. comp. Zool. Harv.* **82**, 1–73.
Romer, A. S. (1941). Notes on the crossopterygian hyomandibular and braincase. *J. Morph.* **69**(**1**), 141–160.
Romer, A. S. (1946). The early evolution of fishes. *Q. Rev. Biol.* **21**(**1**), 33–69.
Romer, A. S. (1947). Review of the Labyrinthodontia. *Bull. Mus. comp. Zool. Harv.* **99**, 7–368.
Romer, A. S. (1955). Herpetichthyes, Amphibioidei, Choanichthyes or Sarcopterygii? *Nature, Lond.* **176**, 126.
Romer, A. S. (1956). "Osteology of the reptiles." University Press, Chicago.
Romer, A. S. (1957). The appendicular skeleton of the Permian embolomerous amphibian *Archeria*. *Contr. Mus. Paleont. Univ. Mich.* **13**, 103–159.
Romer, A. S. (1964). The braincase of the Paleozoic elasmobranch *Tamiobatis*. *Bull. Mus. comp. Zool. Harv.* **131**(**4**), 87–105.
Romer, A. S. (1964a). Problems in early amphibian history. *J. Anim. Morph. Physiol.* **11**(**1**), 1–20.

Romer, A. S. 3rd edn (1966). "Vertebrate paleontology". University Press, Chicago.
Romer, A. S. (1968). "Notes and comments on vertebrate paleontology". University Press, Chicago.
Romer, A. S. (1969). Cynodont reptile with incipient mammalian jaw articulation. *Science, N.Y.* **166**, 881–882.
Romer, A. S. (1969a). A temnospondylous labyrinthodont from the Lower Carboniferous. *Kirtlandia* **6**, 1–20.
Romer, A. S. 4th edn (1970). "The vertebrate body." Saunders, Philadelphia.
Romer, A. S. (1970a). The Chanares (Argentian) Triassic reptile fauna. VI. A chiniquodontid cynodont with an incipient squamosal-dentary jaw articulation. *Breviora* **344**, 1–18.
Romer, A. S. (1972). The vertebrate as a dual animal—Somatic and Visceral. *Evol. Biol.* **6**, 121–156.
Romer, A. S. (1972a). Skin breathing—Primary or Secondary? *Resp. Physiol.* **14**, 183–192.
Romer, A. S. and Price, L. W. (1940). Review of the Pelycosauria. Spec. Pap. geol. Soc. Am. No. 28.
Roofe, P. G. (1935). The endocranial blood vessels of *Amblystoma tigrinum*. *J. comp Neurol.* **61**, 257–293.
Roux, G. H. (1942). The microscopic anatomy of the *Latimeria* scale. *S. afr. J. med. Sci. (Biol. Suppl.)* **7**, 1–18.
Roux, G. H. (1947). The cranial development of certain ethiopian "Insectivores" and its bearing on the mutual affinities of the group. *Acta zool., Stockh.* **28**, 169–397.
Rudebeck, B. (1945). Contributions to the forebrain morphology in Dipnoi. *Acta zool., Stockh.* **26**, 9–156.
Ruge, G. (1897). Ueber das periferische Gebiet des Nervus Facialis bei Wirbelthieren. *Festschr. Für Gegenbaur* **3**, 195–348. Engelmann, Leipzig.
Runnström, J. (1925). Über die Anlage des Parapinealorgans bei *Petromyzon*. *Z. mikrosk.-anat. Forsch.* **31**, 283–294.
Ruud, G. (1920). Über Hautsinnesorgane bei *Spinax niger* Bon. II. Die embryologishce Entwicklung. *Zool. Jb. (Anat.)* **41**, 459–546.
Sagemehl, M. (1883). Beiträge zur vergleichenden Anatomie der Fische. 1. Das Cranium von *Amia calva*. L. *Morph. Jb.* **9**, 177–228.
Sanzo, L. (1911). Distribuzione delle papille cutanee (organi ciatiformi) e suo valore sistematico nei Gobi. *Mitt. zool. Stn Neapel* **20**, 251–328.
Säve-Söderbergh, G. (1932). Preliminary note on Devonian stegocephalians from East Greenland. *Meddr Grønland* **94**, 1–107.
Säve-Söderbergh, G. (1933). The dermal bones of the head and the lateral line system in *Osteolepis macrolepidotus* Ag. With remarks on the terminology of the lateral line system and on the dermal bones of certain other crossopterygians. *Nova Acta R. Soc. Scient. upsal.* (**4**)**9**, 1–129.
Säve-Söderbergh, G. (1934). Some points of view concerning the evolution of the vertebrates and the classification of this group. *Ark. Zool.* **26A**, 1–20.
Säve-Söderbergh, G. (1935). On the dermal bones of the head in labyrinthodont stegocephalians and primitive Reptilia with special reference to Eotriassic stegocephalians from East Greenland. *Meddr Grønland* **98**, 1–211.
Säve-Söderbergh, G. (1936). On the morphology of Triassic stegocephalians from Spitsbergen, and the interpretation of the endocranium in the Labyrinthodontia. *K. svenska VetenskAkad. Handl.* (**3**)**16**, 1–181.
Säve-Söderbergh, G. (1937). On the dermal skulls of *Lyrocephalus, Aphaneramma,* and *Benth-*

osaurus, labyrinthodonts from the Triassic of Spitsbergen and N. Russia. *Bull. geol. Instn Univ. Upsala* **27**, 189–208.

Säve-Söderbergh, G. (1937a). On *Rhynchodipterus elginensis* n.g., n.sp., representing a new group of dipnoan-like Choanata from the Upper Devonian of East Greenland and Scotland. *Ark. Zool.* **29B**, 1–8.

Säve-Söderbergh, G. (1941). On the dermal bones of the head in *Osteolepis macrolepidotus* Ag. and the interpretation of the lateral line system in certain primitive vertebrates. *Zool. Bidr. Upps.* **20**, 523–541.

Säve-Söderbergh, G. (1945). Notes on the trigeminal musculature in non-mammalian tetrapods. *Nova Acta R. Soc. Scient. upsal.* (**4**)**13**, 1–59.

Säve-Söderbergh, G. (1947). Notes on the brain-case in *Sphenodon* and certain Lacertilia. *Zool. Bidr. Upps.* **25**, 489–516.

Säve-Söderbergh, G. (1951). Något om fossila lungfiskar. *Uppsala Univ. Årsskr.* **1951**(**2**), 1–20. (In Swedish.)

Säve-Söderbergh, G. (1952). On the skull of *Chirodipterus wildungensis* Gross, an Upper Devonian dipnoan from Wildungen. *K. svenska VetenskAkad. Handl.* (**4**)**3**, 1–29.

Saxén, L. and Toivonen, S. (1962). "Primary embryonic induction." Logos Press, London.

Scammon, R. E. (1911). Normal plates of the development of *Squalus acanthias. NormTaf. EntwGesch. Wirbeltiere* **12**, 1–140.

Schaeffer, B. (1941). The morphological and functional evolution of the tarsus in amphibians and reptiles. *Bull. Am. Mus. nat. Hist.* **78**, 395–472.

Schaeffer, B. (1952). Rates of evolution in the coelacanth and dipnoan fishes. *Evolution, Lancaster, Pa.* **6**, 101–111.

Schaeffer, B. (1952a). The Triassic coelacanth fish *Diplurus*, with observations on the evolution of the Coelacanthini. *Bull. Am. Mus. nat. Hist.* **99**, 27–78.

Schaeffer, B. (1962). A coelacanth fish from the Upper Devonian of Ohio. *Scient. Publs Cleveland Mus. nat. Hist.* **1**, 5–13.

Schaeffer, B. (1967). Comments on elasmobranch evolution. *In* "Sharks, skates and rays" (Eds P. W. Gilbert *et al.*), 3–35. Johns Hopkins Press, Baltimore.

Schaeffer, B. (1967a). Osteichthyan vertebrae. *In* "Fossil vertebrates" (Eds C. Patterson and P. H. Greenwood). J. Linn. Soc. (Zool.), Vol. 47, 185–195. Academic Press, London.

Schaeffer, B. (1968). The origin and basic radiation of the Osteichthyes. *In* "Current problems of lower vertebrate phylogeny" (Ed. T. Ørvig). Nobel Symp., Vol. 4, 207–222. Almqvist and Wiksell, Stockholm.

Schaeffer, B. (1969). Adaptive radiation of the fishes and the fish-amphibian transition. *Ann. N.Y. Acad. Sci.,* **167**, 5–17.

Schaeffer, B. (1973). Interrelationships of chondrosteans. *In* "Interrelationships of fishes" (Eds P. H. Greenwood *et al.*), 207–226. Academic Press, London.

Schaeffer, B. (1975). Comments on the origin and basic radiation of the gnathostome fishes with particular reference to the feeding mechanism. *Colloques int. Cent. natn. Rech. scient.* **218**, 101–110.

Schaeffer, B. (1977). The dermal skeleton in fishes. *In* "Problems in vertebrate evolution" (Eds S. M. Andrews *et al.*), Linn. Soc. Symp. Ser., Vol. 4, 25–52. Academic Press, London.

Schaeffer, B. and Williams, M. (1977). Relationships of fossil and living elasmobranchs. *Am. Zool.* **17**. 293–302.

Scharrer, E. (1944). The histology of the meningeal myeloid tissue in the ganoids *Amia* and *Lepisosteus. Anat. Rec.* **88**, 291–310.

Schauinsland, H. (1903). Beiträge zur Entwicklungsgeschichte und Anatomie der Wirbeltiere. I–III. *Zoologica, Stuttg.* **16**(**39**), 1–168.

Schauinsland, H. (1905). Die Entwickelung der Wirbelsäule nebst Rippen und Brustbein. *In* "Handbuch der vergleichenden und experimentellen Entwickelungslehre der Wirbeltiere" (Ed. O. Hertwig), Vol. 3(2), 339–572. Fischer, 1906, Jena.

Schaumberg, G. (1978). Neubeschreibung von *Coelacanthus granulatus* Agassiz (Actinistia, Pisces) aus dem Kupferschiefer von Richelsdorf (Perm, W.-Deutschland). *Paläont.Z.* **52**, 169–197.

Schmäh, R. (1934). Die Entwicklung der Unterkieferknochen bei *Polypterus. Gegenbaurs morph. Jb.* **74**, 364–379.

Schmalhausen, J. J. (1910). Die Entwicklung des Extremitätenskelettes von *Salamandra kaiserlingii. Anat. Anz.* **37**, 431–466.

Schmalhausen, J. J. (1912). Die Entwicklung des Skelettes und der Muskulatur der unpaaren Flossen der Fische. *Z. wiss. Zool.* **100**, 509–587.

Schmalhausen, J. J. (1913). Bau und Phylogenese der unpaaren Flossen und insbesonders der Schwanzflosse der Fische. *Z. wiss. Zool.* **104**, 1–80.

Schmalhausen, J. J. (1955). Distribution of the seismosensory organs in urodele Amphibia. *Zool. Zh.* **34**, 1334–1356. (In Russian.)

Schmalhausen, J. J. (1964). "The origin of terrestrial vertebrates." Nauka, Moscow. (In Russian.)

Schmalhausen, J. J. English edn (1968). "The origin of terrestrial vertebrates." Academic Press, New York and London.

Schmidt, L. (1892). Untersuchungen zur Kenntnis des Wirbelbaues von *Amia calva. Z. wiss. Zool.* **54**, 5–21.

Schmidt-Ehrenberg, E. C. (1942). Die Embryogenese des Extremitätenskelettes der Säugetiere. Ein Beitrag zur Frage der Entwicklung der Tetrapodengliedmassen. *Revue suisse Zool.* **49**, 33–131.

Schnitzlein, H. N. and Faucette, J. R. (1969). General morphology of the fish cerebellum. *In* "Neurobiology of cerebellar evolution and development" (Ed. R. Llinás), 77–106. Am. med. Ass., Chicago.

Schöne, G. (1902). Vergleichende Untersuchungen über die Befestigung der Rippen an der Wirbelsäule. *Morph. Jb.,* **57**, 1–43.

Schreiner, K. E. (1902). Einige Ergebnisse über den Bau und die Entwicklung der Occipitalregion von *Amia* und *Lepidosteus. Z. wiss. Zool.* **72**, 467–524.

van Schrick, F. G. (1927). Über den Schwund der praehyalen Visceraltasche bei *Lepidosteus osseus.* Ein Beitrag zur Kenntniss des Kiemendarmes der Wirbeltiere. *Gegenbaurs morph. Jb.* **58**, 197–208.

Schultze, H.-P. (1968). Palaeoniscoidea+Schuppen aus dem Unterdevon Australiens und Kanadas und aus Mitteldevons Spitzbergens. *Bull. Br. Mus. nat. Hist. (Geol.)* **16**, 343–368.

Schultze, H.-P. (1969). *Griphognathus* Gross, ein langschnauziger Dipnoer aus dem Oberdevon von Bergisch-Gladbach (Rheinisches Schiefergebirge) und von Lettland. *Geol. Palaeontol. Marburg* **3**, 21–79.

Schultze, H.-P. (1970). Die Histologie der Wirbelkörper der Dipnoer. *Neues Jb. Geol. Paläont. Abh.* **135**, 311–336.

Schultze, H.-P. (1973). Crossopterygier mit heterozerker Schwanzflosse aus dem Oberdevon Kanadas, nebst Onychodontida-Resten aus dem Mitteldevons Spaniens und aus dem Karbon der USA. *Palaeontographica (A)* **143**, 188–208.

Schultze, H.-P. (1975). Das Axialskelett der Dipnoer aus dem Oberdevon von Bergisch-Gladbach (Westdeutschland). *Colloques int. Cent. natn. Rech. scient.* **218**, 149–157.

Schultze, H.-P. (1977). Ausgangsform und Entwicklung der rhombischen Schuppen der Osteichthyes (Pisces). *Paläont. Z.* **51**, 152–168.

Sedra, S. N. and Michael, M. I. (1957). The development of the skull, visceral arches, larynx and visceral muscles of the South African clawed toad, *Xenopus laevis* (Daudin) during the process of metamorphosis. *Verh. K. ned. Akad. Wet.* (**2**) **51**, 1–80.
Semon, R. (1899). Die Zahnentwickelung des *Ceratodus forsteri*. *In* "Denkschr. med.-naturw. Ges. Jena", Vol. 4(1910), 113–135.
Semon, R. (1901). Über das Verwandtschaftsverhältnis der Dipnoer und Amphibien. *Zool. Anz.* **24**, 180–188.
Senn, D. G. (1970). The stratification in the reptilian central nervous system. *Acta anat.* **75**, 521–552.
Senn, D. G. (1976). Brain structure in *Calamoichthys calabaricus* Smith 1865 (Polypteridae, Brachiopterygii). *Acta zool., Stockh.* **57**, 121–128.
Senn, D. G. (1976a). Notes on the midbrain and forebrain of *Calamoichthys calabaricus* Smith 1865 (Polypteridae, Brachiopterygii). *Acta zool., Stockh.* **57**, 129–135.
Senn, D. G. and Farner, H.-P. (1977). Embryonale Muster in der Anatomie von Amphibien. *Salamandra* **13**, 89–104.
Sensenig, E. C. (1949). The early development of the human vertebral column. *Contr. Embryol.* **33**, 23–40.
Sewertzoff, A. N. (1899). Die Metamerie des Kopfes des electrischen Rochen. *Byull. mosk. Obshch. Ispyt. Prir.* **12**, 197–263, 393–445.
Sewertzoff, A. N. (1902). Zur Entwickelungsgeschichte des *Ceratodus forsteri*. *Anat. Anz.* **21**, 593–608.
Sewertzoff, A. N. (1904). Die Entwickelung der pentadaktylen Extremität der Wirbeltiere. *Anat. Anz.* **25**, 472–494.
Sewertzoff, A. N. (1908). Studien über die Entwickelung der Muskeln, Nerven und des Skeletts der Extremitäten der niederen Tetrapoda. *Byull. mosk. Obshch. Ispyt. Prir.* **21**, 1–430.
Sewertzoff, A. N. (1916–17). Études sur l'évolution des vertébrés inférieurs. *Russk. Arkh. Anat. Gistol. Embriol.* **1**(**1**), 1–104; **1**(**3**), 425–572.
Sewertzoff, A. N. (1923). Die Morphologie des Visceralapparates der Elasmobranchier. *Anat. Anz.* **56**, 389–410.
Sewertzoff, A. N. (1924). The development of the dorsal fin of *Polypterus delhesi*. *J. Morph.* **38**, 551–580.
Sewertzoff, A. N. (1925). Development of bony skull of *Amia calva*. *Byull. mosk. Obshch. Ispyt. Prir.* **34**, 87–127. (In Russian, English summary.)
Sewertzoff, A. N. (1926). Die Morphologie der Brustflossen der Fische. *Jena. Z. Naturw.* **62**, 343–392.
Sewertzoff, A. N. (1926a). Development of the pelvic fins of *Acipenser ruthenus*. New data for the theory of the paired fins of fishes. *J. Morph.* **41**, 547–579.
Sewertzoff, A. N. (1928). The head skeleton and muscles of *Acipenser ruthenus*. *Acta zool., Stockh.* **9**, 193–319.
Sewertzoff, A. N. (1931). "Morphologische Gesetzmässigkeiten der Evolution". Gustav Fischer, Jena.
Sewertzoff, A. N. (1934). Evolution der Bauchflossen der Fische. *Zool. Jb. (Anat.)* **58**, 415–500.
Shelton, P. M. J. (1970). The lateral line system at metamorphosis in *Xenopus laevis* (Daudin). *J. Embryol. exp. Morph.* **24**(**3**). 511–524.
Shishkin, M. A. (1967). [On a basipterygoid articulation in anurans.] *Dokl. Acad. Nauk SSSR* **174**(**6**), 1425–1428. (In Russian.)
Shishkin, M. A. (1968). On the cranial arterial system of the labyrinthodonts. *Acta zool., Stockh.* **49**, 1–22.

Shishkin, M. A. (1970). The origin of Anura and the theory of lissamphibians. *In* "Materials on the evolution of the land vertebrates" (Ed. K. K. Flerov), 30–44. Nauka, Moscow. (In Russian, English title.)

Shishkin, M. A. (1973). "The morphology of the early Amphibia and some problems of the lower tetrapod evolution", 1–260. Nauka, Moscow. (In Russian, English title and table of contents.)

Shishkin, M. A. (1975). Labyrinthodont middle ear and some problems of amniote evolution. *Colloques int. Cent. natn. Rech. scient.* **218**, 337–348.

Shufeldt, R. W. (1885). The osteology of *Amia calva*, including certain references to the skeleton of teleosteans. *Rep. U.S. Commnr Fish* (1883) **11**, 747–878.

Shute, C. C. D. (1956). The evolution of the mammalian eardrum and tympanic cavity. *J. Anat.* **90**, 261–281.

Shute, C. C. D. (1972). The composition of vertebrate and the occipital region of the skull. *In* "Studies in vertebrate evolution" (Eds K. A. Joysey and T. S. Kemp), 21–34. Oliver and Boyd, Edinburgh.

Simons, J. R. (1959). The distribution of the blood from the heart in some Amphibia. *Proc. zool. Soc. Lond.* **132**, 51–63.

Singh-Roy, K. K. (1967). On Goethe's vertebral theory of origin of the skull, a recent approach. *Anat. Anz.* **120**, 250–259.

Slabý, O. (1958). Morphogenesis and evolutionary morphology of the carpus of the sheep. *Čslká. Morf.* **6**, 301–322. (In Czech, English summary.)

Slabý, O. (1967). Die Morphogenese und phylogenetische Morphologie des Carpus der Paarhufer. *Rozpr. čsl. Akad. Věd.* **77**, 1–55.

de Smet, W. (1966). Le développement des sacs aériens des Polyptères. *Acta zool., Stockh.* **47**, 151–183.

Smit, A. L. (1953). The ontogenesis of the vertebral column of *Xenopus laevis* (Daudin). With special reference to the segmentation of the metotic region of the skull. *Annale Univ. Stellenbosch* **29**, 79–136.

Smith, B. G. (1937). The anatomy of the frilled shark *Chlamydoselachus anguineus*. "The Bashford Dean Memorial Volume: Archaic fishes", Vol. 6, 331–505. Am. Mus. Nat. Hist., New York.

Smith, C. L., Rand, C. S., Schaeffer, B. and Atz, J. W. (1975). *Latimeria*, the living coelacanth is ovoviviparous. *Science, N.Y.* **190**, 1105–1106.

Smith, I. C. (1956). The structure of the skin and dermal scales in the tail of Acipenser ruthenus L. *Trans. R. Soc. Edinb.* **58(1)**, 1–14.

Smith, I. C. (1957). New restorations of the heads of *Pharyngolepis oblongus* Kiaer and *Pharyngolepis kiaeri* sp. nov., with a note on their lateral-line systems. *Norsk. geol. Tidsskr.* **37**, 373–402.

Smith, J. L. B. (1939). A living fish of Mesozoic type. *Nature, Lond.* **143**, 455–456.

Smith, J. L. B. (1939a). A surviving fish of the order Actinistia. *Trans. R. Soc. S. Afr.* **27(1)**, 47–50.

Smith, J. L. B. (1940). A living coelacanthid fish from South Africa. *Trans. R. Soc. S. Afr.* **28**, 1–106.

Smith, J. L. B. (1956). "Old fourlegs. The story of the coelacanth." Longmans, London.

Smith, M. M. (1977). The microstructure of the dentition and dermal ornament of three dipnoans from the Devonian of Western Australia. *Phil. Trans. R. Soc. (B)* **281**, 29–72.

Smith, M. M., Hobdell, M. H. and Miller, W. A. (1972). The structure of the scales in *Latimeria chalumnae*. *J. Zool.* **167**, 501–509.

Sobotta, M. J. (1922). Deskriptive Anatomie. III Das Nerven- und Gefässystem und die

Sinnesorgane des Menschen. Lehmanns med. Atlanten No. **4**, 447–774. Lehmann, München.
Sokol, O. M. (1977). A subordinal classification of frogs. (Amphibia: Anura). *J. Zool.* **182**, 505–508.
Špinar, Z. V. (1972). "Tertiary frogs from Central Europe", 1–286. Academia Publishing House, Prague.
Stadtmüller, F. (1924). Über Entwicklung und Bau der papillenförmigen Erhebungen (Filterfortsätze) auf den Branchialbogen der Salamandridenlarven. *Z. Morph. Antrop.* **24**, 125–156.
Stadtmüller, F. (1927). Über das Kiemenfilter der Dipnoer. *Gegenbaurs morph. Jb.* **57**, 489–529.
Stadtmüller, F. (1936). Kranium und Visceralskelett der stegocephalen und Amphibien. *In* "Handbuch der vergleichenden Anatomie der Wirbeltiere," (Eds L. Bolk *et al.*), Vol. 4, 501–698. Urban and Schwarzenberg, Berlin and Wien.
Stadtmüller, F. (1936a). Kranium und Visceralskelett der Säugertiere. *In* "Handbuch der vergleichenden Anatomie der Wirbeltiere" (Eds L. Bolk *et al.*), Vol. 4, 839–1016. Urban and Schwarzenberg, Berlin and Wien.
Stahl, B. J. (1967). Morphology and relationships of the Holocephali with special reference to the venous system. *Bull. Mus. comp. Zool. Harv.* **135**, 141–213.
Starck, D. (1955). "Embryologie." Thieme, Stuttgart. (3rd edn, 1975).
Starck, D. (1963). Die Metamerie des Kopfes der Wirbeltiere. *Zool. Anz.* **170**, 393–428.
Starck, D. (1967). Le crane des Mammifères. *In* "Traité de Zoologie" (Ed. P.-P. Grassé), Vol. 16, 405–549. Masson, Paris.
Steiner, H. (1921). Hand und Fuss der Amphibien, ein Beitrag zur Extremitätenfrage. *Anat. Anz.* **53**, 513–542.
Steiner, H. (1934). Ueber die embryonale Hand- und Fuss-Skelett-Anlage bei den Crocodiliern, sowie über ihre Beziehungen zur Vogel-Flügelanlage und zur ursprünglichen Tetrapoden-Extremität. *Revue suisse Zool.* **41**, 383–396.
Steiner, H. (1935). Beiträge zur Gliedmassentheorie: Die Entwicklung des Chiropterygium aus dem Ichthyopterygium. *Revue suisse Zool.* **42**, 715–729.
Steiner, H. (1942). Der Aufbau des Säugetier-Carpus und -Tarsus nach neueren embryologischen Untersuchungen. *Revue suisse Zool.* **49**, 217–223.
Steiner, H. (1965). Die vergleichend-anatomische und oekologische Bedeutung der rudimentären Anlage eines selbständigen fünften Carpale bei *Tupaia*. *Israel J. Zool.* **14**, 221–233.
Stensiö, E. (1918). Notes on a crossopterygian fish from the Upper Devonian of Spitzbergen. *Bull. geol. Instn Univ. Upsala* **16**, 115–124.
Stensiö, E. (1921). "Triassic fishes from Spitzbergen". I. Adolf Holzhausen, Wien.
Stensiö, E. (1922). Über zwei Coelacanthiden aus dem Oberdevon von Wildungen. *Paläont. Z.* **4**, 167–210.
Stensiö, E. (1922a). Notes on certain crossopterygians. *Proc. zool. Soc. Lond.,* 1241–1271.
Stensiö, E. (1925). Triassic fishes from Spitzbergen. II. *K. svenska VetenskAkad. Handl.* (**3**)**2**, (**1**), 1–261.
Stensiö, E. (1925a). On the head of the macropetalichthyids, with certain remarks on the head of the other arthrodires. *Publs Field Mus. nat. Hist. (Geol.)* **4**(**4**), 91–197.
Stensiö, E. (1925b). Note on the caudal fin of *Eusthenopteron*. *Ark. Zool.* **17**(**B**), 1–3.
Stensiö, E. (1927). The Downtonian and Devonian vertebrates of Spitsbergen. I. Cephalaspidae. Skr. Svalb. Ish. No. **12**, 1–391.
Stensiö, E. (1931). Upper Devonian vertebrates from East Greenland. *Meddr Grønland* **86**, 1–212.
Stensiö, E. (1932). Triassic fishes from East Greenland collected by the Danish Expeditions in 1929–1931. *Meddr Grønland* **83**, 1–305.

Stensiö, E. (1932a). "The cephalaspids of Great Britain." Brit. Mus. (Nat. Hist.), London.
Stensiö, E. (1934). On the Placodermi of the Upper Devonian of East Greenland. I. Phyllolepida and Arthrodira. *Meddr Grønland* **97**(**1**), 1–58.
Stensiö, E. (1935). A new amiid from the Lower Cretaceous of Shantung, China. *Palaeont. sin.* **3**(**1**). 4–48.
Stensiö, E. (1936). On the Placodermi of the Upper Devonian of East Greenland. Supplement to part I. *Meddr Grønland* **97**(**2**), 1–52.
Stensiö, E. (1937a). Notes on the endocranium of a Devonian *Cladodus*. *Bull. Geol. Instn Univ.* dermal skeleton. *K. svenska VetenskAkad. Handl* (**3**) **16**(**4**), 4–56.
Stensiö, E. (1937a). Notes on the endocranium of a Devonian *Cladodus*. *Bull. Geol. Instn Univ. Upsala* **27**, 128–144.
Stensiö, E. (1939). On the Placodermi of the Upper Devonian of East Greenland. Second supplement to part I. *Meddr Grønland* **97**(**3**), 1–33.
Stensiö, E. (1939a). A new anaspid from the Upper Devonian of Scaumenac Bay in Canada, with remarks on the other anaspids. *K. svenska VetenskAkad. Handl.* (**3**)**18**(**1**), 3–25.
Stensiö, E. (1942). On the snout of arthrodires. *K. svenska VetenskAkad. Handl.* (**3**)**20**(**3**), 4–32.
Stensiö, E. (1944). Contributions to the knowledge of the vertebrate fauna of the Silurian and Devonian of Western Podolia. II. Notes on two arthrodires from the Downtonian of Podolia. *Ark. Zool.* **35**(**9**), 1–83.
Stensiö, E. (1945). On the heads of certain arthrodires. II. On the cranium and cervical joint of the Dolichothoraci (Acanthaspida). *K. svenska VetenskAkad. Handl.* (**3**)**22**(**1**), 3–70.
Stensiö, E. (1947). The sensory lines and dermal bones of the cheek in fishes and amphibians. *K. svenska VetenskAkad. Handl.* (**3**)**24**(**3**), 1–195.
Stensiö, E. (1948). On the Placodermi of the Upper Devonian of East Greenland. II Antiarchi: subfamily Bothriolepinae. *Meddr Grønland* **139**, (*Palaeozoologica Groenlandica* **2**), 5–622.
Stensiö, E. (1950). La cavité labyrinthique, l'ossification sclérotique et l'orbite de *Jagorina*. *In* "Paléontologie et Transformisme" (Ed. A. George), 9–43. Albin Michel, Paris.
Stensiö, E. (1958). Les cyclostomes fossiles ou ostracodermes. *In* "Traité de Zoologie" (Ed. P.-P. Grassé), Vol. 13, 173–425. Masson. Paris.
Stensiö, E. (1959). On the pectoral fin and shoulder girdle of the arthrodires. *K. svenska VetenskAkad. Handl.* (**4**)**8**(**1**), 5–226.
Stensiö, E. (1961). Permian Vertebrates. *In* "Geology of the Arctic" (Ed. G. O. Raasch), Vol. 1, 231–247. University of Toronto Press, Toronto.
Stensiö, E. (1963). Anatomical studies on the arthrodiran head. 1. Preface, geological and geographical distribution, the organisation of the arthrodires, the anatomy of the head in the Dolichothoraci, Coccosteomorphi and Pachyosteomorphi. Taxonomic appendix. *K. svenska VetenskAkad. Handl.* (**4**)**9**(**2**), 1–419.
Stensiö, E. (1963a). The brain and the cranial nerves in fossil, lower craniate vertebrates. *Skr. norske Vidensk-Akad. Mat.-naturv. Kl.* **1963**, 1–120.
Stensiö, E. (1964). Les cyclostomes fossiles ou ostracodermes. *In* "Traité de Paléontologie" (Ed. J. Piveteau), Vol. 4(1), 96–382. Masson, Paris.
Stensiö, E. (1968). The cyclostomes with special reference to the diphyletic origin of the Petromyzontida and Myxinoidea. *In* "Current problems of lower vertebrate phylogeny" (Ed. T. Ørvig). Nobel Symp., Vol. 4, 13–71. Almqvist and Wiksell, Stockholm.
Stensiö, E. (1969). Placodermata; Arthrodires. *In* "Traité de Paléontologie" (Ed. J. Piveteau), 4(2), 71–692. Masson, Paris.
Stensiö, E. (1969–71). Anatomie des arthrodires dans leur cadre systématique. *Annls Paléont.* **55**, 151–192; **57**(**1**), 1–39; **57**(**2**), 158–186.
Sterba, G. (1954). Die Physiologie und Histogenese der Schilddrüse und des Thymus beim

Bachneunauge als Grundlagen phylogenetischer Studien über die Evolution der innersekretorischen Kiemendarmderivate. *Wiss. Z. Friedrich Schiller-Univ. Jena* **3**, 239–298.
Sterzi, G. (1904). Die Blutgefässe des Rückenmarks. *Arb. anat. Inst., Wiesbaden (Anatomische Hefte)* **24**, 1–364.
Stokes James, M. (1946). The role of the basibranchial cartilage in the early development of the thyroid of *Hyla regilla. Univ. Calif. Publs Zool.* **51**, 215–228.
Stone, L. S. (1922). Experiments on the development of the cranial ganglia and the lateral line sense organs in *Amblystoma punctatum. J. Exp. Zool.* **35**, 421–495.
Stone, L. S. (1926). Further experiments on the extirpation and transplantation of mesectoderm in *Amblystoma punctatum. J. Exp. Zool.* **44**, 95–131.
Stone, L. S. (1928). Experiments on the transplantation of placodes of the cranial ganglia in the amphibian embryo. *J. comp. Neurol.* **47**, 91–154.
Stone, L. S. (1929). Experiments showing the role of migrating neural crest (mesectoderm) in the formation of the head skeleton and loose connective tissue in *Rana palustris. Arch. EntwMech. Org.* **118**, 40–77.
Strahan, R. (1958). The velum and the respiratory current of *Myxine. Acta zool., Stockh.* **39**, 227–240.
Strahan, R. (1963). The behaviour of myxinoids. *Acta zool., Stockh.* **44**, 73–102.
Strahan, D. (1963a.) The behaviour of *Myxine* and other myxinoids. *In* "The biology of *Myxine*" (Eds A. Brodal and R. Fänge), 22–32. Universitetsforlaget, Oslo.
Streeter, G. L. (1907). On the development of the membranous labyrinth and the acoustic and facial nerves in the human embryo. *Am. J. Anat.* **6**, 139–165.
Streeter, G. L. (1945). Developmental horizons in human embryos. Period of indentation of the lens vesicle. *Contr. Embryol.* **31**, 29–63.
Streeter, G. L. (1949). Developmental horizons in human embryos. A review of the histogenesis of cartilage and bone. *Contr. Embryol.* **33**, 151–167.
Streeter, G. L. (1951). Developmental horizons in human embryos. *Contr. Embryol.* **34**, 167–196.
Swanepoel, J. H. (1970). The ontogenesis of the chondrocranium and of the nasal sac of the microhylid frog *Breviceps adspersus pentheri* Werner. *Annale Univ. Stellenbosch* **45**, 1–119.
Szarski, H. (1962). The origin of the Amphibia. *Q. Rev. Biol.* **37**(3), 189–241.
Szarski, H. (1968). The origin of the vertebrate foetal membranes. *Evolution, Lancaster, Pa.* **22**, 211–214.
Szarski, H. (1977). Sarcopterygii and the origin of tetrapods. *In* "The major patterns of vertebrate evolution" (Eds M. K. Hecht, P. C. Goody and B. M. Hecht). Nato advd Stud. Inst. Ser. (A), Vol. 14, 517–544. Plenum, New York.
Takaya, H. (1953). On the notochord-forming potency of the prechordal plate in *Triturus* gastrulae. *Proc. Japan Acad.* **29**, 374–380.
Tandler, J. (1899). Zur vergleichenden Anatomie der Kopfarterien bei den Mammalia. *Denkschr. Akad. Wiss., Wien.* **67**, 677–784.
Tandler, J. (1902). Zur Entwicklungsgeschichte der Kopfarterien bei den Mammalia. *Morph. Jb.* **30**, 275–372.
Tarlo, L. B. (Halstead) (1960). The invertebrate origins of the vertebrates. *Int. Geol. Congr.*, 21. *Copenhagen.* **22**, 113–122.
Tarlo, L. B. H. (1960a). The Downtonian ostracoderm *Corvaspis kingi* Woodward, with notes on the development of dermal plates in the Heterostraci. *Palaeontology* **3**, 217–226.
Tarlo, L. B. H. (1961). *Rhinopteraspis cornubica* (McCoy) with notes on the classification and evolution of the pteraspids. *Acta palaeont. pol.* **6**, 367–402.

Tarlo, L. B. H. (1964). Psammosteiformes (Agnatha)—A review with descriptions of new material from the Lower Devonian of Poland. I. *Palaeont. pol.* **13**, 1–135.

Tarlo, L. B. H. (1965). Psammosteiformes (Agnatha)—A review with descriptions of new material from the Lower Devonian of Poland. II. *Palaeont. pol.* **15**, 1–168.

Tarlo, L. B. H. (1967). Agnatha. *In* "The fossil record" (Eds W. B. Harland *et al.*). Geol Soc. Lond., 629–636.

Tarlo, L. B. H. and Whiting, H. P. (1965). A new interpretation of the internal anatomy of the Heterostraci (Agnatha). *Nature, Lond.* **206**, 148–150.

Tatarinov, L. P. (1968). Morphology and systematics of the northern Dvina cynodonts (Reptilia, Therapsida; Upper Permian). *Postilla* **126**, 1–51.

Tatarinov, L. P. (1976). "Morphological evolution of the theriodonts and the general problems of phylogenetics." Nauka, Moscow. (In Russian, English title and table of contents.)

Tatarko, K. (1939). The mandibular and hyoid arches and gill cover in polypterini, lepidosteoidei and amioidei. *Dopov. Akad. Nauk URSR.* Papers on animal morphology, **5**, 50–58 (In Russian, English summary).

Taverne, L. (1974). Sur le premier exemplaire complet d'*Enneles audax* Jordan, D. S. and Branner, J. C., 1908 (Pisces, Holostei, Amiidae) du Crétacé supérieur du Brésil. *Bull. Soc. belge Géol. Paléont. Hydrol.* **83**, 61–71.

Tester, A. L. and Nelson, G. J. (1967). *In* "Sharks, skates and rays" (Eds P. W. Gilbert *et al.*), Vol. 34, 503–531. Johns Hopkins, Baltimore.

Thacher, J. K. (1876). Median and paired fins, a contribution to the history of vertebrate limbs. *Trans. Conn. Acad. Arts Sci.* (1876). **3**, 281–310.

Theisen, B. (1970). The morphology and vascularization of the olfactory organ in *Calamoichthys calabaricus* (Pisces, Polypteridae). *Vidensk. Meddr dansk naturh. Foren.* **133**, 31–50.

Theisen, B. (1973). The olfactory system in the hagfish, *Myxine glutinosa. Acta zool., Stockh.* **54**, 271–284.

Theisen, B. (1976). The olfactory system in the Pacific hagfishes *Eptatretus stoutii, Eptatretus deani* and *Myxine circifrons. Acta zool., Stockh.* **57**, 167–173.

Thomson, K. S. (1967). A new genus and species of marine dipnoan fish, from the Upper Devonian of Canada. *Postilla* **106**, 1–6.

Thomson, K. S. (1968). A new Devonian fish (Crossopterygii: Rhipidistia) considered in relation to the origin of the Amphibia. *Postilla* **124**, 1–13.

Thomson, K. S. (1972). New evidence on the evolution of the paired fins of Rhipidistia and the origin of the tetrapod limb, with description of a new genus of Osteolepidae. *Postilla* **157**, 1–7.

Thomson, K. S. (1973). Observations on a new rhipidistian fish from the Upper Devonian of Australia. *Palaeontographica (A)* **143**, 209–220.

Thomson, K. S. (1975). On the biology of cosmine. *Bull. Peabody Mus. nat. Hist.* **40**, 1–59.

Thomson, K. S. (1977). On the individual history of cosmine and a possible elektroreceptive function of the pore–canal system in fossil fishes. *In* "Problems in vertebrate evolution" (Eds S. M. Andrews *et al.*). Linn. Soc. Symp. Ser, Vol. 4, 247–270. Academic Press, London.

Thomson, K. S. and Campbell, K. S. W. (1971). The structure and relationships of the primitive Devonian lungfish—*Dipnorhynchus sussmilchi* (Etheridge). *Bull. Peabody Mus. nat. Hist.* **38**, 1–109.

Thomson, K. S. and Hahn, V. (1968). Growth and form in rhipidistian fishes (Crossopterygii). *J. Zool.* **156**, 199–223.

Thomson, K. S. and Rackoff, J. S. (1974). The shoulder girdle of the Permian rhipidistian fish *Ectosteorhachis nitidus* Cope: Structure and possible function. *J. Paleont.* **48**, 170–179.

Thomson, K. S. and Vaughn, P. P. (1968). Vertebral structure in Rhipidistia (Osteichthyes, Crossopterygii) with description of a new Permian genus. *Postilla* **127**, 1–19.

Thornhill, R. A. (1972). The development of the labyrinth of the lamprey (*Lampetra fluviatilis* L. 1758). *Proc. R. Soc. (B)* **181**, 175–198.

Thorsteinsson, R. (1967). Preliminary note on Silurian and Devonian ostracoderms from Cornwallis and Somerset Islands, Canadian Arctic Archipelago. *Colloques int. Cent. natn. Rech. scient.* **163**, 45–47.

Toerien, M. J. (1963). Experimental studies on the origin of the auditory capsule and columella in *Ambystoma*. *J. Embryol. exp. Morph.* **11**, 459–473.

Toerien, M. J. (1971). The developmental morphology of the chondrocranium of *Podiceps cristatus*. *Annale Univ. Stellenbosch*, No. **46**.

Tokarski, J. (1904). Neue Tatsachen zur vergleichenden Anatomie der Zungenstützorgane der Säugetiere. *Anat. Anz.* **25**, 121–131.

Töplitz, C. (1920). Bau und Entwicklung des Knorpelschädels von *Didelphys marsupialis*. *Zoologica, Stuttgart*. **70**, 1–83.

Torrey, T. W. 3rd edn (1971). "Morphogenesis of the vertebrates." John Wiley, New York.

Trahms, O. K. (1936). Das Geruchsorgan von *Pipa americana*. *Z. Anat. EntwGesch.* **105**, 678–693.

Traquair, R. (1871). Note on the genus *Phaneropleuron* Huxley, with a description of a new species from the Carboniferous formation. *Geol. Mag.* **8**, 530–535.

Traquair, R. (1873). On a new genus (*Ganorhynchus*) of fossil fish of the order Dipnoi. *Geol. Mag.* **10**, 552–555.

Traquair, R. (1875). On the structure and systematic position of the genus *Cheirolepis*. *Ann. Mag. nat. Hist.* 1875, **4**, ser. **15**, 237–249.

Traquair, R. (1878). On the general *Dipterus, Palaedaphus, Holodus* and *Cheirodus*. *Ann. Mag. nat. Hist.* 1878, **5**, ser.**2**, 1–17.

Traquair, R. (1881). On the cranial osteology of *Rhizodopsis*. *Trans. R. Soc. Edinb.* **30**, 167–179.

Traquair, R. (1890). On the fossil fishes found at Achanarras quarry, Caithness. *Ann. Mag. nat. Hist.* **6**, 479–486.

Traquair, R. (1895). The extinct vertebrata of the Moray firth area. *In* "Vertebrate fauna of the Moray Basin", (Eds J. A. Harvie-Brown and T. E. Buckley), Vol. II, 235–285.

Traquair, R. (1899). Report on fossil fishes collected by the Geological Survey of Scotland in Silurian rocks of the south of Scotland. *Trans. R. Soc. Edinb.* **39**, 527–564.

Turner, S. (1973). Siluro-Devonian thelodonts from the Welsh Borderland. *J. Geol. Soc. Lond.* **129**, 557–584.

Turner, S. C. (1967). A comparative account of the development of the heart of a newt and a frog. *Acta zool., Stockh.* **48**, 43–57.

Vaage, S. (1969). The segmentation of the primitive neural tube in chick embryos (*Gallus domesticus*). *Ergebn. Anat. EntwGesch.* **41**(**3**), 1–88.

Vandebroek, G. (1969). "Évolution des vertébrés de leur origine à l'homme." Masson, Paris.

Vandel, A. (1966). Le protée et sa place dans l'embranchement des vertébrés. *Bull. Soc. zool. Fr.* **91**, 171–178.

Vasnetzov, W. (1928). Die Entwicklung des primären Schultergürtels bei *Amia calva*. *Russk. zool. Zh.* **8**, 71–84. (In Russian, German summary.)

Verbout, A. J. (1976). A critical review of the "Neugliederung" concept in relation to the development of the vertebral column. *Acta biotheor.* **25**, 219–258.

Versluys, J. (1898). "Die mittlere und äussere Ohrsphäre der Lacertilia und Rhynchocephalia" 1–247. Fischer, Jena.

Versluys, J. (1936). Kranium und Visceralskelett der Sauropsiden. *In* "Handbuch der vergleichenden Anatomie der Wirbeltiere" (Eds L. Bolk *et al.*), Vol. 4, 699–808. Urban and Schwarzenberg, Berlin and Wien.

Vial. J. L. (1973) (Ed.). "Evolutionary biology of the anurans." University of Missouri Press, Columbia.

de Villiers, C. G. S. (1922). Neue Beobachtungen über den Bau und die Entwickelung des Brustschulterapparates bei den Anuren, insbesondere bei *Bombinator*. *Acta zool., Stockh.* **10**, 153–225.

Vilstrup, T. (1951). "Structure and function of the membranous sacs of the labyrinth in *Acanthias vulgaris*", 1–134. Einar Munksgaard, Copenhagen.

Virchow, H. (1890). Über Spritzlochkieme der Selachier. *Arch. Anat. Physiol.* **1890**, 177–182.

Visser, J. G. J. (1972). Ontogeny of the chondrocranium of the chamaeleon, *Microsauria pumila pumila* (Daudin). *Annale Univ. Stellenbosch* **47(A)**, 1–68.

Vogt, W. (1929). Gestaltungsanalyse am Amphibienkeim mit örtlicher Vitalfärung. II. Gastrulation und Mesodermbildung bei Urodelen und Anuren. *Wilhelm Roux Arch. Entwmech. Org.* **120**, 384–706.

Vorobyeva, E. (1962). [Rhizodont crossopterygians from the Devonian Main Field of the USSR.] *Trudy paleont. Inst.* **104**, 1–108. (In Russian.)

Vorobyeva, E. (1969). Zur Morphologie der Wangenplatte von *Panderichthys rhombolepis* (Gross). *Eesti N S V Tead. Akad. Toim.* **18**, (Geol.), 255–258. (In Russian, German summary.)

Vorobyeva, E. (1971). The ethmoid region of *Panderichthys* and some problems of the cranial morphology of crossopterygians. *In* "Current problems of palaeontology". *Trudy paleont. Inst.*, Vol. **130**, 142–159. Nauka, Moscow. (In Russian, English title.)

Vorobyeva, E. (1973). Einige Besonderheiten im Schädelbau von *Panderichthys rhombolepis* (Gross). *Palaeontographica (A)* **143**, 221–229.

Vorobyeva, E. (1975). Some peculiarities in evolution of the rhipidistian fishes. *Colloques int. Cent. natn. Rech. scient.* **218**, 223–230.

Vorobyeva, E. (1975a). Formenvielfalt und Verwandtschafts-beziehungen der Osteolepidida (Crossopterygii, Pisces). *Paläont. Z* **49**, 44–45.

Vorobyeva, E. (1975b). Bemerkungen zu *Panderichthys rhombolepis* (Gross) aus Lode in Lettland (Gauja-Schichten, Oberdevon). *Neues Jb. Geol. Paläont. Mh.* **1975(5)**, 315–320.

Vorobyeva, E. (1977). "Morphology and nature of evolution of crossopteygian fish" *Trudy paleont. Inst.*, Vol. 163, 1–239. Nauka, Moscow. (In Russian.)

Vorobyeva, E. and Obruchev, D. (1964). Subclass Sarcopterygii. *In* "Fundamentals of Paleontology" (Ed. I. A. Orlow), Vol. 11. English translation, pp. 420–498. Jerusalem, 1967.

Wake, D. B. (1963). Comparative osteology of the plethodontid salamander genus *Aneides*. *J. Morph.* **113**, 77–118.

Wake, D. B. (1966). Comparative osteology and evolution of the lungless salamanders, family Plethodontidae. *Mem. sth. Calif. Acad. Sci.* **4**, 1–111.

Wake, D. B. (1970). Aspects of vertebral evolution in the modern Amphibia. *Forma et Functio* **3**, 33–60.

Wake, D. B. and Lawson, R. (1973). Developmental and adult morphology of the vertebral column in the plethodontid salamander *Eurycea bislineata*, with comments on vertebral evolution in the Amphibia. *J. Morph.* **139**, 251–298.

Wake, M. H. (1968). Evolutionary morphology of the caecilian urogenital system. I. The gonads and the fat bodies. *J. Morph.* **126**, 291–331.

Walls, G. L. (1942). The vertebrate eye and its adaptive radiation. Bull. Cranbrook Inst. Sci., No. 19.

Wängsjö, G. (1944). On the genus *Dartmuthia* Patten, with special reference to the minute structure of the exoskeleton. *Bull. geol. Instn Univ. Upsala* **31**, 349–362.

Wängsjö, G. (1952). Morphologic and systematic studies of the Spitsbergen cephalaspids. *Skr. norsk Polarinst.* **97**, 1–612 + supplementary note.

Warren, J. W. and Wakefield, N. A. (1972). Trackways of tetrapod vertebrates from the Upper Devonian of Viktoria, Aust. *Nature, Lond.* **238**, 469–470.

Watson, D. M. S. (1921). On the coelacanth fish. *Ann. Mag. nat. Hist.* (**9**) **8**, 319–337.

Watson, D. M. S. (1925). The structure of certain palaeoniscids and the relationships of that group with other bony fish. *Proc. zool. Soc. Lond.* **1925**(**14**), 815–870.

Watson, D. M. S. (1926). The evolution and origin of the Amphibia. *Phil. Trans. R. Soc. (B)* **214**, 189–257.

Watson, D. M. S. (1928). On some points in the structure of palaeoniscid and allied fish. *Proc. zool. Soc. Lond.* **1928**(**4**), 49–70.

Watson, D. M. S. (1929). The Carboniferous Amphibia of Scotland. *Palaeont. hung.* **1**, 219–252.

Watson, D. M. S. (1934). The interpretation of Arthrodires. *Proc. zool. Soc. Lond.* **1934**, 437–464.

Watson, D. M. S. (1935). Fossil fishes of the Orcadian Old Red Sandstone. *In* "The geology of the Orkneys" (Eds G. V. Wilson *et al.*). Mem. geol. Surv. U.K., 157–169.

Watson, D. M. S. (1937). Acanthodian fishes. *Phil. Trans. R. Soc. (B)* **228**, 49–146.

Watson, D. M. S. (1938). On *Rhamphodopsis*, a ptyctodont from the Middle Old Red Sandstone of Scotland. *Trans. R. Soc. Edinb.* **59**, 397–410.

Watson, D. M. S. (1940). The origin of frogs. *Trans. R. Soc. Edinb.* **60**, 195–231.

Watson, D. M. S. (1954). On *Bolosaurus* and the origin and classification of reptiles. *Bull. Mus. comp. Zool. Harv.* **111** (**9**), 299–449.

Watson, D. M. S. (1954a). A consideration of ostracoderms. *Phil. Trans. R. Soc. (B)* **238**, 1–25.

Watson, D. M. S. and Day, H. (1916). Notes on some Palaeozoic fishes. *Mem. Proc. Manchr. lit. phil. Soc.* **60**, 1–48.

Watson, D. M. S. and Gill, E. L. (1923). The structure of certain Palaeozoic Dipnoi. *J. Linn. Soc. (Zool.)* **35**, 123–216.

Weber, M. 2nd edn (1927). "Die Säugetiere", Vol. I. Gustav Fischer, Jena.

Wells, G. A. (1917). The skull of *Acanthias vulgaris*. *J. Morph.* **28**, 417–443.

Wenz, S. (1965). Sur un nouveau *Furo, F. normandica,* poisson holostéen du Toarcien de La Caine (Calvados). *C. r. somm. Seanc. Soc. géol. Fr.* **4**, 145–146.

Wenz, S. (1967). Remarques sur les transformations des os dermiques du museau chez les Actinoptérygiens. *Colloques int. Cent. natn. Rech. scient.* **163**, 89–92.

Wenz, S. (1971). Anatomie et position systématique de *Vidalamia* poisson Holostéen du Jurassique supérieur de Montsech (province de Lérida, Espagne). *Annls Paléont. (Vert.)* **57**, 43–62.

Wenz, S. (1977). Le squelette axial et l'endosquelette caudal d'*Enneles audax*, poisson Amiidé du Crétacé de Ceara (Brésil). *Bull. Mus. natn. Hist. nat., Paris* **490**, 341–348.

Werner, C. F. (1930). Das Ohrlabyrinth der Elasmobranchier. *Z. wiss. Zool.* **136**, 485–579.

Werner, C. F. (1960). "Das Gehörorgan der Wirbeltiere und des Menschen." Thieme, Leipzig.

Werner, Y. L. (1971). The ontogenic development of the vertebrae in some gekkonoid lizards. *J. Morph.* **133**, 41–92.

van der Westhuizen, C. M. (1961). The Development of the chondrocranium of *Heleophryne purcelli* Sclater with special reference to the palatoquadrate and the sound-conducting apparatus. *Acta zool., Stockh.* **42**, 3–72.

Westoll, T. S. (1936). On the structures of the dermal ethmoid shield of *Osteolepis*. *Geol. Mag.* **73**, 157–171.
Westoll, T. S. (1937). On the cheek-bones in teleostome fishes. *J. Anat.* **71(3)**, 362–382.
Westoll, T. S. (1937a). The Old Red Sandstone fishes of the north of Scotland, particularly of Orkney and Shetland. *Proc. Geol. Ass.* **48(1)**, 13–45.
Westoll, T. S. (1937b). On a specimen of *Eusthenopteron* from the Old Red Sandstone of Scotland. *Geol. Mag.* **74**, 507–524.
Westoll, T. S. (1938). Ancestry of the tetrapods. *Nature, Lond.* **141**, 127.
Westoll, T. S. (1939). On *Spermatodus pustulosus* Cope, a coelacanth from the "Permian" of Texas. *Am. Mus. Novit.* **1017**, 1–23.
Westoll, T. S. (1940). New Scottish material of *Eusthenopteron*. *Geol. Mag.* **77**, 65–73.
Westoll, T. S. (1943). The origin of the tetrapods. *Biol. Rev.* **18**, 78–98.
Westoll, T. S. (1943a). The origin of the primitive tetrapod limb. *Proc. R. Soc. (B)* **131**, 373–393.
Westoll, T. S. (1943b). The hyomandibular of *Eusthenopteron* and the tetrapod middle ear. *Proc. R. Soc. (B)* **131**, 393–414.
Westoll, T. S. (1944). The Haplolepidae, a new family of late Carboniferous bony fishes. *Bull. Am. Mus. nat. Hist.* **83**, 7–120.
Westoll, T. S. (1945). The paired fins of placoderms. *Trans. R. Soc. Edinb.* **61**, 381–398.
Westoll, T. S. (1945a). A new cephalaspid fish from the Downtonian of Scotland, with notes on the structure and classification of ostracoderms. *Trans. R. Soc. Edinb.* **61**, 341–357.
Westoll, T. S. (1949). On the evolution of the Dipnoi. *In* "Genetics, Paleontology and Evolution" (Eds G. L. Jepson *et al.*), 121–184. University Press, Princeton.
Westoll, T. S. (1958). The lateral fin-fold theory and the pectoral fins of ostracoderms and early fishes. *In* "Studies on Fossil Vertebrates" (Ed. T. S. Westoll), 180–211. Athlone Press, London.
Westoll, T. S. (1961). A crucial stage in vertebrate evolution: Fish to land animal. *Proc. R. Instn Gt Br.* **38**, 600–618.
Westoll, T. S. (1962). Ptyctodontid fishes and the ancestry of Holocephali. *Nature, Lond.* **194**, 949–952.
Westoll, T. S. (1967). *Radotina* and other tesserate fishes. *J. Linn. Soc. (Zool.)* **47**, 83–98.
Westoll, T. S. and Miles, R. S. (1963). On an arctolepid fish from Gemünden. *Trans. R. Soc. Edinb.* **65**, 139–153.
Weston, J. A. (1970). The migration and differentiation of neural crest cells. *In* "Advances in morphogenesis", (Eds M. Abercrombie *et al.*), Vol. 8, 41–114. Academic Press, New York.
White, E. G. (1937). Irrelationships of the elasmobranchs with a key to the order Galea. *Bull. Am. Mus. nat. Hist.* **74**, 26–132.
White, E. I. (1935). The ostracoderm *Pteraspis* Kner and the relationships of the agnathous vertebrates. *Phil. Trans. R. Soc. (B)* **225**, 381–457.
White, E. I. (1939). A new type of palaeoniscoid fish, with remarks on the evolution of the actinopterygian pectoral fins. *Proc. zool. Soc. Lond. (B)* **109**, 41–61.
White, E. I. (1946). *Jamoytius kerwoodi*, a new chordate from the Silurian of Lanarkshire. *Geol. Mag.* **83**, 89–97.
White, E. I. (1950). *Pteraspis leathensis* White a Dittonian zone-fossil. *Bull. Br. Mus. nat. Hist. (Geol.)* **1(3)**, 69–89.
White, E. I. (1952). Australian arthrodires. *Bull. Br. Mus. nat. Hist. (Geol.)* **1(9)**, 251–304.
White, E. I. (1958). On *Cephalaspis lyellie* Agassiz. *Palaeontology* **1(2)**, 99–105.
White, E. I. (1960). Notes on pteraspids from Artois and the Ardenne. *Bull. Mus. r. Hist. nat. Belg.* **36**, 1–16.

White, E. I. (1961). The Old Red Sandstone of Brown Clee Hill and the adjacent area. II. Palaeontology. *Bull. Br. Mus. nat. Hist. (Geol.)* **5(7)**, 243–319.

White, E. I. (1962). A dipnoan from the Assise de Mazy of Hingeon. *Bull. Mus. r. Hist. nat. Belg.* **38**, 1–8.

White, E. I. (1965). The head of *Dipterus valenciennesi*. *Bull. Br. Mus. nat. Hist. (Geol.)* **11(1)**, 3–45.

White, E. I. (1966). Presidential address: A little on lung-fishes. *Proc. Linn. Soc. Lond.* **177**, 1–10.

White, E. I. (1973). Form and growth in *Belgicaspis (Heterostraci)*. *Palaeontographica (A)* **143**, 11–24.

White, E. I. (1978). The larger arthrodiran fishes from the area of the Burrinjuck Dam, N.S.W. *Trans. zool. Soc. Lond.*, **34**, 149–262.

White, E. I. and Toombs, H. A. (1972). The buchanosteid arthrodires of Australia. *Bull. Br. Mus. nat. Hist. (Geol.)* **22(5)**, 379–419.

White, T. E. (1939). Osteology of *Seymouria baylorensis* Broili. *Bull. Mus. comp. Zool. Harv.* **85**, 325–403.

Whiteaves, J. F. (1881). On some remarkable fossil fishes from the Devonian rocks of Scaumenac Bay, Province of Quebec, with descriptions of a new genus and three new species. *Canadian Naturalist,* **10**, 27–35.

Whiteaves, J. F. (1889). Illustrations of the fossil fishes of the Devonian rocks of Canada. II: *Trans. R. Soc. Can.* **6**, 77–96.

Whitehead, P. J. P. (1963). A contribution to the classification of clupeoid fishes. *Ann. Mag. nat. Hist.* **5**, 737–750.

Whiting, H. P. (1972). Cranial anatomy of the ostracoderms in relation to the organisation of larval lampreys. *In* "Studies in vertebrate evolution" (Eds K. A. Joysey and T. S. Kemp), 1–20. Oliver and Boyd, Edinburgh.

Whiting, H. P. (1977). Cranial nerves in lampreys and cephalaspids. *In* "Problems in vertebrate evolution" (Eds S. M. Andrews *et al.*), Linn. Soc. Symp. Ser., Vol. 4, 1–23. Academic Press, London.

Wickbom, T. (1944). Cytological studies on Dipnoi, Urodela, Anura, and *Emys*. *Hereditas* **31**, 241–346.

Wickbom, T. (1949). Further cytological studies on Anura and Urodela. *Hereditas* **35**, 33–48.

Wickstead, J. H. (1969). Some further comments on *Jamoytius kerwoodi* White. *Zool. J. Linn. Soc.* **48**, 421–422.

Wiedersheim, R. (1877). Das Kopfskelet der Urodelen. *Morph. Jb.* **3**, 352–448, 459–546.

Wiedersheim, R. (1892). "Das Gliedmassenskelet der Wirbelthiere mit besonderer Berücksichtigung des Schulter- und Beckengürtels bei Fischen, Amphibien und Reptilien", 1–266. Gustav Fischer, Jena.

Wiedersheim, R. (1909). "Vergleichende Anatomie der Wirbeltiere." 7th edn. Gustav Fischer, Jena.

van Wijhe, J. W. (1882). Ueber die Mesodermsegmente und die Entwickelung der Nerven des Selachierkopfes. *Verh. K. ned. Akad. Wet.* **22**, 1–50.

van Wijhe, J. W. (1882a). Ueber das Visceralskelett und die Nerven des Kopfes der Ganoiden und von *Ceratodus*. *Niederl. Arch. Zool.* **5**, 207–320.

van Wijhe, J. W. (1905). Ueber die Entwicklung des Kopfskeletts bei Selachiern. *C. r. Séanc. Int. Congr. Zool., 6, Berne,* 319–322.

van Wijhe, J. W. (1922). Frühe Entwicklungsstadien des Kopf-und Rumpfskeletts von *Acanthias vulgaris*. *Bijdr. Dierk.* **22**, 271–298.

Wilder, B. G. (1877). On the serrated appendages of the throat of *Amia*. *Proc. Am. Advmt Sci.* **25**, 259–263.

Wiley, E. O. (1976). The phylogeny and biogeography of fossil and recent gars (Actinopterygii: Lepisosteidae). *Misc. Publ Mus. nat. Hist. Univ. Kans.* **64**, 1–111.

Williams, E. E. (1959). Gadow's arcualia and the development of tetrapod vertebrae. *Q. Rev. Biol.* **34**, 1–32.

Willmer, E. N. (1975). The possible contribution of nemertines to the problem of the phylogeny of the protochordates. *In* "Protochordates" (Eds E. J. W. Barrington and R. P. S. Jefferies)., Symp. zool. Soc. Lond., Vol. 36, 319–345. Academic Press, London.

Wills, L. J. (1935). Rare and new ostracoderm fishes from the Downtonian of Shropshire. *Trans. R. Soc. Edinb.* **58(II)**, 427–447.

Winchester, L. and Bellairs, A. d'A. (1977). Aspects of vertebral development in lizards and snakes. *J. Zool.* **181**, 495–525.

Wingstrand, K. G. (1951). "The structure and development of the avian pituitary", 1–316. Gleerup, Lund.

Wingstrand, K. G. (1966). Comparative anatomy and evolution of the hypophysis. *In* "The pituitary gland", (Eds G. W. Harris and B. T. Donovan), Vol. 1, 58–126. Butterworths, London.

Woodward, A. S. (1891). "Catalogue of the fossil fishes in the British Museum (Natural History)", Vol. 2. Br. Mus. Nat. Hist., Taylor and Francis, London.

Woodward, A. S. (1898). "Outlines of vertebrate palaeontology." University Press, Cambridge.

Woodward, A. S. (1900). On a new ostracoderm (*Euphanerops longaevus*) from the Upper Devonian of Scaumenac bay, Province of Quebec, Canada. *Ann. Mag. nat. Hist.* **(7)5**, 416–420.

Woodward, A. S. (1921). Observations on some extinct elasmobranch fishes. *Proc. Linn. Soc. Lond.* **133**, 29–39.

Woodward, A. S. 2nd English edn, revised (1932). Fishes to birds. *In* K. A. von Zittel: "Text-Book of Palaeontology" (Ed. C. R. Eastman) Vol. 2. Macmillan, London.

Woodward, A. S. and White, E. I. (1926). The fossil fishes of the Old Red Sandstone of the Shetland islands. *Trans. R. Soc. Edinb.* **54**, 567–572.

Woskobojnikoff, M. (1936). Die Kiemenatmenapparat bei Dipnoi. *Trudy Inst. Zool. Biol. Kyyiv* **3**, 68–77. (In Russian, German summary.)

Woskobojnikoff, M. (1939). The hyoid arch and branchial arches in lower Gnathostomata. *Dopov. Akad. Nauk URSR. Papers on animal morphology* **5**, 3–16. (In Russian, English summary.)

Wright, R. R. (1885). On the function of the serrated appendages of the throat in *Amia*. *Science, N.Y.* **4**, 511.

Wright, R. R. (1885a). On the hyomandibular clefts and pseudobranchs of *Lepidosteus* and *Amia*. *J. Anat. Physiol., Lond.* **19**, 477–497.

Young, G. C. (1978). A new early Devonian petalichthyid fish from the Taemas/Wee Jasper region of New South Wales. *Alcheringa* **2**, 103–116.

Young, J. Z. (1950). "The life of vertebrates" (2nd edn 1962). Clarendon Press, Oxford.

Zangerl, R. (1973). Interrelationships of early chondrichthyans. *In* "Interrelationships of fishes" (Eds P. H. Greenwood *et al.*), 1–14. Academic Press, London.

Zangerl, R. and Case, G. R. (1973). Iniopterygia, a new order of chondrichthyan fishes from the Pennsylvanian of North America. *Fieldiana Geol. Mem.* **6**, 1–67.

Zangerl, R. and Case, G. R. (1976). *Cobelodus aculeatus* (Cope), An anacanthous shark from Pennsylvanian Black Shales of North America. *Palaeontographica* **154**, 107–157.

Zhang, G.-R. (1978). The antiarchs from the early Devonian of Yunnan. *Vertebr. palasiat.* **16**(**3**), 147–186. (In Chinese, English summary.)
Ziegler, H. E. (1908). Die phylogenetische Entstehung des Kopfes der Wirbeltiere. *Jena. Z. Naturw.* **43**, 653–684.
Zimmermann, K. W. (1891). Ueber die Metamerie des Wirbeltierkopfes. *Verh. anat. Ges., Jena* **5**, 107–113.
Zwick, W. (1897). Beiträge zur Kenntnis des Baues und der Entwicklung der Amphibiengliedmassen, besonders von Carpus und Tarsus. *Z. wiss. Zool.* **63**, 62–114.
Zwilling, E. (1961). Limb morphogenesis. *In* "Advances in morphogenesis" (Eds M. Abercrombie and J. Brachet), Vol. I, 301–330. Academic Press, London.
Zych, W. (1937). *Cephalaspis kozlowskii* n.sp. from the Downtonian of Podole (Poland). *Archwm Tow. nauk. Lwow.* **9**, 50–100.

Subject Index

See also Index in Volume 1. Bold figures denote illustrations